Elements of the Theory of Numbers

Elements of the Theory of Numbers

Joseph B. Dence
University of Missouri
St. Louis, MO

Thomas P. Dence
Ashland University
Ashland, OH

San Diego London Boston
New York Sydney Tokyo Toronto

This book is printed on acid-free paper. ♾

ACADEMIC PRESS
A division of Harcourt Brace & Company
525 B Street, Suite 1900, San Diego, CA 92101-4495, USA
http://www.apnet.com

Academic Press
24–28 Oval Read, London NW1 7DX, UK
http://www.hbuk.co.uk/ap/

Harcourt/Academic Press
200 Wheeler Road, Burlington, MA 01803
http://www.harcourt-ap.com

Library of Congress Cataloging-in-Publication Data

Dence, Thomas P.
Elements of the theory of numbers / Thomas P. Dence, Joseph B. Dence
p. cm.
ISBN 0-12-209130-2 (acid-free paper)
1. Number theory I. Dence, Joseph B. II. Title.
QA241.D425 1998
512′.7—dc21 98-19868
CIP

Printed in the United States of America
98 99 00 01 02 IP 9 8 7 6 5 4 3 2 1

Dedicated

to

the late Harold Tinnappel

and to

Fred Leetch

professors

who motivated us in mathematics

Preface

Our idea in writing this book was to present to the mathematics teaching community a classroom text on number theory that makes greater use of the language and concepts in algebra than is traditionally encountered in introductory texts. We regard it as an important part of the mathematics education of junior and senior mathematics majors that they see the branches of mathematics as interrelated.

All students coming to this course will have completed the standard sequence in the calculus. They may also have completed at least one additional mathematics course, not necessarily in algebra. Consequently, we do not assume any knowledge of this subject on the part of the reader. The elementary material on groups, rings, and fields that is required in this book is introduced at a beginning level when it is needed. The book is thus aimed, in both content and degree of sophistication, somewhere between a traditional slim (200-page) elementary text and the great classic *An Introduction to the Theory of Numbers* by Hardy and Wright.

There is a bit more in this book than can be covered in one semester. This is (1) to allow instructors to make different selections of material according to their personal preferences, and (2) to encourage readers to investigate the material further after completing their course. A standard course could include Chapters 1 through 6, part of Chapter 7, and, in any time that remains, selected sections from any of the later chapters. Some of the later chapters would make suitable reading assignments for extra credit.

Annotated bibliographies appear at the end of each chapter. These bibliographies have been prepared with both the students and the instructors in mind, and we hope that extensive use will be made of them. Each chapter also concludes with a special page entitled Research Problems. These are intended as a source of further stimulation for the more motivated students.

A word about the Problems: There are nearly 900 in the book, ranging from the very elementary to the challenging. A great many are proof-oriented; several others require the writing of computer programs to carry out computations. Problems appear after every section in the book; we recommend that lots of them be assigned as homework, especially since, for many students, this course may be their first in-depth experience with the mechanics of formal proof. An Instructor's Solutions Manual is available to instructors upon request to the publisher.

These individuals gave generously of their time and read selected portions of the manuscript:

George Andrews
Pennsylvania State University

William Donnell
Illinois Wesleyan U.

Ron Dotzel
University of Missouri

Maureen Fenrick
State University Mankato

Wayne McDaniel
University of Missouri

Thomas McLaughlin
Texas Tech University

Carl Pomerance
University of Georgia

Frank DeMeyer
Colorado State University

Kwangil Koh
North Carolina State

Neil Calkin
Georgia Institute of Technology

We thank them for this. Special thanks go to Jesse Hemminger (Huron, Ohio) for his work on the illustrations and graphics in the text. We also thank anonymous reviewers who helped us weed out a great many infelicities at various stages of the manuscript. Any remaining silliness is our fault entirely. Finally, we thank the staff of Academic Press for their professionalism in bringing this project to reality.
We welcome comments from users of the book regarding any aspect of it.

Joseph B. Dence
St. Louis, Missouri

Thomas Dence
Ashland, Ohio

Prologue to the Student

This book covers most of the material on classical number theory that a mathematics major ought to know. For lack of a better designation, classical number theory means roughly the body of concepts and results that were in place up to the time of Gauss and that had been obtained by largely arithmetic methods.

However, beginning in the middle of the 19th century, number theory began a metamorphosis that has changed its very structure. Elements of analysis (the "theory" of the calculus) and of abstract algebra began to be applied to number theory. A mathematics major today must be exposed to some of the connections among the different branches of the discipline. Accordingly, we provide in this book, a glimpse of the exciting role of algebra in number theory in addition to coverage of the classical material.

The book is divided into two parts. Part I consists of fundamental or core material, most of which should be covered in any first course on number theory. It includes primes, congruences, primitive roots, residues, and multiplicative functions. Part II is a collection of more specialized topics, such as a brief look at number fields, recurrence relations, and additive number theory. You are urged to explore on your own as many of these topics as possible. The annotated lists of references at the ends of the chapters should be very useful in this regard.

We assume that the reader has completed the usual three-term sequence in the calculus and that Taylor's Theorem, l'Hôpital's Rule, the Mean Value Theorem, and the common convergence tests for infinite series are familiar results (Ayres, 1978; Leithold, 1990). It is likely that you will also have completed at least one mathematics course beyond the calculus before coming to number theory, one benefit of which will have been to give you some practice in mathematical proof. Direct proof, proof by contrapositive, proof by contradiction, and mathematical induction are all used throughout the present book (Solow, 1990; Kleiner, 1991; Cupillari, 1993). In the presentation of proofs as lemmas, theorems, and corollaries, **Theorem *m.n.*** is the nth theorem in Chapter m, and **Lemma** (or **Corollary)**

m.n.p. is the pth lemma (corollary) attached to Theorem $m.n.$ For general motivation and methodology of doing mathematics, we warmly recommend the stimulating book (Gardiner, 1987).

Although this is not a book on the history of number theory, it is sobering to realize that the subject goes back more than 2000 years (Dickson, 1952; Kramer, 1970; Kline, 1972; Scharlau and Opolka, 1985). Some of its most illustrious practitioners are listed in Table A. You will encounter most of these names in later portions of the book.

The history of number theory has often been the history of work directed toward specific prominent problems. It has usually been the case that these problems became more significant for the research that they sparked than for any intrinsic value of the problems themselves. Thus, the famous Last Theorem of Fermat led to the creation of the theory of algebraic numbers, and the question of how to partition an integer in different ways led to the study of generating functions.

We hope you will enjoy number theory. GOOD LUCK!

Table A Some Luminaries in Number Theory

Name (born–died)	Nationality
Euclid (fl.ca.350 B.C.)	Greek
Diophantus (fl.ca.250 A.D.)	Alexandrian-Greek
Pierre de Fermat (1601–1665)	French
Leonhard Euler (1707–1783)	Swiss
Joseph-Louis Lagrange (1736–1813)	French-Italian
Adrien-Marie Legendre (1752–1833)	French
Carl Friedrich Gauss (1777–1855)	German
Carl Gustav Jacob Jacobi (1804–1851)	German
Peter Gustav Lejeune Dirichlet (1805–1893)	German
Ernst Eduard Kummer (1810–1893)	German
David Hilbert (1862–1943)	German
Leonard Eugene Dickson (1874–1954)	American
Godfrey Harold Hardy (1877–1947)	English
Wacław Sierpiński (1882–1969)	Polish
Viggo Brun (1885–1978)	Norwegian
Srinivasa Ramanujan (1887–1920)	Indian
Ivan M. Vinogradov (1891–1983)	Russian
Derrick Henry Lehmer (1905–1991)	American
Paul Erdös (1913–1996)	Hungarian

References

Ayres, Jr., F., *Differential and Integral Calculus*, 2nd ed., Schaum's Outline Series, McGraw-Hill, New York, 1978. See any parts of this manual for a quick review of the basic theorems in calculus.

Cupillari, A., *The Nuts and Bolts of Proofs*, PWS Publishing Co., Boston, 1993. A short paperback (similar to the book by Solow, below) on the spirit of how to set up proofs.

Dickson, L. E., *History of the Theory of Numbers*, Vols. 1–3, Chelsea Publishing Co., New York, 1952. All you could ever want to know about historical developments in number theory from antiquity to early 20th century is contained in these volumes.

Gardiner, A., *Discovering Mathematics: The Art of Investigation*, Oxford University Press, Oxford, 1987. Very nicely written; should be on the shelf of every mathematics student.

Kleiner, I., "Rigor and Proof in Mathematics: A Historical Perspective," *Math. Mag.*, **64**, 291–314 (1991). Highly recommended article; mathematical proof and rigor have themselves evolved during the course of the evolution of mathematics.

Kline, M., *Mathematical Thought from Ancient to Modern Times*, Oxford University Press, New York, 1972, pp. 813–833. This author's Chapter 34 is a brief treatment of the main developments in number theory during the 19th century.

Kramer, E. E., *The Nature and Growth of Modern Mathematics*, Hawthorn Books, New York, 1970, pp. 497–527. A motivational chapter entitled "East Meets West in the Higher Arithmetic," featuring the eminent **G. H. Hardy** and his brilliant protégé **Ramanujan**.

Leithold, L., *The Calculus of a Single Variable with Analytic Geometry*, 6th ed., Harper Collins College, New York, 1990. A standard, well-written full text on the calculus, with more detail in it than in the preceding reference by Ayres.

Scharlau, W. and Opolka, H., *From Fermat to Minkowski—Lectures on the Theory of Numbers and Its Historical Development*, Springer, New York, 1985. Contains individual chapters on **Fermat, Euler, Lagrange, Legendre, Gauss, Dirichlet**; much mathematics is also included.

Solow, D., *How to Read and Do Proofs*, 2nd ed., John Wiley & Sons, New York, 1990. A short (170 pp) paperback that is a must on the personal bookshelf of every serious student of mathematics.

Contents

3. An Introduction to Congruences

4. Polynomial Congruences

5. Primitive Roots

6. Residues

*Indicates sections that can be omitted without impairing the continuity of the text.

10. Partitions

11. Recurrence Relations

APPENDIX I. Notation

APPENDIX II. Mathematical Tables

APPENDIX III. Sample Final Examinations

Part 1

The Fundamentals

Chapter 1
Introduction: The Primes

1.1 Sets, Logic, and Proof

The concept of a **set** is fundamental in all branches of mathematics (Lipschutz, 1964). In this book sets will generally be denoted by **BOLDFACE CAPITALS**. Some specific sets of frequent occurrence in number theory are indicated in Table 1.1.

The empty set is denoted by $\varnothing$; cardinality of a finite set $\mathbf{S}$ is indicated by $|\mathbf{S}|$. Set membership (or not) is indicated by the symbol $\in$ (or $\notin$)[1]

Example 1.1 Let $\mathbf{S} = \{2, 4, 6, 8, 10\}$. Then $|\mathbf{S}| = 5$ and $3 \notin \mathbf{S}$. ◆

For economy in presentation and clarity in expression, it is convenient to use certain symbols and some of the language from logic (Dence and Dence, 1994). In Table 1.2 let p, q be two propositions (statements) that meaningfully may be separately either true or false.

Lemmas, theorems, and corollaries in the book are generally (although not always) stated in the form "$p \rightarrow q$." This proposition, called a **conditional statement**, may be read as "if p, then q," or "p is a sufficient condition for q," or "q is a necessary condition for p."

The **biconditional statement** "$p \leftrightarrow q$" may be read as "p, if and only if (iff), q," or "p is a necessary and sufficient condition for q."

[1] These symbols, as well as other frequently employed symbols and notation, are gathered together in Appendix I at the rear of the book.

Table 1.1 Symbols for Specific Sets

Set	Illustrative Elements	Symbol
Natural numbers	$1, 2, 3, 4, \ldots,$	$\mathbf{N}$
Integers	$\ldots, -2, -1, 0, 1, 2, \ldots,$	$\mathbf{Z}$
Rational numbers	$1/2, -2/3, -7, 0$	$\mathbf{Q}$
Irrational numbers	$\sqrt{2}, 2 - \sqrt{5}, \sqrt[3]{\sqrt{1 + \sqrt{5}}}, \pi^2$	$\mathcal{I}$
Real numbers	$-3, 3/2, -\sqrt{6}, e^{\pi}$	$\mathcal{R}$
Complex numbers	$i, 2 - i\sqrt{2}, i\pi, 1$	$\mathcal{C}$

Table 1.2 Notation from Logic

Symbol	Use	Meaning
$\sim$	$\sim p$	Negation of proposition p
$\rightarrow$	$p \rightarrow q$	Proposition p implies proposition q
$\leftrightarrow$	$p \leftrightarrow q$	Propositions p, q imply each other (are **equivalent**)

Example 1.2

p: "$\sqrt{x}$ is rational."
q: "x is rational."
$p \rightarrow q$: "A sufficient condition for x to be rational is that $\sqrt{x}$ is rational."
This proposition is true. ◆

Example 1.3

p: "$\sqrt{x}$ is irrational."
q : "x is irrational."
$p \leftrightarrow q$: "A necessary and sufficient condition for x to be irrational is that $\sqrt{x}$ is irrational."
This proposition is false. ◆

The **converse** of the proposition "$p \rightarrow q$" is the proposition "$q \rightarrow p$." The **contrapositive** of the proposition "$p \rightarrow q$" is the proposition "$\sim q \rightarrow \sim p$." In Aristotelian logic the following hold:

Law of Contradiction: The propositions p, $\sim p$ cannot be true simultaneously.

Law of Contrapositive: The propositions $p \rightarrow q$, $\sim q \rightarrow \sim p$ are equivalent.

On the other hand, if "$p \rightarrow q$" is true, the converse may or may not be true.

Example 1.4

p: "x is a negative number"
q: "x^2 is a positive number"

$p \to q$: "If x is a negative number, then x^2 is a positive number." TRUE
$\sim q \to \sim p$: "If x^2 is not a positive number, then x is not a negative number." TRUE
$q \to p$: "If x^2 is a positive number, then x is a negative number." FALSE ◆

Some theorems are proved, beginning with the hypothesis (p), in a sequence of conditional statements, that is, by **direct proof**:

$$p \to p' \to p'' \to p''' \to q.$$

Other theorems are more conveniently proved using **proof by contrapositive**, that is, by using the Law of Contrapositive; in this procedure one begins with the negation of the conclusion ($\sim q$) :

$$\sim q \to q' \to q'' \to q''' \to \sim p.$$

A very common method of proof in number theory is **proof by contradiction**. In this technique one assumes (a) the hypothesis (p), (b) some other proposition (r) that may be a consequence of p or may be an axiom or some previously proven theorem, and (c) the negation of the conclusion ($\sim q$). From these assumptions one then shows that $\sim r$ follows. In view of the Law of Contradiction, it follows that $\sim q$ must be false (i.e., q is true).

Example 1.5 Let $x \neq 0$. Prove that if $x > 0$ (p), then $1/x > 0$ (q).

Here we assume the truth of this proposition r:

r: The product of a positive number and a negative number is always negative.

Suppose, then, that $x > 0$ but $1/x < 0$. By the definition of a reciprocal, we must have $x \cdot (1/x) = 1$. But now we have arrived at $\sim r$. Since $\sim r$, r cannot both hold (and because we really do believe r), we conclude that $1/x < 0$ is false, that is, $1/x > 0$ (because $1/x = 0$ is impossible; why?). ◆

Many results in number theory are obtained using **proof by mathematical induction** (Sominsky, 1975). The method is based on the following general principle (Problem 1.37):

Theorem 1.1 (Principle of Finite Induction). Let $\mathbf{S} \subset \mathbf{N}$ and suppose that $\mathbf{S}$ has the properties

(i) $1 \in \mathbf{S}$
(ii) whenever $k \in \mathbf{S}$, then $k + 1 \in \mathbf{S}$.

Then $\mathbf{S} = \mathbf{N}$.

There are a number of variants of the theorem (Problem 1.20).

For every positive integer k let P(k) be a proposition that is either true or false. Suppose that (i) P(1) is true and (ii) whenever P(k) is true, then the proposition P($k + 1$) is also true. Let **S** represent the set of positive integers k for which P(k) is true. Applying Theorem 1.1 to this set, we conclude that P(k) is true for all $k \in \mathbf{N}$.

A point of confusion regarding mathematical induction is the following: In the process of implementing the previous statement (ii), referred to sometimes as the **induction hypothesis**, aren't we actually *assuming* that P(k) holds for all k, whereas this is what we are trying to *prove*? The answer is no. What we are doing is showing that *if* P(k) *happens to be true,* then P($k + 1$) is true as a logical consequence. This is not at all the same as supposing that P(k) is true for all k.

Example 1.6 Prove the formula

$$1 \cdot 2 + 2 \cdot 3 + 3 \cdot 4 + \ldots + n \cdot (n + 1) = \frac{n(n + 1)(n + 2)}{3}.$$

Let P(k) be the proposition that the formula holds for $n = k$. We see that P(1) is true since $1 \cdot 2 = [1(1 + 1)(1 + 2)]/3 = 2$. Now we must show that if P(k) is true, then so is P($k + 1$). But,

$$\begin{aligned}\sum_{j=1}^{k+1} j(j + 1) &= \sum_{j=1}^{k} j(j + 1) + (k + 1)(k + 2) \\ &= \frac{k(k + 1)(k + 2)}{3} + (k + 1)(k + 2) \\ &= \frac{(k + 1)(k + 2)(k + 3)}{3},\end{aligned}$$

so P($k + 1$) is true. By the Principle of Finite Induction the proposition P(n) is true for all positive integers n. ◆

Example 1.7 Show that $\int_0^\infty x^{n-1}e^{-ax}\,dx = (n - 1)!/a^n$ for $a > 0$ and all $n \in \mathbf{N}$.

The proposition P(1) is true since for $n = 1$ we have

$$\int_0^\infty e^{-ax}\,dx = \left.\frac{e^{-ax}}{(-a)}\right|_0^\infty = 0 - \frac{1}{(-a)} = \frac{1}{a} = \frac{0!}{a^1}\ (a > 0).$$

Assume that the proposition is true for $n = k$. Then

$$\int_0^\infty x^k e^{-ax}\,dx = \frac{x^k e^{-ax}}{(-a)}\bigg|_0^\infty - \int_0^\infty \frac{e^{-ax}}{(-a)} \cdot kx^{k-1}\,dx$$

$$= 0 - 0 + \frac{k}{a}\int_0^\infty x^{k-1}e^{-ax}\,dx$$

$$= \frac{k}{a} \cdot \frac{(k-1)!}{a^k}$$

$$= \frac{k!}{a^{k+1}}, \text{ so } \mathrm{P}(k+1) \text{ is true.}$$

Hence, by the Principle of Finite Induction, the proposition P(n) is true for all positive integers n. ◆

Finally, one simple method of proof should never be overlooked. A statement can sometimes be shown to be false as stated using **proof by counterexample**. Only one exception to the statement is necessary for this.

Example 1.8 If the lengths of two sides of a triangle are irrational, then the third side must be irrational also.

Consider ΔABC where $m(\angle B) = 90°$, $AB = \sqrt{2}$ and $BC = \sqrt{2}$. Then $AC = 2$, and we have proved the preceding statement to be false. ◆

Problems[2]

1.1. Translate these statements into words:

(a) $x \in \mathbf{S} \cap \mathbf{T}$.
(b) $\mathbf{Q} \cup \mathcal{I} = \mathcal{R}$.
(c) $x \in \mathcal{R}$ or $y \in \mathcal{C}$.
(d) $N(m) = |\mathbf{S}|, \mathbf{S} = \{x\colon x \in \mathbf{N}, x^2 < m\}$.
(e) $x \in \mathbf{Q}$ and $y \in \mathcal{I} \rightarrow xy \in \mathcal{I}$.

1.2. Translate these statements into mathematical language:

(a) If x is an element of set **S**, then it is also an element of set **T**.
(b) The cube of a rational number is a rational number.

[2] Hints and answers to selected problems are given in Appendix IV in the rear of the book.

(c) A necessary condition that x lie in the intersection of sets **S**, **T** is that x be an element of **S** and x be an element of **T**.
(d) An integer is even iff it is divisible by 2.
(e) The empty set is a subset of every set.

1.3. Let p, q be the propositions

$$p: \text{“}\sqrt{x} \text{ is an irrational number”}$$

$$q: \text{“}x \text{ is a rational number.”}$$

State the following propositions in words:

(a) $p \to q$. (b) $\sim\sim q$. (c) $\sim q \to \sim p$. (d) $p \leftrightarrow q$.

1.4. In a truth table various propositions are assigned truth values of T (true) or F (false). For example, the truth table for the basic operation of negation is

p	$\sim p$
T	F
F	T

How does this show that $p \leftrightarrow \sim\sim p$?

1.5. The **conjunction** of two propositions p, q is the proposition “p and q.” It is true only when both p, q are true.

p	q	p and q
T	T	T
T	F	F
F	T	F
F	F	F

The **disjunction** of two propositions p, q is the proposition “p or q.” It is false only when both p, q are false. Show that $\sim(p \text{ and } q) \leftrightarrow (\sim p \text{ or } \sim q)$. Illustrate this result with a concrete example.

1.6. Use direct proof to show that $\mathbf{W} = \mathbf{S} \cap \mathbf{T} \to \mathbf{W} \subset \mathbf{S}$.

1.7. Use direct proof to show that $\mathbf{S} \subset \mathbf{T}$ and $\mathbf{T} \subset \mathbf{W} \to \mathbf{S} \subset \mathbf{W}$.

1.8. A finite set **S** has n elements. How many subsets of **S** are possible? Illustrate with $\mathbf{S} = \{1, 2, 3, 4\}$.

1.9. Use direct proof to show that a point on the perpendicular bisector of line segment AB is equidistant from the ends of AB.

1.10. Use proof by contrapositive to show that if $x > 0$ and $x \in \mathcal{I}$, then $\sqrt{x} \in \mathcal{I}$.

1.11. Differentiate clearly between the techniques of proof by contrapositive and proof by contradiction.

1.12. Use proof by contradiction to show that if $x \neq 0$, $x \in \mathbf{Q}$, and $y \in \mathcal{I}$, then $xy \in \mathcal{I}$.

1.13. Use proof by contradiction to show that if a planar polygon has exactly three right angles, then it is not a quadrilateral.

1.14. Use mathematical induction to show that $x - y$ is a factor of $x^n - y^n$ for all $n \in \mathbf{N}$.

1.15. Use mathematical induction to show that

$$\int (\ln x)^n \, dx = (-1)^n n! \, x \sum_{r=0}^{n} (-\ln x)^r / r! + C$$

for all $n \in \mathbf{N}$.

1.16. The **Stirling's numbers of the second kind**, $s(n, k)$, are defined by the recurrence relation

$$s(n, k) = ks(n - 1, k) + s(n - 1, k - 1),$$

together with the specifications $s(n, n) = 1$ for $n \geq 0$, $s(n, 0) = 0$ for $n > 0$, $s(n, k) = 0$ for $k > n$. Use mathematical induction to show that $s(n, n - 1) = n(n - 1)/2$.

1.17. The Principle of Finite Induction need not be initialized at $k = 1$.

(a) How would you reword it so that it could be applied to a problem where the initial value of the integral variable is some arbitrary positive integer k_0?
(b) Use (a) to prove that $1/k! < 3^{1-k}$ (specify conditions). Illustrate the result.

1.18. **Bernoulli's Inequality** says that if $x > -1, x \neq 0, n \in \mathbf{N}, n > 1$, then $(1 + x)^n > 1 + nx$. Prove it. Illustrate the result with $x = -0.1, n = 5$.

1.19. Prove that the inequality $(1 + n^{-1})^n > 9/4$ holds for all integers n beyond a certain point. [Hint: Confirm and use

$$[1 + (n + 1)^{-1}]^{n+1} = \left(\frac{n + 2}{n + 1}\right)\left(\frac{n + 1}{n}\right)^n [1 - (n + 1)^{-2}]^n.]$$

1.20. An important variant of the Principle of Finite Induction is the **Second Principle of Induction** (or **Strong Mathematical Induction**). Let $\mathbf{S} \subset \mathbf{N}$ and suppose that $\mathbf{S}$ has the properties

(i) $n_0 \in \mathbf{S}$
(ii) whenever $n_0, n_0 + 1, n_0 + 2, \ldots, k \in \mathbf{S}$, then $k + 1 \in \mathbf{S}$.

Then $\mathbf{S} = \{n: n \in \mathbf{N}, n \geq n_0\}$.

Use the Second Principle for the following problem. The **Fibonacci numbers**, F_n, are defined recursively by $\mathrm{F}_1 = \mathrm{F}_2 = 1$, $\mathrm{F}_n = \mathrm{F}_{n-1} + \mathrm{F}_{n-2}$ for $n > 2$. Show that for $n \in \mathbf{N}$, $\mathrm{F}_{n+1} < [(1 + \sqrt{5})/2]^n$. Illustrate the result with $n = 6, 10, 15$.

1.21. The **Chebyshev polynomials**, $\mathrm{T}_n(x)$, are important in analysis and statistics. They are defined by $\mathrm{T}_0(x) = 1$, $\mathrm{T}_1(x) = x$, $\mathrm{T}_n(x) = 2x\mathrm{T}_{n-1}(x) - \mathrm{T}_{n-2}(x)$ for $n \geq 2$.

(a) Deduce $\mathrm{T}_4(x)$, $\mathrm{T}_5(x)$.
(b) Prove that the sum of the coefficients in $\mathrm{T}_n(x)$ is 1.

1.22. Let the symbol $\sigma(n)$ denote the sum of the positive integral divisors of the positive integer n. For example, $\sigma(10) = 1 + 2 + 5 + 10 = 18$. The following theorem is conjectured: $\sigma(n) < 5n/2$. Prove that this proposition is false.

1.2 REAL NUMBERS AND THE WELL-ORDERING PROPERTY

In this book we take the existence of real numbers for granted. We assume that you have already been exposed in college algebra (or elsewhere) to the popular set of axioms for $\mathfrak{R}$, first given by the German mathematician **David Hilbert** (1862–1943) in 1899. We view these axioms as providing a characterization of $\mathfrak{R}$, including the common arithmetic operations on elements of $\mathfrak{R}$ (the Field axioms) and an ordering among these elements (the Order axioms).

One special axiom for $\mathfrak{R}$ that is needed when one deals with sets of real numbers is the following:

COMPLETENESS AXIOM

Every nonempty set $\mathbf{S}$ of real numbers that is bounded above (or below) has a least upper bound (or a greatest lower bound).

This axiom has many wonderful consequences that form part of the subject matter for a first course in real analysis.

Our interest in the Completeness Axiom is that from it one can prove rather easily the very plausible Well-Ordering Property for integers. We will use the Well-Ordering Property several times in the book. We give it here without proof in order that we may proceed expeditiously.

Theorem 1.2 (The Well-Ordering Property). If **S** is a nonempty, finite, or infinite set of nonnegative integers, it contains a smallest member.

Example 1.9 Let **S** be the set of all positive odd integers n such that $n + 27$ is divisible by 11. Then **S** is nonempty because $17 \in \mathbf{S}$; in fact, **S** is infinite. By Theorem 1.2, **S** has a smallest member; checking shows that 17 is this smallest member. ◆

Example 1.10 Let $N(j, m)$ denote the number of times the digit j, $j = 0, 1, 2, \ldots, 9$, appears in the rational approximation of π, written out to the first m decimal places. The set of integers $\mathbf{S} = \{N(0, 10^{100}), N(1, 10^{100}), \ldots, N(9, 10^{100})\}$ has a smallest member by Theorem 1.2, although its identity is probably not known. ◆

An important application of the Well-Ordering Property is to the demonstration of the Division Algorithm (Section 1.3), one of the cornerstones of arithmetic in **Z**. At this point you may be tempted to view Theorem 1.2 as rather simplistic; this is wrong. The theorem can be shown to be equivalent to the Principle of Finite Induction, which as you know is used extensively throughout mathematics as a method of proof. Further, the theorem has a generalization that is applicable to much wider sets of numbers than nonnegative integers, but this is outside the scope of the present book (Lipschutz, 1964; Wilder, 1983).

Problems (mainly for those who have seen the axioms for $\mathcal{R}$)

1.23. State in words the particular Field axiom for $\mathcal{R}$ that justifies each of the following statements:

(a) $2^{961} + 3^{107} = 3^{107} + 2^{961}$
(b) $\pi \cdot (-7 + 5^{110}) = \pi \cdot (-7) + \pi \cdot (5^{110})$
(c) $\pi \cdot (2^{961} \cdot 3^{107}) = [\pi \cdot (2^{961})] \cdot 3^{107}$
(d) $\sqrt{3} \cdot (\sqrt{3})^{-1} = 1.$

1.24. Using Field axioms for $\mathfrak{R}$ prove the **Cancellation Law of Addition:** if $x, y, z \in \mathfrak{R}$ and $x + y = x + z$, then $y = z$.

1.25. Prove that if $x \in \mathfrak{R}$, then $-x$ (the additive inverse of x) is unique. [Hint: Make use of Problem 1.24.]

1.26. Prove that if $x \in \mathfrak{R}$, then $x \cdot 0 = 0$. [Hint: Begin with $x \cdot 0 + 0 = x \cdot 0 = x \cdot (0 + 0)$.]

1.27. Prove that 0 has no multiplicative inverse.

1.28. Prove that $(-1) \cdot x = -x$. [Hint: Multiply both sides of the equation $1 + (-1) = 0$ by x.]

1.29. The Order axioms for $\mathfrak{R}$ can be stated in various ways. One way is in terms of the primitive concept of **positivity**:

I. Some elements of $\mathfrak{R}$ have the property of being positive.
II. For any $x \in \mathfrak{R}$, only one of the following three propositions is true:
  (a) $x = 0$
  (b) x is positive
  (c) $-x$ is positive.
III. The sum and product of any two positive numbers are each positive.

We make the definition that a real number x is **negative** iff $-x$ is positive. Explain how it then follows from this definition and the axioms above that the product of two negative numbers is a positive number. [Hint: See Problem 1.28.]

1.30. The symbols $<$ and $>$ ("less than" and "greater than") are defined by

$$\begin{cases} x < y \text{ iff } y - x \text{ is positive} \\ x > y \text{ iff } x - y \text{ is positive.} \end{cases}$$

Prove that $x < y \rightarrow x + z < y + z$ for any $x, y, z \in \mathfrak{R}$.

1.31. Suppose that $x, y, z \in \mathfrak{R}$ and $x < 0$, $y < z$. Prove that $xy > xz$.

1.32. Explain why the equation $x^2 + 2 = 0$ has no real root.

1.33. One of the wonderful consequences of the Completeness Axiom is the **Archimedean Property**.[3] It states: If $x, y > 0$, then there is an $n \in \mathbf{N}$ such that $nx > y$. Now suppose that $x, y \in \mathfrak{R}$ and that $y > x$. Prove that there is a rational number between x and y. Illustrate the result using $x = \pi$, $y = 22/7$.

[3] Named after **Archimedes of Syracuse** (in Sicily) (287–212 B.C.), who attributed a related statement to **Eudoxus** (ca. 408—ca. 355 B.C.).

1.34. Let x, y be two distinct real numbers. Prove that there is an irrational number between x and y. Illustrate with the case $x = 10.00$, $y = 10.01$. The results of this problem and the previous one may be phrased by saying, respectively, that the rational numbers and the irrational numbers are each **dense** in the system of real numbers.

1.35. How could you reword the Well-Ordering Property to apply to nonempty sets of nonpositive integers?

1.36. Let **S** be the set of integers n such that $n \ln n - 2n$ is positive. Does the Well-Ordering Property apply? If so, find the smallest member of **S**.

1.37. Look again at the statement of the Principle of Finite Induction (Theorem 1.1). Prove it from the Well-Ordering Property. [Hint: Suppose that there is a nonempty subset $\mathbf{S}' \subset \mathbf{N}$ for which the Principle fails.]

1.38. Show that the Second Principle of Induction (see Problem 1.20) follows from the Well-Ordering Property.

1.39. A review problem on use of the Second Principle of Induction: The **Chebyshev polynomials of the second kind** are defined recursively by

$$\begin{cases} U_0(x) = 1 \\ U_1(x) = 2x \\ U_n(x) = 2xU_{n-1}(x) - U_{n-2}(x), \; n \geq 2. \end{cases}$$

Formulate a conjecture on the values of $U_n(1)$, and then prove it.

1.3 The Division Algorithm

Divisibility is one of the principal areas of classical number theory (Allenby and Redfern, 1989; Hardy and Wright, 1979; Rosen, 1993). The main theorem of this section occupies a key place in the arithmetic in **Z**.

Theorem 1.3 (The Division Algorithm). If $a, b \in \mathbf{Z}, b > 0$, then there exist unique integers q, r with the properties $a = bq + r, 0 \leq r < b$.

Proof Let $a, b \in \mathbf{Z}$ be fixed, and consider the set of nonnegative integers $\mathbf{S} = \{a - bn : n \in \mathbf{Z}, a - bn \geq 0\}$. The Archimedean Property (Problem 1.33) guarantees that **S** is nonempty.

By the Well-Ordering Property (Theorem 1.2), **S** contains a smallest integer, $r = a - bq$, corresponding to some $q \in \mathbf{Z}$. To see that $r < b$, assume that $r \geq b$. Then $r - b \geq 0$ and $r - b \in \mathbf{S}$ because $r - b = (a - bq) - b =$

$a - b(q + 1)$, $q + 1 \in \mathbf{Z}$. But r is minimal, and yet $r - b < r$, a contradiction.

To show uniqueness of q, r, assume that $a = bq + r = bq' + r'$, where $q \geq q'$ and $0 \leq r' < b$. We obtain $b(q - q') = r' - r$. If $q > q'$ holds strictly, then the left side is at least b. But $r' - r$ cannot be at least b because both r and r' are nonnegative and less than b. This contradiction implies $q' = q$, so the right side is 0 and $r' = r$. ◆

If $b < 0$, we write $a = |b|q + r = b(-q) + r$, in which case $0 \leq r < |b|$. In this way, the Division Algorithm is applicable to all nonzero $b \in \mathbf{Z}$.

The Division Algorithm has a simple geometric visualization. Along the real line $\mathcal{R}$ in Figure 1.1 we locate the point corresponding to the rational number a/b. The Division Algorithm then identifies the unique point corresponding to the integer q that is closest to and lies to the left of the point a/b. The line segment of length Δ, when multiplied by the integer b, gives the unique integer $r = b\Delta = b[(a/b) - q] = a - bq$.

Theorem 1.3 is an example of an **existence theorem**; it asserts the existence of one or more objects without explicitly producing them. Theorem 1.2 is also an existence theorem. However, the technique of the proof of Theorem 1.3 is algorithmic in nature. We actually construct r by subtracting enough (but a finite number of) multiples of b from a to obtain a number that is minimal and nonnegative.

DEF. If a, $b \in \mathbf{Z}$, then b is said to **divide** a, written $b \mid a$, iff in the Division Algorithm one has $r = 0$, that is, iff $a = bq$ for some $q \in \mathbf{Z}$.

It follows that if $a \neq 0$, then division by $b = 0$ is impossible since no q, r can be found. If $a = b = 0$, we shall again say that division is impossible since q is not unique.

For every nonzero $a \in \mathbf{Z}$, one has $a \mid a$ (take $q = 1$) and $1 \mid a$ (take $q = a$). Note also that if $a \neq 0$ and $b \mid a$, then $|a| \geq |b|$. For suppose $|b| > |a|$; then $|b| - |a| = |b| - |b||q| = |b|\{1 - |q|\}$. The left-hand side is now positive, but the right-hand side is either 0 or negative, there being no integers q between 0 and ± 1.

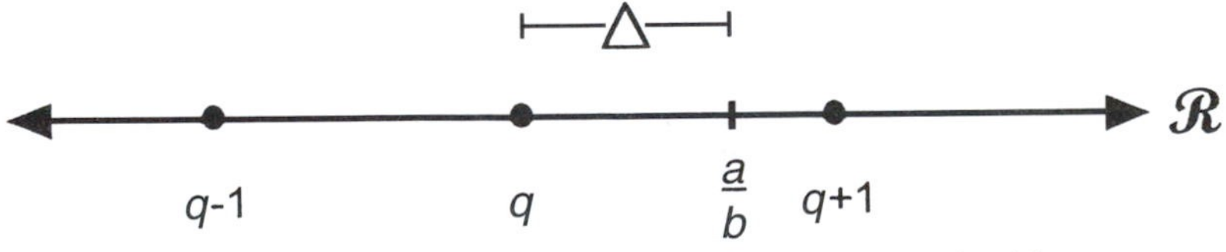

Figure 1.1 Geometric illustration of the Division Algorithm

Theorem 1.4 Let $a, b, c \in \mathbf{Z}$. Then

(a) If $a \mid b$ and $b \mid c$, then $a \mid c$.
(b) If $a \mid b$ and $a \mid c$, then $a \mid (bx + cy)$ for all $x, y \in \mathbf{Z}$.

Both parts of Theorem 1.4 are extendable by induction in obvious ways.

Example 1.11 Consider any integer of the form $5n^3 + 12m$, where $m, n \in \mathbf{Z}$ and m is a multiple of 10. Then $10 \mid m$ and $5 \mid 10$, so by part (a) of Theorem 1.4 we have $5 \mid m$. In part (b) of Theorem 1.4 make the identifications $a = 5$, $b = 5$, $c = m$, $x = n^3$ and $y = 12$. Then $5 \mid b$ and $5 \mid c$, so $5 \mid (5n^3 + 12m)$. ◆

Problems

1.40. Prove both parts of Theorem 1.4. Illustrate the parts with numerical examples.

1.41. Let n_1, n_2 be integers that satisfy $0 < n_1 < n_2$. Apply the Division Algorithm to obtain q_1, r_1. Now replace n_2 by n_1 and n_1 by r_1, and apply the Division Algorithm again to obtain q_2, r_2. If this procedure is repeated until it is no longer possible to do so, what should be the final r?

1.42. Let $a, b \in \mathbf{N}$; prove that there exist unique nonnegative integers c_0, c_1, c_2 such that $a = c_0 b^2 + c_1 b + c_2$, $0 \le c_1, c_2 < b$. Illustrate the result with the choice $a = 191$, $b = 12$.

1.43. Let the positive integer N have the form

$$N = 100a + 10b + c, \qquad a + b + c = 3n,$$

where a, b, c are digits, $a \neq 0$. Prove that $3 \mid$ N. This is the **divisibility rule by 3** for three-digit numbers. Illustrate the rule with some examples. How do you think the rule could be extended to positive integers of more than three digits?

1.44. Let N be a four-digit number. Form the difference between the sum of the odd-place digits in it and the sum of the even-place digits. Prove that if 11 divides this difference, then 11 divides N. This is the **divisibility rule by 11** for four-digit numbers. Illustrate the rule with some examples. Discuss, with examples, whether the rule could be extended to positive integers of more than four digits.

The usual base for the representation of integers is the base 10 (see the next problem). Do you see any connection between this choice of base and the divisibility rule by 11 just given? What happens if other bases are selected?

1.45. The base-10 representation of an integer $N > 0$ is an expansion of the form

$$N = A_k \cdot 10^k + A_{k-1} \cdot 10^{k-1} + \ldots + A_1 \cdot 10^1 + A_0,$$

where the integer A_k satisfies $1 \leq A_k \leq 9$ and the remaining A_i's satisfy $0 \leq A_i \leq 9$. The integer N is then written in shorthand as $A_k A_{k-1} \ldots A_1 A_0$. The **digital sum** of N, written DS(N), is defined by

$$\text{DS}(N) = \sum_{i=0}^{k} A_i.$$

Prove that $\text{DS}(3^n) < 5n$ for all integers $n > 0$.

1.4 The Primes

The prime integers (or primes) are the building blocks for **Z** (Beyer, 1984). The precise meaning of this is given shortly, but first let's define what we mean by prime integers.

> **DEF.** An integer $p \neq \pm 1$ is a **prime integer** (**prime**) iff the only nonzero integers that divide p are ± 1, $\pm p$. An integer n that is not prime has a nonzero divisor x other than ± 1 and $\pm n$, and n is then termed **composite**.[4]

The integers 2, 3, -5, -7 are primes; the integers 4, 6, -8, -9 are composite. Note that the definition of composite merely refers to x as a divisor of n; it does not identify x as a prime necessarily. The connection between prime and composite is supplied by Theorem 1.5 shortly.

Theorem 1.5 is a global result in the sense that it applies to a very wide class of objects. Frequently, a way to prove a global result is to assume the existence of a nonempty set **S** of objects that are exceptions, and then to obtain a contradiction. In the present case, we analyze the set **S** with the aid of the Well-Ordering Property.

[4] The definition of prime here permits primes to be either positive or negative; also the definition leads naturally to a generalization that we shall make later in the book (Section 9.7). However, in all that follows we shall (unless stipulated otherwise) take the words prime or divisor to refer to *positive* integers.

Theorem 1.5 Every integer $n > 1$ is either a prime or a product of (positive) primes.

Proof Assume, to the contrary, that $\mathbf{S} \subset \mathbf{N}$, the set of positive integers that are neither prime nor products of primes, is nonempty. By the Well-Ordering Property, $\mathbf{S}$ has a smallest member $n_{\min}$.

Because $n_{\min}$ is not prime, it is composite: $n_{\min} = xy$, where $x, y < n_{\min}$. Since now $x, y \notin \mathbf{S}$, then x is either a prime or a product of primes, and similarly for y. But then $n_{\min}$ is just a product of two (or possibly more) primes, a contradiction of $n_{\min} \in \mathbf{S}$. Hence, $\mathbf{S}$ cannot be nonempty. ◆

Theorem 1.5 provides no guidance on how to determine if an integer is prime or on how to decompose a composite integer into prime factors. Clearly, a table of primes would be useful. Beyond 2, the primes must be sought from among the odd integers. According to one scheme, the technique of counting is all we need in order to determine if an odd integer N is prime or not.

First write down (figuratively) the odd integers from 3 to N:

$$3, 5, 7, 9, 11, 13, 15, 17, 19, 21, 23, 25, 27, 29,$$
$$31, 33, 35, 37, 39, 41, 43, 45, \ldots, N.$$

We accept that 3 is the smallest odd prime. Next, *by counting*, remove from the list every third integer beginning with the 3:

$$\_, 5, 7, \_, 11, 13, \_, 17, 19, \_, 23, 25, \_, 29,$$
$$31, \_, 35, 37, \_, 41, 43, \_, \ldots, N.$$

We retain the empty locations in our display. The smallest integer that remains (5) is the next prime (i.e., $p_1 = 2, p_2 = 3, p_3 = 5$).

Continuing, we remove from the display any integer that is in every fifth location beginning with the 5. Some locations, of course, may already have been empty because of previous deletions. We now obtain

$$\_, \_, 7, \_, 11, 13, \_, 17, 19, \_, 23, \_, \_, 29,$$
$$31, \_, \_, 37, \_, 41, 43, \_, \ldots, N.$$

The smallest integer that now remains is the next prime after 5 ($p_4 = 7$). The procedure is continued.

If N is not deleted during any of the sets of deletions that are initialized by integers less than $\sqrt{N}$, then N is a prime. During all of these maneuvers, nothing more sophisticated than counting has been used, although the legitimacy of the procedure demands certain facts from us.

It is Theorem 1.5 that guarantees that this method will produce a correct listing of primes up to N. With but minor modification the method is the **Sieve of Eratosthenes**, which was given by the Alexandrian mathematician, poet, and philosopher **Eratosthenes** (ca. 284–ca. 192 B.C.). It is obvious, though, that use of this method rapidly becomes laborious as N increases, at least if done by hand. Most of the labor can be removed by running the procedure on computer (Problem 1.49). Also, there are more recent sieve methods that are more efficient than that of Eratosthenes. A partial listing of primes is given in Table C of Appendix II.

Example 1.12 By use of the Sieve of Eratosthenes (as implemented on computer) one finds that 647 is prime, but 649 is not. Trial and error shows that $649 = 11 \cdot 59$. ◆

Problems

1.46. Find two primes larger than 10,000.

1.47. **Pierre de Fermat** (1601–1665) conjectured that all integers of the form $2^{2^n} + 1$, where n is a nonnegative integer, are prime. Investigate this for $n = 0, 1, 2, 3, 4$.

1.48. Of what significance is the decomposition $4{,}294{,}967{,}297 = 641 \cdot 6{,}700{,}417$?

1.49. Write a computer program to implement the Sieve of Eratosthenes, and then use it to determine how many primes there are less than 10,000.

1.50. Prove that every integer greater than 11 is the sum of two composite integers.[5]

1.51. Some elementary properties of integers are demonstrable if one represents integers in simple algebraic forms. Explain why any integer is of one of these three forms: $3n$, $3n-1$, $3n+1$. Then prove that the product of any three consecutive positive integers is divisible by 6.

1.52. Show that if $n \in \mathbf{N}$ is odd and $n > 3$, then $(n^2-1)/4$ is integral and composite.

1.53. Show that if $n, m \in \mathbf{N}$ and $2 \le m \le n$, then $n! + m$ is composite.

[5] A similarly worded statement asserts that any even integer greater than 3 can be written as the sum of two primes. This is **Goldbach's Conjecture** (after **Christian Goldbach**, 1690–1764); it is still unproven.

1.54. Show that if a positive integer is of the form $6k + 5$, $k \geq 0$, then it is also of the form $3k' - 1$, $k' \geq 1$. Is the converse true?

1.55. **Twin primes** are integral pairs $(p, p + 2)$ in which both members are prime. Examples are (5, 7), (11, 13), and (107, 109). It is believed that there are infinitely many twin primes, but as yet the matter is undecided. Explain carefully why it would be fruitless to look for **triple primes** $(p, p + 2, p + 4)$ beyond (3, 5, 7).

1.56. Do a computer search and find a pair of twin primes larger than 10,000.

1.57. Prove that if $n > 1$ is a positive integer, then the integer $n^4 + 4$ cannot be prime. [Hint: Consider factoring.]

1.58. To determine whether 71 is prime, one need only test its divisibility by three integers: 3, 5, and 7. To determine whether 233 is a prime, one need only test its divisibility by five integers: 3, 5, 7, 11, and 13. Explain these observations. On the basis of your answer, write a computer program alternative to the one in Problem 1.49 to locate all the primes from 2 to N. Prepare such a list for $N =$ 10,000.

1.5 Infinitude of the Primes

We come now to one of the truly beautiful theorems in all of mathematics, one that's beautiful because of its wide generality yet simplicity of proof. The theorem appeared as Proposition 20 in Book IX of the *Elements* of **Euclid**. It showed the impossibility of the Sieve of Eratosthenes as a device to list *all* of the primes.

The proof of Theorem 1.6 uses the fact that any finite set **S** of real numbers has a maximum element. This follows because there is an ordering relation ($<$) on the elements of $\mathfrak{R}$, and the relation can certainly be applied a sufficient, but finite, number of times in order to arrange the members of **S** from low to high.

Theorem 1.6 (Euclid). The number of primes is infinite.

Proof Assume, to the contrary, that the set **S** of primes is finite; let P be maximal in **S**. Now form the number N, defined as

$$N = 2 \cdot 3 \cdot 5 \cdot 7 \cdot 11 \ldots P + 1,$$

where the product includes all of the primes from 2 to P. Clearly, $N > P$, and if N is not composite, then we have found a new prime larger than P, a contradiction.

If N is not prime, then it is a product of primes (Theorem 1.5).

$$N = p_1 p_2 \cdots p_k$$

None of the p_i's can be among the list 2, 3, 5, . . . , P because the Division Algorithm shows that each of these leaves a remainder of $r = 1$ when considered as a divisor of N. It follows that each of $p_1, p_2, \ldots, p_k$ must exceed P, again a contradiction. It follows that **S** is not finite. ◆

This is essentially the original proof of **Euclid**; there are a number of variations of it (Problem 1.59). Because **Euclid** knew of no formula for the nth prime, he was unable to generate the complete set of primes. Hence, he chose indirect proof; this is sometimes the easiest way to show that a certain set is infinite. The construction of the number

$$N = 2 \cdot 3 \cdot 5 \cdot 7 \cdot 11 \ldots P + 1$$

was, of course, a stroke of intuition (or possibly just plain trial and error) on the part of **Euclid**.

The symbol $\pi(n)$ is conventionally used to indicate the discrete function that gives the number of primes that do not exceed n: $\pi(2) = 1$, $\pi(3) = 2$, $\pi(20) = 8$, and so forth.[6] Then Euclid's theorem is equivalent to the statement

$$\lim_{n \to \infty} \pi(n) = \infty.$$

The function $\pi(n)$ has been the subject of intense research for nearly 200 years. We touch on one fundamental topic in this direction in Section 1.8, but there are a great many other such topics of inherent interest. For example, it was once an hypothesis that

$$\pi(2n) - \pi(n) \geq 1$$

for $n \geq 2$ (**Bertrand's Postulate**). This was first proved in 1852 by the Russian mathematician **Pafnuti Chebyshev** (1821–1894). A similar conjecture that is still undecided is

$$\pi(p_{n+1}^2) - \pi(p_n^2) \geq 4$$

for $p_n \geq 3$. These conjectures seem eminently plausible; that doesn't mean that they are easy to prove.

Example 1.13 Suppose 11 is imagined to be the largest prime. Then $2 \cdot 3 \cdot 5 \cdot 7 \cdot 11 + 1 = 2311$. Since $\sqrt{2311} \approx 48$, we find that 2311 is not divisible

[6] The symbol π here has nothing to do with the *number* π.

by any of the odd integers up through 47. Hence, 2311 is a prime and exceeds 11. Note that this example mimics the proof of Theorem 1.6. ◆

Example 1.14 (a) $\pi(20) - \pi(10) = 8 - 4 = 4 \geq 1$.
(b) $\pi(7^2) - \pi(5^2) = 15 - 9 = 6 \geq 4$. ◆

It is interesting to inquire when the quantity $N = 2 \cdot 3 \cdot 5 \ldots P + 1$ of Theorem 1.6 is prime. Templer found by computer that for $P \leq 1031$, there were only nine cases where N was prime; the cases $P = 379$, 1019, and 1021 yielded integers that were not previously known to be prime (Templer, 1980). The record in 1994 for the largest prime P for which N is also prime was $P =$ 13,649 (Ribenboim, 1994). It is clear that although Euclid's method of proof provides a foolproof recursive way of generating an infinite set of primes, it is highly impractical.

Problems

1.59. This proof of Theorem 1.6 is due to **E.E. Kummer** (1810–1893). Suppose that there are only k primes: $p_1 < p_2 < \ldots < p_k$. Define the quantity

$$N = \prod_{j=1}^{k} p_j.$$

The integer $N - 1$ is necessarily odd (why?) and must be composite (why?). Hence, it has an odd prime p_i as a common factor with N itself. It follows that p_i divides $N - (N - 1)$, that is, $p_i \mid 1$. Write out the proof in nice form.

1.60. Do the primes that are produced in succession by using Euclid's recursive formula increase continually?

1.61. Consider the following alleged proof by mathematical induction of Theorem 1.6: "Accept that $p_1 = 2$ and that there is another prime larger than p_1, namely $p_2 = 3$. Now suppose p_k is a prime. Form the extended product of it with all of the previously discovered primes, and add 1,

$$N = p_1 p_2 \ldots p_k + 1.$$

Since none of the primes $p_1, p_2, \ldots p_k$ divides N because each leaves a remainder of 1, then N must be a prime; set $p_{k+1} = N$. By the Principle of Finite Induction there is a larger prime for each succeeding $k \in \mathbf{N}$, so the set of primes is infinite." Comment on this proof.

1.62. State clearly in your own words the content of Bertrand's Postulate. Show that the postulate is true for $n = 5, 8, 12$.

1.63. Do one of these two computer explorations:

(a) Write a computer program to hunt for a prime between n^2 and $(n + 1)^2$, $n \in \mathbf{N}$. Test it out on several cases.
(b) Use your computer program from either Problem 1.49 or Problem 1.58 to confirm the conjecture $\pi(p_{27}^2) - \pi(p_{26}^2) \geq 4$.

1.6 Remarks on the Distribution of the Primes

In view of Theorem 1.6, it is natural to ask how the primes are distributed along the positive half of the real line. The data in Table 1.3 suggest that over long intervals the density of primes tends to decrease as one approaches larger and larger integers.

When relatively small intervals are examined, the density of primes varies erratically. For example, in intervals of length 200 from 10,000,000 to 10,001,000

Table 1.3 Distribution of Prime Numbers in Various Intervals

Interval	Number of Primes	Interval	Number of Primes
1–10,000	1,229	50,001–60,000	924
10,001–20,000	1,033	60,001–70,000	878
20,001–30,000	983	70,001–80,000	902
30,001–40,000	958	80,001–90,000	876
40,001–50,000	930	90,001–100,000	879

Interval	Number of Primes	% Primes
1–100,000	9,592	9.6
1–1,000,000	78,498	7.8
1–10,000,000	664,579[a]	6.6
1–100,000,000	5,761,455[a]	5.8
1–1,000,000,000	50,847,534[a]	5.1
1–10,000,000,000	455,052,511[a,b]	4.6
1–1,000,000,000,000	37,607,912,018[b]	3.8

[a] D.H. Lehmer, "On the Exact Number of Primes Less than a Given Limit," *Illinois J. Math.*, **3**, 381–388 (1959).

[b] J.C. Lagarias, V.S. Miller, and A.M. Odlyzko, "Computing $\pi(x)$: The Meissel–Lehmer Method," *Math. Comp.*, **44**, 537–560 (1985).

the numbers of primes are 8, 12, 9, 17, 15. On the average, in each block of 200 integers here only about 6% of the integers are prime.

As if to emphasize the thinness of occurrence of the primes, it is possible to find regions of any arbitrary length wherein no primes at all occur (Pegg, Jr., 1997). The idea of the demonstration of this is to basically use Theorem 1.4(b) to generate a sequence of integers of the form $bx + cy$, for each of which there is some corresponding integer $a > 1$ that divides bx and cy.

Theorem 1.7 There are arbitrarily long blocks of consecutive composite numbers.

Proof Let p be a prime and consider the set $\mathbf{S}$ of $p - 1$ integers, where

$$\mathbf{S} = \{N + 2, N + 3, N + 4, \ldots, N + p\},$$

$N = 2 \cdot 3 \cdot 4 \ldots p = p!$. These integers are divisible by 2, 3, 4, 5, . . . , p, respectively, and so are composite. By Theorem 1.6, there is no limit to the size of p, and thus there is no limit to $|\mathbf{S}|$. ◆

In contrast, there are no arbitrarily long blocks of consecutive odd integers that are prime (why?).

One way to examine the density of occurrence of certain subsets of integers among all of $\mathbf{N}$ is to look at their series of reciprocals. For example, the set of all even positive integers yields a series of reciprocals that diverges,

$$\sum_{n=1}^{\infty} \frac{1}{2n} = \frac{1}{2} \sum_{n=1}^{\infty} \frac{1}{n} = \infty.$$

On the other hand, the Taylor series expansion for e^x

$$e^x = 1 + \sum_{n=1}^{\infty} \frac{x^n}{n!} \qquad (x \in \mathfrak{R})$$

implies that the series of reciprocals of the factorials converges:

$$\sum_{n=1}^{\infty} \frac{1}{n!} = e - 1.$$

What about the series of reciprocals of the primes? It is certainly true that $p_n > 2n$ for $n \geq 5$, so that $1/p_n < 1/2n$ for $n \geq 5$. Then

$$\sum_{n=5}^{\infty} \frac{1}{p_n} \leq \sum_{n=5}^{\infty} \frac{1}{2n},$$

and if the primes are "scarce enough," the series on the left *might* converge.

On the other hand, examination of tables suggests that $p_n < n!$ holds for $n \geq 3$ ($p_3 = 5$, $3! = 6$; $p_8 = 19$, $8! = 40320$; $p_{9500} \approx 10^5$, $9500! \approx 10^{33658}$). So, possibly, $1/n! < 1/p_n$, and

$$\sum_{n=3}^{\infty} \frac{1}{p_n} \geq \sum_{n=3}^{\infty} \frac{1}{n!}$$

and if the primes are not scarce enough, the series on the left *might* diverge.

We don't have a formula for the nth prime, so the idea of somehow using the Comparison Test to decide the convergence of $\Sigma_{n=1}^{\infty} p_n^{-1}$ seems hopeless. It was therefore an interesting analytical result when the matter was settled in the mid-18th century. Our proof uses the concept of squarefree integers. The term **squarefree** signifies an integer that is not divisible by any integral square larger than 1^2. The primes are squarefree, as are integers such as 6, 10, and 15. Any positive integer n can, when factored, be written uniquely as the product of a square j^2 times a squarefree integer k.[7] For example, one has $12 = 2^2 \cdot 3$, $45 = 3^2 \cdot 5$, and $216 = 6^2 \cdot 6$.

The expansion of $e - 1$ as a series of reciprocals of factorials that was given earlier suggests that we might look at a similar expansion that yields reciprocals of the primes (and other terms, as well). For example, we observe that

$$\begin{aligned} e^{1/p_1} e^{1/p_2} &= \left[1 + \frac{1}{p_1} + \ldots\right]\left[1 + \frac{1}{p_2} + \ldots\right] \\ &= 1 + \left[\frac{1}{p_1} + \frac{1}{p_2}\right] + \left[\frac{1}{p_1 p_2}\right] + \text{other terms} \end{aligned}$$

and among the terms on the right-hand side are those of the form $1/k$, where k is squarefree and does not exceed $p_1 p_2$. The idea of the proof that follows is to set up a pair of contradictory statements involving a series of reciprocal squarefree integers, $\Sigma\, 1/k$, and the series of reciprocal primes.

Theorem 1.8 The series $\Sigma_{n=1}^{\infty} p_n^{-1}$ diverges.

Proof Let N be a positive integer and define the partial sum $S_N = \Sigma_{n=1}^{N} p_n^{-1}$. Then

$$e^{S_N} = \prod_{n=1}^{N} e^{1/p_n} > \prod_{n=1}^{N} \left[1 + \frac{1}{p_n}\right].$$

[7] We borrow this result as a consequence of the Fundamental Theorem of Arithmetic, which we prove in the next chapter.

If the right-hand side is multiplied out, it yields exclusively terms with numerator 1 and denominators that are all of the squarefree integers not exceeding $p_1 p_2 \cdots p_N$. Hence, we can write, upon summation over all squarefree k not exceeding N,

$$e^{S_N} > \sum_{k \le N} \frac{1}{k}.$$

The function e^x is an increasing function. Let us assume, contrary to what is to be proved, that $\lim_{N\to\infty} S_N = S < \infty$. Then $S > S_N$, and we obtain

$$e^S > \sum_{k \le N} \frac{1}{k}, \qquad (*)$$

for any positive integer N.

On the other hand, since any positive integer n can be written uniquely as $n = j^2 k$, where k is squarefree, we have the inequality

$$\left(\sum_{k \le N} \frac{1}{k}\right)\left(\sum_{j=1}^{N} \frac{1}{j^2}\right) > \sum_{n=1}^{N} \frac{1}{n} \qquad (N > 1).$$

The inequality follows because every n on the right-hand side can be found as a denominator somewhere on the left-hand side.

Now the series $\Sigma_{j=1}^{\infty}\ 1/j^2$ converges (use the Integral Test), say, to J. Hence, the preceding inequality leads to

$$\sum_{k \le N} \frac{1}{k} > J^{-1} \sum_{n=1}^{N} \frac{1}{n}.$$

But the harmonic series diverges, so we can choose N large enough so that

$$\sum_{n=1}^{N} \frac{1}{n} > Je^S.$$

We obtain, consequently, the inequality

$$\sum_{k \le N} \frac{1}{k} > e^S,$$

which contradicts (*). We conclude that $\lim_{N\to\infty} S_N$ is not finite. ◆

Theorem 1.8 was first proved (but differently) in 1737 by the marvelous Swiss mathematician **Leonhard Euler** (1707–1783) (Burckhardt, 1983). Our proof is patterned after Niven (1971). Several other proofs, each with certain advantages, are known (Leavitt, 1979; Vanden Eynden, 1980).

Problems

1.64. Find a block of 10 consecutive composite numbers.

1.65. What is the longest possible string of squarefree integers? What is the longest possible string of consecutive cubefree integers? Give two such strings.

1.66. From the data in Table 1.3 make a plot of $\ln \pi(N)$ versus $\ln N$ for $N = 10^4$ to 10^{12}. If your plot suggests that some other (but related) graph might be more revealing, prepare such a graph.

1.67. Consider the set of integers $\mathbf{S} = \{2^0 0!, 2^1 1!, 2^2 2!, 2^3 3!, \ldots\}$. What do we get when we sum its series of reciprocals? Answer the same question for the set $\mathbf{S}' = \{3, 7, 11, 15, 19, \ldots\}$.

1.68. Verify, as asserted in the proof of Theorem 1.8, that the series $\Sigma_{j=1}^{\infty} j^{-2}$ is convergent.

1.69. Illustrate numerically with a particular choice of N that $\Pi_{n=1}^{N} (1 + p_n^{-1}) > \Sigma_{k \leq N} 1/k$ holds, as indicated in the proof of Theorem 1.8.

1.70. Use the computer to arrive at a numerical estimate of the number J in the proof of Theorem 1.8.

1.71. Discuss the convergence or divergence of the series

$$\sum_{n=1}^{\infty} [\exp\frac{1}{p_n} - 1]$$

where $\{p_n\}_{n=1}^{\infty}$ is the sequence of the primes. Discuss the convergence or divergence of the series $\Sigma_{n=1}^{\infty} p_n^{-2}$.

1.72. An alternative but accessible proof of Theorem 1.8 is given in Clarkson (1966). Consult this and work through the details.

1.7 Primes of Various Forms

Various polynomial forms can be made to fit certain subsets of the primes. The simplest such forms are linear forms. Thus, some primes such as 7 and 13 are of the form $3k + 1$. Obviously, not all integers of the form $3k + 1$ are prime, nor are all primes of the form 3k+1. No single polynomial in one variable can accommodate all primes, although some such polynomials can yield several primes (Boston and Greenwood, 1995). Interestingly, it is possible to construct more complex polynomials, all of whose positive values are the primes. One such polynomial, shown explicitly in Jones, Sato, Wada, and Wiens (1976), is a polynomial of algebraic degree 25 in 26 integral variables. Still other functions are known, that are not of a polynomial nature, but yield all the primes (Willans, 1964). Nevertheless, we can ask about the distribution of primes of a given representational type. The following simple theorem illustrates one particular linear form.

Theorem 1.9 The number of primes of the form $4k + 3$ is infinite.

Proof All primes greater than 2 are of the form $4k + 1$ or $4k + 3$ since integers of the form $4k$, $4k + 2$ are even and hence composite. Also, observe that the product of any two integers of the form $4k + 1$ is another integer of that form:

$$(4k + 1)(4m + 1) = 16km + 4(k + m) + 1 = 4(4km + k + m) + 1.$$

Assume, to the contrary, that P is the largest possible $(4k + 3)$-prime, and consider the integer N, defined as

$$N = 2^2 \cdot 3 \cdot 5 \cdot 7 \cdot 11 \ldots P - 1,$$

where the extended product is over all primes from 2 to P. This integer is of the form $4k + 3$ since it can be written as

$$\begin{aligned} N &= (2^2 \cdot 3 \cdot 5 \cdot 7 \cdot 11 \ldots P - 4) + (4 - 1) \\ &= 4(3 \cdot 5 \cdot 7 \cdot 11 \ldots P - 1) + 3. \end{aligned}$$

If N is prime, then we have found a new $(4k + 3)$-prime that exceeds P, a contradiction. If N is composite, then at least one of its prime divisors must be a $(4k + 3)$-prime since (as we just saw) multiplication of only $(4k + 1)$-primes yields integers of the same form.

A $(4k + 3)$-prime that divides N cannot be any of the $(4k + 3)$-primes in the list $3, 7, 11, \ldots, P$ because any such division leaves a remainder. Hence, the $(4k + 3)$-prime factor must lie outside this list; that is, it must exceed P, (again, a contradiction). The conclusion is that no P exists that is the largest possible $(4k + 3)$-prime, and the set of such primes is infinite. ◆

You can see that the preceding proof is patterned very closely after our proof of Euclid's theorem (Theorem 1.6). In almost the same way, we can prove the following (Problem 1.75):

Table 1.4 Primes of the Form $11k + 3$

k	$11k + 3$	Prime ?	k	$11k + 3$	Prime ?
2	25	No	100	1103	Yes
4	47	Yes	200	2203	Yes
6	69	No	300	3303	No
8	91	No	500	5503	Yes
10	113	Yes	1000	11003	Yes
14	157	Yes	2000	22003	Yes

Theorem 1.10 The number of primes of the form $6k + 5$ is infinite.

Let us randomly take another linear form and examine it (see Table 1.4).

It seems plausible to conjecture that there are an infinite number of primes of the form $11k + 3$. More generally, it seems plausible to conjecture that there are an infinite number of primes of any linear form $Ak + B$, where A and B have no common divisor (for if A and B did, any such number $Ak + B$ would be composite!). Unfortunately, the proofs of Theorems 1.9 and 1.10 employ special algebraic devices in order to get started, and these devices do not appear to be generalizable to other linear forms. The general result, stated shortly as Theorem 1.11, was proved by the German mathematician **Peter G.L. Dirichlet** (1805–1893) in 1837, but the proof requires considerable analysis and is too difficult to be included here. No simple proof is known. Also, one might well ask about primes of higher forms (e.g., quadratic, cubic). Very little is known about them; for example, it is not known if there are infinitely many primes of the form $k^2 + 1$ or $k^2 + n$ or $k^2 + k + 1$.

Theorem 1.11 (Dirichlet's Theorem). Let $A, B \in \mathbf{N}$, and suppose that A, B have no common divisor other than ± 1. Then the arithmetic progression $Ak + B, k = 0, 1, 2, \ldots$ contains an infinite number of primes.

Example 1.15 There are an infinite number of integers of the form $4k^2 + 4k + 1$ that are squares of primes since $4k^2 + 4k + 1 = (2k + 1)^2$, and Theorem 1.11 applies to $2k + 1$.

Problems

1.73. Find six primes (a) of the form $6k + 1$, (b) of the form $7k + 5$, (c) of the form $12k + 7$.

1.74. Find six primes that are simultaneously of the form $4k + 3$ and $5k' + 1$.

1.75. Prove Theorem 1.10.

1.76. Find six squarefree integers of the form $12k + 10$. How do you know that there is actually an infinite supply of them?

1.77. Write a computer program to check the following polynomial functions as generators of primes: (a) $2n^2 + 29$, (b) $n^2 - n + 41$, (c) $n^2 - 81n + 1681$, (d) $n^2 - 79n + 1601$, (e) $9n^2 - 231n + 1523$ (Higgins, 1981–1982). In each case restrict n to values in $\mathbf{N}$.

1.78. In searching for primes of the form $n^2 + n + 1$, why should one omit as choices of n those integers that are perfect squares?

1.79. Prove: If an odd prime p is of the form $7k + 2$, then it is actually of the form $14k' + 9$. Similarly, show that if the prime $p > 3$ is of the form $7k + 3$, then it is actually either of the form $21k' + 10$ or of the form $21k' + 17$.

1.80. There are only seven positive integers C such that the quadratic $x^2 + x + C$ yields exclusively primes for all x in $0 \leq x \leq C - 2$ (Fendel, 1985). All of these C's are less than 100. Write a computer program to discover them. Why do we exclude $x = C - 1$?

1.81. It is an automatic deduction from Dirichlet's Theorem that the number of $(4k + 1)$-primes is infinite. Discuss what problems arise if you try to prove this by mimicking the proof of Theorem 1.9.

1.82. Dirichlet's Theorem says that the sequence of integers $\{Ak + B\}$, where A, B have no common divisor other than ± 1, contains infinitely many primes. It does not say that all such numbers $Ak + B$ are prime, nor does it say how many integers in such a sequence are themselves *primes in an arithmetic progression.*

(a) Find four primes in arithmetic progression. Find five primes in arithmetic progression.
(b) If you wish to discover six or more primes in arithmetic progression $\{Ak + B\}$, where $k = 0, 1, 2, 3, 4, 5, \ldots$, why must A be divisible by 10?
(c) Find six primes in arithmetic progression.

In 1983 an arithmetic progression of 18 primes was discovered; at that time this was a record (Pritchard, 1983). Slightly longer progressions are now known (Pritchard, Moran, and Thyssen, 1995).

1.8 Other Theorems and Conjectures about Primes

Dirichlet's Theorem was one of the crowning achievements of prime number theory. A more stunning accomplishment was the **Prime Number Theorem**. To explain it, we need to present a definition.

> **DEF.** A function f(x) is said to be **asymptotic** to g(x) as $x \to x_0$ if $\lim_{x \to x_0}$ f(x)/g(x) $= 1$, and we write f(x) $\sim$ g(x) as $x \to x_0$.

A simple case involves polynomials. A polynomial $P(x) = \Sigma_{k=0}^{n} a_k x^{n-k}$, $a_0 \neq 0$, is asymptotic to its leading term $a_0 x^n$ as x $\to \infty$ because

$$\lim_{x\to\infty} \frac{P(x)}{a_0 x^n} = \lim_{x\to\infty} \left[1 + \frac{a_1}{a_0}\frac{1}{x} + \frac{a_2}{a_0}\frac{1}{x^2} + \ldots + \frac{a_n}{a_0}\frac{1}{x^n} \right]$$

$$= 1.$$

Example 1.16 Let g(x) $= 1 + x$ and f(x) $= e^x$. Then we obtain

$$\lim_{x\to 0} \frac{f(x)}{g(x)} = \lim_{x\to 0} \frac{e^x}{1+x} = 1,$$

so $e^x \sim (1 + x)$ as $x \to 0$.

In some cases it may be necessary to use l'Hôpital's Rule.

Example 1.17 Let g(x) $= x^{-1}$ and f(x) $=$ csc x. Then from l'Hôpital's Rule we have

$$\lim_{x\to 0} \frac{\csc x}{x^{-1}} = \lim_{x\to 0} \frac{x}{\sin x} = \lim_{x\to 0} \frac{1}{\cos x} = 1,$$

and so csc $x \sim x^{-1}$ as $x \to 0$.

Qualitatively, the definition says that as $x \to x_0$, the function f(x) tends to "look like" the function g(x). The concept of asymptotic equivalence and its look-alike interpretation hold even when f(x), g(x) are not defined at $x = x_0$. This is the case with Example 1.17. The asymptotic equivalence there is shown graphically in Figure 1.2.

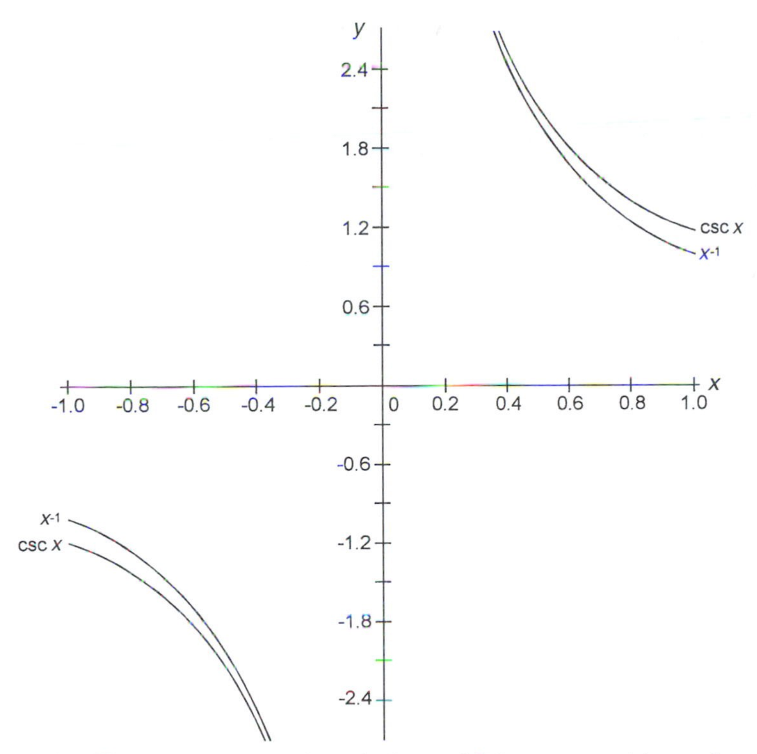

Figure 1.2 Showing the asymptotic equivalence of $f(x) = \csc x$ to $g(x) = x^{-1}$ as $x \to 0$

Now recall the discrete function $\pi(n)$ mentioned in Section 1.5. When n is large, $\pi(n)$ itself is large (e.g., $\pi(10^6) = 78{,}498$) and changes very slowly with n. That is, for large n, $\pi(n)$ behaves almost like a continuous function. Which continuous function?

Adrien-Marie Legendre (1752–1833) and **Carl Friedrich Gauss** (1777–1855) conjectured independently around 1800 that $\pi(x)$ increases somewhat like $x/\ln(x)$. Their conjectures were based just on numerical evidence (Goldstein, 1973; Apostol, 1996), as neither possessed a proof. Finally, in 1896 the French mathematician **Jacques Hadamard** (1865–1963) and the Belgian mathematician **Charles J.G.N. de la Vallée Poussin** (1866–1962) independently proved Theorem 1.12.

Theorem 1.12 (The Prime Number Theorem). If n is a positive integral variable, then $\pi(n) \sim n/\ln(n)$ as $n \to \infty$.

The proofs by **Hadamard** and **de la Vallée Poussin** employed very advanced analysis and are too difficult to be included here. More than 50 years later (in 1949) the brilliant and eccentric **Paul Erdös** (1913–1996) (Bollobás, 1996, 1998) and **Atle Selberg** (1917–) showed how Theorem 1.12 could be proved without most of the heavy analytical machinery; regrettably, their proofs are still very long and difficult.

Before the Prime Number Theorem was proved in 1896, **Chebyshev** had deduced a related but simpler result. He proved in 1852 that there exist positive constants A, B, not greatly different from 1, such that

$$\mathrm{A}\frac{n}{\ln(n)} < \pi(n) < \mathrm{B}\,\frac{n}{\ln(n)}$$

for all $n > 1$. What this says is that the "order of magnitude" of $\pi(n)$ is $n/\ln(n)$. This, of course, was stronger than the mere numerical conjectures of **Legendre** and **Gauss**, but did not go quite as far as Theorem 1.12. Table 1.5 gives some illustrative data. It is known that for all integers $n \geq 11$, one has $n/\ln\, n < \pi(n)$.

A great many questions about primes to this day remain undecided. We list a few of the best-known ones here.

Conjecture 1 (Goldbach's Conjecture). Every even integer ≥ 4 can be written as the sum of two odd primes.

It is known that every integer n from some point on is the sum of no more than four odd primes. Also, the conjecture has been verified by computer for $N \leq 2 \times 10^{10}$ (Silverman, 1991). These are steps forward; there is still a long way to go, but the conjecture seems plausible (Problem 1.95).

Table 1.5 Illustration of Chebyshev's Result on the Order of Magnitude of $\pi(n)$

n	$\pi(n)$	$n/\ln(n)$	% Error
10,000	1,229	1,086	12
100,000	9,592	8,686	9
1,000,000	78,498	72,382	8
10,000,000	664,579	620,421	7
100,000,000	5,761,455	5,428,681	6

Conjecture 2 (Fermat). There are infinitely many primes of the form

$$F_n = 2^{2^n} + 1, \qquad n = 0, 1, 2, \ldots.$$

To date, only five Fermat primes have been found, those for $n = 0$ to 4. The Fermat numbers for $n = 5$ to 22 are known to be composite, and factorizations are known for most of them (Crandall et al., 1995; Pomerance, 1996).

Conjecture 3 (Mersenne). There are only finitely many primes of the form $M_p = 2^p - 1$, where p is itself a prime.

At least 36 such Mersenne primes M_p are known, a recent one being $2^{2,976,221} - 1$, which has 895,932 digits. Mersenne primes lead to what are called perfect numbers; we say more on Mersenne primes in Chapter 4 and on perfect numbers in Chapter 7. The scarcity of Mersenne primes is illustrated by the fact that there are only two Mersenne exponents p between 100,000 and 139,268 and none between 216,092 and 353,620 (Colquitt and Welsh, 1991).

Conjecture 4 (Twin Primes Conjecture). There are infinitely many pairs of primes of the form $(p, p + 2)$.

This seems to be supported by numerical evidence. There are 8169 such pairs below 1,000,000. However, the density of twin primes cannot be very high because the Norwegian mathematician **Viggo Brun** (1885–1978) proved in 1919 that the series

$$\sum_p \left[\frac{1}{p} + \frac{1}{p + 2}\right]$$

converges, where the sum is taken over all primes p for which $p + 2$ is also prime. The limit of this series (Problem 1.97) is now called **Brun's constant**, B (Ribenboim, 1996). We know from Theorem 1.8 that the preceding series diverges if p is taken over *all* primes. A 1996 record (but see Problem 1.98) for the largest known pair of twin primes was $570{,}918{,}348 \times 10^{5,120} \pm 1$ (Ribenboim, 1996).

Problems

1.83. Show that the important function $f(x) = x/\ln x$, which occurs in the Prime Number Theorem, is an increasing function for $x > e$.

1.84. If we accept Theorem 1.14 as true, what is the value of

$$\lim_{n\to\infty} \frac{\pi(n)}{n}?$$

What can you say about

$$\lim_{n\to\infty} \frac{\pi(n)}{\sqrt{n}}?$$

1.85. The following exact but impractical formula for $\pi(n)$ appears in (Ribenboim, 1996):

$$\pi(n) = \sum_{k=2}^{n} \left\lfloor \frac{(k-1)!+1}{k} - \left\lfloor \frac{(k-1)!}{k} \right\rfloor \right\rfloor.$$

Here, the brackets $\lfloor \cdot \rfloor$ mean **greatest integer function**, that is, $\lfloor 10.6 \rfloor = 10$, $\lfloor 0.91 \rfloor = 0$, and $\lfloor 3 \rfloor = 3$. Verify the formula for $n = 13$.

1.86. An alternative version of the Prime Number Theorem is that $\pi(x)$ is asymptotic to the function $\mathrm{li}(n) = \int_2^n (1/\ln t)\, dt$ as $n \to \infty$. The integral is called the **logarithmic integral**; values of it are calculated numerically and tabulated in reference sources. Consult such a source and determine how well li(10,000) approximates $\pi(10{,}000)$ (Abramowitz and Stegun, 1965; Jahnke and Emde, 1945).

1.87. Suppose p is a prime and let $\{p_i\}$ be the primes in the interval $[2, \sqrt{p})$. A proposition due to **Legendre** says that the number N of primes in the interval $(\sqrt{p}, p]$ is

$$N = \left\{ p - \sum_i \left\lfloor \frac{p}{p_i} \right\rfloor + \sum_{i,j} \left\lfloor \frac{p}{p_i p_j} \right\rfloor - \sum_{i,j,k} \left\lfloor \frac{p}{p_i p_j p_k} \right\rfloor + \ldots \right\} - 1.$$

The summations are over all primes p_i, products of all distinct pairs p_i, p_j of primes (order of multiplication is unimportant), products of all distinct triples p_i, p_j, p_k of primes, and so forth. Program this on the computer and check out how well it works for several choices of p.

1.88. Let P_n be the probability that an integer drawn randomly from the set of all positive n-digit numbers is a prime. A theorem due to the German mathematician **Edmund Landau** (1877–1938) says that $P_n \sim 1/(n \ln 10)$ as $n \to \infty$. Consult a listing of primes and check out how close this is for $n = 2, 3, 4, 5$.

1.89. A conjecture by the French mathematician **Joseph-Louis Lagrange** (1736–1813) states that every $(4k + 3)$-prime (≥ 7) is the sum of a $(4k + 1)$-prime and twice some other $(4k + 1)$-prime. For example, $419 = 193 +$

2(113). Write a computer program to check out Lagrange's conjecture for some challenging primes.

1.90. Corroborate the stated number of digits in the Mersenne prime $2^{2,976,221} - 1$.

1.91. Write a computer program to express an even integer of six digits or fewer as the sum of two primes. Test it out on 1492, 24,680, 123,456.

1.92. The following was once proposed as a criterion to determine if an integer $n > 1$ is a prime: "n is a prime if n divides $2^n - 2$." Write a computer program to check the validity of this for several n.

1.93. A conjecture by the English mathematician **G.H. Hardy** (1877–1947) says that if $P_2(n)$ is the number of *pairs* of twin primes (see Problem 1.55) in the interval $[3, n + 2]$, then

$$P_2(n) \sim \frac{2Cn}{(\ln n)^2}$$

as $n \to \infty$, where

$$C = \prod_{p \geq 3}^{\infty} \left\{1 - \frac{1}{(p - 1)^2}\right\} \approx 0.66016.$$

Write a computer program and check out how well this alleged asymptotic formula works for $n = 1000, 2000, 5000, 10{,}000$. See Rubinstein (1993) for a heuristic proof of the conjecture.

1.94. Let $\mathrm{p}(n) = 60n^2 - 1710n + 12{,}150$. Investigate the occurrence of twin primes, $\{\mathrm{p}(n) - 1, \mathrm{p}(n) + 1\}$, for various $n \in \mathbf{Z}$.

1.95. An elementary combinatorial approach (Clarke and Shannon, 1983) to Goldbach's Conjecture shows that if primes less than 10^6 are used, the probability of a violation of the conjecture is less than 10^{-2911}! Read this article and make a brief report on its contents.

1.96. Not every assertion about primes is hard to prove. Here's an easy one: Prove that there are an infinite number of primes that end in 999. Also show that if p is such a prime, then 37 does not divide $p - 999$.

1.97. Estimate Brun's constant to one significant figure. [Actually, this is quite hard to do at this point, and you will not be able to be very rigorous in this effort, but try anyway.]

1.98. The search for larger and larger twin primes goes on (moral: mathematical records are continually being broken). The pair $6{,}797{,}727 \times 2^{15{,}328} \pm 1$ has been reported (Forbes, 1997). How many digits are in these primes?

RESEARCH PROBLEMS

1. It is plausible that the density of twin primes might decrease as one moves farther out on the positive real axis. We already know that this behavior is true of the primes themselves. Investigate the behavior of the ratio of the density of twin primes to the density of the primes.

2. Refer to Problem 1.86 for the definition of the logarithmic integral. This is of interest in number theory only for large n. Investigate ways for the numerical evaluation of the integral. For background you might wish to look at a chapter on integration in a numerical analysis text.

3. Appropriately constructed systems of polynomials have some properties analogous to the system of integers $\mathbf{Z}$. Let $\mathbf{Z}_3[x]$ be the system of polynomials in x in which the coefficients are 0, 1, or 2. Let addition and multiplication of the coefficients be defined by these tables:

$\oplus$	0	1	2
0	0	1	2
1	1	2	0
2	2	0	1

$\otimes$	0	1	2
0	0	0	0
1	0	1	2
2	0	2	1

The analog in $\mathbf{Z}_3[x]$ of primes in $\mathbf{Z}$ are those polynomials with leading coefficient 1 that are not factorable (reducible) into products of simpler polynomials in $\mathbf{Z}_3[x]$. For example, $x^2 + 1$ is such an irreducible polynomial, but $x^2 + 2 = (x + 1)(x + 2)$ is not. Investigate if an analog of Theorem 1.6 can hold in $\mathbf{Z}_3[x]$.

References

Abramowitz, M., and Stegun, I.A. (eds.), *Handbook of Mathematical Functions*, Dover, New York, 1965. See Chap. 5, expecially Table 5.2, on the exponential integral and related functions (such as the logarithmic integral).

Allenby, R.B.J.T., and Redfern, E.J., *Introduction to Number Theory with Computing*, Edward Arnold, London, 1989. Chapters 0, 1, 2 on introductory matters, divisibility, and the primes somewhat parallel our first two chapters; interesting reading.

Apostol, T.M., "A Centennial History of the Prime Number Theorem," *Engineering & Science* (California Institute of Technology), **59** (4), 18–28 (1996). Very nicely written, elementary synopsis of the story of the Prime Number Theorem.

Beyer, W.H. (ed.), *CRC Standard Mathematical Tables*, 27th ed., CRC Press, Boca Raton, 1984. See pp. 84–91 for a table of primes up to 99,991.

Bollobás, B., "A Life of Mathematics—Paul Erdös, 1913–1916," *Focus*, **16** (6), 1–6 (1996). An obituary of this seminal figure in 20th-century mathematics.

Bollobás, B., "To Prove and Conjecture: Paul Erdös and His Mathematics," *Amer. Math.*

Monthly, **105**, 209–237 (1998). Compared to the previous cite, this is a more detailed look at this 20th-century giant, whose motto was "Another roof, another proof." His career began in 1932 with an independent proof of Bertrand's Postulate, which prompted **N. Fine** some years later to pen the lines, "Chebyshev said, and I say it again, There's always a prime between n and $2n$." Nice photos in this article.

Boston, N. and Greenwood, M.L., "Quadratics Representing Primes," *Amer. Math. Monthly*, **102**, 595–599 (1995). Fascinating article.

Burckhardt, J.J., "Leonhard Euler, 1707–1783," *Math. Mag.*, **56**, 262–273 (1983). Interesting article on the 200th anniversary of Euler's death. Devoted mainly to brief summaries of the areas in which **Euler** worked. The man was a writing machine.

Clarke, J.H., and Shannon, A.G., "A Combinatorial Approach to Goldbach's Conjecture," *Math. Gaz.*, **67**, 44–46 (1983). Elementary but persuasive!

Clarkson, J.A., "On the Series of Prime Reciprocals," *Proc. Amer. Math. Soc.*, **17**, 541 (1966). Very short and elementary.

Colquitt, W.N., and Welsh, L., Jr., "A New Mersenne Prime," *Math. Comp.*, **56**, 867–870 (1991). Skim-read this to get an idea of what is involved computationally in discovering Mersenne primes. The Lucas–Lehmer test, stated in this article, will be discussed in our Section 4.4.

Crandall, R., Doenias, J., Norrie, C., and Young, J., "The Twenty-Second Fermat Number Is Composite," *Math. Comp.*, **64**, 863–868 (1995). Hence, F_4 is still the largest known Fermat prime.

Dence, J.B., and Dence, T.P., *A First Course of Collegiate Mathematics*, Krieger Publishing Co., Malabar, FL, 1994. Read Chapter 2, "Logical Argumentation," for background material on formal logic.

Fendel, D., "Prime-producing Polynomials and Principal Ideal Domains," *Math. Mag.*, **58**, 204–210 (1985). This article may require more algebraic background than you have presently, but skim-read it anyway.

Forbes, T., "A Large Pair of Twin Primes," *Math. Comp.*, **66**, 451–455 (1997). High-level computational work, such as described in this article, is a highly specialized part of contemporary number theory.

Goldstein, L.J., "A History of the Prime Number Theorem," *Amer. Math. Monthly*, **80**, 599–615, 1115 (1973). The Prime Number Theorem has had a fascinating history, which is told here excellently; there is more detail here than in Apostol's article.

Hardy, G.H., and Wright, E.M., *An Introduction to the Theory of Numbers*, 5th ed., Oxford University Press, Oxford, 1979. **Hardy** was a master expositor. Read his Chapters 1 and 2 (pp. 1–22) in order to catch the flavor of his writing.

Higgins, O., "Another Long String of Primes," *J. Recreational Math.*, **14** (3), 185 (1981–1982). See if you can discover why the two prime-producing polynomials $9x^2 - 231x + 1523$ and $9x^2 - 471x + 6203$ are interesting.

Jahnke, E., and Emde, F., *Tables of Functions*, 4th ed., Dover, New York, 1945. Material on the logarithmic integral is contained in Chapter 1.

Jones. J.P., Sato, D., Wada, H., and Wiens, D., "Diophantine Representation of the Set of Prime Numbers," *Amer. Math. Monthly*, **83**, 449–464 (1976). We dare you to program the authors' polynomial and evaluate it for several choices of the 26 independent variables.

Leavitt, W.G., "The Sum of the Reciprocals of the Primes," *Two Year Coll. Math. J.*, **10**, 198–199 (1979). An alternative, neat proof of our Theorem 1.8.

Lipschutz, S., *Set Theory and Related Topics*, Schaum's Outline Series, McGraw–Hill, New York, 1964, ppl 179–184. Read Chapters 1 through 3 on sets and Chapter 12 on "Axiom of Choice, Zorn's Lemma, Well-Ordering Theorem."

Niven, I., "A Proof of the Divergence of $\Sigma(1/p)$," *Amer. Math. Monthly*, **78**, 272–273 (1971). A very nice proof of this basic result.

Pegg, Jr., E., "A Long Sequence of Composite Numbers," *College Math. J.*, **28**, 121 (1997). According to MAPLE the 1411 consecutive integers between the primes $131 \times 10^{55} - 51$ and $131 \times 10^{55} + 1361$ may constitute the longest known computer-verified sequence of composite integers.

Pomerance, C., "A Tale of Two Sieves," *Notices Amer. Math. Soc.*, **43**, 1473–1485 (1996). Skim-read this summary of contemporary work on the factoring of large numbers, such as the Fermat numbers.

Pritchard, P., Moran, A., and Thyssen, A., "Twenty-two Primes in Arithmetic Progression," *Math. Comp.*, **64**, 1337–1339 (1995). $28{,}383{,}220{,}937{,}263 + 1{,}861{,}263{,}814{,}410\,k$ is prime for $0 \leq k \leq 21$.

Pritchard, P.A., "Eighteen Primes in Arithmetic Progression," *Math. Comp.*, **41**, 697 (1983). Perhaps you would care to confirm the primality of the members of this sequence!

Ribenboim, P., "Prime Number Records," *College Math. J.*, **25**, 280–290 (1994). Filled with neat facts about primes.

Ribenboim, P., *The New Book of Prime Number Records*, Springer, New York, 1996, pp. 181, 259–265. The first citation is to a formula for $\pi(n)$, and the second is a short summary of some facts about twin primes.

Rosen, K.H., *Elementary Number Theory and Its Applications*, 3rd ed., Addison–Wesley, Reading, MA, 1993. See Chapters 1 and 2 (pp. 4–118) on divisibility, greatest common divisor, and prime factorization.

Rubinstein, M., "A Simple Heuristic Proof of Hardy and Littlewood's Conjecture B," *Amer. Math. Monthly*, **100**, 456–460 (1993). Uses simple combinatorial arguments plus a theorem by **Dirichlet** that we have not included in this book. Otherwise, this article should be accessible to you.

Silverman, R.D., "A Perspective on Computational Number Theory," *Notices Amer. Math. Soc.*, **38**, 562–568 (1991). A very interesting article that summarizes contemporary results obtained using large computers.

Sominsky, I.S., *The Method of Mathematical Induction*, Mir Publishers, Moscow, 1975. This is an excellent little (62-page) booklet, which we strongly recommend to undergraduate readers.

Templer, M., "On the Primality of $k! + 1$ and $2 \cdot 3 \cdot 5 \ldots p + 1$," *Math. Comp.*, **34**, 303–304 (1980). Imagine that $2 \cdot 3 \cdot 5 \ldots 1021 + 1$ is prime! Care to guess how many digits are in this prime?

Vanden Eynden, C., "Proofs That $\Sigma\ 1/p$ Diverges," *Amer. Math. Monthly*, **87**, 394–397 (1980). Summary of several other proofs alternative to ours. The proof by **Erdös** is worth studying for its utter simplicity.

Wilder. R.L., *Introduction to the Foundations of Mathematics*, 2nd ed., Krieger Publishing Co., Malabar, FL, 1983, pp. 114–123. A discussion of well-ordered sets and of Zermelo's Well-Ordering Theorem; advanced.

Willans, C.P., "On Formulae for the *N*th Prime Number," *Math. Gaz.*, **48**, 413–415 (1964). Neat article; uses some material from our Chapter 3.

Chapter 2
The Fundamental Theorem of Arithmetic and Its Consequences

2.1 The Fundamental Theorem of Arithmetic

We shall use the term **factorization** to refer to a representation of a composite integer as a product completely of positive primes. Although Theorem 1.5 has told us that all positive integers greater than 1 that are not prime must be factorable into products of primes, we do not as yet know if such factorizations are unique. Theorem 2.1, which follows shortly, is the cornerstone of all arithmetic in **N**.

Lemma 2.1.1 The prime p is in a factorization of the composite positive integer n iff $p \mid n$.

Proof ($\rightarrow$) Suppose p is in a factorization of n: $n = pp_1p_2 \ldots p_k$. Hence, in the Division Algorithm we have $n = pq + r$, where $q = p_1p_2 \ldots p_k$ and $r = 0$. By our definition in Section 1.3, this says that $p \mid n$.

($\leftarrow$) Suppose $p \mid n$; then $n = qp + 0 = qp$. If q is a prime, then we have a factorization of n. If q is composite, then by Theorem 1.5 it is expressible as a product of primes: $q = p_1p_2 \ldots$, and again we have a factorization of n that contains p. ◆

Note that Lemma 2.1.1 does not require n to possess unique factorization. The lemma may be extended, but with care, since we do not yet know unique factorization to be valid.

Lemma 2.1.2 Suppose that the positive integer n possesses unique factorization. Then the distinct primes $p_1, p_2, \ldots, p_r$ appear in this factorization iff $p_1 \mid n, p_2 \mid n, \ldots, p_r \mid n$.

This generalization is obtainable from Theorem 1.5 and Lemma 2.1.1 by mathematical induction (Problem 2.6).

Example 2.1 Suppose that 540 has a unique factorization. The primes 2, 3, 5 are all divisors of 540. Hence, 2, 3, 5 appear in in the factorization of 540. In fact, we can write $540 = 2^2 \cdot 3^3 \cdot 5^1$.

Suppose that 189 has the unique factorization: $189 = 3^3 \cdot 7^1$. Hence, $3 \mid 189$ and $7 \mid 189$. ◆

We come now to the main result of this section, the **Fundamental Theorem of Arithmetic**. In some treatments the Fundamental Theorem is deduced as a consequence of Euclid's Lemma (our Corollary 2.12.3). But this requires considerable development of the derivative topic of the greatest common divisor. We prefer to arrive at the Fundamental Theorem from consideration of the primitive notion of factorization together with use of the very basic Well-Ordering Property. Euclid's Lemma will then appear much farther down the line as a by-product of the Fundamental Theorem.

Theorem 2.1 (The Fundamental Theorem of Arithmetic).[1] The factorization of a positive integer is unique except possibly for the ordering of the factors.

Proof Let $n \in \mathbf{N}$ be composite and arbitrary; by Theorem 1.5 it is factorable into a product of primes in at least one way. Let $\mathbf{S} \subset \mathbf{N}$ be the set of composite positive integers that are factorable into primes in *more than one* way. Assume $\mathbf{S}$ is nonempty. By the Well-Ordering Property (Theorem 1.2) $\mathbf{S}$ has a smallest member $n_{\min}$,

$$n_{\min} = p_1 p_2 \cdots p_k = q_1 q_2 \cdots q_m,$$

where the p_i's, q_i's are primes.

No p_i can be a q_j and vice versa. For example, if $p_1 = q_1$ held, then from Lemma 2.1.1 both factorizations of $n_{\min}$ would be divisible by p_1, leaving $p_2 p_3 \cdots p_k = q_2 q_3 \cdots q_m$ as two factorizations of a new element in $\mathbf{S}$ that is smaller than $n_{\min}$. This is a contradiction.

Now arrange the two factorizations of $n_{\min}$ so that p_1, q_1 are the smallest of the p_i's, q_j's, respectively. Then $p_1 \le p_2$ and multiplication of both sides by p_1 gives $p_1^2 \le p_1 p_2 \le n_{\min}$. Similarly, we obtain $q_1^2 \le n_{\min}$ and combination with the previous inequality gives $p_1^2 q_1^2 < n_{\min}^2$, or $p_1 q_1 < n_{\min}$, strictly, because $p_1 \ne q_1$.

[1] The first proof of this theorem that is acceptable by modern standards appears to have been given by **Gauss** (who did not attach the modifier "Fundamental"), although the content of the theorem in restricted form appeared as Proposition 14 in Book IX of Euclid's *Elements*. Worthwhile comments about this are given in Knorr (1976). Our proof here comes from (Lindemann, 1933).

Next, consider the positive integer

$$N = n_{\min} - p_1 q_1.$$

This integer does not belong to **S** because it is less than $n_{\min}$. Since $p_1 \mid n_{\min}$ and $p_1 \mid p_1 q_1$, then by Theorem 1.4(b) $p_1 \mid N$. Similarly, $q_1 \mid N$. Because $N \in \mathbf{S}$, it has a unique factorization. By Lemma 2.1.2 the primes p_1, q_1 appear in the factorization of N since each divides N: $N = p_1 q_1 p_a p_b \ldots p_n$. Thus, in the Division Algorithm $N = (p_1 q_1)q + r$, where $q = p_a p_b \ldots p_n$ and $r = 0$, so $p_1 q_1 \mid N$. Again, from Theorem 1.4(b) it follows that $p_1 q_1 \mid n_{\min}$ or, equivalently, that $q_1 \mid (n_{\min}/p_1)$.

But $(n_{\min}/p_1) < n_{\min}$ and thus does not belong to **S**. It has the unique factorization $p_2 p_3 \ldots p_k$. Since q_1 is not one of these p_i's, then $q_1 \mid (n_{\min}/p_1)$ is a contradiction. We conclude that **S** is empty. ◆

Theorem 2.1 is an elegant, global statement about *all* of **N**. The following example, due in spirit to **David Hilbert**, shows what can happen in the case of a proper subset of **N**.

Consider the following set, on which multiplication is **closed** (the product of any two numbers in **S** is another number in **S**):

$$\mathbf{S} = \{x: x = 3n - 2, n \in \mathbf{N}\}.$$

For this system we define a prime p as any member of **S** that has no factors from **S** other than 1 and p. The following are some primes and some composite integers in **S**.

Primes	*Composites*
4	$16 = 4 \cdot 4$
7	$28 = 4 \cdot 7$
10	$40 = 4 \cdot 10$
13	$49 = 7 \cdot 7$
19	$52 = 4 \cdot 13$

The composites shown have unique factorizations in **S**. However, we soon discover that $100 = 4 \cdot 25 = 10 \cdot 10$, where 4, 10, 25 are primes in **S**. So although Theorem 1.5 still holds, the Fundamental Theorem of Arithmetic has broken down in **S**.

Theorem 1.5, involving only the concept of a prime, makes use only of the arithmetic operation of *multiplication*. On the other hand, the proof of Theorem 2.1 requires the *subtraction* $\mathbf{N} = \boldsymbol{n_{\min} - p_1 q_1}$. This operation in **N** still leaves one in **N**, but in **S** subtraction may take one outside of that system. In fact, $100 - 4 \cdot 10 = 60$ and $60 \notin \mathbf{S}$. In effect, the Fundamental Theorem fails in **S** because **S** is *not large enough* to provide unique prime factors for every composite integer. The integer 100 in **S** has been termed **S-irregular** (Schreiber, 1978).

The Fundamental Theorem of Arithmetic allows us to identify the complete set of positive divisors of a composite positive integer n as 1 and all possible multiplicative combinations of the primes and their powers in the factorization of n.

Example 2.2 The factorization of 12 is $2^2 \cdot 3^1$. Hence, the divisors of 12 are the members of the set

$$\{1,\ 2^1,\ 3^1,\ 2^2,\ 2^1 \cdot 3^1,\ 2^2 \cdot 3^1\},$$

six in number. This number six is intimately connected with the exponents (2 and 1) in the factorization of 12. See if you can work out the connection in this Example and discover the generalization. ◆

Naturally, some integers have more divisors than others. Intuitively, we might suppose that one integer is "rounder" than another near it if the first integer possesses more divisors than the second. This leads to an interesting classification of integers (Ratering, 1991).

> **DEF.** A **highly composite integer** is a positive integer that has more positive divisors than all lesser positive integers. A **special highly composite integer** is a highly composite integer that is a divisor of all larger highly composite integers.

The first 15 highly composite integers are 1, 2, 4, 6, 12, 24, 36, 48, 60, 120, 180, 240, 360, 720, and 840. The first 5 special highly composite integers are 1, 2, 6, 12, and 60.

Example 2.3 Find the next highly composite integer (HCI) after 60.

Since $60 = 2^2 \cdot 3^1 \cdot 5^1$, it has $(2 + 1)(1 + 1)(1 + 1) = 12$ divisors, if we reason from Example 2.2. The next HCI must have at least 13 divisors:

13: smallest integer with 13 divisors $= 2^{12} = 4096$

14: smallest integer with 14 divisors $= 2^6 \cdot 3^1 = 192$

15: smallest integer with 15 divisors $= 2^4 \cdot 3^2 = 144$

The integer 16 factors as $8 \cdot 2$ or $4 \cdot 4$ or $4 \cdot 2 \cdot 2$ or $2 \cdot 2 \cdot 2 \cdot 2$. Hence, potential HCI's include the following:

$8{\cdot}2$: $2^7{\cdot}3^1 = 384$ $\quad$ $4{\cdot}2{\cdot}2$: $2^3{\cdot}3^1{\cdot}5^1 = 120$

$4{\cdot}4$: $2^3{\cdot}3^3 = 216$ $\quad$ $2{\cdot}2{\cdot}2{\cdot}2$: $2{\cdot}3{\cdot}5{\cdot}7 = 210.$

A few simple trials show that no integers with 17 or more divisors can be less than $2^3 \cdot 3^1 \cdot 5^1$. Hence, 120 is the next HCI. ◆

Problems

2.1. Find the prime factorizations of the following composite numbers:

(a) 2673, (b) 8450, (c) 2257, (d) 17,017, (e) 12,167, (f) 56,721.

2.2. Find the prime factorizations of the following composite numbers:

(a) $2^{2^5} - 1$, (b) $3^{20} - 1$, (c) $11^4 - 5^4$, (d) $71^2 + 2(71) + 1$.

2.3. It is conjectured that any positive integer N expressible in the form $N = n^2 + m^2$ is not factorable and so must be prime. Comment on this.

2.4. A **repunit** is a positive integer consisting only of 1's in its base-10 expansion.

(a) Find two repunits that are divisible by 7.
(b) Find two repunits that are divisible by 41.
(c) Find a repunit that is divisible by both 7 and 41.

2.5. Let us write the prime factorization of a positive integer n in this standard way:

$$n = \prod_i p_i^{\alpha_i},$$

where the p_i's are distinct. If the ordering of the prime powers were important, how many factorizations would there be of 61,740?

2.6. Extend Lemma 2.1.1 by mathematical induction on the number of primes to give Lemma 2.1.2.

2.7. **Fermat's Factorization Method** is an elementary technique for factoring positive, composite integers. It says, "The odd integer $N \geq 3$ is composite iff there are nonnegative integers n, m such that $n - m > 1$ and $N = n^2 - m^2$."

(a) Prove both directions of this proposition. [Hint: If $N = rs$, then

$$N = [(r + s/2]^2 - [(r - s)/2]^2.]$$

(b) Apply Fermat's Factorization Method to $N = 39, 161, 737$.

2.8. Consider the set **S** $\{x: x = 6n - 5, n \in \mathbf{N}\}$.

(a) Is the set **S** closed under multiplication?
(b) Show that the Fundamental Theorem of Arithmetic fails in **S** by producing a member of **S** that is **S**-irregular.

2.9. Let p be a prime and $n > p$ be a positive integer. Show that the number of powers of p contained in the factorization of n! is given by

$$\sum_{k=1} \left\lfloor \frac{n}{p^k} \right\rfloor,$$

where the summation terminates as soon as $\lfloor n/p^k \rfloor = 0$. Apply the result to a particular example.

2.10. Show that there is a bijection[2] from the set $\mathcal{P}$ of primes to the set **S** of positive integers that have only three positive divisors.

2.11. Show that the only exponent n for which the base-10 representation of 3^n consists only of 9's is $n = 2$.

2.12. A **Niven number** (named after the Canadian mathematician **Ivan Niven**, 1915–) is an integer N that is divisible by the sum of its digits. For example, 27 is Niven since $(2 + 7) \mid 27$.

(a) Recall from Problem 1.45 the digital sum function of an integer N. Prove that $\mathrm{DS}(N) \leq 9 \log_{10} N + 9$.
(b) Let the factorization of N be given as $N = p_1 p_2 p_3 \ldots p_n$ without the use of exponents, where $p_1 \leq p_2 \leq p_3 \leq \ldots \leq p_n$. Prove that if N is Niven and N is not a power of 10, then $p_1 \leq \mathrm{DS}(N)$. Is the converse true?
(c) Let k be a fixed positive integer, and consider the integer p^k, where p is any prime. Prove that if p^k is Niven, then

$$\frac{p - 9}{\log_{10} p} \leq 9_k$$

must hold, and therefore to a given k there can be only a finite number of primes p such that p^k is Niven (Liu, 1983).
(d) Let $k = 3$. What is the upper bound on p if p^3 is to be Niven? In fact, how many cubes of primes are Niven?

2.13. The following proof of Theorem 1.6 was given by the Hungarian mathematician **George Pólya** (1888–1985), but probably goes back as far as **C. Goldbach**. The numbers $F_n = 2^{2^n} + 1$ in Conjecture 2 in Section 1.8 are called **Fermat numbers**. Let $m, n \in \mathbf{N}$ and $n > m$.

(a) Show that $(F_n - 2)/F_m$ is an integer.
(b) Assume that there is a prime p in the factorization of both F_n, F_m. Show that this leads to a contradiction.

[2] Recall that a function is a **bijection** if *each* element of the range is related to *exactly one* element of the domain.

(c) Hence, there must be at least n distinct primes. Now let $n \to \infty$. Write out the proof in detail.

2.14. A set **S** of numbers is said to be **countable** if it is possible to define a one-to-one function f: $\mathbf{S} \to \mathbf{N}$. Use Theorems 1.6 and 2.1 to show that **Q** is countable.

2.15. Consult the paper by Ratering listed in the References, study his Lemma 1, and then add some additional highly composite integers to his table.

2.2 A Theorem of Euclid

The two strands of divisibility of one integer by another and of the factorization of a composite integer come together in the Fundamental Theorem of Arithmetic. The next theorem continues the merger of these two strands; it appeared as Proposition 30 in Book VII of the *Elements* of **Euclid.**[3]

Theorem 2.2 (Euclid). If p is a prime and $a, b \in \mathbf{N}$, then $p \mid ab$ implies $p \mid a$ or $p \mid b$.

Proof Suppose, arbitrarily, that $p \nmid a$. Then by Lemma 2.1.1 the prime p is not in the factorization of a. That is, a is expressible as $a = p_1 p_2 \ldots$, where none of the p_i's is p. Let b have the prime factorization $q_1 q_2 \ldots q_n$; hence, we can write

$$ab = p_1 p_2 \ldots p_k q_1 q_2 \ldots q_n.$$

By the Fundamental Theorem of Arithmetic, the preceding is the unique factorization of the number ab. Since $p \mid ab$, then by Lemma 2.1.1 p is contained in this factorization. As p is not one of the p_i's, it must be one of the q_j's. Finally, from the lemma a third time, p in the factorization of b implies that $p \mid b$. ◆

Theorem 2.2 has something of the flavor of Lemma 2.1.2. It presupposes less than that lemma because we now have the Fundamental Theorem of Arithmetic. Mathematical induction, as expected, can be used to extend Theorem 2.2 (Problem 2.18).

Example 2.4 Assume $\sqrt[3]{3}$ is rational, p/q, and reduced to lowest terms. Then $p^3 = 3q^3$ and Theorem 2.2 then yields $3 \mid p$ or $3 \mid p^2$; hence, $3 \mid p$ holds. Let

[3] **Hardy** refers to the theorem as Euclid's First Theorem, perhaps because he thought it was the first theorem on number theory in the *Elements* to have real substance. We shall not use this designation.

$p = 3k$; we now have $\sqrt[3]{3} = 3k/q$, or $q^3 = 9k^3$. A second application of Theorem 2.2 gives $3 \mid q$, a contradiction since p/q was stipulated to be in lowest terms. It follows that there are no integers p, q such that $\sqrt[3]{3} = p/q$. ◆

Let us write the prime factorization of a positive integer n in the standard way given in Problem 2.5:

$$n = \prod_i p_i^{\alpha_i},$$

where the p_i's are distinct. Either Theorem 2.2 or the Fundamental Theorem of Arithmetic itself can then be used to show that n^2 is uniquely represented as

$$n^2 = \prod_i p_i^{2\alpha_i},$$

and consequently

Lemma 2.3.1 Every exponent in the prime factorization of a square is an even integer.

This lemma can be used to show that $\sqrt{2}$ is not rational (Problem 2.21). A less immediate result is the following interesting proposition.

Theorem 2.3 For any integer $n > 1$, $n!$ is not a square.

Proof The proposition is trivially true for $n = 2$, so assume $n > 2$. Let p be the largest prime that does not exceed n. It must be that $2p > n$, for if $2p \leq n$ held, then by Bertrand's Postulate (see Section 1.5) there would exist a prime p' satisfying $p < p' < 2p \leq n$; this is a contradiction. If n itself is a prime, then p is n. In this case the integer N, defined by

$$N = \frac{n!}{p} = 1 \cdot 2 \cdot 3 \ldots (p - 1)$$

is not divisible by p because $1, 2, 3, \ldots, p - 1$ are all less than p (contrapositive of Theorem 2.2).

If n is composite, then $p < n$ and now

$$N = 1 \cdot 2 \cdot 3 \ldots (p - 1)(p + 1)(p + 2) \ldots n.$$

The prime p cannot divide any of the first $p - 1$ factors; it also cannot divide $p + 1, p + 2, \ldots, n$ because each such division leaves a remainder $(1, 2, \ldots, k$, where $n = p + k$ and $k < p)$. Hence, again from the contrapositive of Theorem 2.2 we have $p \nmid N$.

Thus, in either case $p^2 \nmid n$, so from Lemma 2.3.1 we conclude that $n!$ is not a square because the exponent of p in the prime factorization of $n!$ is 1. ◆

Problems

2.16. We observe that $6 \mid 9{\cdot}4$, but neither $6 \mid 9$ nor $6 \mid 4$ holds. What is wrong here? Is this an exception to Theorem 2.2?

2.17. Apply Theorem 2.2 to these cases:

(a) $p = 17,\ a = 51,\ b = 103.$ (c) $p = 31,\ a = 63,\ b = 464.$
(b) $p = 43,\ a = 89,\ b = 731.$ (d) $p = 127,\ a = 2159,\ b = 5207.$

2.18. Extend Theorem 2.2 to this: Let $M = \Pi_{i=1}^{r} n_i$, where each $n_i \in \mathbf{N}$, and let p be a prime. Then $p \mid M$ implies p divides at least one of the n_i's.

2.19. Use Theorem 2.2 to show that there are no nonzero rational numbers x, y such that $x + y\sqrt[5]{7}$ is rational.

2.20. Confirm Lemma 2.3.1 for the integers (a) 21,609, (b) 4,840,000.

2.21. Use Lemma 2.3.1 explicitly to show that $\sqrt{2}$ is not a rational number.

2.22. Consider again the set **S** in Problem 2.8. Does Euclid's theorem apply there? Prove or give a counterexample.

2.23. Let p be a prime and $n \geq 2$ be a positive integer. Show that $\sqrt[n]{p}$ is not rational.

2.24. Prove that $\log_2 5$ is not rational.

2.25. Let $n > 1$ be a positive integer. Prove that

$$S = \sum_{k=2}^{n} \frac{1}{k}$$

is not an integer. [Hint: Assume it is, and look at $n!S$.]

2.26. Let p be a prime and let n be any integer satisfying $1 \leq n \leq p - 1$. Prove that p divides the **binomial coefficient**

$$\binom{p}{n} = \frac{p!}{(p - n)!\ n!}.$$

2.3 Groups

Some topics in this book, including those in Sections 2.4 and 2.7, will be discussed using concepts from abstract algebra (Ayres, 1965; Fraleigh, 1994). Algebra deals with general systems of elements that possess various properties. One such property is possession of a binary operation.

DEF. Let **S** be a set of mathematical objects. An operation $(\cdot)$ is called a **binary operation**[4] on **S** if for any $x, y \in \mathbf{S}$ one has $x \cdot y \in \mathbf{S}$.

An especially important algebraic system is that of a group (Kleiner, 1986); a group has just one binary operation defined on it.

DEF. A set **G** of elements, together with a binary operation $(\cdot)$ on **G** (called, generically, **multiplication**), is a **group** $\langle \mathbf{G}, \cdot \rangle$ if

1. There is an element $\mathbf{I} \in \mathbf{G}$ such that $\mathbf{I} \cdot g = g \cdot \mathbf{I} = g$ for every $g \in \mathbf{G}$.
2. Given $g \in \mathbf{G}$, there is an element $h \in \mathbf{G}$ such that $h \cdot g = g \cdot h = \mathbf{I}$.
3. $(\cdot)$ obeys the associative law on **G**; that is, $x \cdot (y \cdot z) = (x \cdot y) \cdot z$ for any $x, y, z \in \mathbf{G}$.

The preceding element **I** is called an **identity element** of the group $\langle \mathbf{G}, \cdot \rangle$. In 2, the element **h** is called the **inverse** (g^{-1}) of the element g.

Example 2.5 Let $\mathbf{G} = \mathfrak{R} - \{0\}$, the set of all nonzero real numbers. Let $(\cdot)$ be ordinary multiplication $(\times)$, and let $\mathbf{I} = 1$ and g^{-1} be the arithmetic reciprocal of g. Then $\langle \mathfrak{R} - \{0\}, \times \rangle$ is a group. ◆

Example 2.6 Let $\mathbf{G} = M_2(\mathfrak{R})$, the set of all invertible 2×2 matrices defined over the real numbers. Let $(\cdot)$ be matrix multiplication, and let $\mathbf{I} = \left(\begin{smallmatrix} 1 & 0 \\ 0 & 1 \end{smallmatrix}\right)$ and g^{-1} be the matrix inverse of g. As matrix multiplication is known to be associative, then $\langle M_2(\mathfrak{R}), \cdot \rangle$ is a group. ◆

The definition of a group states that there is at least one element in **G** that can serve as an identity for $\langle \mathbf{G}, \cdot \rangle$. More can be asserted (cf. Problem 1.25).

Theorem 2.4 Let $\langle \mathbf{G}, \cdot \rangle$ be a group. Then the identity element **I** is unique.

Proof Assume, to the contrary, that $\mathbf{I}'$ is another (distinct) identity element of $\langle \mathbf{G}, \cdot \rangle$. Since group multiplication by an identity is commutative, we have

[4] Possession of a binary operation on **S** is synonymous with **S** being **closed** (Section 2.1) under the operation $(\cdot)$.

$\mathbf{I} \cdot \mathbf{I}' = \mathbf{I}' \cdot \mathbf{I}$. Regarding **I** on the left side as the identity, we obtain $\mathbf{I} \cdot \mathbf{I}' = \mathbf{I}'$. However, regarding $\mathbf{I}'$ on the right side as the identity, we obtain $\mathbf{I}' \cdot \mathbf{I} = \mathbf{I}$. The equality $\mathbf{I} \cdot \mathbf{I}' = \mathbf{I}' \cdot \mathbf{I}$ thus reduces to $\mathbf{I} = \mathbf{I}'$, which contradicts the assumption that $\mathbf{I}, \mathbf{I}'$ are distinct. It follows that $\langle \mathbf{G}, \cdot \rangle$ contains only one identity, **I**. ◆

In a similar way one can show that inverses are unique in a group (Problem 2.27).

A group with only a finite number of elements is called a **finite group**. The number of elements is the **order** of the group, written $|\mathbf{G}|$. All that can be known about a finite group is encoded in its **multiplication table**. By convention, the left column of the table consists of multipliers and the top row consists of multiplicands. The multiplication table for a group of order 3, for example, must be

$\cdot$	$\mathbf{I}$	a	b
$\mathbf{I}$	$\mathbf{I}$	a	b
a	a	b	$\mathbf{I}$
b	b	$\mathbf{I}$	a

It is easy to see that all entries in a column of a group multiplication table must be distinct. Suppose, to the contrary, that for some distinct $y_1, y_2 \in \mathbf{G}$ one has from the table

$$x \cdot y_1 = x \cdot y_2$$

for some $x \in \mathbf{G}$. Then premultiplication of both sides by x^{-1} yields $(x^{-1} \cdot x) \cdot y_1 = (x^{-1} \cdot x) \cdot y_2$, or $y_1 = y_2$, a contradiction. Hence, distinct y_1, y_2 must yield distinct values of $x \cdot y_1, x \cdot y_2$ for any $x \in \mathbf{G}$. Similarly, all entries in a row of a group multiplication table must be distinct.

The preceding multiplication table shows an interesting feature. Notice that $a^1 = a, a^2 = a \cdot a = b, a^3 = a \cdot a^2 = a \cdot b = \mathbf{I}$. All elements of the group are generated by the first n positive powers of the single element a.

DEF. An element a of a group $\langle \mathbf{G}, \cdot \rangle$ generates **G**, or is a **generator** for **G**, if all elements of **G** are expressible as positive powers of a. A group $\langle \mathbf{G}, \cdot \rangle$ is called a **cyclic group** if **G** has a generator.

The group of order 3 is a cyclic group. In general, the generator for a cyclic group is not unique (Problem 2.33).

When we come to groups of order 4 a new feature arises. One finds that there are two possible multiplication tables as shown in Fig. 2.1. The tables are certainly

$\bullet$	$\mathbf{I}$	a	b	c
$\mathbf{I}$	$\mathbf{I}$	a	b	c
a	a	$\mathbf{I}$	c	b
b	b	c	$\mathbf{I}$	a
c	c	b	a	$\mathbf{I}$

(a)

$\bullet$	$\mathbf{I}$	A	B	C
$\mathbf{I}$	$\mathbf{I}$	A	B	C
A	A	B	C	$\mathbf{I}$
B	B	C	$\mathbf{I}$	A
C	C	$\mathbf{I}$	A	B

(b)

Figure 2.1 Multiplication table for groups of order 4

distinct (look at the diagonals). Tedious checking shows that the associative law is upheld in both tables. No other tables can be constructed (try it!). Therefore, there are two, and only two, distinct groups of order 4.

The table in Fig. 2.1(b) is that for a cyclic group because A is a generator for the group. The group in Fig. 2.1(a) is noncyclic; it is called the **Klein 4-group** (or the *Vierergruppe*), and is usually symbolized $\mathbf{V}$.

There is only one distinct group of order 5. Its multiplication table is shown in Fig. 2.2. The group is also cyclic. You should experiment a bit to convince yourself that no other multiplication table is possible for a group of order 5.

As the order increases, it becomes progressively more tedious to work out group multiplication tables. For an **infinite group** (a group with an infinite number of elements, such as the groups in Examples 2.5 and 2.6), it is impossible to even write down the multiplication table.

In some groups, whether finite or infinite, the following phenomenon can be observed. The identity $\mathbf{I}$, of course, commutes with all other elements of a

$\bullet$	$\mathbf{I}$	a	b	c	d
$\mathbf{I}$	$\mathbf{I}$	a	b	c	d
a	a	b	c	d	$\mathbf{I}$
b	b	c	d	$\mathbf{I}$	a
c	c	d	$\mathbf{I}$	a	b
d	d	$\mathbf{I}$	a	b	c

Figure 2.2 Multiplication table for the group of order 5

group. However, in Fig. 2.1(a) one also has $a \cdot b = b \cdot a$, $a \cdot c = c \cdot a$, and $b \cdot c = c \cdot b$. In Fig. 2.1(b) one has $A \cdot B = B \cdot A$, $A \cdot C = C \cdot A$, and $B \cdot C = C \cdot B$. Finally, in Fig. 2.2 similar equalities hold throughout that table.

DEF. A group $\langle \mathbf{G}, \cdot \rangle$ is termed **Abelian**[5] if $x \cdot y = y \cdot x$ for every $x, y \in \mathbf{G}$.

The cyclic groups displayed in Figs. 2.1 and 2.2 are all Abelian groups. The following speculation seems warranted.

Theorem 2.5 Every cyclic group is Abelian.

Proof Let $\langle \mathbf{G}, \cdot \rangle$ be a cyclic group with generator a. Let $g_1, g_2 \in \mathbf{G}$ be arbitrary. Then there exist positive integers m, n such that $g_1 = a^m$ and $g_2 = a^n$. Hence, upon multiplication we easily obtain $g_1 \cdot g_2 = a^m \cdot a^n = a^{m+n} = a^{n+m} = a^n a^m = g_2 \cdot g_1$. ◆

The converse of Theorem 2.4 is false (Problem 2.32).

Certain subsets of the elements in a group can have significance. These subsets, called subgroups, are actually of central importance in group structure.

DEF. Let $\langle \mathbf{G}, \cdot \rangle$ be a group, and let $\mathbf{S} \subset \mathbf{G}$. Then $\mathbf{S}$, together with the operation $(\cdot)$ defined on $\mathbf{G}$, is a **subgroup** $\langle \mathbf{S}, \cdot \rangle$ if $(\cdot)$ is a binary operation on $\mathbf{S}$ and all of the group postulates hold for $\mathbf{S}$.

At the outset this means that one must verify that $(\cdot)$ is a binary operation on $\mathbf{S}$. If $(\cdot)$ is a binary operation on $\mathbf{S}$, then associativity is automatically inherited from $\mathbf{G}$. Next, $\mathbf{S}$ must contain the identity $\mathbf{I}$ of $\mathbf{G}$; this element, of course, is unique (Theorem 2.4). Finally, $\mathbf{S}$ must contain all of its inverses; these are also unique.

Example 2.7 In Fig. 2.1(a), the Klein 4-group has five subgroups, which consist of the sets $\{\mathbf{I}\}$, $\{\mathbf{I}, a\}$, $\{\mathbf{I}, b\}$, $\{\mathbf{I}, c\}$, and $\{\mathbf{I}, a, b, c\}$. ◆

Example 2.8 The group $\langle \mathbf{Z}, + \rangle$ is a cyclic subgroup of $\langle \mathfrak{R}, + \rangle$. ◆

[5] After the Norwegian mathematician **Niels Henrik Abel** (1802–1829), who died tragically at the early age of 26 from tuberculosis (Ore, 1974).

DEF. If $\langle \mathbf{G}, \cdot \rangle$ is a group, then the subgroup containing all of $\mathbf{G}$ is the **improper subgroup** of $\langle \mathbf{G}, \cdot \rangle$. Any other subgroup is called a **proper subgroup**. The subgroup $\langle \mathbf{I}, \cdot \rangle$ is termed the **trivial subgroup** of $\langle \mathbf{G}, \cdot \rangle$. All other subgroups are referred to as **nontrivial subgroups**.

The subgroup of Example 2.8 is a nontrivial proper subgroup.

We note one more thing here. The orders of the five subgroups in Example 2.7 are 1, 2, 2, 2, and 4, all of which are divisors of the order of the group itself (whose order is 4). This is no accident: we say more on this later.

Problems

2.27. Prove that the inverse of any element g in a group is unique.

2.28. Are these binary operations on the indicated sets $\mathbf{S}$?

(a) $\mathbf{S} = \mathbf{Z} - \{0\}$, $a \cdot b = a + b$.
(b) $\mathbf{S} = \mathfrak{R}$, $a \cdot b = (a - b)^2$.
(c) $\mathbf{S} =$ the set of all quadratic polynomials $c_1x^2 + c_2x + c_3$, with each $c_i \in \{0, 1\}$, and $(\cdot) =$ termwise addition defined by

$$c_i + c_j = \begin{cases} c_i + c_j & \text{if } c_i + c_j < 2 \\ 0 & \text{if } c_i + c_j = 2. \end{cases}$$

2.29. Decide which of the following are groups:

(a) $\mathbf{G} = \mathbf{Q}$; $(\cdot) =$ ordinary multiplication.
(b) $\mathbf{G} = \mathbf{Z}$; $(\cdot) =$ ordinary addition.
(c) $\mathbf{G} =$ the set $\{1, -1, i, -i\}$; $(\cdot) =$ complex multiplication.
(d) $\mathbf{G} =$ the positive rationals; $a \cdot b = a/b$.
(e) $\mathbf{G} =$ the set of all numbers of the form $(1 + 2m)/(1 + 2n)$, where $m, n = 0, \pm 1, \pm 2, \ldots$; $(\cdot) =$ ordinary multiplication.
(f) $\mathbf{G} =$ the set of all 11×11 matrices defined over $\mathfrak{R}$; $(\cdot) =$ matrix multiplication.
(g) $\mathbf{G} =$ the set of all nonzero numbers of the form $a + b\sqrt[3]{2} + c\sqrt[3]{4}$, where $a, b, c \in \mathbf{Q}$; $(\cdot) =$ ordinary multiplication.

2.30. A set $\mathbf{S}$ has exactly three elements. How many different binary operations could conceivably be defined on $\mathbf{S}$?

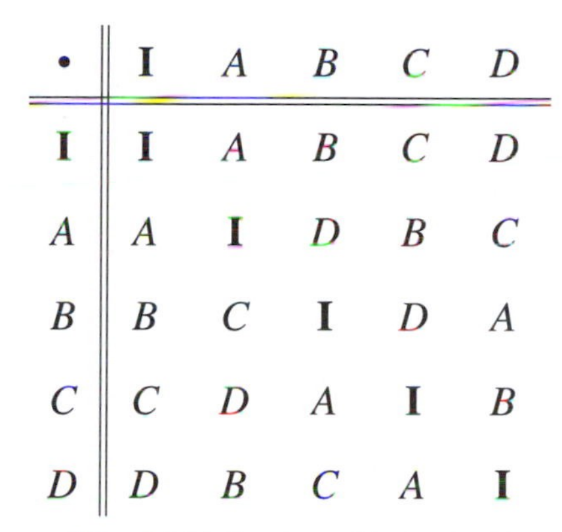

•	**I**	A	B	C	D
I	**I**	A	B	C	D
A	A	**I**	D	B	C
B	B	C	**I**	D	A
C	C	D	A	**I**	B
D	D	B	C	A	**I**

Figure 2.3 Multiplication table for Problem 2.34

2.31. A binary operation $(\cdot)$ on **N** is defined by $x \cdot y = 2^{xy}$. Is $(\cdot)$ associative?

2.32. Find an Abelian group that is not cyclic. Give an example of an infinite group that is not Abelian.

2.33. Show that a cyclic group need not have a unique generator.

2.34. Examine the multiplication table shown in Fig. 2.3. Is it the multiplication table of a group?

2.35. The multiplication table for a group of order 6 is shown in Fig. 2.4. Find the nontrivial proper subgroups.

2.36. Let $\langle \mathbf{G}, \cdot \rangle$ be a finite group. Show that there cannot be more than $2^{|\mathbf{G}|-1}$ subgroups.

•	**I**	a	b	c	d	e
I	**I**	a	b	c	d	e
a	a	b	**I**	e	c	d
b	b	**I**	a	d	e	c
c	c	d	e	**I**	a	b
d	d	e	c	b	**I**	a
e	e	c	d	a	b	**I**

Figure 2.4 Multiplication table for Problem 2.35

2.37. If x, y are elements of a group, then $(x \cdot y)^2$ is generally equal to $x^2 \cdot y^2$. Prove, or give a counterexample.

2.38. Show that if $\langle \mathbf{G}, \cdot \rangle$ is a group and $g_1, g_2 \in \mathbf{G}$, then $(g_1 \cdot g_2)^{-1} = g_2^{-1} \cdot g_1^{-1}$.

2.39. Show that if $\langle \mathbf{G}, \cdot \rangle$ is a group and if $(x \cdot y)^2 = x^2 \cdot y^2$ for every $x, y \in \mathbf{G}$, then $\langle \mathbf{G}, \cdot \rangle$ is Abelian.

2.40. A group element x which satisfies $x^2 = x$ is called **idempotent**. Prove that a group has only one idempotent element. Identify the idempotent element in the groups of Examples 2.7 and 2.8.

2.41. Let $\langle \mathbf{G}, \cdot \rangle$ be a finite group. An element $g \in \mathbf{G}$ is said to have **order k** if k is the smallest positive integer such that $g^k = \mathbf{I}$. Prove that g, g^{-1} have the same order.

2.42. Show that if $g_1, g_2, g_1 \cdot g_2$ are each of order 2, then g_1, g_2 commute.

2.43. Show that if the group $\langle \mathbf{G}, \cdot \rangle$ is of even order, then $\mathbf{G}$ contains an odd number of elements of order 2.

2.44. Show that if $\langle \mathbf{G}, \cdot \rangle$ is a finite group, then for each $g \in \mathbf{G}$ there is at least one corresponding positive integer n such that $g^n = \mathbf{I}$. [Hint: Consider the set $\{g, g^2, g^3, \ldots, g^{|\mathbf{G}|+1}\}$.]

2.45. Let $\langle \mathbf{G}, \cdot \rangle$ be Abelian, and suppose $\langle \mathbf{H}, \cdot \rangle$ and $\langle \mathbf{K}, \cdot \rangle$ are both subgroups of $\langle \mathbf{G}, \cdot \rangle$. Let $\mathbf{HK} = \{x: x = h \cdot k, h \in \mathbf{H}, k \in \mathbf{K}\}$. Then show that $\langle \mathbf{HK}, \cdot \rangle$ is a subgroup of $\langle \mathbf{G}, \cdot \rangle$.

2.46. Let $\langle \mathbf{G}, \cdot \rangle$ be Abelian, and define $\mathbf{S} = \{x: x \cdot x = \mathbf{I}, x \in \mathbf{G}\}$. Then show that $\langle \mathbf{S}, \cdot \rangle$ is a subgroup of $\langle \mathbf{G}, \cdot \rangle$.

2.47. Do the 10th roots of unity constitute a group $\mathbf{G}$? If so, what is its order? Does $\mathbf{G}$ have any proper subgroups?

2.4 Greatest Common Divisor

The concept of the greatest common divisor is already familiar to you from earlier studies. Thus, the greatest common divisor of 12 and 18 is 6. The arithmetic definition of the greatest common divisor is simple but not very illuminating.

DEF. Let $x, y \in \mathbf{Z}$, with at least one of these being nonzero. The largest positive integer d that divides both x, y is termed the **greatest common divisor** of x, y and we write $d = (x, y)$.

Shortly, we shall see how the definition can be rephrased in the language of group theory.

That two nonzero integers x, y should even have a greatest common divisor is a consequence of the Well-Ordering Property (Theorem 1.2). Let $M = \max\{|x|, |y|\}$ and $d > 0$ denote any common divisor of x, y; clearly, $d \leq M$. Define $\mathbf{S}$ to be the set of nonnegative integers

$$\mathbf{S} = \{n: n = M - d, d \mid x, d \mid y\}.$$

This set is nonempty since $M - 1 \in \mathbf{S}$. By Theorem 1.2 it has a minimum n, so there is a maximum common divisor d.

The Fundamental Theorem of Arithmetic allows us to identify this maximum d. Let us employ infinite product notation and write the factorization of an integer $x > 1$ in the form

$$x = \prod_{i=1}^{\infty} p_i^{a_i} = p_1^{a_1} p_2^{a_2} p_3^{a_3} \ldots,$$

where $p_1 = 2, p_2 = 3, p_3 = 5$, and so forth. If a certain prime p_k does not actually appear in the factorization of x, the corresponding exponent a_k is set equal to 0. Thus, $90 = 2^1 \cdot 3^2 \cdot 5^1$ and $a_1 = 1, a_2 = 2, a_3 = 1, a_k = 0$ for $k \geq 4$. The use of an infinite product (a concept that we haven't even defined properly) is merely a notational convenience. Any actual integer is, of course, only a finite product of prime powers. The following theorem shows how (x, y) can be computed from the factorizations of x and y.

Theorem 2.6 Let x, y be integers whose absolute values are factorable as

$$|x| = \prod_{i=1}^{\infty} p_i^{a_i}, \qquad |y| = \prod_{i=1}^{\infty} p_i^{b_i}.$$

Then $(x, y) = \Pi_{i=1}^{\infty} p_i^{n_i} > 0$, where $n_i = \min\{a_i, b_i\}$ for all $i \in \mathbf{N}$.

Proof From Lemma 2.1.1 we can deduce that if a power of a prime, p^n, divides both x and y, then p^n must appear in both of their factorizations. Hence, for each p_i the exponent n_i in the factorization of (x, y) cannot exceed the lesser of a_i, b_i. Since one of $\min\{a_i, b_i\} - a_i$, $\min\{a_i, b_i\} - b_i$ is 0 and the other is nonnegative, this forces n_i to be exactly $\min\{a_i, b_i\}$. ◆

DEF. If for two integers x, y we have $(x, y) = 1$, then x, y are said to be **relatively prime** or **coprime**. By convention, 1 is coprime to any other positive integer.

Example 2.9 Let $x = 588$, $y = 3080$. Their factorizations are $x = 2^2 \cdot 3^1 \cdot 7^2$, $y = 2^3 \cdot 5^1 \cdot 7^1 \cdot 11^1$. The greatest common divisor is thus

$$
\begin{aligned}
(588, 3080) &= 2^{\min\{2,3\}} \cdot 3^{\min\{1,0\}} \cdot 5^{\min\{0,1\}} \cdot 7^{\min\{2,1\}} \cdot 11^{\min\{0,1\}} \\
&= 2^2 \cdot 3^0 \cdot 5^0 \cdot 7^1 \cdot 11^0 \\
&= 28.
\end{aligned}
$$
◆

Although Theorem 2.6 provides a way of computing the greatest common divisor of two integers, it is not efficient because factorizations require too much labor. We'll see in Section 2.6 how this can be improved.

The greatest common divisor has the following set of elementary properties. We prove part (d), leaving the other parts for you.

Theorem 2.7 (a) $(x, y) = (y, x)$.
(b) $(x, (y, z)) = ((x, y), z)$.
(c) $(x, 1) = (1, x) = 1$.
(d) $(xz, yz) = |z|\,(x, y)$.

Proof (d) We appeal to the preceding theorem and to the fact that the greatest common divisor is a positive integer. Let

$$|x| = \prod_{i=1}^{\infty} p_i^{a_i}, \quad |y| = \prod_{i=1}^{\infty} p_i^{b_i}, \quad |z| = \prod_{i=1}^{\infty} p_i^{n_i}.$$

Then the greatest common divisor of xz, yz is

$$
\begin{aligned}
(xz, yz) &= \prod_{i=1}^{\infty} p_i^{\min\{a_i+n_i, b_i+n_i\}} \\
&= \prod_{i=1}^{\infty} p_i^{n_i} \prod_{i=1}^{\infty} p_i^{\min\{a_i, b_i\}} \\
&= |z|(x, y).
\end{aligned}
$$
◆

Example 2.10 Let $x = -28$, $y = 35$, $z = 10$, and $xz = 280 = 2^3 \cdot 5^1 \cdot 7^1$, $yz = 350 = 2^1 \cdot 5^2 \cdot 7^1$.

$$\begin{aligned}(-280,\ 350) &= 2^{\min\{3,1\}} \cdot 5^{\min\{1,2\}} \cdot 7^{\min\{1,1\}} \\ &= 2^1 \cdot 5^1 \cdot 7^1 \\ &= 70\end{aligned}$$

$$|z| = 10 \quad \text{and} \quad 28 = 2^2 \cdot 7, \qquad 35 = 5^1 \cdot 7^1$$

$$\begin{aligned}(-28,\ 35) &= 2^{\min\{2,0\}} \cdot 5^{\min\{0,1\}} \cdot 7^{\min\{1,1\}} \\ &= 2^0 \cdot 5^0 \cdot 7^1 \\ &= 7\end{aligned}$$

$$(xz,\ yz) = (-280,\ 350) = 10(-28,\ 35),$$

in accordance with Theorem 2.7(d). ◆

The group interpretation of the greatest common divisor $d = (x, y)$ is based on the observation that x is a multiple of d, y is a multiple of d, and by Theorem 1.4(b) any linear combination of x, y is a multiple of d.

We let the symbol $k\mathbf{Z}$ stand for the subset of $\mathbf{Z}$ consisting of all multiples of the given fixed nonzero integer k:

$$k\mathbf{Z} = \{kz\colon z \in \mathbf{Z}\}.$$

Theorem 2.8 For any integer k, the set $k\mathbf{Z}$ together with the operation of addition is a subgroup of $\langle \mathbf{Z}, +\rangle$.

Proof Let $x_1, x_2 \in k\mathbf{Z}$; then there are integers z_1, z_2 such that $x_1 = kz_1$ and $x_2 = kz_2$. Hence, $x_1 + x_2 = kz_1 + kz_2 = k(z_1 + z_2)$, and since $z_1 + z_2 \in \mathbf{Z}$, we see that $x_1 + x_2 \in k\mathbf{Z}$. Thus, $(+)$ is a binary operation on k$\mathbf{Z}$.

Take $z = 0$, so $kz = 0$ and thus the identity of $\mathbf{Z}$ is also in $k\mathbf{Z}$. Let $x = kz \in k\mathbf{Z}$. The group inverse of z is $-z$, and so $k(-z) = -(kz)$. Then $kz + \{-(kz)\} = kz + k(-z) = k[z + (-z)] = k \cdot 0 = 0$, so $k(-z)$ is the inverse of kz and lies in $k\mathbf{Z}$. That is, $k\mathbf{Z}$ contains all of its inverses. Finally, $k\mathbf{Z}$ inherits associativity from $\mathbf{Z}$, and this completes the proof that $k\mathbf{Z}$ is a group, a subgroup of $\mathbf{Z}$.[6] ◆

[6] Since the group operation on $\mathbf{Z}$ has already been identified as ordinary addition, we drop from here on the complete symbol $\langle \mathbf{Z}, +\rangle$ for the group in favor of just $\mathbf{Z}$.

Theorem 2.9 Every subgroup $\mathbf{H} \subset \mathbf{Z}$ is of the type $\mathbf{H} = k\mathbf{Z}$ for some integer k.

Proof The case $\mathbf{H} = \{0\}$ is trivial, so we assume that $\mathbf{H}$ is a nontrivial subgroup of $\mathbf{Z}$. Any such subgroup $\mathbf{H}$ must contain both positive and negative numbers since $\mathbf{H}$ contains all its inverses. From the Well-Ordering Property (Theorem 1.2), there is a smallest positive integer $N_{\min}$ among the positive members of $\mathbf{H}$.

Now consider $N_{\min}\mathbf{Z}$, a subgroup of $\mathbf{Z}$ (Theorem 2.8), and let $N_{\min}z \in N_{\min}\mathbf{Z}$, where z is a positive integer. We have

$$N_{\min}z = \underbrace{N_{\min} + N_{\min} + \ldots + N_{\min}}_{z \text{ terms}},$$

and since $(+)$ is a binary operation on $\mathbf{H}$, then $N_{\min}z \in \mathbf{H}$. So is $N_{\min}(-z) = -(N_{\min}z)$ since $\mathbf{H}$ contains all its inverses. Finally, $N_{\min} \cdot 0 = 0 \in \mathbf{H}$. Therefore, every element in $N_{\min}\mathbf{Z}$ is also in $\mathbf{H}$; that is, $N_{\min}\mathbf{Z} \subset \mathbf{H}$.

On the other hand, let $h \in \mathbf{H}$. From the Division Algorithm (Theorem 1.3) there exist unique integers q, r such that

$$h = N_{\min}q + r, \qquad 0 \leq r < N_{\min}.$$

Since $h \in \mathbf{H}$ and also $N_{\min}q$ belongs to $\mathbf{H}$ (see the argument of the preceding paragraph), then $r = h + [-(N_{\min}q)]$ is in $\mathbf{H}$ too. But $N_{\min}$ was designated as the smallest positive element in $\mathbf{H}$, and yet $0 \leq r < N_{\min}$. This forces $r = 0$, so $h = N_{\min}q$ and this says $h \in N_{\min}\mathbf{Z}$. Thus, $h \in \mathbf{H}$ has implied $h \in N_{\min}\mathbf{Z}$, so $\mathbf{H} \subset N_{\min}\mathbf{Z}$.

But if both $N_{\min}\mathbf{Z} \subset \mathbf{H}$ and $\mathbf{H} \subset N_{\min}\mathbf{Z}$ hold, then $\mathbf{H} = N_{\min}\mathbf{Z}$; that is, $\mathbf{H}$ is of the form $k\mathbf{Z}$. ◆

It is easy to regard Theorem 2.9 too lightly. This is wrong, however, because $\mathbf{Z}$ is not typical of most groups. For most groups with nontrivial subgroups, not all of these subgroups will be cyclic. Theorem 2.9 says that every subgroup of $\mathbf{Z}$ is cyclic, and that each (nontrivial) one is generated by its smallest positive element.

Example 2.11 The numbers $\{\ldots, -14, -7, 0, 7, 14, 21, \ldots\}$ are a subgroup of $\mathbf{Z}$. It is cyclic and is generated by 7. ◆

Clearly, there are infinitely many cyclic subgroups of $\mathbf{Z}$, each of infinite order, and each generated by a single generator k. But, for the moment, we can imagine a subgroup of $\mathbf{Z}$ that is generated by two integers a, b. We define the symbol $a\mathbf{Z} + b\mathbf{Z}$ to mean

$$a\mathbf{Z} + b\mathbf{Z} = \{az_1 + bz_2 \colon z_1, z_2 \in \mathbf{Z}\}.$$

To prove that this set is a subgroup is an easy exercise for you (Problem 2.61). But now Theorem 2.9 tells us that the subgroup $a\mathbf{Z} + b\mathbf{Z}$ must be of the form $k\mathbf{Z}$, where k is some integer.

Theorem 2.10 Let k be the generator for the subgroup $a\mathbf{Z} + b\mathbf{Z}$. Then

(a) One has $k = az_1 + bz_2$, for some integers z_1, z_2.
(b) $k \mid a$ and $k \mid b$.
(c) If x is an integer such that $x \mid a$ and $x \mid b$ hold, then $x \mid k$ also.

Proof (a) This follows merely because $k \in a\mathbf{Z} + b\mathbf{Z}$.

(b) In $az_1 + bz_2$, take $z_1 = 1$ and $z_2 = 0$. Hence, $a \in a\mathbf{Z} + b\mathbf{Z}$. Similarly, $b \in a\mathbf{Z} + b\mathbf{Z}$. But since k generates $a\mathbf{Z} + b\mathbf{Z}$, it must be that $\underbrace{k + k + k + \ldots + k}_{r \text{ terms}} = a$, or $kr = a$ for some integer r. So k | a. Similarly $\underbrace{k + k + k + \ldots + k}_{s \text{ terms}} = b$, or $ks = b$ for some integer s; thus $k \mid b$.

(c) Now let $x \in \mathbf{Z}$ be such that $x \mid a$ and $x \mid b$. This means that a, b belong to the subgroup $x\mathbf{Z}$. Hence, since $(+)$ is a binary operation on $x\mathbf{Z}$, any number $t = az_1 + bz_2$ is also a member of $x\mathbf{Z}$. Thus, $(a\mathbf{Z} + b\mathbf{Z}) \subset x\mathbf{Z}$, that is, $x\mathbf{Z}$ contains all the same elements as $k\mathbf{Z}$, and possibly more. In particular, x must generate $k\mathbf{Z}$, so $x \mid k$. ◆

We may summarize Theorems 2.9 and 2.10 by saying that if a, b are two integers, not both zero, then the unique positive integer that is the generator for the group $a\mathbf{Z} + b\mathbf{Z}$ is the largest integer that divides a, b. Hence, the following definition is equivalent to our previous one for the greatest common divisor:

DEF. Let $a, b \in \mathbf{Z}$, with at least one of these being nonzero. The unique positive integer d that generates the subgroup $a\mathbf{Z} + b\mathbf{Z}$ is termed the **greatest common divisor** of a, b.

The greatest common divisor can be generalized to finite sets of more than two integers. Inductively, we define the greatest common divisor of k integers to be $(x_1, (x_2, x_3, \ldots, x_k))$. In view of Theorem 2.7(a, b) the inner parentheses are redundant and the order of listing within the outer parentheses is immaterial. Hence, the greatest common divisor is unambiguously specified by the symbol $(x_1, x_2, x_3, \ldots, x_k)$. For example, $(45, 27, 90, -18) = 9$. Again, if $(x_1, x_2, x_3, \ldots, x_k) = 1$, then $x_1, x_2, x_3, \ldots, x_k$ are termed **relatively prime**, even if particular pairs of them are not relatively prime. Thus, 2, 3, 16 are relatively prime, even though $(2, 16) = 2$.

Do you see, however, that if one pair of integers from a set of k integers is relatively prime, then the whole set is relatively prime (Problem 2.53)? This allows us to extend the idea in the preceding paragraph to *infinite* sets of integers.

DEF. An **infinite coprime sequence** is an infinite sequence of integers, usually arranged in ascending magnitude, in which the greatest common divisor of any two members is unity.

Of course, one obvious infinite coprime sequence is the sequence of the primes themselves,

$$2, 3, 5, 7, 11, 13, 17, 19, 23, \ldots .$$

However, construction of this example of an infinite coprime sequence depends on our knowledge of the sequence of primes. We have scant knowledge of incredibly large primes so there is a question here of just what is known. We can, though, construct other infinite coprime sequences without a knowledge of the sequence of primes (Edwards, 1964).

Theorem 2.11 Let p be any prime and suppose $(p, u_1) = 1$, $u_1 > p$. Then the sequence $\{u_n\}_{n=1}^{\infty}$ defined by $u_n = p + u_{n-1}(u_{n-1} - p)$ is an infinite coprime sequence.

Proof From $u_n = p + u_{n-1}(u_{n-1} - p)$ and $u_{n-1} = p + u_{n-2}(u_{n-2} - p)$, we obtain $u_n = p + u_{n-1}u_{n-2}(u_{n-2} - p)$. By induction, this leads ultimately to

$$\begin{aligned} u_n &= p + u_{n-1}u_{n-2}u_{n-3} \ldots u_1(u_1 - p) \\ &= p + U(u_1 - p), \qquad U = \prod_{i=1}^{n-1} u_i. \end{aligned}$$

Now suppose $k \geq 1$ is any integer such that $k \mid u_n$ and $k \mid U$. Then k must divide $u_n - U(u_1 - p) = p$, and hence k is either 1 or p. Assume it is p; then from Problem 2.18, p must divide one of $u_1, u_2, \ldots, u_{n-1}$. Let it divide u_j, $1 \leq j \leq n - 1$. Now we have

$$u_j = p + u_{j-1}u_{j-2} \ldots u_1(u_1 - p)$$

and $p \mid u_j$, $p \mid p$ imply p divides one of $u_1, u_2, \ldots, u_{j-1}$. We continue in this manner until we reach the conclusion that p divides u_1; this is a contradiction, so $k = 1$ and the finite set $\{u_1, u_2, \ldots, u_n\}$ is relatively prime in pairs. Since n was arbitrary, this conclusion holds for the infinite sequence $\{u_n\}_{n=1}^{\infty}$, so by definition we have an infinite coprime sequence. ◆

Example 2.12 Let $p = 3$, $u_1 = 4$. Then the sequence

$$\{3, 4, 7, 31, 871, 756031, \ldots\}$$

is an infinite coprime sequence. Observe that some elements of this sequence are composite (4, 871). ◆

Problems

2.48. We have used the Fundamental Theorem of Arithmetic to identify the greatest common divisor of two integers. But is the existence of the greatest common divisor dependent upon the validity of the Fundamental Theorem? Consider again the set **S** in Problem 2.8. In that set what is the value of (1375, 3025)?

2.49. Use Theorem 2.6 to determine the greatest common divisor of the following sets of integers:

(a) 204, 498.
(b) 1055, 1056.
(c) 12832, 205664.
(d) 83853, 160083.
(e) 236, 362, 632.
(f) 9633, 415233, 1341015.

2.50. Let N be any positive integer. Are N, $N + 1$ necessarily relatively prime? How about N and $N^2 + 1$?

2.51. Show that if $(x, y) = 1$ and $(x, z) = 1$, then $(x, yz) = 1$.

2.52. Show that if $n \geq 8$ is an even integer, then it can be written as the sum of two integers greater than 1 that are relatively prime.

2.53. Show that if one pair from the set of integers $\{x_1, x_2, \ldots, x_k\}$ is coprime, then the whole set is relatively prime.

2.54. Prove Theorem 2.7(b).

2.55. In this problem, d stands for (x, y); assume $1 < x < y$.

(a) If $d > 1$, is $(x/d, y) = 1$?
(b) If $d = 1$, and $xy = z^5$, are x, y both fifth powers?
(c) If $d > 1$, is $(x + 1, y + 1) = 1$?

2.56. Recall the Fibonacci numbers (Problem 1.20). Explain why $(F_n, F_{n+1}) = 1$ holds for all $n \in \mathbf{N}$.

2.57. Prove that if $(x, y) = 1$, then $(x + y, x - y) = 1$ or 2.

2.58. Show that the integers $3k + 1$ and $5k + 2$, $n \in \mathbf{N}$, are relatively prime.

2.59. Find all y satisfying $0 < y \leq 50$ such that $(y - 8, y + 20) = 14$.

2.60. Is it possible for two subgroups of $\langle \mathbf{Z}, + \rangle$ to have only two elements in common? Only three elements in common?

2.61. Let a, b be integers, at least one of which is nonzero. Show that the set $a\mathbf{Z} + b\mathbf{Z} = \{az_1 + bz_2\colon z_1, z_2 \in \mathbf{Z}\}$ is a subgroup of $\mathbf{Z}$.

2.62. Let P_n be the probability that two integers x, y drawn at random from $1, 2, 3, \ldots, n$ have $(x, y) = 1$. The Italian mathematician **Ernesto Cesàro** (1859–1906) and the English mathematician **James J. Sylvester** (1814–1897) showed around 1883 that $P_n \sim 6/\pi^2$ as $n \to \infty$. Write a computer program to check out how well the asymptotic formula works for some large n.

2.63. Let $u_1 > 1$ be an odd integer and define the sequence $\{u_n\}_{n=1}^{\infty}$ inductively by $u_n = u_{n-1}^2 - 2$. Show that $\{u_n\}_{n=1}^{\infty}$ is an infinite coprime sequence.

2.5 Application to Linear Diophantine Equations

A single equation in more than one unknown whose solution is sought in integers is called a **Diophantine equation**, after the Alexandrian mathematician **Diophantus** (ca. 250 A.D.) (Heath, 1964). An equation of the first degree in each unknown such as $7x + 3y = 1$ is a **linear Diophantine equation**. The greatest common divisor plays an important role in the solution of such equations.

Theorem 2.12 Given any two integers a, b, there exist integers x, y that satisfy $ax + by = 1$ iff $(a, b) = 1$.

Proof ($\rightarrow$) Suppose integers x_0, y_0 exist such that $ax_0 + by_0 = 1$. Let $d = (a, b)$; then $d \mid a$ and $d \mid b$. By Theorem 1.4(b), $d \mid (ax_0 + by_0)$ and hence $d \mid 1$. This is possible only if $d = (a, b) = 1$.

($\leftarrow$) If $(a, b) = 1$, then by the definition on page 59, the subgroup $a\mathbf{Z} + b\mathbf{Z}$ is generated by 1, and hence there are $x_0, y_0 \in \mathbf{Z}$ such that $ax_0 + by_0 = 1$. ◆

Notice how quickly the sufficiency part ($\leftarrow$) of the theorem follows. This is a nice illustration of the power of the algebraic approach.

Example 2.13 (a) Consider $17x - 35y = 1$. Since $(17, -35) = 1$, the equation has a solution by Theorem 2.12. In fact, $x = -2$ and $y = -1$ is one such solution, obtained by inspection.

(b) The integers $a = 4$, $x = -12$, $b = 7$, $y = 7$ give $ax + by = 1$. We observe, in accordance with Theorem 2.12, that $(a, b) = (4, 7) = 1$. ◆

Corollary 2.12.1 Given any two integers a, b, there exist integers x, y such that $ax + by = (a, b)$.

Proof Define the integers $c = a/(a, b)$ and $d = b/(a, b)$. By the definition on page 55, it follows that $(c, d) = 1$. Theorem 2.12 then guarantees the existence of integers x_0, y_0 such that $cx_0 + dy_0 = 1$. Multiplication by (a, b) then yields $ax_0 + by_0 = (a, b)$. ◆

Example 2.14 Consider $6x + 8y = 2$, for which $(6, 8) = 2$. By Corollary 2.12.1 it has a solution; $x = -5$, $y = 4$ is such a solution. The original equation is, of course, equivalent to $3x + 4y = 1$, to which Theorem 2.12 applies. ◆

Corollary 2.12.2 If $(a, b) \mid n$, then there exist integers x, y satisfying $ax + by = n$.

Proof Define the integers $n^* = n/(a, b)$, $c = a/(a, b)$, and $d = b/(a, b)$. By Theorem 2.12 there are integers x_0, y_0 such that $cx_0 + dy_0 = 1$. Multiplication by $n^*(a, b)$ yields

$$[c(a, b)](n^*x_0) + [d\ (a, b)](n^*y_0) = n^*(a, b)$$

or $a(n^*x_0) + b(n^*y_0) = n$. Hence, $x = n^*x_0$ and $y = n^*y_0$ satisfy $ax + by = n$. ◆

Example 2.15 Consider $12x + 33y = 6$. Here, $(12, 33) = 3$ and $n^* = 6/3 = 2$. The associated equation $cx + dy = 1$ in Corollary 2.12.2 is $4x + 11y = 1$. By easy inspection, it has a solution $x_0 = 3$, $y_0 = -1$. Hence, a solution to the original Diophantine equation is $x = x_0 n^* = 3(2) = 6$, $y = y_0 n^* = (-1)(2) = -2$. ◆

Incidentally, the Diophantine equations in each of the three preceding examples have been simple enough to solve by casual inspection. For more complicated equations, systematic procedures are available. We'll see one such method in the next section and another method in Section 3.11.

We come now to another famous result ascribed to **Euclid**. It appeared in Book VII of his *Elements*.

Corollary 2.12.3. (Euclid's Lemma). Let x, y, z be nonzero integers, and suppose $x \mid yz$. If $(x, y) = 1$, then $x \mid z$.

Proof Since $(x, y) = 1$, then from Corollary 2.12.1 we can write $xa + yb = 1$ for some integers a, b. Multiplication of both sides by z gives $(xz)a + (yz)b = z$. But $x \mid xz$ clearly, and $x \mid yz$ by hypothesis, so from Theorem 1.4(b) we must have $x \mid z$. ◆

This last corollary is a generalization of Theorem 2.2. It reduces to that theorem if x is a prime p that is not in the factorization of y.

Example 2.16 We have $27 \mid 21{,}060$. If we write $21{,}060 = 52 \cdot 405$, then $(27, 52) = 1$ implies $27 \mid 405$. Or if we write $21{,}060 = 130 \cdot 162$, then $(27, 130) = 1$ implies $27 \mid 162$. ◆

This section has been your first introduction to Diophantine equations, a vast subject. We'll meet them again in the next section and again briefly in Section 3.11 in the guise of linear congruences, and once more in Chapter 8.

Problems

2.64. Find a particular solution (if it exists) to each of the following Diophantine equations:

(a) $21x + 35y = 26$.
(b) $718x + 10y = 52$.
(c) $-17x + 31y = 11$.
(d) $97x + 129y = 1$.
(e) $84x - 357y = -105$.
(f) $209x + 114y = -221$.

2.65. Express the greatest common divisor of 12,832 and 205,664 as a linear combination of them.

2.66. Supply reasons for the steps in the following proof:

Theorem: If x_0, y_0 is a particular solution to the Diophantine equation $ax + by = c$, then all solutions to the equation have the form $x = x_0 + (b/d)k, y = y_0 - (a/d)k, d = (a, b), k \in \mathbf{Z}$.

Proof:

Statements	*Reasons*
1. The numbers x, y as given are a family of solutions to the equation $ax + by = c$.	1. Why?
2. Assume that $x = x' \neq x_0$, $y = y' \neq y_0$ is any solution of the equation. Then we have $\frac{x' - x_0}{y' - y_0} = \frac{b}{-a} = \frac{b/d}{-a/d}$.	2. Why?
3. The quantity on the right is a quotient of two integers and is in lowest terms.	3. Why?
4. It follows that there is an integer k such that $x' - x_0 = (b/d)k$ and $y' - y_0 = (-a/d)k$.	4. Why?
5. Hence, x', y' have the general form just given, so all solutions have this form.	5. Why?

2.67. Suppose $(a, b) = 1$, where $a, b > 0$. Prove that the Diophantine equation $ax + by = ab$ has no solution in *positive* integers.

2.68. Show that $14x^2 + 15y^2 = 7^{1990}$ has no solution in integers.

2.69. A particular solution to the Diophantine equation $2x - 3y + 5z = 17$ is $x_0 = 2$, $y_0 = 4$, $z_0 = 5$. See if you can modify the statement of the theorem in Problem 2.66 so as to provide five more particular solutions to this Diophantine equation.

2.70. Consult the reference Davis and Shisha (1981) and write up a synopsis in your own words of the five proofs of Euclid's Lemma given therein.

2.6 The Euclidean Algorithm

Euclid showed in Book VII of the *Elements* a systematic way to determine the greatest common divisor of two integers x, y without having to factor them (a tedious task for large integers). His algorithm is nothing more than a sequence of applications of the Division Algorithm.

We illustrate the **Euclidean Algorithm** with the pair of integers $\{x, y\} = \{72, 120\}$.

STEP 1. Attempt division of the integer of greater magnitude by the one of smaller magnitude (assumed nonzero).

$$120 = 72 \cdot 1 + 48$$

STEP 2. Attempt division of the former divisor (72) by the former remainder (48).

$$72 = 48 \cdot 1 + 24$$

STEP 3. Repeat STEP 2 until it terminates.

$$48 = 24 \cdot 2 + 0$$

STEP 4. The last nonzero remainder (24) is the greatest common divisor of the two integers (72 and 120).

Theorem 2.13 (The Euclidean Algorithm). Let $a, b \in \mathbf{N}$ with $a > b$ and $b \nmid a$. Define q_1, r_1 by $a = bq_1 + r_1$, $0 < r_1 < b_1$. Define the sequences $\{a_k\}$, $\{b_k\}$, $\{q_k\}$, $\{r_k\}$ by

$$\begin{cases} a_k = b_{k-1} \\ b_k = r_{k-1} \\ a_k = b_k q_k + r_k, \qquad 0 \leq r_k < b_k. \end{cases}$$

Then if $r_n > 0$ but $r_{n+1} = 0$ for some $n \in \mathbf{N}$, then $r_n = (a, b)$.

Proof The algorithm yields $r_k < r_{k-1} < r_{k-2} < \ldots < r_1$. The string of inequalities must terminate since the r_k's are nonnegative integers. Thus, for some $n \in \mathbf{N}$, one has $r_n > 0$ and $r_{n+1} = 0 < r_n$:

$$\begin{cases} b_{n-1} = a_n = b_n q_n + r_n, & 0 < r_n < b_n \\ b_n = a_{n+1} = b_{n+1} q_{n+1} + r_{n+1} & \\ \qquad\qquad = r_n q_{n+1} + 0. & \end{cases}$$

Then working from the bottom equation up to the first equation we have $r_n \mid b_n \to r_n \mid b_{n-1} \to r_n \mid b_{n-2} \to \ldots \to r_n \mid a,\ r_n \mid b_n$. The implications follow by repeated application of Theorem 1.4. Hence, $r_n \le (a, b) = d$.

On the other hand, $d \mid a$ and $d \mid b \to d \mid r_1 \to d \mid b_2$. Then as $b = a_2 = b_2 q_2 + r_2$, we obtain $d \mid b_2 \to d \mid r_2 \to \ldots \to d \mid r_n$. Hence, $d \le r_n$, and the two inequalities imply that $d = (a, b) = r_n$. ◆

The Euclidean Algorithm can be used to determine the greatest common divisor of more than two integers. We know from Theorem 2.7 that the greatest common divisor of x, y, z, for example, depends only on these numbers and not on the order in which they are written nor on the way they are grouped. So if we write $d = (x, y, z)$ and $d' = (y, z)$, then $d = (x, d')$. The Euclidean Algorithm is used once to compute d', and then again to compute d.

Example 2.17 Find the greatest common divisor of the integers 861, 2706, 9061.

We first compute $d' = (861, 2706)$ as follows:

$$2706 = 3 \cdot 861 + 123$$

$$861 = 7 \cdot 123 + 0. \qquad \therefore d' = 123.$$

Next, we compute $d = (123, 9061)$ as follows:

$$9061 = 73 \cdot 123 + 82$$

$$123 = 1 \cdot 82 + 41$$

$$82 = 2 \cdot 41 + 0. \qquad \therefore d = 41. \quad ◆$$

Example 2.18 Does the Diophantine equation

$$861x + 2706y + 9061z = 204$$

have a solution?

ERRATA

Location		*Replace*	*By*
p 34.	Line 1	1.14	1.12
p. 67.	Line 20	first =	—
p. 92.	2nd line of Problem 3.37(d)	xz – wy	xz + wy
p. 94.	1st line of Example 3.14	2	Q
p. 121.	Line 6	d_0,	d_0;
p. 179.	Line 7 of Example 5.11	8, 7, 11	8, 11, 7
p. 192.	Line 7	ap	p
p. 321.	Line 16	(1939)	(1939b)
p. 377.	Line 4 of Example 10.28	$F_{n-1} + F_{n-2}$	$\hat{F}_{n-1} + \hat{F}_{n-2}$
p. 410.	Line 10	Delete }	
p. 414.	Line 3 of Example 11.5	1	0
p. 441.	Line 2 of Problem 11.60	n; = 0	n = 0
p. 451.	8th line from the bottom	note of **II**,	Note on the **II**,
p. 455.	Line 3	incongruents	incongruent
p. 501.	11.51.	No.	Yes.

Dence/Dence, Elements of the Theory of Numbers,
0-12-209130-2

A necessary condition for there to be integers x, y, z that satisfy the equation is that $d = (861, 2706, 9061)$ shall divide 204. But from Example 2.17, $d = (861, 2706, 9061) = 41$ and $41 \nmid 204$. Thus, no solution is possible. ◆

If regarded too narrowly, the Euclidean Algorithm might seem to be good only for finding greatest common divisors. Look again, however, at the sequence of equations that defines the algorithm. The sequence actually provides a systematic method for finding a solution to the Diophantine equation $ax + by = c$, assuming that one exists.

Example 2.19 Solve $15x - 42y = -12$.

This has a solution since $(15, 42) \mid (-12)$. The Euclidean Algorithm gives the set of equations

$$\begin{aligned} 42 &= 15 \cdot 2 + 12 & & \\ 15 &= 12 \cdot 1 + 3 & \rightarrow \quad 42 &= 15(2 + 1) - 3 \\ 12 &= 3 \cdot 4 & 12 &= 3 \cdot 4. \end{aligned}$$

We eliminate the first remainder (12) from between the two equations grouped in the first loop. Then eliminate the second remainder (3) that is contained in the second loop:

$$42(4) + 12 = 15[4(2 + 1)],$$

or

$$15[-4(2 + 1)] = 42(-4) = -12.$$

It follows that a solution is $x = -12$, $y = -4$. The complete set of solutions is obtainable using Problem 2.66.

Problems

2.71. Rework Problem 2.49, but use Theorem 2.13.

2.72. Find the complete solution sets of these equations:

(a) $77x - 23y = 6$. (b) $300x + 105y = 1110$.
(c) $-17x + 32y = 11$. (d) $84x - 354y = -105$.

2.73. What would happen if you applied the Euclidean Algorithm to the pair of integers 25, 55 in the set **S** of Problem 2.8?

2.74. The Euclidean Algorithm is not the only algorithm for finding the greatest common divisor d of two positive integers. The following procedure employs no divisions, only subtractions and multiplications. We have for $x > y \geq 1$,

$$(x, y) = \begin{cases} 2(x/2, y/2) & \text{if } x, y \text{ are both even} \\ (x/2, y) & \text{if } x \text{ is even, } y \text{ is odd} \\ (x - y, y) & \text{if } x, y \text{ are both odd, } x > y. \end{cases}$$

The preceding is repeated enough times until (d, d) is obtained. Program this on a computer and use it to find (262584, 527773). Also program the Euclidean Algorithm and apply it to the same pair of integers. See if you can decide which algorithm is faster.

2.75. Start with the arbitrary pair of Fibonacci numbers F_{n+1}, F_n (Problem 1.20) and apply the Euclidean Algorithm to it. Deduce a series formula for the product $F_{n+1}F_n$.

2.7 Least Common Multiple

Given two nonzero integers x, y, it is always possible to find a positive integer that is divisible by both of them. For example, $|xy|$ is such an integer. Let $\mathbf{S}_{xy}$ be the set of all positive integers that are divisible by both x and y. Then $\mathbf{S}_{xy}$ is nonempty and is bounded below by 0. By the Well-Ordering Property, $\mathbf{S}_{xy}$ must contain a least element. For example, if $x = 14$ and $y = 35$, then 70 is the least positive integer that is divisible by 14, 35.

DEF. Let $x, y \in \mathbf{Z}$, with both being nonzero. The smallest positive integer M that is divisible by both x, y is termed the **least common multiple** of x, y and we write $M = [x, y]$.

Example 2.20 Suppose $x = p_1$, $y = p_2$, where p_1, p_2 are distinct primes; then $[x, y] = p_1p_2$. Suppose $x = 14p_1$, $y = 35p_2$, where p_1, p_2 are distinct primes larger than 7; then $[x, y] = 70p_1p_2$. ◆

If x, y are not too small, it is not easy to tell at a glance what is the value of $[x, y]$. How could you proceed? Let us suppose for simplicity that $1 < x < y$, and suppose you could identify one prime divisor p of either x or y. Then p has to occur in the factorization of $[x, y]$. We would then examine just those multiples of p, beginning at y or at the first such multiple that exceeds y, until we find the smallest multiple of p that is divisible by both x and y.

Code for a computer program to do this work would be quite short.

```
10  ENTER X, Y. COMMENT: 1 < X < Y.
20  ENTER P.   COMMENT: P DIVIDES X OR Y.
30  M = P * (INT( (Y − .1)/P) + 1).
40  DO X, Y BOTH DIVIDE M ?
50  IF NO, M := M + P. GOTO 40.
60  IF YES, PRINT M.
70  END
```

Example 2.21 Let $X = 69$, $Y = 299$. We quickly identify $P = 3$ as a prime divisor of X. Initialize M at $M = 3 * (\lfloor 298.9/3 \rfloor + 1) = 300$. We now check the sequence of integers $\{300, 303, 306, 309, \ldots\}$ until we find the first one that is divisible by 69 and by 299. Execute this algorithm on the computer. ◆

The preceding procedure is safe but brute-force. Checking of fewer multiples would be required if we could at the outset identify *two* distinct prime divisors p_1, p_2 of x or y. Then it would only be necessary to examine those multiples of $p_1 \cdot p_2$ beginning at y or at the first such multiple that exceeds y. In the computer program we enter P1, P2 instead of P, and we modify lines 30, 50 by replacing P by P1 · P2. Try this out on an example of your own.

There is no reason to limit our considerations to only two prime divisors of x or y. If the preceding ideas are extended, then we would arrive at the following theorem, which is the analog of Theorem 2.6, and is again a numerical consequence of the Fundamental Theorem of Arithmetic.

Theorem 2.14 Let x, y be nonzero integers whose absolute values are factorable as

$$|x| = \prod_{i=1}^{\infty} p_i^{a_i}, \qquad |y| = \prod_{i=1}^{\infty} p_i^{b_i}.$$

Then

$$[x, y] = \prod_{i=1}^{\infty} p_i^{n_i}, \qquad \text{where } n_i = \max\{a_i, b_i\} \text{ for all } i \in \mathbf{N}.$$

Example 2.22 Refer to Example 2.9. The least common multiple of 588, 3080 is therefore

$$\begin{aligned}[588, 3080] &= 2^{\max\{2,3\}} \cdot 3^{\max\{1,0\}} \cdot 5^{\max\{0,1\}} \cdot 7^{\max\{2,1\}} \cdot 11^{\max\{0,1\}} \\ &= 2^3 \cdot 3^1 \cdot 5^1 \cdot 7^2 \cdot 11^1 \\ &= 64{,}680.\end{aligned}$$

Unfortunately, Theorem 2.14 suffers from the same problem as does Theorem 2.6: it requires time-consuming factorizations. However, we notice in the preceding example that $(588, 3080)[588, 3080] = 28(64{,}680) = 1{,}811{,}040$, and that $xy = 588(3080) = 1{,}811{,}040$. This is no accident.

Theorem 2.15 Let x, y be two nonzero integers. Then

$$(x, y)[x, y] = |xy|.$$

Proof The proof (Problem 2.81), which follows from Theorems 2.6 and 2.14, is left to the reader. ◆

Now Theorem 2.15 is really useful for the computation of the least common multiple. No factorizations are needed, and (x, y) is obtained quickly from the Euclidean Algorithm.

Example 2.23 Compute [861, 2706].

The Euclidean Algorithm gives $(861, 2706) = 123$ (see Example 2.17). Hence, $[861, 2706] = 861(2706)/123 = 18{,}942$. ◆

The least common multiple has the following set of elementary properties. We prove part (b), leaving the other parts for you.

Theorem 2.16

(a) $[x, y] = [y, x]$.
(b) $[x, [y, z]] = [[x, y], z]$.
(c) $[x, 1] = [1, x] = x$.
(d) $[xz, yz] = |z|\,[x, y]$.

Proof (b) We apply the result in Theorem 2.14 to the left side. Let

$$|x| = \prod_{i=1}^{\infty} p_i^{a_i}, \qquad |y| = \prod_{i=1}^{\infty} p_i^{b_i}, \qquad |z| = \prod_{i=1}^{\infty} p_i^{c_i}.$$

Then we have

$$\begin{aligned}[x, [y, z]] &= \prod_{i=1}^{\infty} p_i^{\max\{a_i, \max\{b_i, c_i\}\}} \\ &= \prod_{i=1}^{\infty} p_i^{\max\{\max\{a_i, b_i\}, c_i\}} \\ &= [[x, y], z]. \quad \blacklozenge\end{aligned}$$

It is clear from Theorem 2.16 (a, b) that the inner brackets in $[[x, y], z]$ are redundant and that $[x, y, z]$ is uniquely determined by x, y, z.

Example 2.24 We compute $[6, 7, 9] = [[6, 7], 9] = [42, 9] = 126$. ◆

As with the greatest common divisor, the least common multiple can be given an algebraic treatment that provides additional insight. The following lemma is needed.

Lemma 2.17.1 Let $\mathbf{H}$, $\mathbf{K}$ be subgroups of $\mathbf{G}$. Then $\mathbf{H} \cap \mathbf{K}$ is also a subgroup of $\mathbf{G}$.

Proof Let $x, y \in \mathbf{H} \cap \mathbf{K}$; then $x \cdot y \in \mathbf{H}$ because $\mathbf{H}$ is a group. Also $x \cdot y \in \mathbf{K}$ because $\mathbf{K}$ is a group; hence, $x \cdot y \in \mathbf{H} \cap \mathbf{K}$, so $(\cdot)$ is a binary operation on $\mathbf{H} \cap \mathbf{K}$. Similarly, the identity $\mathbf{I}$ of $\mathbf{G}$ resides in both $\mathbf{H}$, $\mathbf{K}$ and so is also in $\mathbf{H} \cap \mathbf{K}$. By parallel reasoning we see that if $x \neq \mathbf{I}$ is in $\mathbf{H} \cap \mathbf{K}$, then x^{-1} resides in both $\mathbf{H}$, $\mathbf{K}$ and so is also in $\mathbf{H} \cap \mathbf{K}$. Thus, $\mathbf{H} \cap \mathbf{K}$ contains all its inverses. Finally, $\mathbf{H} \cap \mathbf{K}$ inherits associativity from $\mathbf{G}$. Thus, $\mathbf{H} \cap \mathbf{K}$ meets all the requirements of a group, so it is a subgroup of $\mathbf{G}$. ◆

Example 2.25 Let $\mathbf{G} = \mathbf{Z}$, $\mathbf{H} = 6\mathbf{Z}$, and $\mathbf{K} = 8\mathbf{Z}$. The elements of $\mathbf{H} \cap \mathbf{K}$ are those integers that are simultaneously multiples of 6 and 8, that is, $\mathbf{H} \cap \mathbf{K} = 24\mathbf{Z}$. By Theorem 2.8 this is a subgroup of $\mathbf{Z}$. ◆

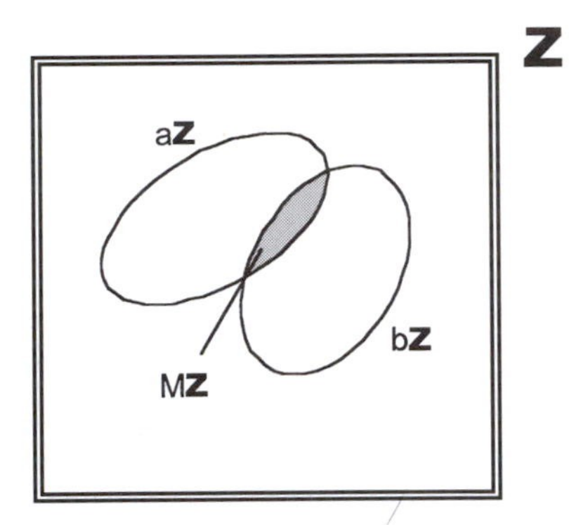

Figure 2.5 The intersection of the subgroups $a\mathbf{Z}$, $b\mathbf{Z}$ is generated by the least common multiple of a, b.

Theorem 2.17 Let a, b (both nonzero) be generators for the subgroups $a\mathbf{Z}$, $b\mathbf{Z}$, respectively. Then their intersection has as its generator M, where

(a) $a \mid M$ and $b \mid M$.
(b) if x is an integer such that $a \mid x$ and $b \mid x$ hold, then $M \mid x$ also.

Proof (a) From Lemma 2.17.1 $a\mathbf{Z} \cap b\mathbf{Z}$ is a subgroup of $\mathbf{Z}$, while from Theorem 2.9 we have that $a\mathbf{Z} \cap b\mathbf{Z}$ is of the form $M\mathbf{Z}$, for some integer M. But if $M \in a\mathbf{Z} \cap b\mathbf{Z}$, then $M \in a\mathbf{Z}$, which is generated by a. This means $az_1 = M$ for some $z_1 \in \mathbf{Z}$, or $a \mid M$. Similarly, $M \in b\mathbf{Z}$ means $bz_2 = M$ for some $z_2 \in \mathbf{Z}$, or $b \mid M$.

(b) If $x \in \mathbf{Z}$ is such that $a \mid x$, $b \mid x$ hold (a, $b \neq 0$), then this says that x belongs simultaneously to the subgroups generated by a, b. That is, $x \in a\mathbf{Z} \cap b\mathbf{Z}$; but this group is generated by M. So, $x \in M\mathbf{Z}$ and, therefore, $x = Mz$ for some $z \in \mathbf{Z}$, or $M \mid x$. ◆

We may summarize Theorem 2.17 by saying that if a, b are two nonzero integers, then the unique positive integer that is the generator for the group $a\mathbf{Z} \cap b\mathbf{Z}$ (Fig. 2.5) is the smallest positive integer that is divisible by both a and b.

Hence, the following definition is equivalent to the more familiar arithmetic one given earlier for the least common multiple:

DEF. Let $a, b \in \mathbf{Z}$, both of these being nonzero. The unique positive integer M that generates the subgroup $a\mathbf{Z} \cap b\mathbf{Z}$ is termed the **least common multiple** of a, b, $M = [a, b]$.

The algebraic picture permits some facts about the least common multiple to appear naturally (Problem 2.85).

Problems

2.76. Let $\mathbf{S} \subset \mathbf{N}$ be closed under multiplication. Is it possible for every pair of integers in $\mathbf{S}$ to have a least common multiple, even if the Fundamental Theorem of Arithmetic is invalid in $\mathbf{S}$?

2.77. Suppose that the only information we have about two integers x, y is that $x = 2p_1$, $y = 3p_2$, and p_1, p_2 are distinct primes. Can we conclude that $[x, y] = 6p_1p_2$? Explain.

2.78. Do what Example 2.21 requests.

2.79. Use Theorem 2.14 explicitly in order to compute the following:

(a) [78, 204].
(b) [100, 117, 405].
(c) [1573, 1859].
(d) [982, 8726].

2.80. You wish to add $1/6 + 13/34 + 5/44 + 1/1089$. What is the best choice for a common denominator? Explain.

2.81. Prove Theorems 2.14 and 2.15.

2.82. Prove the following distributive-like laws of the greatest common divisor and the least common multiple:

$$[(x, y), z] = ([x, z], [y, z])$$

$$([x, y], z) = [(x, z), (y, z)].$$

Illustrate these results with concrete examples.

2.83. Let $x = 20$ and suppose $[x, y] = 140$. Find all (positive) values for y.

2.84. Evaluate:

(a) $[F_n, F_{n+1}]$ (the F_n's are Fibonacci numbers).
(b) $[2^{2^n} + 1, 2^{2^m} + 1]$, $n \neq m$.
(c) $[u_n, u_m]$, $n \neq m$ (see Theorem 2.11).
(d) [118726, 530982].

2.85. Explain how the algebraic interpretation of the least common multiple allows you to see parts (a), (b), and (c) of Theorem 2.16 readily.

2.86. Let x, y, z be nonzero integers. Does Theorem 2.15 extend to this: $(x, y, z)\ [x, y, z] = |xyz|$?

RESEARCH PROBLEMS

1. A special kind of divisor $d > 0$ of a positive integer N is one for which $(d, N/d) = 1$. Let us designate such a divisor by d_u. Thus, the complete set of d_u's for $N = 12$ is $\{1, 3, 4, 12\}$. For any $N > 1$ we always have $\Sigma_{d_u|N}\, d_u > N$. Investigate by computer if there are any even $N > 6$ such that $\Sigma_{d_u|N}\, d_u = 2N$.

2. Let $\mathbf{Q}[x]$ denote the set of all polynomials in x with rational coefficients. Investigate if it is possible to formulate for $\mathbf{Q}[x]$ an analog of the Euclidean Algorithm for finding a (the ?) greatest common divisor (definition ?) of two polynomials in $\mathbf{Q}[x]$. Consider a proof and consider programming what you discover.

3. Let $\mathbf{Z}[\sqrt{-d}]$, $d \in \mathbf{N}$, be the system of numbers of the form $a + b\sqrt{-d}$, $a, b \in \mathbf{Z}$. We define the "size" of $a + b\sqrt{-d}$ to be $a^2 + db^2$. A prime $a + b\sqrt{-d}$ in $\mathbf{Z}[\sqrt{-d}]$ is a number that cannot be factored in $\mathbf{Z}[\sqrt{-d}]$ as $a + b\sqrt{-d} = (A + B\sqrt{-d})(C + D\sqrt{-d})$ and for which $A^2 + dB^2, C^2 + dD^2 > 1$. For example, $1 + 3\sqrt{-2}$ is a prime in $\mathbf{Z}[\sqrt{-2}]$. Investigate by computer if a statement analogous to Theorem 2.1 appears to be valid for the systems corresponding to $d = 2, 5, 13, 17$.

References

Ayres, Jr., F., *Modern Algebra*, Schaum's Outline Series, McGraw–Hill, New York, 1965. A very readable coverage of elementary material on groups, rings, fields, and vector spaces. See especially Chapters 9–11.

Davis, D., and Shisha, O., "Simple Proofs of the Fundamental Theorem of Arithmetic," *Math. Mag.*, **54**, 18 (1981). This paper is important for you to read.

Edwards, A.W.F., "Infinite Coprime Sequences," *Math. Gaz.*, **48**, 416–422 (1964). Interesting article with several examples of such sequences.

Fraleigh, J.B., *A First Course in Abstract Algebra*, 5th ed., Addison–Wesley, Reading, MA, 1994. A fine book; see Chapter 1 on groups.

Heath, T.L., *Diophantus of Alexandria*, Dover, New York, 1964. Written by an eminent authority on ancient Greek mathematics.

Kleiner, I., "The Evolution of Group Theory: A Brief Survey," *Math. Mag.*, **59**, 195–215 (1986). Group theory has had a glorious history; its roots lie in classical algebra, number theory, geometry, and analysis. Read about it here so that you can begin to appreciate why some of it is in the present book.

Knorr, W., "Problems in the Interpretation of Greek Number Theory: Euclid and the 'Fundamental Theorem of Arithmetic'," *Studies Hist. Phil. Sci.*, **7**, 353–368 (1976). The author warns against a too narrow interpretation of ancient written Greek mathematics. For example, **Euclid** may have understood the Fundamental Theorem more fully than its statement in the *Elements* might suggest.

Lindemann, F.A., "The Unique Factorization of a Positive Integer," *Quart. J. Math.* (Oxford), **4**, 319–320 (1933). The source of our proof of the Fundamental Theorem of Arithmetic. **Frederick Alexander Lindemann** (1886–1957) was originally trained as a physicist; he worked for a while in Berlin in the laboratory of **Hermann Walther Nernst**. Later, he came to England and became a don at Oxford. During World War II he was a scientific advisor to **Churchill**. Eventually, Lindemann was made a peer (Lord Cherwell, 1st Viscount). Interesting comments on him can be found in C.P. Snow, *Science and Government*, Harvard University Press, Cambridge, MA, 1961.

Liu, L.–C., "Niven Numbers and Products and Multiples of Primes," M.S. thesis, Central Missouri State University, Warrensburg, 1983. Our material in Problem 2.12 on integers p^k that are Niven is taken from pp. 10–12.

Ore, O., *Niels Henrik Abel*, Chelsea Publishing Co., New York, 1974. The life of the brilliant Norwegian mathematician **Abel** is a powerful human-interest story. Read about it here.

Ratering, S., "An Interesting Subset of the Highly Composite Numbers," *Math. Mag.*, **64**, 343–346 (1991). Straightforward proofs of theorems involving special highly composite numbers.

Schreiber, M., "Irregular Integers," *Amer. Math. Monthly*, **85**, 165–172 (1978). Skim-read this.

Chapter 3
An Introduction to Congruences

3.1 Elementary Properties of Congruences

Can you imagine writing nearly a whole book on the topic of division? Seems incredible, but in the summer of 1801 the 24-year-old **Carl Friedrich Gauss** published just such a work, *Disquisitiones Arithmeticae*. Written in Latin and dedicated to Gauss's patron, the Duke of Brunswick, the book is divided into seven chapters (Bühler, 1981). It proved to be one of the most influential books ever in the history of mathematics.

Of course, there was much more to the book than just mere division of integers. A new notation for an ancient concept played no small role in leading **Gauss** to deep results about integers.

DEF. The notation $a \equiv b \pmod{n}$, read "a **is congruent to** b **modulo** n," means that $n \mid (a - b)$.

For example, we have $7 \equiv 2 \pmod{5}$ and $3 \equiv 11 \pmod{4}$. Clearly, if $a \equiv b \pmod{n}$ holds, then $a \equiv b \pmod{-n}$ is also valid and vice versa. We shall therefore always assume moduli to be positive integers.

Let us now review from earlier mathematical studies the concept of an equivalence relation. We recall that a binary relation $\mathfrak{R}$ on a set **S** is a correspondence rule that connects various pairs of elements in **S**.

DEF. Let $\mathfrak{R}$ be a binary relation between mathematical objects in some set **S**. Then $\mathfrak{R}$ is termed an **equivalence relation** if for any $x, y, z \in \mathbf{S}$:

1. $x \,\mathfrak{R}\, x$ holds (reflexive property).
2. If $x \,\mathfrak{R}\, y$ is true, then $y \,\mathfrak{R}\, x$ also holds (symmetric property).
3. If $x \,\mathfrak{R}\, y$ and $y \,\mathfrak{R}\, z$ hold, then $x \,\mathfrak{R}\, z$ also holds (transitive property).

When we wish to indicate notationally that the objects x, y are equivalent with respect to some relation $\mathfrak{R}$ on a set **S**, we write $x \sim y$.

Equality between real numbers is an obvious equivalence relation.

Theorem 3.1 Modulo a given integer $n > 1$, congruence is an equivalence relation.[1]

Proof The proof (Problem 3.4), which follows from the preceding definitions, is left to the reader. ◆

Being an equivalence relation is not the only point of similarity between equalities and congruences. The very notation of a congruence suggests that some manipulations with equalities may have their counterparts among those with congruences. Look at these sample calculations.

Example 3.1 We have $11 \equiv 5 \pmod 3$ and $-2 \equiv 7 \pmod 3$; then subtraction gives $13 \equiv -2 \pmod 3$ (true), and multiplication gives $-22 \equiv 35 \pmod 3$ (also true). Squaring $-2 \equiv 7 \pmod 3$ yields the valid result $4 \equiv 49 \pmod 3$.

On the other hand, attempted subtraction of $8 \equiv 1 \pmod 7$ and $11 \equiv 5 \pmod 3$ leads to $-3 \not\equiv -4$ (mod 4, 7, or 3). Similarly, attempted multiplication leads to $88 \not\equiv 5$ (mod 21, 7, or 3). ◆

The following comprehensive theorem gives the correct formulation of the basic arithmetic of congruences.

Theorem 3.2 Suppose $a \equiv b \pmod n$ and $c \equiv d \pmod n$. Then

(1) $ka \equiv kb \pmod{kn}$ for any integer $k \neq 0$.
(2) $ka \equiv kb \pmod n$.

[1] Any two integers are congruent modulo 1; this is uninteresting. Henceforth, we assume $n > 1$ for any modulus.

(3) $a \pm c \equiv b \pm d \pmod{n}$.
(4) $ka + k'c \equiv kb + k'd \pmod{n}$ for any integers k, k'.
(5) $ac \equiv bd \pmod{n}$.
(6) $a^m \equiv b^m \pmod{n}$ for any $m \in \mathbf{N}$.
(7) If $n = pq, p, q \in \mathbf{N}$, then $a \equiv b \pmod{p}$ and $a \equiv b \pmod{q}$.

Proof Part (4) follows immediately from parts (2) and (3). To obtain (5), write $a = b + jn$ and $c = d + j'n$. Then $ac = bd + n(jd + j'b + jj'n)$, so $ac \equiv bd \pmod{n}$. For part (7) we have that $pq|(a - b)$ implies $p|(a - b)$, so $a \equiv b \pmod{p}$, and similarly for q. Parts (1) through (3) and (6) are left as exercises. ◆

Thus, we see from Theorem 3.2 that congruences to the same modulus can be validly added, subtracted, multiplied, and raised to positive integral powers. However, not every operation with equalities has a precise counterpart among congruences. Notice what happens in these examples.

Example 3.2 We have $72 \equiv 12 \pmod{6}$, but attempted division by 12 leads to $6 \not\equiv 1 \pmod{6}$. Similarly, $-35 \equiv 25 \pmod{15}$, but attempted division by -5 leads to $7 \not\equiv -5 \pmod{15}$.

On the other hand, we have $72 \equiv 12 \pmod{5}$ and division by 3 gives $9 \equiv 4 \pmod{5}$, a valid result. ◆

Division of a congruence by a nonzero integer can be salvaged if we make a change in the modulus.

Theorem 3.3 If $ka \equiv kb \pmod{n}$, then $a \equiv b \pmod{[n/(n, k)]}$.

Proof Let $n/(n, k) = n_1$ and $k/(n, k) = k_1$. Then $k(a - b) = jn, j \in \mathbf{N}$, or $k_1(n, k)(a - b) = j(n, k)n$, that is, $k_1(a - b) = jn_1$. Since $(n_1, k_1) = 1$, then $n_1|k_1(a - b)$ implies $n_1|(a - b)$ by Corollary 2.12.3. Putting this in congruence notation, we obtain $a \equiv b \pmod{n_1}$. ◆

Corollary 3.3.1 If $(k, n) = 1$, then $ka \equiv kb \pmod{n}$ implies $a \equiv b \pmod{n}$.

Proof This follows immediately from the theorem. ◆

Thus, whereas $72 \equiv 12 \pmod{6}$ did not imply $6 \equiv 1 \pmod{6}$, the congruence $72 \equiv 12 \pmod{5}$ does imply $6 \equiv 1 \pmod{5}$ because the divisor 12 and the modulus 5 are relatively prime. Notice the difference between this corollary and part (7) of Theorem 3.2. In the corollary the congruent integers are reduced, but the modulus is left intact. In Theorem 3.2, part (7), the congruent integers are left intact, but the modulus is reduced.

Parts (1) through (6) of Theorem 3.2 can be generalized in an obvious way to polynomials. We let the symbol $\mathbf{Z}[x]$ denote the set of all polynomials in the single indeterminate x with coefficients in $\mathbf{Z}$. For example, $17, x + 4, -2x^3 - x + 1$ are elements of $\mathbf{Z}[x]$.

Theorem 3.4 Suppose $f(x) \in \mathbf{Z}[x]$ and $a \equiv b \pmod{n}$. Then $f(a) \equiv f(b) \pmod{n}$.

Proof Let $f(x) = \sum_{k=1}^{r} c_k x^k$, where each $c_k \in \mathbf{Z}$. Then from part (6) of Theorem 3.2 we have $a^k \equiv b^k \pmod{n}$ for $k = 1, 2, \ldots, r$. Part (2) then gives us $c_k a^k \equiv c_k b^k \pmod{n}$ for $k = 1, 2, \ldots, r$. Repeated use of part (4) now yields $\sum_{k=1}^{r} c_k a^k \equiv \sum_{k=1}^{r} c_k b^k \pmod{n}$, that is, $f(a) \equiv f(b) \pmod{n}$. ◆

Example 3.3 Show that the square of an odd integer is congruent to 1 modulo 8.

Let $n = 2m + 1$. Then $n^2 = 4m^2 + 4m + 1 = 4m(m + 1) + 1$. One of m, $m + 1$ must be even, so in congruence language $m(m + 1) \equiv 0 \pmod{2}$. By Theorem 3.2, part (1), we have $4m(m + 1) \equiv 0 \pmod{8}$, so by part (3) addition of $4m(m + 1) \equiv 0 \pmod{8}$ to $1 \equiv 1 \pmod{8}$ gives $n^2 \equiv 1 \pmod{8}$.

Example 3.4 Let $f(x) = 2x^3 - x - 5$. Since $7 \equiv -2 \pmod{9}$, then $f(7) = 2(7)^3 - 7 - 5 = 674$, $f(-2) = 2(-2)^3 + 2 - 5 = -19$, and $674 \equiv -19 \pmod{9}$, as $9 \mid (674 + 19)$. ◆

A congruence can be interpreted as a statement about equality of remainders for two separate division problems. Thus, $a \equiv b \pmod{n}$ is equivalent to saying a/n and b/n yield the same remainder r (nothing is said about their quotients) if we insist that r lie in the range $0 \leq r < n$. For example, $27 \equiv 17 \pmod{5}$ and the divisions 27/5 and 17/5 produce the common remainder $r = 2$. In the special case when b already lies in the interval [0, n), then the congruence $a \equiv b \pmod{n}$ means that the single division a/n gives the remainder b.

Example 3.5 4^{100} divided by 7 gives a remainder of 4.

This statement is equivalent to $4^{100} \equiv 4 \pmod{7}$. To verify this congruence, we note that $4^3 = 64 \equiv 1 \pmod{7}$. Hence, it follows from Theorem 3.2, part (6), that

$$4^{99} = (4^3)^{33} \equiv 1^{33} \equiv 1 \pmod{7},$$

and multiplication of both sides of the congruence by 4 (use Theorem 3.2, part (2)) gives $4^{100} \equiv 4 \pmod{7}$. ◆

An algorithm for the determination of the remainder when a power a^k is divided by n is easily programmed on a computer or a contemporary graphing calculator. After inputting the integers a, k, n, we begin by initializing a product variable P to 1. The program then runs through a FOR loop k times, and reassigns P each time to the product aP, with adjustments made if $P \geq n$. The following code demonstrates this:

```
FOR dummy := 1 to k DO
  P := a * P
  IF P ≥ n THEN P := P − n * INT(P/n)
END.
```

The function INT, common on Texas Instruments calculators, represents the greatest integer function. The final value of P, after finishing the FOR loop, then represents the desired remainder when a^k is divided by n.

Using this program we can quickly verify

$$131^{76} \equiv 12 \pmod{37}$$

$$58^{95} \equiv 32 \pmod{100}.$$

With this latter congruence, we now know that if 58^{95} were multiplied out (to yield a 168-digit integer), its last two digits would be 3 and 2. Furthermore, in the first congruence, Theorem 3.2, part (6), allows us to reduce the base value of 131 to 20, and thereby examine the equivalent congruence $20^{76} \equiv 12 \pmod{37}$.

One advantage of the method just described for establishing $a^k \equiv b \pmod{n}$ is its simplicity. The code consists of only a few lines. On the other hand, a disadvantage is the potential for a large number of multiplications that need to be executed (maximum of k).

An alternative method that greatly reduces the number of multiplications requires that we initially express k in binary form. The 1's in the binary expansion of k will serve to dictate which (and how many) multiplications are necessary. This method has been called the **Method of Successive Squaring** (Silverman, 1997).

Example 3.6 Show that $37^{549} \equiv 14 \pmod{79}$.

Because $549 = 512 + 32 + 4 + 1 = 1000100101_2$, we then compute successive doubling powers of 37,

$$
\begin{aligned}
37 &\equiv 37 \pmod{79} & 37^{32} &\equiv 20^2 \equiv 5 \pmod{79}\\
37^2 &\equiv 26 \pmod{79} & 37^{64} &\equiv 5^2 \equiv 25 \pmod{79}\\
37^4 &\equiv 26^2 \equiv 44 \pmod{79} & 37^{128} &\equiv 25^2 \equiv 72 \pmod{79}\\
37^8 &\equiv 44^2 \equiv 40 \pmod{79} & 37^{256} &\equiv 72^2 \equiv 49 \pmod{79}\\
37^{16} &\equiv 40^2 \equiv 20 \pmod{79} & 37^{512} &\equiv 49^2 \equiv 31 \pmod{79}.
\end{aligned}
$$

From here we only need to multiply four terms because

$$
\begin{aligned}
37^{549} &= 37^{512} \cdot 37^{32} \cdot 37^4 \cdot 37^1\\
&\equiv 31 \cdot 5 \cdot 44 \cdot 37 \pmod{79}\\
&\equiv 31 \cdot 5 \cdot 48 \pmod{79}\\
&\equiv 31 \cdot 3 \pmod{79}\\
&\equiv 14 \pmod{79}. \quad \blacklozenge
\end{aligned}
$$

The interested student may wish to program this method and compare its speed of execution with that obtained by the first method.

Problems

3.1. Is congruence between triangles an equivalence relation? Is similarity between triangles an equivalence relation? Is asymptotic equivalence (Section 1.8) an equivalence relation between functions? Support your answers.

3.2. Define the binary relation $x \mathfrak{R} y$ on the set of nonzero real numbers to mean that $x/y > 0$. Is $\mathfrak{R}$ an equivalence relation? Define the binary relation $x \mathfrak{R} y$ on $\mathbf{Z}$ to mean that at least one of $3 \mid (x - y)$, $3 \mid (x + y)$ is true. Is $\mathfrak{R}$ an equivalence relation?

3.3. Define the binary relation $p \mathfrak{R} q$ on the set of positive, odd primes to mean that there is an integer x satisfying $0 \le x < q$ such that $x^2 \equiv p \pmod{q}$ holds. Is $\mathfrak{R}$ an equivalence relation?

3.4. Prove Theorem 3.1. Illustrate this theorem with specific examples.

3.5. Find at least one solution x of these congruences:

(a) $16x \equiv 64 \pmod{24}$.
(b) $(5x)^3 \equiv 8 \pmod{7}$.
(c) $2x + 3 \equiv 9 - 5x \pmod{13}$.
(d) $3x^2 + 19x \equiv 3 \pmod{11}$.

3.6. Prove parts (1) through (3) and (6) of Theorem 3.2.

3.7. Suppose $a \equiv b \pmod{n}$ holds and $|2a| < n, |2b| < n$. What can you say about a, b?

3.8. If $a^m \equiv b^m \pmod{n}$ holds for some $m \in \mathbf{N}$, does it follow that $a \equiv b \pmod{n}$? Prove, or give a counterexample.

3.9. Determine which of the following are true:

(a) $37^{73} \equiv 11 \pmod{53}$.
(b) $133^{50} \equiv 28 \pmod{29}$.
(c) $1997^{1997} \equiv 8 \pmod{99}$.
(d) $2001^{30} \equiv 84 \pmod{111}$.

3.10. If the number 243^{72} were multiplied out completely, what would be its units digit? Its tens digit? Its hundreds digit? Incidentally, how many digits does 243^{72} have?

3.11. Use the Method of Successive Squaring to calculate b, where $177^{345} \equiv b \pmod{217}$, and $0 \leq b < 217$.

3.12. Show that if $p > 3$ is a prime, then $p^2 \equiv 1 \pmod{3}$.

3.13. Let a be odd and b be even. Prove that $a^2 + b^2 \equiv 1 \pmod{4}$.

3.14. Show that $a^3 \equiv a \pmod{6}$ for every $a \in \mathbf{Z}$.

3.15. Prove that if n is a positive integer, then

$$10^n \equiv \begin{cases} -1 \pmod{11} & \text{if } n = \text{odd} \\ +1 \pmod{11} & \text{if } n = \text{even.} \end{cases}$$

3.16. A positive integer is written in base-9 notation (see Problem 1.45 for the meaning of the analogous base-10 notation). Deduce a criterion for its divisibility by 6.

3.17. Suppose that m is the units digit of an integer $M > 0$ written in base-10 notation.

(a) Show that $M \equiv m \pmod{10}$.
(b) Show that $M^4 \equiv 0, 1, 5,$ or $6 \pmod{10}$. Some data to support this can be found in Table B of Appendix II.

3.18. If DS(N) denotes the digital sum function of a positive integer N (Problem 1.45), then prove that $N \equiv \mathrm{DS}(N) \pmod{9}$. Hence, show that for $n \geq 2$, $9 \mid \mathrm{DS}(3^n)$.

3.19. Verify, without calculator, that $2^{70} + 3^{70} \equiv 0 \pmod{13}$.

3.20. Show that $a^7 \equiv a \pmod{42}$ for every $a \in \mathbf{Z}$.

3.21. Prove that if $x^2 + y^2 \equiv 0 \pmod 7$, then $x \equiv 0 \pmod 7$ and $y \equiv 0 \pmod 7$. Is this result true if 7 is replaced by any other prime?

3.22. The **digital root** function $\mathrm{DR}(N)$ of an integer $N > 0$ is the digital sum function composed with itself enough times to give a value between 1 and 9, inclusive; for example, $\mathrm{DR}(84) = \mathrm{DS}\{\mathrm{DS}(84)\} = \mathrm{DS}(12) = 3$. The digital root function has many interesting properties (Dence, 1982).

(a) How does $N \equiv \mathrm{DR}(N) \pmod 9$ follow from Problem 3.18?
(b) If $N, M > 0$, show that $\mathrm{DR}(N + M) \equiv \mathrm{DR}(N) + \mathrm{DR}(M) \pmod 9$.

3.23. Let integers $N, M > 0$ be arbitrary.

(a) Show that $\mathrm{DR}(NM) \equiv \mathrm{DR}(N) \cdot \mathrm{DR}(M) \pmod 9$.
(b) Prove that the digital roots of the powers of 2 are periodic. That is, find the smallest integer $k > 0$ such that $\mathrm{DR}(2^{n+k}) = \mathrm{DR}(2^n)$ for all $n \in \mathbf{N}$.

3.24. Prove that if the positive integer N is a perfect square, then its digital root is 1, 4, 7, or 9.

3.25. Write a computer program to print out the digital roots of the squares of the first 5000 positive integers. Continuing Problem 3.24, investigate the frequencies of occurrence of 1, 4, 7, and 9.

3.26. Modify your program in Problem 3.25 to print out the digital roots of the first 1000 primes. Formulate a conjecture.

3.27. Prove that in contrast to the behavior in Problem 3.25, the digital roots of the cubes of the integers appear periodically. Consult Table B in Appendix II for supporting data.

3.28. Show that $2(4^n - 1) \equiv 3n(3n - 1) \pmod{54}$ for any $n \in \mathbf{N}$.

3.29. In a deck of 52 cards the initial locations of the cards are designated as positions $0, 1, \ldots, 51$. Now define a **perfect shuffle** as one where the deck is cut into two (equal) halves and the halves are perfectly interlaced. Designate the original sequence of cards as $C_0, C_1, \ldots, C_{51}$. After one perfect shuffle the new sequence is now $C_0, C_{26}, C_1, C_{27}, C_2, C_{28}, \ldots, C_{25}, C_{51}$. Cards C_0, C_{51} remain fixed in any perfect shuffle.

(a) After one perfect shuffle, the card initially in position r will now be in what new position?
(b) After k perfect shuffles, the card initially in position r will now be in what new position?

(c) Write the congruence that gives the condition for the sequence of cards after k perfect shuffles to be the same as the initial sequence.
(d) Solve this congruence for the minimum k and thereby find the least number of perfect shuffles needed to restore the deck to its original order.

This problem, although elementary in nature, has some advanced extensions (Kolata, 1982).

3.2 Residue Classes

Let us return again to the subject of equivalence relations. Suppose that we have a set **S** on which there is defined an equivalence relation $\mathcal{R}$. We make this definition on the construction of certain subsets of **S**.

DEF. For any $x \in \mathbf{S}$ define the set $\mathbf{S}_x$ by

$$\mathbf{S}_x = \{y : y \in \mathbf{S},\ x \sim y\}.$$

The set $\mathbf{S}_x$ is called the **equivalence class** of **S** given by $\mathcal{R}$ and containing the element x.

Clearly, if $\mathbf{S}_x = \mathbf{S}_y$, then $x \sim y$. Also, the union of all the equivalence classes of **S** is **S** since every element in **S** is in some equivalence class.

We need to know more about equivalence classes than this, if they are to be useful to us in number theory. Consider this simple example. Congruence between triangles is an equivalence relation (Problem 3.1). So triangles that are congruent to each other are in the same equivalence class. It is geometrically clear that a given triangle cannot simultaneously be in two distinct equivalence classes (Fig. 3.1).

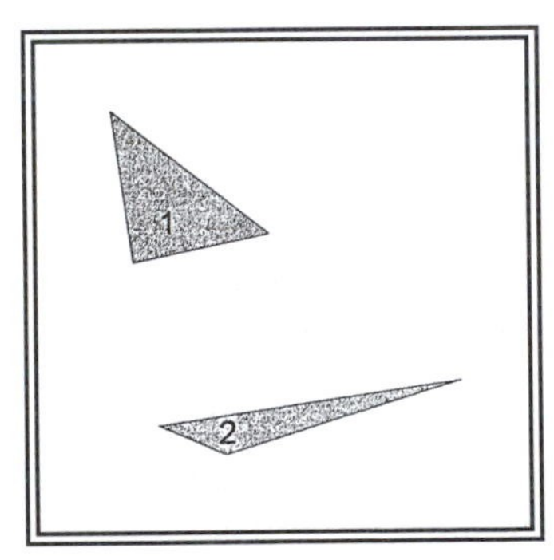

Figure 3.1 Neither triangle 1 nor triangle 2 can belong to the equivalence class of the other

The generalization that this observation suggests is this:

Theorem 3.5 The equivalence classes of a set **S** are disjoint.

Proof Suppose that $\mathbf{S}_x$, $\mathbf{S}_y$ are equivalence classes of **S** and that $\mathbf{S}_x \neq \mathbf{S}_y$, and assume that $\mathbf{S}_x \cap \mathbf{S}_y \neq \varnothing$. Our object is to reach a contradiction. Let $z \in \mathbf{S}_x \cap \mathbf{S}_y$; then $z \in \mathbf{S}_x$ and $z \in \mathbf{S}_y$, so $x \sim z$ and $y \sim z$. By the Symmetric Property $z \sim y$, and by the Transitive Property $x \sim y$. Let $w \in \mathbf{S}_x$; then $x \sim w$ and combination of this with $x \sim y$ yields $w \sim y$, that is, $w \in \mathbf{S}_y$. So $w \in \mathbf{S}_x \rightarrow \mathrm{w} \in \mathbf{S}_y$, that is, $\mathbf{S}_x \subset \mathbf{S}_y$. The interchange of x and y throughout the argument leads to $\mathbf{S}_y \subset \mathbf{S}_x$. It follows that $\mathbf{S}_x = \mathbf{S}_y$, a contradiction, so $\mathbf{S}_x \cap \mathbf{S}_y = \varnothing$. ◆

Thus, whereas the statement $\mathbf{S}_x = \mathbf{S}_y \rightarrow \mathrm{x} \sim \mathrm{y}$ follows from the definition of equivalence class, the converse statement $x \sim y \rightarrow \mathbf{S}_x = \mathbf{S}_y$ is a consequence of Theorem 3.5.

Any decomposition of a set **S** into disjoint subsets whose union is all of **S** is called a **partition** of **S**. Theorem 3.5 says that an equivalence relation defined on a set **S** induces a partition of **S**. Since congruence modulo a given positive integer n is an equivalence relation on **Z** (Theorem 3.1), then congruence modulo n induces a partition of **Z**.

DEF. The equivalence classes of **Z** given by congruence modulo $n > 1$ are termed the **residue classes** modulo n.

The word *residue* derives from the prose connotation of something left over after a certain action. Thus, if $a > 0$ and n is a modulus, then from the Division Algorithm there are unique integers q, r such that $0 \leq r < n$ and $a = qn + r$, or in congruence language, $a \equiv r \pmod{n}$. The remainder r is a "residue" left over after attempted division of a by n. We generalize this as follows.

DEF. If a, b are any integers satisfying $a \equiv b \pmod{n}$, then b is called a **residue** of a modulo n.

Let $k \in \mathbf{Z}$; then $0 \equiv kn \pmod{n}$. Combination of this with $a \equiv b \pmod{n}$ gives from Theorem 3.2, part (3),

$$a \equiv b + kn \pmod{n}.$$

This shows that not only is b a residue of a modulo n , but also any integer obtained by adding multiples of n to b is a residue of a modulo n.

Example 3.7 Since $7 \equiv 1 \pmod{3}$, the residues of 7 modulo 3 are all of the integers in the residue class

$$\mathbf{S}_1 = \{\ldots, -5, -2, 1, 4, 7, 10, 13, \ldots\}. \quad \blacklozenge$$

Any member of $\mathbf{S}_1$ in the preceding example is called a **representative** of that residue class. The element $1 \in \mathbf{S}_1$ is called the **least nonnegative residue** of 7 modulo 3.

Corresponding to any modulus n there can be only n residue classes because from the Division Algorithm one has for any $x \in \mathbf{Z}$ unique integers q, r such that $x = nq + r$, where $r = 0, 1, 2, \ldots,$ or $n - 1$ (n possible values).

Example 3.8 There are just two other residue classes modulo 3 besides the one shown in Example 3.7:

$$\mathbf{S}_0 = \{\ldots, -6, -3, 0, 3, 6, 9, 12, \ldots\}.$$

$$\mathbf{S}_2 = \{\ldots, -4, -1, 2, 5, 8, 11, 14, \ldots\}.$$

Theorem 3.5 tells us that $\mathbf{S}_0 \cap \mathbf{S}_1 = \mathbf{S}_0 \cap \mathbf{S}_2 = \mathbf{S}_1 \cap \mathbf{S}_2 = \varnothing$, and $\mathbf{S}_0 \cup \mathbf{S}_1 \cup \mathbf{S}_2 = \mathbf{Z}$. ◆

We shall see shortly that the residue classes possess algebraic structure. To prepare for this, some additional background in algebra will be presented in the next two sections.

Problems

3.30. The following is a residue class of what modulus?

$$\{\ldots, -18, -11, -4, 3, 10, 17, 24, \ldots\}$$

Write a congruence with this modulus that has 3 as its least nonnegative residue. Indicate two other residue classes of the same modulus.

3.31. Let $n > 2$ be a modulus and suppose that b, c are both residues of a modulo n. Show that there are positive integers B, C, both less than n, such that $Bb + Cc$ is also a residue of a modulo n.

3.32. Find the least nonnegative residue as indicated:

(a) Of 2^{50} modulo 13.
(b) Of 12! modulo 17.
(c) Of $2^{100} - 1$ modulo 33.

3.33. Find the least nonnegative residue as indicated:

(a) Of 6! modulo 7.
(b) Of 10! modulo 11.
(c) Of 12! modulo 13.
(d) Formulate a conjecture for the general case.

3.34. Construct an example of a finite set **S** and an equivalence relation $\mathfrak{R}$ defined on it that has the property that not all of the equivalence classes are of the same size.

3.35. Let $\mathbf{S}_0$, $\mathbf{S}_1$, $\mathbf{S}_2$, $\mathbf{S}_3$, $\mathbf{S}_4$, $\mathbf{S}_5$, $\mathbf{S}_6$ be the residue classes modulo 7. What do you get when

(a) You multiply a member of $\mathbf{S}_3$ by a member of $\mathbf{S}_4$?
(b) You add a member of $\mathbf{S}_2$ to a member of $\mathbf{S}_6$?

3.3 Rings

The next algebraic structure that we need to look at briefly is a ring; it has two binary operations defined on it.[2]

DEF. A set **R** of elements, together with two binary operations on **R**, denoted $(+)$ and $(\cdot)$, is a **ring** $\langle \mathbf{R}, +, \cdot \rangle$ if

1. $\langle \mathbf{R}, + \rangle$ is an Abelian group.
2. $(\cdot)$ obeys the associative law on **R**.
3. The two distributive laws are obeyed:

$$x \cdot (y + z) = x \cdot y + x \cdot z$$

$$(y + z) \cdot x = y \cdot x + z \cdot x.$$

The binary operations are called, generically, **addition** and **multiplication**. Two cautionary notes: (1) A ring must have an additive identity (a **zero**), symbolized 0, but it need not have a multiplicative identity (a **unity**), which would be symbolized by 1 and which would obey $x \cdot 1 = 1 \cdot x = x$ for any $x \in \mathbf{R}$.

[2] In the discussion that follows, we use the notation **R** to stand for the elements of a ring. This symbol should not be confused with $\mathfrak{R}$, which has previously been used either for the set of all real numbers or for a binary relation. In case of doubt about any symbol, consult Appendix I in the rear of the book.

(2) The operation $(+)$ must be commutative, but the operation $(\cdot)$ need not. A ring that has a 1 is called a **ring with identity**, and a ring in which $(\cdot)$ is commutative is called a **commutative ring**. If a ring has a unity, it must be unique (Problem 3.42).

Analogous to the concept of a subgroup (see Section 2.3), there is the concept of a subring.

> **DEF.** Let $\langle \mathbf{R}, +, \cdot \rangle$ be a ring, and let $\mathbf{S} \subset \mathbf{R}$. Then $\mathbf{S}$, together with the operations $(+)$, $(\cdot)$ defined on $\mathbf{R}$, is a **subring** $\langle \mathbf{S}, +, \cdot \rangle$ if $(+)$, $(\cdot)$ are binary operations on $\mathbf{S}$ and all of the ring postulates hold for $\mathbf{S}$.

It is automatic that if $\langle \mathbf{S}, +, \cdot \rangle$ is a subring of $\langle \mathbf{R}, +, \cdot \rangle$, then $\langle \mathbf{S}, + \rangle$ is a subgroup of $\langle \mathbf{R}, + \rangle$ (Problem 3.38).

Example 3.9 Let $\mathbf{R} = \mathbf{Z}$, and let $(+)$, $(\cdot)$ be ordinary addition and multiplication. The additive identity is the zero integer. Then $\langle \mathbf{Z}, +, \times \rangle$ is a ring; in fact, it is a commutative ring with identity. If $\mathbf{Z}$ is replaced by the reals $\mathfrak{R}$, then we also have a commutative ring with identity and the first ring is a **subring** of this latter. ◆

Example 3.10 Let $\mathbf{R}$ be the set of all functions $f\colon \mathfrak{R} \to \mathfrak{R}$. Define $(+)$, $(\cdot)$ for any $f, g \in \mathbf{R}$ by

$$(f + g)(x) = f(x) + g(x)$$

$$(f \cdot g)(x) = f(x)g(x).$$

$\langle \mathbf{R}, + \rangle$ is easily seen to be an Abelian group whose additive identity is the zero function $\mathbf{O}(\mathrm{x}) = 0$, and where the inverse of f is $(-f)(x) = -f(x)$. Associativity of $(\cdot)$ is easily shown, as are the distributive laws. Thus, $\langle \mathbf{R}, +, \cdot \rangle$ is a ring, a commutative ring with identity. ◆

Example 3.11 Let $\mathbf{R}$ be the set of all 3×3 matrices defined over $\mathfrak{R}$. Define $(+)$, $(\cdot)$ by the usual matrix addition and multiplication. For example,

$$\begin{pmatrix} 0 & 1 & 2 \\ 3 & 4 & 5 \\ 2 & 1 & 0 \end{pmatrix} + \begin{pmatrix} 6 & -1 & 1 \\ 2 & 0 & 0 \\ 3 & 3 & 3 \end{pmatrix} = \begin{pmatrix} 6 & 0 & 3 \\ 5 & 4 & 5 \\ 5 & 4 & 3 \end{pmatrix}$$

$$\begin{pmatrix} 0 & 1 & 2 \\ 3 & 4 & 5 \\ 2 & 1 & 0 \end{pmatrix} \cdot \begin{pmatrix} 6 & -1 & 1 \\ 2 & 0 & 0 \\ 3 & 3 & 3 \end{pmatrix} = \begin{pmatrix} 8 & 6 & 6 \\ 41 & 12 & 18 \\ 14 & -2 & 2 \end{pmatrix}.$$

The additive identity is the matrix $\begin{pmatrix} 0 & 0 & 0 \\ 0 & 0 & 0 \\ 0 & 0 & 0 \end{pmatrix}$ and the additive inverse of $\begin{pmatrix} a & b & c \\ d & e & f \\ g & h & i \end{pmatrix}$ is the matrix $\begin{pmatrix} -a & -b & -c \\ -d & -e & -f \\ -g & -h & -i \end{pmatrix}$. So, $\langle \mathbf{R}, + \rangle$ is easily seen to be an Abelian group. Matrix multiplication is known to be associative, and the distributive laws are easily verified in the general case. Thus, $\langle \mathbf{R}, +, \cdot \rangle$ is a ring, in fact, a noncommutative ring with identity $\begin{pmatrix} 1 & 0 & 0 \\ 0 & 1 & 0 \\ 0 & 0 & 1 \end{pmatrix}$. ◆

Example 3.12 Let $\mathbf{R} = \{a + b\sqrt{2}\colon a, b \in \mathbf{Q}\}$. This is a ring; we denote it by $\mathbf{Q}(\sqrt{2})$ (Problem 3.36). ◆

The following theorem gives some elementary properties of members of any ring. We prove the first two parts and leave the third as a routine exercise.

Theorem 3.6 Let $\mathbf{R}$ be a ring[3] and let $x, y \in \mathbf{R}$. Then

(1) $0 \cdot x = x \cdot 0 = 0$
(2) $x \cdot (-y) = (-x) \cdot y = -(x \cdot y)$
(3) $(-x) \cdot (-y) = x \cdot y$.

Proof (1) By definition of the additive identity, we can write

$$0 = 0 + 0.$$

Premultiply both sides by x and use a distributive law.

$$x \cdot 0 = x \cdot 0 + x \cdot 0$$

Add $-(x \cdot 0)$ to both sides of this to obtain $0 = x \cdot 0$. A repetition of the whole argument but with a replacement of "premultiplication by x" by "postmultiplication by x" leads to $0 = 0 \cdot x$. (2) By definition of the additive identity, we can write

$$(-x) \cdot y = (-x) \cdot y + 0.$$

[3] We continue here the practice, initiated earlier with groups, of letting the simpler symbol for an algebraic structure replace the more complete notation in those instances where there is no possibility of confusion.

Again, by definition of the additive identity, we can write

$$x \cdot y + [-(x \cdot y)] = 0.$$

Substitution of the second equation into the first gives

$$\begin{aligned}(-x) \cdot y &= (-x) \cdot y + \{x \cdot y + [-(x \cdot y)]\} \\ &= \{(-x) \cdot y + x \cdot y\} + [-(x \cdot y)] \\ &= \{(-x) + x) \cdot y + [-(x \cdot y)] \\ &= 0 \cdot y + [-(x \cdot y)] \\ &= 0 + [-(x \cdot y)] \\ &= -(x \cdot y).\end{aligned}$$

A repetition of the whole argument but with "$(-x) \cdot y$" in line 1 replaced by "$x \cdot (-y)$" leads to $x \cdot (-y) = -(x \cdot y)$. ◆

The results in Theorem 3.6 seem familiar to us from high-school algebra; the theorem shows that they hold in any ring (not just in $\mathfrak{R}$). Note especially that the multiplicative property of the additive identity (zero) is a provable result and not a convention (recall Problem 1.26).

An interesting phenomenon can be observed for some rings. Let us simplify Example 3.11 slightly to the ring **R** of 2×2 matrices over $\mathfrak{R}$, for which the additive identity is the matrix $\begin{pmatrix} 0 & 0 \\ 0 & 0 \end{pmatrix}$. Now consider the following multiplication of two specific members of **R**:

$$\begin{pmatrix} 1 & 0 \\ 0 & 0 \end{pmatrix} \cdot \begin{pmatrix} 0 & 0 \\ 1 & 1 \end{pmatrix} = \begin{pmatrix} 0 & 0 \\ 0 & 0 \end{pmatrix}.$$

The product is the zero element, and yet neither multiplier is the zero element.

DEF. Let **R** be a ring, and let $x \in \mathbf{R}$, $x \neq 0$. Then x is called a **right** (or **left**) **divisor of zero** if there is a $y \in \mathbf{R}$, $y \neq 0$, such that $x \cdot y = 0$ (or $y \cdot x = 0$). Either a right or a left divisor of zero is called simply a **divisor of zero**.

In the preceding example, $\begin{pmatrix} 1 & 0 \\ 0 & 0 \end{pmatrix}$ is a right divisor of zero and $\begin{pmatrix} 0 & 0 \\ 1 & 1 \end{pmatrix}$ is a left divisor of zero. Of course, in a commutative ring the distinction between a right and a left divisor of zero disappears. This prompts a definition.

DEF. A commutative ring **R** with identity and no divisors of zero is called an **integral domain D**.

Example 3.13 We have seen in Example 3.9 that **Z** is a commutative ring. Further, in **Z** if $x \cdot y = 0$, then at least one of x, y must be 0. Thus, **Z** is an integral domain. ◆

The next theorem gives one reason why integral domains are important.

Theorem 3.7 (Cancellation Theorem). Let **D** be an integral domain, and suppose $x, y, z \in \mathbf{D}$ and $x \neq 0$. If $x \cdot y = x \cdot z$, then $y = z$.

Proof Addition of $-(x \cdot z)$ to both sides of $x \cdot y = x \cdot z$ gives

$$x \cdot y + [-(x \cdot z)] = 0.$$

From Theorem 3.6, part (2), we can rewrite this as

$$x \cdot y + x \cdot (-z) = 0,$$

and a distributive law now yields

$$x \cdot [y + -(z)] = 0.$$

Since $x \neq 0$ and we are in an integral domain, then it must be that $y + -(z) = 0$. The additive inverse of $-z$ is z; addition of this to both sides gives, finally,

$$\begin{aligned} [y + -(z)] + z &= 0 + z \\ y + [-(z) + z] &= 0 + z \\ y + 0 &= 0 + z \\ y &= z. \end{aligned}$$ ◆

Theorem 3.7 may be viewed as a cancellation law for multiplication in an integral domain. Note that the law fails in arbitrary rings because they may contain divisors of zero.

If a ring contains a unity (multiplicative identity), it must be unique (Problem 3.42). The unity possesses the following elementary properties:

Theorem 3.8 Let **R** be a ring with identity 1. Then if $x \in \mathbf{R}$, one has

1. $(-1) \cdot x = -x$.
2. $(-1) \cdot (-1) = 1$.

Proof The proof of these two assertions is left to the reader. ◆

Problems

3.36. Verify all aspects of ring definition for $\mathbf{Q}(\sqrt{2})$ (see Example 3.12).

3.37. Decide which of the following are rings:

(a)	$\mathbf{R}$ = the set of all numbers of the form $a + b\sqrt{2} + c\sqrt{6}$, $a, b, c \in \mathbf{Q}$	$(+)$ = ordinary addition $(\cdot)$ = ordinary multiplication
(b)	$\mathbf{R}$ = the set of all vectors in xyz-space	$(+)$ = vector addition $(\cdot)$ = vector (cross-) multiplication
(c)	$\mathbf{R}$ = the set of all even integers	$(+)$ = ordinary addition $(\cdot)$ = ordinary multiplication
(d)	$\mathbf{R}$ = the set of all ordered pairs (x, y) of real numbers	$(x, y) + (w, z) = (x + w, y + z)$ $(x, y) \cdot (w, z) = (xw - yz, xz + wy)$
(e)	$\mathbf{R} = \mathscr{C} - \{0\}$, the set of all nonzero numbers of the form $a + bi$, $a, b \in \mathscr{R}$	$(+)$ = complex multiplication $(a + bi) \cdot (c + di) = ac + (bc + ad)i$

3.38. Why is it automatic that if $\langle \mathbf{S}, +, \cdot \rangle$ is a subring of $\langle \mathbf{R}, +, \cdot \rangle$, then $\langle \mathbf{S}, + \rangle$ is a subgroup of $\langle \mathbf{R}, + \rangle$? Does it make sense to ask if $\langle \mathbf{S}, \cdot \rangle$ is then a subgroup of $\langle \mathbf{R}, \cdot \rangle$?

3.39. Which of the rings in Problem 3.37 are actually integral domains? Do any of the rings that are not integral domains contain a unity?

3.40. Prove part (3) of Theorem 3.6.

3.41. In the proof of Theorem 3.7 why couldn't one simply premultiply both sides of $x \cdot y = x \cdot z$ by $1/x$, make use of the associative law for $(\cdot)$, and thereby arrive quickly at $y = z$?

3.42. Prove that if a ring contains a unity, it is unique.

3.43. Let $\mathbf{R}$ be a ring, and let n be any nonnegative integer and $x \in \mathbf{R}$. Define $n \cdot x$ by

$$n \cdot x = \begin{cases} 0 & \text{if } n = 0 \\ x & \text{if } n = 1 \\ \underbrace{x + x + x + \cdots + x}_{n \text{ times}} & \text{if } n > 1. \end{cases}$$

Further, define x^k to be $x \cdot x^{k-1}$ for any $k \in \mathbf{N}, k > 1$. If **R** is commutative and $x, y \in \mathbf{R}$, then show that

$$(x + y)^2 = x^2 + 2 \cdot x \cdot y + y^2$$

$$(x + y)^3 = x^3 + 3 \cdot x^2 \cdot y + 3 \cdot x \cdot y^2 + y^3.$$

Do the results still hold if **R** is not commutative?

3.44. Let **R** be a ring and let $w, x, y, z \in \mathbf{R}$. Prove that

$$(w + x)(y + z) = wy + wz + xy + xz.$$

3.45. Let **R** be a ring with identity, and let $x \in \mathbf{R}$ be nonzero. Prove that there is *at most* one element $y \in \mathbf{R}$ such that $x \cdot y = y \cdot x = 1$.

3.46. The ring $\mathbf{R} = \{0\}$ is called the **trivial ring**. Let **R** be a ring with identity which is not the trivial ring. Show that the zero and the unity are distinct. [Hint: Proof by contradiction.]

3.47. An element x of a ring **R** is called **nilpotent** if $x^n = 0$ for some positive integer n. Prove that if x, y are nilpotent in a commutative ring **R**, then $x + y$ is also nilpotent.

3.48. Let **R** be a ring and suppose that 0 is the only solution of $x^2 = 0$. Show that **R** has no nonzero nilpotent elements.

3.49. Suppose that **S**, **T** are subrings of the ring **R**.

(a) Show that $\mathbf{S} \cap \mathbf{T}$ is also a subring of **R**.
(b) Illustrate this result for the case where **S**, **T** are two specific subrings of the ring **Z**, and $(+)$, $(\cdot)$ are ordinary addition and multiplication.
(c) Is $\mathbf{S} \cup \mathbf{T}$ also a subring of **R**, generally?

3.4 Fields

Rings are richer structures than groups. Integral domains are still richer. Even richer than integral domains are fields. Fields arise when one considers algebraic systems in which elements have multiplicative inverses.

DEF. Let **R** be a ring. If **R** contains a unity and if $x, y \in \mathbf{R}$ are such that $x \cdot y = 1$, then x and y are said to be **multiplicative inverses** of each other; x and y are referred to collectively as **units**.

A ring might have no units, one unit (the unity), two units, or several units (Problem 3.55). We let the symbol $\mathbf{R}^{\times}$ denote the subset of $\mathbf{R}$ that consists of all the units of $\mathbf{R}$. For example, $\mathbf{Z}^{\times} = \{1, -1\}$ (Problem 3.57) and $\mathbf{Q}^{\times} = \mathbf{Q} - \{0\}$.

DEF. If every nonzero element of **R**, a ring with identity, is a unit, then **R** is a **division ring**. A **field** is a commutative division ring.

We can write out for convenience all of the defining characteristics of a field **F**:

1. **F** is a collection of elements on which there are defined two binary operations: field addition $(+)$ and field multiplication $(\cdot)$.
2. $(+)$, $(\cdot)$ are commutative and associative.
3. The two distributive laws hold.
4. **F** contains a zero (0) and a unity (1).
5. Every element in **F** has an additive inverse.
6. Every element in **F** except the 0 has a multiplicative inverse.

Example 3.14 $\mathcal{R}$, $\mathcal{Q}$, $\mathcal{C}$ are fields under the usual addition and multiplication operations. In contrast, the integral domain **Z** fails to be a field because of the preceding property 6. ◆

Example 3.15 The ring in Example 3.12 is actually a field. The multiplicative inverse of $x = a + b\sqrt{2}$ is

$$x^{-1} = \frac{a}{(a^2 - 2b^2)} + \left[\frac{b}{(2b^2 - a^2)}\right]\sqrt{2}$$

provided $a^2 - 2b^2 \neq 0$ (Problem 3.63). ◆

So what is the connection between a field and an integral domain? The next theorem completes the hierarchy of group, ring, integral domain, and field.

Theorem 3.9 Every field **F** is an integral domain.

Proof The one concept not mentioned in the definition of a field is that of divisors of zero. We need to show that a field contains no divisors of zero. Suppose $x, y \in \mathbf{F}, x \neq 0, x \cdot y = 0$. As we are in a field, x^{-1} exists. Multiplication of both sides of $x \cdot y = 0$ by x^{-1} gives

$$x^{-1} \cdot 0 = 0 = x^{-1} \cdot (x \cdot y) = (x^{-1} \cdot x) \cdot y = 1 \cdot y = y.$$

Since y is forced to be 0 and x was arbitrary, then **F** contains no divisors of zero and is therefore a field. ◆

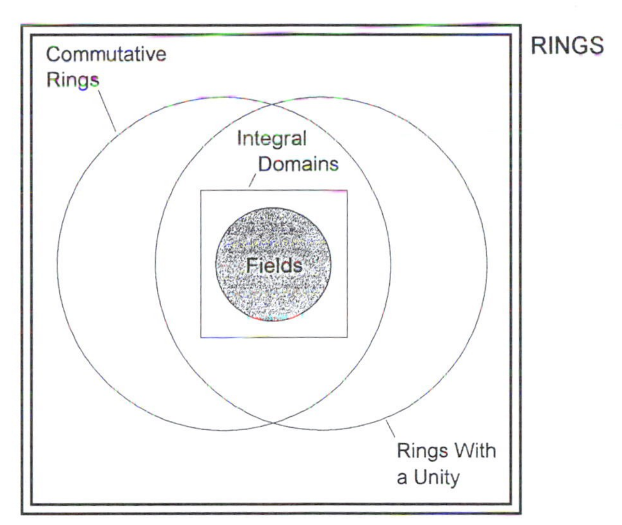

Figure 3.2 The hierarchy of rings, integral domains, and fields

Figure 3.2 is a helpful way for you to remember the connection between rings, integral domains, and fields. Although not all integral domains are fields (e.g., **Z**), the following is almost the converse of Theorem 3.9.

Theorem 3.10 Every finite integral domain **D** is a field.

Proof Since **D** is finite, we can list its elements: $0, 1, x_1, x_2, \ldots, x_n$. What we have to show is that each of the x_k's is a unit (the 1 is trivially a unit). Let $x \in \{1, x_1, x_2, \ldots, x_n\}$ be arbitrary and consider the products: $x \cdot 1$, $x \cdot x_1, x \cdot x_2, \ldots, x \cdot x_n$. The products, of course, all reside in **D**. No two of the products are the same, for if $x \cdot x_i = x \cdot x_j$ were true for some $i \neq j$, then Theorem 3.7 could be applied to give $x_i = x_j$. This is a contradiction.

Further, none of the products is 0 since, by definition, **D** has no divisors of zero. It follows that the $n + 1$ quantities $\{x \cdot 1, x \cdot x_1, \ldots, x \cdot x_n\}$ must be the $n + 1$ nonzero elements of **D**. One of these nonzero elements is the unity. Hence, either $x \cdot 1 = 1$ (in which case, $x = 1$) or $x \cdot x_j = 1$ for some j (in which case, x_j is the multiplicative inverse of x). Thus, x is a unit, and since x was arbitrary, **D** must be a field. ◆

Problems

3.50. Let $a + bi$ be a nonzero element of the field $\mathscr{C}$. What is its additive inverse? Its multiplicative inverse? The operations $(+)$, $(\cdot)$ here are the usual complex addition and complex multiplication.

3.51. Let **F** be the set of all vectors in three-dimensional Cartesian space, and let $(+)$, $(\cdot)$ be the usual vector addition and vector (cross-) multiplication. Is $\langle \mathbf{F}, +, \cdot \rangle$ a field?

3.52. Let **F** be the set $\{0, 1, 2, 3, 4, 5, 6\}$, and let $(+)$, $(\cdot)$ be addition modulo 7 and multiplication modulo 7. Is $\langle \mathbf{F}, +, \cdot \rangle$ a field?

3.53. Let x, y be any two elements in the **F** of Problem 3.52. Prove that $(x + y)^7 = x^7 + y^7$ in **F**.

3.54. Let **S** be a given set and let **F** be the set of all subsets of **S**. For **V**, $\mathbf{W} \in \mathbf{F}$ define $(+)$, $(\cdot)$ by

$$\mathbf{V} + \mathbf{W} = \mathbf{V} \cup \mathbf{W}$$

$$\mathbf{V} \cdot \mathbf{W} = \mathbf{V} \cap \mathbf{W}.$$

Is $\langle \mathbf{F}, +, \cdot \rangle$ a field?

3.55. Give an example of a ring that has more than one unit but is not a division ring. Give an example of a nontrivial ring that has no units.

3.56. Let $\mathbf{F} = \{0, 1, A, B\}$ and let the binary operations $(\oplus)$, $(\cdot)$ be defined by the tables

$\oplus$	0	1	A	B
0	0	1	A	B
1	1	0	B	A
A	A	B	0	1
B	B	A	1	0

$\cdot$	0	1	A	B
0	0	0	0	0
1	0	1	A	B
A	0	A	B	1
B	0	B	1	A

Is $\langle \mathbf{F}, \oplus, \cdot \rangle$ a field?

3.57. Prove that the ring **Z** has only two units.

3.58. Let **R** be a ring with identity, and let $x, y \in \mathbf{R}^{\times}$. Then prove that

(a) $x^{-1} \in \mathbf{R}^{\times}$.
(b) $x \cdot y \in \mathbf{R}^{\times}$ and $(x \cdot y)^{-1} = y^{-1} \cdot x^{-1}$.

3.59. Show that the units of a ring form a group under the operation of ring multiplication. [Hint: Use the preceding problem.]

3.60. A field has exactly three elements. Write out the addition and multiplication tables for the field.

3.61. A field has exactly five elements. Show that it cannot contain a subfield of only two elements (see Problem 3.64).

3.62. A number α in the field $\mathbf{F} = \mathfrak{R}$ is said to be **algebraic** over **Q** if there is a polynomial $p(x) = \sum_{k=0}^{n} c_k x^k$, with each $c_k \in \mathbf{Q}$, such that $p(\alpha) = 0$. Show that $\alpha = \sqrt{2} + \sqrt{3}$ is algebraic over **Q**.

3.63. Discuss the qualifying clause in Example 3.15.

3.64. If $\mathbf{H} \subset \mathbf{F} \subset \mathbf{K}$ and $\mathbf{H}$, $\mathbf{F}$, $\mathbf{K}$ are all fields, then we say that $\mathbf{H}$ is a **subfield** of $\mathbf{F}$ and $\mathbf{K}$ is an **extension field** of $\mathbf{F}$. A function $f(x)$ has a **zero** in $\mathbf{F}$ if for some $\alpha \in \mathbf{F}$ one has $f(\alpha) = 0$. Produce an example of a function that simultaneously has no zero in some field $\mathbf{H}$, one zero in a field $\mathbf{F}$, and three zeros in an extension field $\mathbf{K}$. We will explore extension fields in much more detail in Chapter 9.

3.5 The Algebra of Residue Classes

Consider the modulus $n = 5$, for which we have the following five residue classes:

$$\mathbf{S}_0 = \{, \ldots, -20, -15, -10, -5, 0, 5, 10, 15, 20, \ldots\}$$
$$\mathbf{S}_1 = \{, \ldots, -19, -14, -9, -4, 1, 6, 11, 16, 21, \ldots\}$$
$$\mathbf{S}_2 = \{, \ldots, -18, -13, -8, -3, 2, 7, 12, 17, 22, \ldots\}$$
$$\mathbf{S}_3 = \{, \ldots, -17, -12, -7, -2, 3, 8, 13, 18, 23, \ldots\}$$
$$\mathbf{S}_4 = \{, \ldots, -16, -11, -6, -1, 4, 9, 14, 19, 24, \ldots\}.$$

Now observe the following additions and multiplications:

$$\begin{array}{ccccccc} -18 & + & 9 & = & -9 & \qquad & (-3) \times (-6) = 18 \\ \mathbf{S}_2 & & \mathbf{S}_4 & & \mathbf{S}_1 & & \mathbf{S}_2 \qquad \mathbf{S}_4 \qquad \mathbf{S}_3 \end{array}$$

and and

$$\begin{array}{ccccccc} 7 & + & (-11) & = & -4 & \qquad & 17 \times (-1) = -17 \\ \mathbf{S}_2 & & \mathbf{S}_4 & & \mathbf{S}_1 & & \mathbf{S}_2 \qquad \mathbf{S}_4 \qquad \mathbf{S}_3 \end{array}$$

It appears that any member from $\mathbf{S}_2$ plus (times) any member from $\mathbf{S}_4$ gives an element in $\mathbf{S}_1(\mathbf{S}_3)$ (recall Problem 3.35).

DEF. An operation on pairs of sets defined by means of representatives from the sets is said to be **well defined** if the result is independent of the choice of representatives.

Example 3.16 Let $\mathbf{O}$ stand for the set of odd integers and $\mathbf{E}$ stand for the set of even integers. The following example of class addition and multiplication is familiar from elementary arithmetic.

$\oplus$	O	E
O	E	O
E	O	E

$\otimes$	O	E
O	O	E
E	E	E

◆

If we wish to define addition and multiplication of residue classes by means of their representatives, then it is necessary to know Theorem 3.11.

Theorem 3.11 Addition and multiplication of residue classes modulo n are well defined.

Proof Let $x \in \mathbf{S}_i$, $y \in \mathbf{S}_j$, and suppose $x + y \equiv z \in \mathbf{S}_k$ and $x \cdot y \equiv w \in \mathbf{S}_m$, where $\mathbf{S}_i$, $\mathbf{S}_j$, $\mathbf{S}_k$, $\mathbf{S}_m$ are residue classes modulo n. Let $x' \in \mathbf{S}_i$ and $y' \in \mathbf{S}_j$. Then we have

$$x' \equiv x \pmod{n}$$

$$y' \equiv y \pmod{n}.$$

From Theorem 3.2, part (3), one has

$$x' + y' \equiv x + y \pmod{n}.$$

But $x + y \equiv z \pmod{n}$ and since congruence modulo n is an equivalence relation (Theorem 3.1), then

$$x' + y' \equiv z \pmod{n},$$

that is, $x' + y' \in \mathbf{S}_k$. This shows that addition of residue classes is well defined. The case of multiplication follows analogously, upon using Theorem 3.2, part (5). ◆

The broad implications for the set of residue classes $\{\mathbf{S}_0, \mathbf{S}_1, \mathbf{S}_2, \ldots, \mathbf{S}_{n-1}\}$ now follow easily from Theorem 3.11 if one considers a corresponding set of representatives such as $\{0, 1, 2, \ldots, n-1\}$. Addition and multiplication of the classes are commutative, and the associative and distributive laws also hold since these laws are valid for real numbers under modular arithmetic. The residue class $\mathbf{S}_0$ is the additive identity,

$$\mathbf{S}_0 \oplus \mathbf{S}_i = \mathbf{S}_i.$$

The additive inverse of the class $\mathbf{S}_i$ is the class $\mathbf{S}_{n-i}$ since $i + (n - i) = n \equiv 0 \pmod{n}$. Finally, the residue class $\mathbf{S}_1$ is the multiplicative identity: $\mathbf{S}_1 \otimes \mathbf{S}_i = \mathbf{S}_i$.

Corollary 3.11.1 The residue classes modulo n constitute a finite commutative ring with identity, where addition and multiplication are defined by means of representatives of the classes.

Example 3.17 The addition and multiplication tables for the residue classes modulo 5 are as follows:

$\oplus$	$\mathbf{S}_0$	$\mathbf{S}_1$	$\mathbf{S}_2$	$\mathbf{S}_3$	$\mathbf{S}_4$
$\mathbf{S}_0$	$\mathbf{S}_0$	$\mathbf{S}_1$	$\mathbf{S}_2$	$\mathbf{S}_3$	$\mathbf{S}_4$
$\mathbf{S}_1$	$\mathbf{S}_1$	$\mathbf{S}_2$	$\mathbf{S}_3$	$\mathbf{S}_4$	$\mathbf{S}_0$
$\mathbf{S}_2$	$\mathbf{S}_2$	$\mathbf{S}_3$	$\mathbf{S}_4$	$\mathbf{S}_0$	$\mathbf{S}_1$
$\mathbf{S}_3$	$\mathbf{S}_3$	$\mathbf{S}_4$	$\mathbf{S}_0$	$\mathbf{S}_1$	$\mathbf{S}_2$
$\mathbf{S}_4$	$\mathbf{S}_4$	$\mathbf{S}_0$	$\mathbf{S}_1$	$\mathbf{S}_2$	$\mathbf{S}_3$

$\otimes$	$\mathbf{S}_0$	$\mathbf{S}_1$	$\mathbf{S}_2$	$\mathbf{S}_3$	$\mathbf{S}_4$
$\mathbf{S}_0$	$\mathbf{S}_0$	$\mathbf{S}_0$	$\mathbf{S}_0$	$\mathbf{S}_0$	$\mathbf{S}_0$
$\mathbf{S}_1$	$\mathbf{S}_0$	$\mathbf{S}_1$	$\mathbf{S}_2$	$\mathbf{S}_3$	$\mathbf{S}_4$
$\mathbf{S}_2$	$\mathbf{S}_0$	$\mathbf{S}_2$	$\mathbf{S}_4$	$\mathbf{S}_1$	$\mathbf{S}_3$
$\mathbf{S}_3$	$\mathbf{S}_0$	$\mathbf{S}_3$	$\mathbf{S}_1$	$\mathbf{S}_4$	$\mathbf{S}_2$
$\mathbf{S}_4$	$\mathbf{S}_0$	$\mathbf{S}_4$	$\mathbf{S}_3$	$\mathbf{S}_2$	$\mathbf{S}_1$

◆

The boxed entries in the multiplication table show that each residue class other than $\mathbf{S}_0$ has a multiplicative inverse; for example, $\mathbf{S}_2^{-1} = \mathbf{S}_3$. Hence, for $n = 5$ the set of residue classes is actually a field (Section 3.4), and is a special case of a commutative ring with identity (recall Fig. 3.2).

In contrast, we show in Fig. 3.3 the multiplication table for the residue classes modulo 6. We see that only $\mathbf{S}_1$ and $\mathbf{S}_5$ have multiplicative inverses. Why is the modulus 6 different from the modulus 5? Before reading on, you can experiment a bit here. Construct the multiplication tables for some other moduli in order to see which ones behave like modulus 5 and which behave like modulus 6. We suggest that you include moduli 4, 7, 9, 11 in your investigation.

Theorem 3.12 Let $\mathbf{Z}_n$ denote the finite ring of the residue classes modulo n, that is, $\mathbf{Z}_n = \{\mathbf{S}_0, \mathbf{S}_1, \mathbf{S}_2, \ldots, \mathbf{S}_{n-1}\}$. Then $\mathbf{Z}_n$ is a field iff n is a prime p.

Proof ($\leftarrow$) Suppose n is a prime p; $\mathbf{Z}_p$ is immediately a commutative ring with identity. Then if $\mathbf{S}_j \neq \mathbf{S}_0$ and $\mathbf{S}_k \neq \mathbf{S}_0$, one has $\mathbf{S}_j \otimes \mathbf{S}_k \neq \mathbf{S}_0$. This follows because if $x \in \mathbf{S}_j$ and $y \in \mathbf{S}_k$, then $xy = cp$ for some integer $c \neq 0$, that

$\otimes$	S_0	S_1	S_2	S_3	S_4	S_5
S_0	S_0	S_0	S_0	S_0	S_0	S_0
S_1	S_0	S_1	S_2	S_3	S_4	S_5
S_2	S_0	S_2	S_4	S_0	S_2	S_4
S_3	S_0	S_3	S_0	S_3	S_0	S_3
S_4	S_0	S_4	S_2	S_0	S_4	S_2
S_5	S_0	S_5	S_4	S_3	S_2	S_1

Figure 3.3 Multiplication table for the residue classes modulo 6

is, $p|xy$. But this is impossible from Theorem 2.2 because $(x, p) = (y, p) = 1$. So $\mathbf{Z}_p$ is really an integral domain, and Theorem 3.10 applies.

($\rightarrow$) Conversely, suppose n is composite. Then there are integers r, s such that $1 < r, s < n$ and $n = rs$. So $rs \equiv 0 \pmod{n}$, even though neither $r \equiv 0 \pmod{n}$ nor $s \equiv 0 \pmod{n}$ holds. Hence, by Theorem 3.11, $\mathbf{S}_r$ and $\mathbf{S}_s$ are divisors of zero and so $\mathbf{Z}_n$ is not a field. ◆

Theorem 3.12 provides us with an infinite family of finite fields that are important in number theory. There are still other finite fields not covered by this theorem.

Incidentally, we have also proved the following very useful corollary:

Corollary 3.12.1 If p is a prime and $a \not\equiv 0 \pmod{p}$, then $ax \equiv 1 \pmod{p}$ always has one, and only one, solution x, $0 < x < p$.

The corollary follows because $\mathbf{Z}_p$ is a field, so every nonzero element in $\mathbf{Z}_p$ has a multiplicative inverse, and this inverse is unique. Hence, if $a \not\equiv 0 \pmod{p}$, then a^{-1} exists and multiplication of $ax \equiv 1 \pmod{p}$ by a^{-1} gives $x \equiv a^{-1} \pmod{p}$.

Example 3.18 Solve $3x \equiv 1 \pmod{5}$.

From Example 3.17 we see that $\mathbf{S}_3^{-1} = \mathbf{S}_2$.

Hence, $x \equiv 2 \pmod{5}$. ◆

Example 3.19 Solve $8x \equiv 1 \pmod{29}$.

The congruence is equivalent to finding positive integers x, k that satisfy $8x = 1 + 29k$. Let k be, successively, $1, 2, 3, \ldots$, until the quantity $1 + 29k$ becomes an integer divisible by 8. We find $k = 3$, so $x \equiv 11 \pmod{29}$. ◆

Problems

3.65. In Example 3.16 are sets **O**, **E** residue classes? Are they equivalence classes of **Z**?

3.66. If the Corollary to Theorem 3.11 is to hold, then the residue classes modulo n must obey the associative law with respect to multiplication. Demonstrate this with several examples drawn from Fig. 3.3.

3.67. Let p be a prime. What kind of algebraic structure does the set $\{\mathbf{S}_1, \mathbf{S}_2, \ldots, \mathbf{S}_{p-1}\}$ possess? Prove it.

3.68. Find the solution to the congruences:

(a) $23x \equiv 1 \pmod{113}$. (b) $1 - 48x \equiv 0 \pmod{67}$.
(c) $16x \equiv 1 \pmod{71}$. (d) $-1 \equiv -31x \pmod{83}$.

3.69. Regarding the corollary to Theorem 3.12, if $p = n$ is not prime and $a \not\equiv 0 \pmod{n}$, can $ax \equiv 1 \pmod{n}$ possess solutions? For each of the following choices of n, determine how many values of a in the interval $1 \leq a < n$ there are such that $ax \equiv 1 \pmod{n}$ has a solution: (a) $n = 6$, (b) $n = 9$, (c) $n = 10$. See if you can formulate a generalization that answers this same question for arbitrary n.

3.70. Theorem 3.12 has told us that $\mathbf{Z}_4$ is not a field. Could the following tables be the abstract addition and multiplication tables for a field of four elements?

$\oplus$	0	1	A	B
0	0	1	A	B
1	1	0	B	A
A	A	B	0	1
B	B	A	1	0

$\otimes$	0	1	A	B
0	0	0	0	0
1	0	1	A	B
A	0	A	B	1
B	0	B	1	A

3.6 Residue Systems

The following terminology is widely used in number theory in any discussion involving residue classes.

DEF. Let $n > 1$ be a modulus. Any set of integers $\{r_1, r_2, \ldots, r_n\}$ formed by choosing one representative from each of the n residue classes is termed a **complete residue system modulo n**.

An example of a complete residue system modulo n is the set $\mathbf{S} = \{0, 1, 2, \ldots, n - 1\}$; this set contains all of the least nonnegative representatives of the residue classes. The modifier *complete* connotes the fact that for any $k \in \mathbf{Z}$ there is one, and only one, r_i such that $k \equiv r_i \pmod{n}$ (Problem 3.72). In view of Theorem 3.5 it is clear that r_i, r_j are **incongruent** if $i \neq j$, that is, $r_i \not\equiv r_j \pmod{n}$. We also know from Theorem 3.12 that a complete residue system modulo n is a field if n is a prime, and is a commutative ring with identity otherwise.

When n is composite, the complete residue system contains some divisors of zero. For example, if $n = 6$ then the divisors of zero present in the complete residue system $\{0, 1, 2, 3, 4, 5\}$ are 2, 3, 4 (why ?). It is convenient, therefore, to define a set of residues reduced in size from that of the complete residue system.

DEF. A **reduced residue system modulo *n*** is any set of integers obtained from a complete residue system modulo n by removal of (a) all of the divisors of zero and (b) the unique member that is congruent to 0 modulo n.

Example 3.20 A complete residue system modulo 9 is the set $\mathbf{S} = \{9, 10, 2, 3, 13, 5, 15, -2, 8\}$ (verify!). The element 9 is congruent to 0 modulo 9; delete it. The divisors of zero are 3, 15 (verify!); delete them. Hence, a reduced residue system modulo 9 is the set $\mathbf{S}' = \{10, 2, 13, 5, -2, 8\}$. ◆

What are the divisors of zero in a complete residue system modulo n? We notice from Example 3.20 that they were the elements in $\mathbf{S}$ incongruent to 0 (mod 9) and not relatively prime to 9. Here's another example.

Example 3.21 A complete residue system modulo 10 is the set $\mathbf{S} = \{0, 1, 2, 3, 4, 5, 6, 7, 8, 9\}$. The divisors of zero are 2, 4, 5, 6, 8 (verify!); these are not coprime to 10. A reduced residue system modulo 10 is therefore $\mathbf{S}' = \{1, 3, 7, 9\}$. ◆

Theorem 3.13 Let $n > 3$ be composite. Then a nonzero element x of the complete residue system $\mathbf{Z}_n$ is a divisor of zero iff $(x, n) > 1$.

Proof ($\rightarrow$) Let x be a divisor of zero in the complete residue system $\mathbf{Z}_n = \{0, 1, \ldots, n - 1\}$. By definition, there is then a $y \in [1, n - 1]$ such that $xy \equiv 0 \pmod{n}$. If $(x, n) = 1$ were true, then by Corollary 2.12.3, $n|y$ would hold. This is impossible because $y < n$. It follows that $(x, n) > 1$.

($\leftarrow$) Suppose the nonzero element $x \in \mathbf{Z}_n$ has $(x, n) = d > 1$. Then $y = n/d$ is an integer and $1 < y < n$. As x/d is also an integer, we have $y \cdot x = (x/d) \cdot n \equiv 0 \pmod{n}$, so x is a divisor of zero. ◆

Example 3.22 In Examples 3.20 and 3.21 the sets of divisors of zero are precisely the sets of nonzero elements that are not relatively prime to n in the respective complete residue systems. ◆

A closer look at Example 3.22 reveals another important feature. The divisors of zero in the set $\mathbf{S}$ of Example 3.21 are also the elements of $\mathbf{Z}_{10}$ that do not possess multiplicative inverses. That is, the congruences such as $2x \equiv 1 \pmod{10}$ and $6x \equiv 1 \pmod{10}$ have no solutions. This is not a coincidence.

Corollary 3.13.1 A nonzero element x in a complete residue system $\mathbf{Z}_n$ is a divisor of zero iff x is not a unit of $\mathbf{Z}_n$.

Proof $(\rightarrow)$ Suppose $x \in \mathbf{Z}_n$ is a unit; then x^{-1} exists. Now consider $x \cdot y \equiv 0 \pmod{n}$. Multiplication by x^{-1} gives $x^{-1} \cdot (x \cdot y) \equiv (x^{-1} \cdot x) \cdot y \equiv y \equiv 0 \pmod{n}$, so x is not a divisor of zero.

$(\leftarrow)$ Suppose the nonzero element $x \in \mathbf{Z}_n$ is not a divisor of zero. By Theorem 3.13 we have $(x, n) = 1$, or from Theorem 2.12 there are integers a, b such that $ax + bn = 1$. Hence, $ax \equiv 1 \pmod{n}$, so $a \equiv x^{-1} \pmod{n}$ and x is a unit. ◆

Corollary 3.13.2 A reduced residue system modulo n is a multiplicative group.

Proof By Theorem 3.13 and the preceding corollary we see that a reduced residue system modulo n consists precisely of the units of the ring $\mathbf{Z}_n$. By Problem 3.59 this is a group. ◆

Example 3.23 The group multiplication table for the set $\mathbf{S}'$ in Example 3.21 is:

$\cdot$	1	3	7	9
1	1	3	7	9
3	3	9	1	7
7	7	1	9	3
9	9	7	3	1

◆

The corollaries imply that any nonzero $k \in \mathbf{Z}$ that is coprime to a modulus n is congruent to one, and only one, element of a reduced residue system modulo n (Problem 3.76).

In the construction of complete or reduced residue systems, it is possible to require that they possess certain additional characteristics. The following theorem, as applied to complete residue systems, shows one such possibility.

Theorem 3.14 Suppose $(m, n) = 1$ and $r \in \mathbf{Z}$. Then the n integers

$$\mathbf{S} = \{r, r + m, r + 2m, \ldots, r + (n - 1)m\}$$

form a complete residue system modulo n.

Proof It is sufficient to show that any pair of integers in **S** is incongruent. Assume to the contrary that

$$r + km \equiv r + k'm \pmod{n}$$

for some distinct $k, k' \in \{0, 1, 2, \ldots, n - 1\}$, or equivalently,

$$m(k - k') \equiv 0 \pmod{n}.$$

Since $(m, n) = 1$, Corollary 3.3.1 applies, and we obtain

$$k - k' \equiv 0 \pmod{n},$$

or $n|(k - k')$. But this is impossible since $0 < |k - k'| < n$. Hence, $r + km \not\equiv r + k'm \pmod{n}$, and since **S** contains n elements, these must be representatives from each of the n residue classes modulo n. ◆

This theorem says that we can construct a complete residue system modulo n that begins at any arbitrary integer r and whose members are in arithmetic progression with a common difference that is any integer coprime to n.

Example 3.24 Find a complete residue system modulo 8 whose largest member is 11 and whose members form an arithmetic progression with a common difference of 5.

In Theorem 3.14, set $n = 8$, $m = 5$, and $r + (n - 1)m = 11$; this gives $r = -24$. The required residue system is thus

$$\mathbf{S} = \{-24, -19, -14, -9, -4, 1, 6, 11\}.$$

These integers are congruent modulo 8, respectively, to 0, 5, 2, 7, 4, 1, 6, 3. ◆

There are statements more general than that of Theorem 3.14 for both complete and reduced residue systems.

Problems

3.71. Which of the following are complete residue systems?

(a) $\{1, 3, 5, 7, 9, 11, 13, 15, 17, 19, 21\}$.
(b) $\{1, 5, 7, 11, 13, 17\}$.
(c) $\{5, 6, 7, 8, 19\}$.
(d) $\{0, 1, 2, 4, 5, 6, 7, 12\}$.

3.72. The following is a complete residue system modulo n:

$$\mathbf{S} = \{-28, -9, -3, 3, 10, 14, 26, 35, 58, 79\}.$$

For each integer k indicated next, determine which element $r_i \in \mathbf{S}$ satisfies $k \equiv r_i \pmod{n}$:

(a) $k = 11$.
(b) $k = 104$.
(c) $k = -78$.
(d) $k = -1001$.

3.73. Regarding the complete residue system given in Problem 3.72,

(a) Which element is the additive identity?
(b) Which element is the multiplicative identity?
(c) What are the multiplicative inverses of 3, 26?

3.74. Which of the following are reduced residue systems modulo 10?

(a) $\{2, 4, 8, 10\}$.
(b) $\{3, 9, 21, 27\}$.
(c) $\{-9, -7, -3, -1\}$.
(d) $\{13, 40, 91, 117\}$.

3.75. If $m \neq n$, is it possible for a reduced residue system modulo m to have the same number of elements as a reduced residue system modulo n?

3.76. Regarding the complete residue system given in Problem 3.72,

(a) Extract from it the reduced residue system, $\mathbf{S}'$, modulo n.
(b) For each integer k indicated, determine which element $r_i \in \mathbf{S}'$ satisfies $k \equiv r_i \pmod{n}$: i) $k = -83$, (ii) $k = 57$, (iii) $k = 111$, (iv) $k = 3^{10}$.

3.77. Write a computer program that will take any given modulus $n > 1$ and construct a reduced residue system modulo n.

3.78. What are the divisors of zero in the complete residue system modulo 30 given by $\mathbf{S} = \{0, 1, 2, \ldots, 29\}$? What is the size of the reduced residue system modulo 30? Write out the multiplication table for the reduced residue system obtainable from $\mathbf{S}$, and verify that it is the multiplication table for a group.

3.79. Let $n = 14$ be a modulus.

(a) Construct a complete residue system modulo 14, all of whose members are divisible by 3.
(b) Construct another one, all of whose members are congruent to 1 (mod 5).
(c) Construct a third complete residue system modulo 14 that has a positive maximum element that is 4 times the minimum element.

3.80. Would the proof of Theorem 3.13 fail if a complete residue system different from the one there ($\mathbf{Z}_n = 0, 1, 2, \ldots, n - 1$) were examined?

3.81. For each of the following complete residue systems, identify the set of divisors of zero:

(a) $\{0, 7, 9, -3, 4, 35\}$.
(b) $\{0, 1, 2, 3, 4, 5, 6\}$.

(c) $\{2, 1, 0, -1, -2, -3, 6, 7, 8, 9, 10, 17, 18, 19\}$.
(d) $\{-1, 2, 5, 8, 11, 14, 17, 20, 23, 26\}$.

3.82. Find a complete residue system modulo 20 in arithmetic progression and with minimum and maximum elements of 50 and 183, respectively.

3.83. Let $(m, n) = 1$ and let u run through a reduced residue system modulo n. Show that the integers of the form mu also form a reduced residue system modulo n. Illustrate this theorem for the case $n = 9, m = 4$.

3.84. Prove that for any prime p there exists a complete residue system modulo p, all of whose members are primes. Illustrate this for the case $p = 7$. Is the result still true if p is not prime? For any composite integer $n > 3$ does there exist a reduced residue system modulo n, all of whose members are primes?

3.85. Suppose $(m, n) = 1$, $(M, n) = 1$, $r \in \mathbf{Z}$, and u runs through a complete residue system modulo n while v runs through a complete residue system modulo m. Show that the mn integers of the form

$$r + Mmu + nv$$

form a complete residue system modulo mn.

3.7 Introduction of the Euler ϕ-Function

In the previous Section we saw that a reduced residue system modulo n is a maximal set of incongruent integers that are relatively prime to n. The following function is therefore important not only here, but also in many other places in number theory. It was introduced by **Euler** (in another connection) in 1760.

DEF. The number of positive integers equal to or less than n that are relatively prime to n is denoted by $\phi(n)$, where ϕ is called the **Euler ϕ-function**. By convention, we take $\phi(1) = 1$.

Thus, we find that $\phi(6) = 2$, $\phi(25) = 20$, and $\phi(29) = 28$. We are now going to prove a result that is an analog of Problem 3.85 and that leads to an unexpected bonus regarding the Euler ϕ-function.

Theorem 3.15 Suppose $(m, n) = 1$ and u runs through a reduced residue system modulo n while v runs through a reduced residue system modulo m. Then the set of integers of the form $mu + nv$ is a reduced residue system modulo mn.

Proof Problem 3.85 has indicated (taking $r = 0$, $M = 1$ there) that if u, v actually run through complete residue systems modulo n, m, respectively, then the integers $mu + nv$ form a complete residue system modulo mn. This system will then automatically contain $\phi(mn)$ integers that are coprime to mn. Call this set of $\phi(mn)$ integers **S**; what we have to show is that the elements of **S** are necessarily formed by allowing u, v to run through only the elements of the reduced residue systems modulo n, m, respectively.

For $mu + nv \in \mathbf{S}$, we have

$$(mu + nv, mn) = 1 \quad \leftrightarrow \quad \begin{cases} (mu + nv, m) = 1 \\ (mu + nv, n) = 1. \end{cases}$$

The direction ($\leftarrow$) follows from Problem 2.51. To see ($\rightarrow$), suppose $(mu + nv, m) = d > 1$ and let the prime p divide d. Then $p \mid (mu + nv)$, and $p \mid m \rightarrow p \mid mn$; hence, $(mu + nv, mn) \geq p > 1$.

Since $m \mid mu$, $n \mid nv$, and $(m, n) = 1$, we have the further implications

$$\begin{cases} (mu + nv, m) = 1 & \text{iff } (nv, m) = 1 \\ (mu + nv, n) = 1 & \text{iff } (mu, n) = 1. \end{cases}$$

But $(nv, m) = 1$ if and only if v assumes any of the $\phi(m)$ values in a reduced residue system modulo m. Similarly, $(mu, n) = 1$ if and only if u assumes any of the $\phi(n)$ values in a reduced residue system modulo n. This proves the theorem. ◆

Example 3.25 Let $m = 5$ and $n = 6$, and let the values of u be the members of a reduced residue system modulo n and the values of v be the members of such a system modulo m:

$$u: \{1, 5\}$$
$$v: \{1, 2, 3, 4\}.$$

Hence, a reduced residue system modulo 30, as given by Theorem 3.15, is $\{5u + 6v\}$, or

$$\{11, 17, 23, 29, 31, 37, 43, 49\}. \quad ◆$$

The promised bonus is a quick but significant observation that we can make about the Euler ϕ-function.

Corollary 3.15.1 If $(m, n) = 1$, then $\phi(mn) = \phi(m)\,\phi(n)$.

Proof The corollary is immediate since the $\phi(mn)$ integers coprime to mn are linear combinations of $\phi(m)$ v's and $\phi(n)$ u's. ◆

A function with the property shown in the Corollary is called a **multiplicative function**. Let us note that the two functions f_1 and f_2 defined by $f_1(n) = n$ and $f_2(n) = 1/n$ are each multiplicative, whereas the functions $f_3(n) = n + 1$ and $f_4(n) = 2/n$ are not multiplicative. There are several important multiplicative functions in number theory, and we shall consider some of them in detail in Chapter 7.

Problems

3.86. (a) Without resorting to tedious counting, determine the values of $\phi(37)$, $\phi(89)$, $\phi(22)$.

(b) Do the same for $\phi(121)$, $\phi(64)$.

3.87. Construct a reduced residue system modulo 21 that has minimum and maximum elements of 31, 95, respectively.

3.88. Prepare a computer program that will print out the values of $\phi(n)$ for all $n \in [3, 100]$. Perhaps you can modify your program from Problem 3.77. Formulate any conjectures, if possible.

3.89. If you did not previously do Problem 3.85, do it now.

3.90. Prove that the number N of nonintegral, positive rational numbers that are reduced to lowest terms, are less than 1, and have denominators not exceeding n is given by

$$N = \sum_{k=2}^{n} \phi(k).$$

Illustrate this with $n = 10$.

3.91. (a) Show that in $\mathbf{Z}_n$, if k is relatively prime to n, then $n - k$ is also relatively prime to n.

(b) Use the result of part (a) to prove that in $\mathbf{Z}_n$ the sum of the integers that are relatively prime to n is $n\ [\phi(n)/2]$. For example, in $\mathbf{Z}_8$ one has $1 + 3 + 5 + 7 = 8\ [\phi(8)/2] = 8(4)/2$.

(c) We notice in part (b) that $8 \mid (1 + 3 + 5 + 7)$. What simple fact would permit us to say that for any $n > 2$, the modulus n divides the sum of all the integers in $\mathbf{Z}_n$ relatively prime to it? How does this simple fact actually follow from part (a) via an elementary counting argument?

3.8 Fermat's Theorem

The result known as Fermat's Theorem (and sometimes also as Fermat's Little Theorem[4]) resulted during an attempt by **Fermat** to solve another problem. A fellow countryman and number theory enthusiast, **Bernard Frénicle de Bessy** (1612?–1675), challenged **Fermat** to find a perfect number of 20 digits or the next largest perfect number (Fletcher, 1989). During the course of this investigation in 1640, **Fermat** discovered the theorem to be given shortly. He claimed to have a proof of it but never revealed it (Mahoney, 1994). We shall give two proofs of this important theorem.

Theorem 3.16 (Fermat's Theorem). If the prime p does not divide the integer a, then $a^{p-1} \equiv 1 \pmod{p}$.

Proof Since $p \nmid a$, then $(p, a) = 1$. In Theorem 3.14 take $r = 0$; then $\{0, a, 2a, \ldots, (p-1)a\}$ is a complete residue system modulo p. Hence, the members of the set $\{a, 2a, \ldots, (p-1)a\}$ are congruent modulo p in some order to members of the set $\{1, 2, 3, \ldots, p-1\}$. Now make repeated use of Theorem 3.2, part (5), to write

$$a(2a)(3a) \ldots (p-1)a \equiv 1(2)(3) \ldots (p-1) \pmod{p}.$$

Since $(p, (p-1)!) = 1$, Corollary 3.3.1 holds and we obtain upon division

$$a^{p-1} \equiv 1 \pmod{p}. \quad ◆$$

This proof was first given by the Scottish mathematician **James Ivory** (1765–1842) in 1806, but other proofs existed earlier, including an unpublished one by **G.W. Leibniz** (1646–1716) before 1683.

Example 3.26 Consider again Problem 3.32(a). Fermat's Theorem gives immediately $2^{12} \equiv 1 \pmod{13}$, so from Theorem 3.2, part (6), $(2^{12})^4 \equiv 2^{48} \equiv 1 \pmod{13}$. Since $2^2 \equiv 4 \pmod{13}$, we obtain from Theorem 3.2, part (5), that $2^{48} \cdot 2^2 = 2^{50} \equiv 4 \pmod{13}$.

Corollary 3.16.1 If p is a prime and a is any integer, then $a^p \equiv a \pmod{p}$.

Proof If $p \mid a$, then $p \mid a^p$ also, and $p \mid (a^p - a)$ follows from Theorem 1.4(b). When $(p, a) = 1$, Theorem 3.16 holds and multiplication of both sides of the congruence there by a yields $a^p \equiv a \pmod{p}$. ◆

[4] To distinguish **Fermat's Theorem** from **Fermat's Last Theorem**, the somewhat pejorative adjective *little* has frequently been appended to the former. But Fermat's Theorem is a significant result in number theory, so we shall delete the adjective.

We are now going to develop an alternative proof of Fermat's Theorem, one that brings out underlying algebraic structure. The following is a new kind of subset of the elements of a group.

DEF. Let $\mathbf{H}$ be a subgroup of a finite group $\mathbf{G}$, and let g be an arbitrary fixed member of $\mathbf{G}$. The subset $g\mathbf{H} = \{gh : h \in \mathbf{H}\}$ is called the **left coset** of $\mathbf{H}$ that contains g.[5]

Clearly, every element $g \in \mathbf{G}$ is in some left coset of $\mathbf{H}$ because $g \in g\mathbf{H}$.

Example 3.27 Refer to the multiplication table in Fig. 2.1(a) for the Klein 4-group. Let $\mathbf{H}$ be $\{\mathbf{I}, a\}$, and choose arbitrarily $g = b$. Then the left coset $b\mathbf{H} = \{b \cdot \mathbf{I}, b \cdot a\} = \{b, c\}$. Observe that $b\mathbf{H}$ is not a subgroup of $\mathbf{G}$ since $\mathbf{I} \notin b\mathbf{H}$. ◆

But what do the cosets really look like? We show next that belonging to a coset is an equivalence relation, but first here is a preliminary note. Now $g' \in g\mathbf{H}$ is the same as $g' = gh$ for some $h \in \mathbf{H}$ or, equivalently, $g^{-1}g' = h$ and this is the same as $g^{-1}g' \in \mathbf{H}$. So saying that g' belongs to the same coset of $\mathbf{H}$ as does g is equivalent to saying $g^{-1}g'$ belongs to $\mathbf{H}$. We use this idea in what follows.

Theorem 3.17 Coset membership is an equivalence relation.

Proof Let the binary relation $x \mathfrak{R} y$ mean that x, y belong to the same left coset of $\mathbf{H}$, where $x, y \in \mathbf{G}$.
Reflexive. If $x \in \mathbf{G}$, then $x^{-1}x = \mathbf{I} \in \mathbf{H}$; hence, $x \mathfrak{R} x$.
Symmetric. Suppose $x \mathfrak{R} y$; then $y^{-1}x \in \mathbf{H}$. But $\mathbf{H}$ is a group, so it contains all its inverses, that is, $(y^{-1}x)^{-1} = x^{-1}y \in \mathbf{H}$. In turn, this says $y \in x\mathbf{H}$ or $y \mathfrak{R} x$.
Transitive. Suppose $x \mathfrak{R} y$ and $y \mathfrak{R} z$. Then $y^{-1}x \in \mathbf{H}$ and $z^{-1}y \in \mathbf{H}$. But $\mathbf{H}$ is a group, so it is closed under the group operation, that is, $(z^{-1}y)(y^{-1}x) = (z^{-1}x) \in \mathbf{H}$. This is equivalent to $x \in z\mathbf{H}$ or $x \mathfrak{R} z$. It follows that $\mathfrak{R}$ is an equivalence relation $\sim$ on $\mathbf{G}$. ◆

Corollary 3.17.1 The left cosets of $\mathbf{H}$ induce a partition of $\mathbf{G}$.

Proof This follows immediately from Theorems 3.5 and 3.17. ◆

[5] Similarly, the **right coset** of $\mathbf{H}$ that contains g is $\mathbf{H}g = \{hg : h \in \mathbf{H}\}$. If $\mathbf{G}$ is Abelian, then $g\mathbf{H}$, $\mathbf{H}g$ are just called **cosets**. If $\mathbf{G}$ is not Abelian, as in Fig. 2.4, then $g\mathbf{H} \neq \mathbf{H}g$.

A question that remains is that of the sizes of the various cosets of **H**. We saw in Example 3.27 that the cosets **H**, b**H** were the same size. However, the group there was small so there were not many possibilities for coset sizes. Also, the equivalence classes formed when an equivalence relation $\mathcal{R}$ partitions a finite set **S** are not necessarily the same size (Problem 3.34). However, consider Theorem 3.18.

Theorem 3.18 Let **H** be a subgroup of a finite group **G**. Then the left (right) cosets of **H** are of the same size as **H**.

Proof Let g**H** be an arbitrary left coset of **H**. We define the following mapping from elements of **H** into elements of g**H**:

$$f : \mathbf{H} \rightarrow g\mathbf{H}$$

where $f(h) = gh$. The mapping is actually *onto* g**H** because g**H** is defined as the set of all elements gh, where $h \in \mathbf{H}$.

Now let $h_1, h_2 \in \mathbf{H}$ and suppose $f(h_1) = f(h_2)$. This means $gh_1 = gh_2$, and premultiplication of both sides by g^{-1} gives $h_1 = h_2$, that is, contrapositively, $h_1 \neq h_2$ implies $f(h_1) \neq f(h_2)$. This says that the mapping is *one-to-one* (distinct elements in the domain have distinct images).

A mapping that is one-to-one and onto is a bijection between two sets. Since the sets here (**H**, g**H**) are finite, they must then be of the same size. ◆

We can now get a picture of the set of left cosets of a subgroup **H** of a group **G**. The cosets are all the same size as **H**, they have no elements in common, and they completely exhaust **G** (Fig. 3.4). The following theorem, which is one of the cornerstones of group theory, now seems almost like a triviality. We alluded to it at the very end of Section 2.3.

Theorem 3.19 (Lagrange's Theorem). Let **H** be a subgroup of a finite group **G**. Then the order of **H** divides the order of **G**.

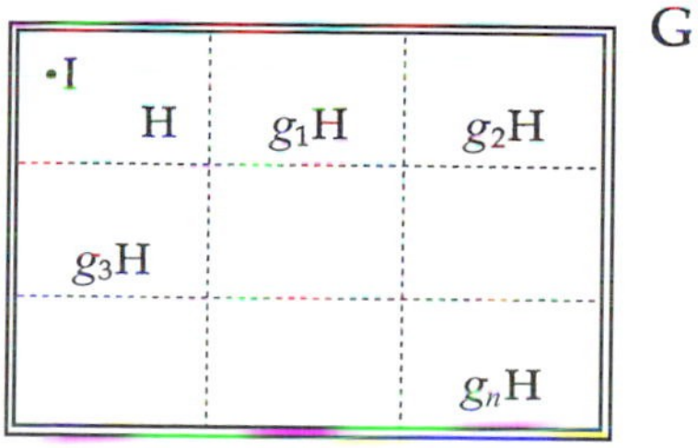

Figure 3.4 All (left) cosets of a given subgroup **H** have order $|\mathbf{H}|$

Proof From Theorem 3.18 let $|\mathbf{H}|$ be the common order of all of the left cosets of $\mathbf{H}$ in $\mathbf{G}$, and let $|\mathbf{G}|$ be the order of $\mathbf{G}$. Suppose there are n such cosets. Then from Corollary 3.17.1 we can write $|\mathbf{G}| = n|\mathbf{H}|$. Since $|\mathbf{G}|$, n, $|\mathbf{H}|$ are all integers, we have $|\mathbf{H}|$ divides $|\mathbf{G}|$. ◆

Refer to Problem 2.41 where we defined the **order of an element** g in $\mathbf{G}$. If $g^k = \mathbf{I}$, then $g^{k-1} = g^{-1}$, $g^{k-2} = (g^2)^{-1}$, $g^{k-3} = (g^3)^{-1}, \ldots, g^1 = (g^{k-1})^{-1}$. So the k elements $\{g^1, g^2, g^3, \ldots, g^k\}$ all contain their own inverses as well as the identity $\mathbf{I}$, multiplication is closed, and associativity is inherited from $\mathbf{G}$. Therefore, this set of elements is a group; it is the cyclic subgroup of $\mathbf{G}$ that is generated by g. The definition of order of the element g that was given in Problem 2.41 is thus equivalent to the definition, "The order of the element $g \in \mathbf{G}$ is the order of the cyclic subgroup generated by g."

Corollary 3.19.1 In a finite group $\mathbf{G}$ the order of any element g divides the order of the group.

Proof The element g generates a subgroup $\mathbf{H}$ of $\mathbf{G}$. Suppose the order of g is m; then $|\mathbf{H}| = m$. By Theorem 3.19 we have $|\mathbf{H}| \,|\, |\mathbf{G}|$. It follows that $m \,|\, |\mathbf{G}|$. ◆

Example 3.28 Refer to the multiplication table for the group of order 6 shown in Problem 2.35. Elements a, b are of order 3; elements c, d, e are of order 2; and $\mathbf{I}$ is (always) of order 1.

Corollary 3.19.2 (Fermat's Theorem). If p is a prime and $a \not\equiv 0 \pmod{p}$, then $a^{p-1} \equiv 1 \pmod{p}$.

Proof A complete residue system $\mathbf{S}$ modulo a prime p is a field (Theorem 3.12), and the subset $\mathbf{S}' \subset \mathbf{S}$ of elements incongruent to 0 (mod p) is a group of order $p - 1$ under modular multiplication (Corollary 3.13.2). Let $a \in \mathbf{S}'$ be of order m, so by Corollary 3.19.1, $mk = p - 1$ for some $k \in \mathbf{N}$. Hence, $a^{p-1} = (a^m)^k = \mathbf{I}^k = \mathbf{I}$, or in congruence notation, $a^{p-1} \equiv 1 \pmod{p}$. ◆

The upshot of the algebraic development is that Fermat's Theorem holds because when p is a prime, then the elements of a complete residue system modulo p that are incongruent to 0 are a group and any member a of this group generates a cyclic subgroup whose order is a divisor of $p - 1$ by Lagrange's Theorem. This gives, immediately, $a^{p-1} = \mathbf{I}$.

If p is not a prime, then Fermat's Theorem must be modified. The generalization required is called Euler's Theorem, and we prove it in Section 3.10 by a method analogous to that of Corollary 3.19.2.

Example 3.29 Show that 5 divides $n^{17} - n$ for every $n \in \mathbf{Z}$.

If $5 \mid n$, then $5 \mid (n^{17} - n)$ follows. If $5 \nmid n$, then consider $n^{16} - 1$. By Fermat's Theorem we have $n^4 \equiv 1 \pmod 5$, so from Theorem 3.2, part (6), $(n^4)^4 \equiv 1^4 \pmod 5$. Hence, $5 \mid (n^{16} - 1)$ and the theorem follows from Theorem 3.2, part (2). Example: $3^{17} - 3 = 129{,}140{,}160 = 5(25{,}828{,}032)$. ◆

Problems

3.92. In each case find the least positive residue by making use of Fermat's Theorem:

(a) $4^{24} \pmod{11}$. (b) $4^{20} \pmod{37}$. (c) $31^{31} \pmod{29}$. (d) $3^{100} \pmod{47}$.
(e) $5^{999{,}999} \pmod 7$ (attributed to Fermat, ca. 1635).
(f) $(-16)^{253} \pmod{251}$.

3.93. Show that

(a) $2^{11{,}213} - 1$ is not divisible by 11.
(b) $n^{33} - n$ is divisible by 15 for any $n \in \mathbf{Z}$.

3.94. Select one of the two subgroups of order 3 of the group $\mathbf{G}$ whose multiplication table is given in Problem 2.35. Work out the left cosets of the subgroup. Is Corollary 3.17.1 upheld?

3.95. A certain group $\mathbf{G}$ of order 40 has a subgroup $\mathbf{H}$ of order 4. How many left cosets of $\mathbf{H}$ are there? How many of these cosets are subgroups of $\mathbf{G}$? Could $\mathbf{G}$ have a subgroup of order 6?

3.96. Let $\mathbf{G}_p$ be the cyclic group of order p, where $p > 2$ is a prime. Show that if $g \in \mathbf{G}_p$, $g \neq \mathbf{I}$, then the order of g is p.

3.97. Prove that if p is a prime, then

$$p \mid (1^{p-1} + 2^{p-1} + \cdots + (p-1)^{p-1} + 1).$$

It has been conjectured that the divisibility holds only if p is a prime. This is known to be true for integers $\leq 10^{1000}$ (Sierpiński, 1964).

3.98. Let p be an odd prime and consider the quantity $C = \prod_{k=1}^{p-1} k^{p-2}$.

(a) Show that $C \cdot (p-1)! \equiv 1 \pmod p$.
(b) Find the least positive residue of C for the cases $p = 5, 7, 11$.
(c) Formulate a conjecture.

3.99. Fermat's Theorem can also be proved by an elementary combinatorial argument (Golomb, 1956). Read this short note and prepare a summary of it in your own words.

3.9 Pseudoprimes*

The determination of whether an integer is prime or not is a difficult affair if the integer is large. There are no *easy* tests that one can apply. A number of exact criteria do exist for primality of an integer, but these criteria are not generally practical for large integers.

Fermat's Theorem is the basis for an inexact but, nevertheless, useful test for primality, and also for the introduction of the very interesting topic of pseudoprimes. The converse of Theorem 3.16 is not generally true, but it is observed to hold in a great many cases.

DEF. If $n > 3$ is a composite integer that divides $2^n - 2$, then n is called a **pseudoprime**.[6]

The definition permits both odd and even values for n. When n is odd, the divisibility requirement is equivalent to $n|(2^{n-1} - 1)$. Pseudoprimes occur infrequently, much less frequently than primes. Hence, if an odd integer n does not divide $2^{n-1} - 1$, it cannot be a prime, and if it does divide $2^{n-1} - 1$, then there is a high probability that it is a prime and a lesser probability that it is composite.

The first four odd pseudoprimes are 341, 561, 645, and 1105. In the interval [3, 1105] there are, in contrast, 184 primes. However, despite their scarcity, the odd pseudoprimes still constitute an infinite set. This result was established by **E. Malo** in 1903 and **M. Cipolla** in 1904.

Theorem 3.20 The number of odd pseudoprimes, psp(2), is infinite.

Proof Let k be an odd pseudoprime; this means $k|(2^{k-1} - 1)$ or, equivalently, $Ak = 2^{k-1} - 1$ for some $A \in \mathbf{N}$. Further, by definition, k is composite, $k = ab$ and $a > 1, b > 1$.

We now claim that 2^{k-1} is also a pseudoprime. First observe that $2^k - 1 = 2^{ab} - 1$ is composite since $2^a - 1$ divides $2^{ab} - 1$. Next, we show that $(2^k - 1)|\{2^{(2^k-1)-1} - 1\}$, that is, $(2^k - 1)|(2^{2Ak} - 1)$. Let $x = 2^k$; then we must show that $x - 1$ divides $x^{2A} - 1$. But this follows from the Factor Theorem since $x^{2A} - 1 = 0$ implies $x = 1$ is one of the roots, so $x - 1$ is a

* Sections marked with an asterisk can be skipped without impairing the continuity of the text.

[6] More precisely, we should say "pseudoprime to the base 2." In general, n is a **pseudoprime to base a**, psp(a), if n is composite and $n|(a^n - a)$. The smallest even psp(2) is 161,038, which was discovered by the American number theorist **Derrick H. Lehmer** (1905-1991) in 1950 (Brillhart, 1992). The next year **N.G.W. Beeger** proved that there are infinitely many even psp(2)'s. The psp(2)'s are sometimes called **Poulet numbers** because **P. Poulet** (in 1938) tabulated all odd ones less than 10^8.

factor of $x^{2A} - 1$. Thus, we now have a new pseudoprime $2^k - 1 > k$ generated from a prior smaller one. Clearly there is no end to this process. ◆

Theorem 3.20 does not imply that all odd pseudoprimes must be of the form $2^k - 1$. For example, none of 341, 561, 645, 1105 is of this form. Not much is known about what types of representations are possible for pseudoprimes. In answer to a question of whether there are any pseudoprimes of the form $2^k - 2$ (Rotkiewicz, 1972), the number $2^{465,794} - 2$ was exhibited as such a pseudoprime (McDaniel, 1989). This even pseudoprime has 140,218 digits; it is not known if this is the smallest such pseudoprime.

The definition given previously for a psp(a), $a > 2$, does not preclude the possibility that $(a, n) > 1$, where n is the pseudoprime. Thus, when $a = 5$ we have

$$5^{15} - 5 \equiv 0 \pmod{15},$$

so 15 is a psp(5). Permitting $(a, n) > 1$ greatly lengthens the list of psp(a)'s and thus removes some of the interest in them. Accordingly, some writers define a psp(a) as any composite integer $n \geq a + 1$ for which $a^{n-1} \equiv 1 \pmod{n}$. Any n that satisfies this is automatically coprime to a. By this definition, the smallest psp(5) is now 217.

Example 3.30 Find the smallest psp(7).

The first few odd candidates are $n = 9, 15, 21, 25$. Checking shows that $7^{25-1} \equiv 1 \pmod{25}$. Among the first few even candidates greater than 7, none satisfies $7^{n-1} \equiv 1 \pmod{n}$. Hence, the smallest psp(7) is 25. ◆

Scarcer than pseudoprimes to base a are those integers that are simultaneously a psp(a) and a psp(b). Such an integer is 1105, that is, 1105 is composite and since $(1105, 2) = (1105, 3) = 1$, then $1105 | (2^{1104} - 1)$ and $1105 | (3^{1104} - 1)$. Thus, if an odd integer n divides both $2^{n-1} - 1$ and $3^{n-1} - 1$ and is coprime to 3, there is a higher likelihood that it is a prime than if it only divides $2^{n-1} - 1$. Of course, here we already know that 1105 is not a prime; it is a pseudoprime since it is composite.

Some software systems (e.g., MAPLE) do primality testing by looking for **strong pseudoprimes to base** a, spsp(a). These are odd, composite integers n, $0 < a < n$, $(a, n) = 1$ such that if $n - 1 = K \cdot 2^t$ (K = odd), then either $a^K \equiv 1 \pmod{n}$ or $a^{2^r} \equiv -1 \pmod{n}$ holds for some r, $0 \leq r < t$. If n is a prime, it also satisfies these conditions. The smallest composite integer that is simultaneously an spsp(2), an spsp(3), and an spsp(5) is 25,326,001. The software system goes further and tests with several bases, say 2, 3, 5, 7, 11, 13; if n passes all six tests, it is declared "probably prime." Interestingly, around 1992 the software system found that $n = 3{,}474{,}749{,}660{,}383$ is "probably prime," and yet

Table 3.1 Distribution of Carmichael Numbers

x	C(x)	x	C(x)	x	C(x)
10^3	1	10^8	255	10^{12}	8241
10^4	7	10^9	646	10^{13}	19279
10^5	16	10^{10}	1547	10^{14}	44706
10^6	43	2.5×10^{10}	2163	10^{15}	105212
10^7	105	10^{11}	3605	10^{16}	246683

$n = 1303(16{,}927)(157{,}543)$ (Pinch, 1993). This showed that the computer algorithm is not foolproof.[7]

Still scarcer than any set of psp(a)'s are integers n that are pseudoprime to all bases a for which $1 < a < n$ and $(a, n) = 1$. That such integers exist was discovered in 1909 by the American mathematician **Robert D. Carmichael** (1879–1967). They are called **Carmichael numbers** (or **absolute pseudoprimes**); there are only seven of them less than 10,000 (Granville, 1992). The smallest Carmichael number is 561 (Problem 3.105). Table 3.1 gives the distribution of Carmichael numbers up to 10^{16}; C(x) stands for the number of such numbers that do not exceed x. The first 48 Carmichael numbers are given in Table I(6) in Appendix II.

Any hope of preparing a *complete* list of Carmichael numbers as a useful aid in primality testing was dashed in 1992. It was proved by **W.R. Alford**, **A. Granville**, and **C. Pomerance** of the University of Georgia that $\lim_{x\to\infty} C(x) = \infty$. They also proved that $C(x) > x^{2/7}$ for all x sufficiently large; an exponent greater than 2/7 is probably true also. This shows that the Carmichael numbers eventually become plentiful.

Example 3.31 Show that 1105 is a Carmichael number.

First, we note that $1105 = 5 \cdot 13 \cdot 17$. Then if $a > 1$ is any base such that $(a, 1105) = 1$, then $(a, 5) = (a, 13) = (a, 17) = 1$. Fermat's Theorem now gives $a^{1104} = (a^4)^{276} \equiv 1^{276} \equiv 1 \pmod{5}$, $a^{1104} = (a^{12})^{92} \equiv 1^{92} \equiv 1 \pmod{13}$, and $a^{1104} = (a^{16})^{69} \equiv 1^{69} \equiv 1 \pmod{17}$, so finally $a^{1104} \equiv 1 \pmod{5 \cdot 13 \cdot 17}$ because 5, 13, 17 are pairwise coprime. ◆

Problems

3.100. Show that 561 is a psp(5).

3.101. Suppose the integer n is both a psp(a) and a psp(a'). Prove that it is also a psp(aa').

[7] This problem has since been corrected by a newer version of MAPLE.

3.102. Recall the **Fermat numbers**, defined in Problem 2.13. Prove that a composite Fermat number is a psp(2).

3.103. Refer to Table 3.1. Prepare a plot of $\log_{10} C(x)$ versus $\log_{10} x$. Formulate an empirical expression for the counting function $C(x)$.

3.104. Let $a > 1$ be a base, and let p be an odd prime that does not divide $a(a^2 - 1)$. There are an infinite number of such primes (why?). Define the number $m = (a^{2p} - 1)/(a^2 - 1)$. Then m is composite; show this.

Next, show that $2p$ divides $(m - 1)$ or, equivalently, that $m = 1 + 2kp$ for some $k \in \mathbf{N}$. This gives $a^{2p} \equiv 1 \pmod{m}$ and finally, $a^{m-1} \equiv 1 \pmod{m}$, which says that m is a psp(a). Thus, we have proved that there are an infinite number of psp(a) for any $a > 1$. Fill in the gaps and write out the complete proof. Illustrate the result with $a = 3, p = 5$.

3.105. Prove that 561 is a Carmichael number.

3.106. Write a computer program and uncover the smallest spsp(2).

3.107. A reliable "quick and dirty" test for primality of an integer n, where $n < 2.5 \times 10^{10}$, is given in (Pomerance, Selfridge, and Wagstaff, Jr., 1980). Consult Table 7 there and write a summary of the test.

3.10 Euler's Theorem

Euler considered the extension of Fermat's Theorem to the case where the modulus m is composite. Is it even reasonable that there should exist an exponent $k > 0$ such that $a^k \equiv 1 \pmod{m}$, for $(a, m) = 1$?

Consider the set

$$\mathbf{S} = (a^1, a^2, a^3, \ldots, a^{m-1}),$$

reduced modulo m. If these were all incongruent, then one of them would have the value 1. If, on the other hand, $a^i \equiv a^j \pmod{m}$ and $i > j$, then $a^{i-j} \equiv 1 \pmod{m}$. Thus, for $(a, m) = 1$ there is always some power of a that is congruent to $1 \pmod{m}$. **Euler** showed how easy it is to compute this power. Our approach (unlike Euler's) is algebraic in spirit, in sympathy with the proof of Corollary 3.19.1.

Theorem 3.21 (Euler's Theorem). If $m > 1$ and $(a, m) = 1$, then $a^{\phi(m)} \equiv 1 \pmod{m}$, where ϕ is the Euler ϕ-function.

Proof Corollary 3.13.2 shows that $\mathbf{S}'$, a reduced residue system modulo m, is a group of order $\phi(m)$. Let $a \in \mathbf{S}'$ and suppose a has order k. Then $a^k = \mathbf{I}$, where $\mathbf{I}$ is the identity of the group. By Corollary 3.19.1 we must have $kr = \phi(m)$ for some $r \in \mathbf{N}$. It follows that

$$a^{\phi(m)} = (a^k)^r = \mathbf{I}^r = \mathbf{I}, \quad \text{or} \quad a^{\phi(m)} \equiv 1 \pmod{m}. \quad \blacklozenge$$

The proof of Euler's Theorem is a carbon copy of that of Corollary 3.19.2. An immediate application of Theorem 3.21 is to the solution of linear congruences.

Example 3.32 Solve $13x \equiv 24 \pmod{30}$.

We find easily that $\phi(30) = 8$. Multiply both sides of the congruence by $13^{\phi(30)-1} = 13^7$:

$$13^{\phi(30)}\, x \equiv 24 \cdot 13^7 \pmod{30}$$

or

$$\begin{aligned} x &\equiv 24 \cdot 13^7 \pmod{30} \\ &\equiv 24 \cdot 7 \pmod{30} \\ &\equiv 18 \pmod{30}. \quad \blacklozenge \end{aligned}$$

Notice that the solution of Example 3.32 depended upon the fact that $(13, 30) = 1$. Observe also that Example 3.32 is a reformulation in congruence language of the Diophantine equation $13x + 30y = 24$. A particular solution of this is (from the preceding) $x_0 = 18$, $y_0 = [24 - 13(18)]/30 = -7$; the general solution is obtained from Problem 2.66.

Problems

3.108. Compute $\phi(m)$: (a) $m = 42$. (b) $m = 54$. (c) $m = 100$. (d) $m = 520$.

3.109. Suppose that k is the order of a modulo m: $a^k \equiv 1 \pmod{m}$, $(a, m) = 1$. Explain how you know that $k \mid \phi(m)$.

3.110. Find the order k of each integer a modulo m:

(a) $a = 8$, $m = 35$. (b) $a = 27$, $m = 50$.
(c) $a = 12$, $m = 49$. (d) $a = 2$, $m = 49$.

3.111. Let $m = 18$, and let $\mathbf{S}'$ be the set of $\phi(18)$ least positive integers that are relatively prime to m. Work out the multiplication table for the members of $\mathbf{S}'$, and convince yourself that it is the multiplication table of a group. Is the group a cyclic group?

3.112. Solve these congruences by making use of Euler's Theorem:

(a) $22x \equiv 3 \pmod{35}$.
(b) $63x \equiv 1 \pmod{100}$.
(c) $-13x \equiv 2 \pmod{28}$.
(d) $100x \equiv 33 \pmod{49}$.
(e) $-81x \equiv 80 \pmod{106}$.
(f) $44x \equiv 18 \pmod{51}$.

3.113. Let $m > 2$ be a modulus. By Corollary 3.13.2 the reduced residue system modulo m is a group. Prove that the order of this group is even. [Hint: See Problem 3.91.]

3.114. Let $\mathbf{S}'$ be a reduced residue system modulo m, $m \geq 3$, and assume that $\mathbf{S}'$ is cyclic. Suppose that $a \in \mathbf{S}'$ is a generator for $\mathbf{S}'$. Show that

$$\sum_{n=0}^{\phi(m)-1} a^n \equiv 0 \pmod{m}.$$

3.115. Write a computer program that will solve linear congruences such as those in Problem 3.112 by explicitly using Euler's Theorem.

3.11 Solving Linear Congruences; Finite, Simple Continued Fractions

Look again at Example 3.32. It began improperly by not even inquiring if a solution should exist. Solving congruences is analogous to solving equations; three questions should be asked of any congruence, and in the following order:

1. Does a solution exist?
2. How many solutions are there?
3. How can I obtain these solutions?

Also, as with equations, there is no general strategy that works for all congruences. In this section we look only at some simple cases.

Theorem 3.22 Let $(a, m) = d$. Then the congruence $ax \equiv b \pmod{m}$ has no solutions if $d \nmid b$ and has d incongruent solutions if $d \mid b$.

Proof Suppose $d \nmid b$; since $d \mid a$, then $d \nmid (ax - b)$. So there is no x such that $ax \equiv b \pmod{d}$, and thus no x such that $ax \equiv b \pmod{m}$, from Theorem 3.2, part (7).

On the other hand, if $d|b$, then Theorem 3.3 gives

$$a'x \equiv b' \pmod{m'},$$

where $a' = a/d$, $b' = b/d$ and $m' = m/d$, and $(a', m') = 1$. By Theorem 3.13 and Corollary 3.13.1 we deduce that a' is a unit in $\mathbf{Z}_{m'}$, so a'^{-1} exists. Hence, the preceding congruence has the solution

$$x' \equiv a'^{-1}\, b' \pmod{m'},\ 1 \leq a'^{-1}\, b' < m'.$$

The integers $x = a'^{-1}\, b' + km', k = 0, 1, 2, \cdots, d - 1$, are all solutions of $ax \equiv b \pmod{m}$, as can be seen by direct substitution. Also, no two of them are congruent modulo m, so they constitute a set of d incongruent solutions of $ax \equiv b \pmod{m}$. ◆

Theorem 3.22 is due to **Gauss**. In the case where m is a prime p, $p \nmid a$, and $b = 1$, Theorem 3.22 reduces to Corollary 3.12.1.

Example 3.33 Find all solutions of $8x \equiv 8 \pmod{28}$.

We have $d = (8, 28) = 4$ and $4|8$, so we first solve

$$2x' \equiv 2 \pmod{7}.$$

A nonnegative solution of this is $x' \equiv 1 \pmod{7}$. Now form the set $\{1 + 0 \cdot 7, 1 + 1 \cdot 7, 1 + 2 \cdot 7, 1 + 3 \cdot 7\}$. The complete set of solutions of $8x \equiv 8 \pmod{28}$ is therefore $x \equiv 1 \pmod{28}$, $x \equiv 8 \pmod{28}$, $x \equiv 15 \pmod{28}$, and $x \equiv 22 \pmod{28}$. ◆

We solved $2x' \equiv 2 \pmod{7}$ in Example 3.33 by inspection since it was an easy case. But suppose we needed to solve, instead, $113x \equiv 51 \pmod{173}$. This is not easy to do by inspection.

A systematic method was given in Example 3.32. We simply multiply both sides of $113x \equiv 51 \pmod{173}$ by $113^{\phi(173)-1}$ and then reduce modulo 173. Although straightforward, this method also requires a bit of labor. The following alternative method gives us an opportunity to look at continued fractions, an interesting topic in classical number theory.

In general, a **continued fraction** is any expression (finite or infinite) of the form (Olds, 1963)

$$d_0 + \cfrac{n_1}{d_1 + \cfrac{n_2}{d_2 + \cfrac{n_3}{d_3 + \cdots}}}$$

where n_i's and d_i's are real or complex numbers. This notation is cumbersome. A better one from the standpoint of printing is

$$d_0 + \frac{n_1}{d_1 +} \frac{n_2}{d_2 +} \frac{n_3}{d_3 +} \cdots.$$

When $n_1, n_2, n_3, \ldots$ are all 1's and $d_1, d_2, d_3, \ldots$ are all positive integers, the continued fraction is termed **simple**. For this special case a still briefer notation is $[d_0; d_1, d_2, d_3, \ldots]$. We note here that all the terms are followed by a comma, except the first term, d_0.

DEF. The elements of the sequence

$$\left\{d_0, d_0 + \frac{n_1}{d_1}, d_0 + \frac{n_1}{d_1 +} \frac{n_2}{d_2}, d_0 + \frac{n_1}{d_1 +} \frac{n_2}{d_2 +} \frac{n_3}{d_3}, \ldots\right\}$$

are called the **convergents** of order 0, 1, 2, 3, . . . of the continued fraction.

Example 3.34 Find the convergents to the simple continued fraction representation of the rational number 355/113. We have $d_0 = [355/113] = 3$. Then

$$\frac{355}{113} = 3 + \frac{16}{113} = 3 + \frac{1}{\frac{113}{16}} = 3 + \frac{1}{7 + \frac{1}{16}},$$

so

$$d_0 + \frac{n_1}{d_1} = 3 + \frac{1}{7} = \frac{22}{7}.$$

Finally, in the last step, we have simply

$$d_0 + \frac{n_1}{d_1 +} \frac{n_2}{d_2} = 3 + \frac{1}{7 +} \frac{1}{16} = 3\frac{16}{113} = \frac{335}{113}.$$

The first convergent is just the rough value for π that you learned in high school. The second-order convergent is a still better approximation, its value being 3.14159292. ◆

For present purposes we shall require only a few facts about simple continued fractions. The first will involve a convenient way to evaluate such fractions.

Theorem 3.23 Let the sequences $\{P_n\}$, $\{Q_n\}$ be recursively defined by

$$\begin{cases} P_0 = d_0 \\ P_1 = d_1 d_0 + 1 \\ P_n = d_n P_{n-1} + P_{n-2} \end{cases} \qquad \begin{cases} Q_0 = 1 \\ Q_1 = d_1 \\ Q_n = d_n Q_{n-1} + Q_{n-2} \end{cases}$$

$$(2 \leq n) \qquad\qquad (2 \leq n)$$

Then the convergents of the simple continued fraction are given by $[d_0; d_1, d_2, d_3, \ldots, d_n] = P_n/Q_n$.

Proof The proof is by induction. The theorem is true for $n = 0, 1$ since

$$P_0/Q_0 = d_0/1 = [d_0] \quad \text{and} \quad P_1/Q_1 = (d_1 d_0 + 1)/d_1 = d_0 + \frac{1}{d_1} = [d_0; d_1].$$

Assume the theorem is true for $n = k$ and that the continued fraction of interest has a defined convergent of order $k + 1$. Then, by definition,

$$\begin{aligned} \frac{P_{k+1}}{Q_{k+1}} &= \frac{d_{k+1}P_k + P_{k-1}}{d_{k+1}Q_k + Q_{k-1}} \\ &= \frac{d_{k+1}(d_k P_{k-1} + P_{k-2}) + P_{k-1}}{d_{k+1}(d_k Q_{k-1} + Q_{k-2}) + Q_{k-1}} \\ &= \frac{P_{k-1}(d_k d_{k+1} + 1) + d_{k+1}P_{k-2}}{Q_{k-1}(d_k d_{k+1} + 1) + d_{k+1}Q_{k-2}} \\ &= \frac{P_{k-1}(d_k + 1/d_{k+1}) + P_{k-2}}{Q_{k-1}(d_k + 1/d_{k+1}) + Q_{k-2}} \\ &= \left[d_0; d_1, d_2, \ldots, d_{k-1}, d_k + \frac{1}{d_{k+1}}\right] \end{aligned}$$

by the induction hypothesis. But from the expanded form of a simple continued fraction we see that

$$[d_0; d_1, d_2, \ldots, d_{k+1}] = \left[d_0; d_1, d_2, \ldots, d_{k-1}, d_k + \frac{1}{d_{k+1}}\right].$$

Hence, $P_{k+1}/Q_{k+1} = [d_0; d_1, d_2, \ldots, d_{k+1}]$, and the theorem follows by the Principle of Finite Induction. ◆

Theorem 3.23 provides a straightforward algorithmic procedure for calculating the convergents up to finite order of a simple continued fraction. It is easily programmed on a computer or a graphing calculator that has programming capabilities.

Example 3.35 The software system MAPLE contains the function *cfrac* in its number theory package. Applying this function to $\sqrt{3}$ we find that $\sqrt{3} = [1; 1, 2, 1, 2, 1, 2, \ldots]$. Find the sixth order convergent to $\sqrt{3}$.

Even though this continued fraction is infinite, we can still apply the preceding procedure for calculating the first few convergents. We compute, successively,

$$\begin{aligned}
P_0 &= 1 & Q_0 &= 1\\
P_1 &= 1(1) + 1 = 2 & Q_1 &= 1\\
P_2 &= 2(2) + 1 = 5 & Q_2 &= 2(1) + 1 = 3\\
P_3 &= 1(5) + 2 = 7 & Q_3 &= 1(3) + 1 = 4\\
P_4 &= 2(7) + 5 = 19 & Q_4 &= 2(4) + 3 = 11\\
P_5 &= 1(19) + 7 = 26 & Q_5 &= 1(11) + 4 = 15\\
P_6 &= 2(26) + 19 = 17 & Q_6 &= 2(15) + 11 = 41.
\end{aligned}$$

Hence, $\sqrt{3} \approx 71/41 \approx 1.7317$. ◆

Theorem 3.24 Any rational number can be represented by a finite simple continued fraction.

Proof Without loss of generality we consider only positive numbers x. If $x \in \mathbf{N}$, then $d_0 = x$ and this is the only element in the sequence of convergents. If $x \in \mathbf{Q}$ and is not integral, then $x = a/b = d_0 + r_0/b$ from the Division Algorithm, where $a, b, d_0, r_0 \in \mathbf{N}$ and $0 < r_0 < b$. Since $r_0 \neq 0$, then define $x_1 = b/r_0$, and again from the Division Algorithm we obtain, uniquely, $x_1 = d_1 + r_1/r_0$, $0 \leq r_1 < r_0$. If $r_1 = 0$, then $x = d_0 + r_0/b = d_0 + (b/r_0)^{-1} = d_0 + x_1^{-1} = d_0 + 1/d_1 = [d_0; d_1]$. If $r_1 \neq 0$, then set $x_2 = r_0/r_1$ and obtain $x_2 = d_2 + r_2/r_1$, $0 \leq r_2 < r_1$. This gives, if $r_2 = 0$, $x = d_0 + [d_1 + x_2^{-1}]^{-1} = [d_0; d_1, d_2]$. The sequence $\{r_0, r_1, r_2, \ldots\}$ is thus a strictly decreasing sequence of integers bounded below by 0. The sequence is therefore finite and converges to $r_n = 0$ for some $n \in \mathbf{N}$. Hence, we have $x = [d_0; d_1, d_2, \ldots, d_n]$. ◆

Converting a rational number to its finite simple continued fraction form is an easy task for today's graphing calculators. It merely requires a succession of keystrokes that alternate between subtraction and reciprocal. We illustrate this (on a

Texas Instruments calculator) by showing that the rational number 117/31 is represented by [3; 1, 3, 2, 3].

Key Strokes		Integer Portion	Continued Fraction
117/31	ENTER	3	[3]
Ans − 3	ENTER		
x^{-1}	ENTER	1	[3; 1]
Ans − 1	ENTER		
x^{-1}	ENTER	3	[3; 1, 3]
Ans − 3	ENTER		
x^{-1}	ENTER	2	[3; 1, 3, 2]
Ans − 2	ENTER		
x^{-1}	ENTER	3	[3; 1, 3, 2, 3]

Furthermore, this same method can be used if we wish to determine the first n terms in the expansion of a given irrational number.

You will recognize that the procedure outlined in Theorem 3.24 is identical to that described in the Euclidean Algorithm (Section 2.6). The greatest common divisor of a, b $(x = a/b)$ is simply the r_{n-1} we just noted.

Theorem 3.25 If $\{P_n\}$, $\{Q_n\}$ are defined as in Theorem 3.23, then

$$P_nQ_{n-1} - P_{n-1}Q_n = (-1)^{n-1} \quad \text{for all } n \geq 1.$$

Proof From the definitions of the P_n's and Q_n's, we have

$$\begin{aligned} P_nQ_{n-1} - P_{n-1}Q_n &= (d_nP_{n-1} + P_{n-2})Q_{n-1} - P_{n-1}(d_nQ_{n-1} + Q_{n-2}) \\ &= -(P_{n-1}Q_{n-2} - P_{n-2}Q_{n-1}). \end{aligned}$$

Let n be replaced by $n - 1$ on both sides; this gives

$$P_{n-1}Q_{n-2} - P_{n-2}Q_{n-1} = -(P_{n-2}Q_{n-3} - P_{n-3}Q_{n-2}),$$

so the first equation is now

$$P_nQ_{n-1} - P_{n-1}Q_n = (-1)^2(P_{n-2}Q_{n-3} - P_{n-3}Q_{n-2}).$$

Now continue this sequence of steps, replacing n successively by $n - 2, n - 3, \ldots, 2$. A total of $n - 1$ replacements are made in this way. At the last stage we obtain the equation

$$\begin{aligned} P_nQ_{n-1} - P_{n-1}Q_n &= (-1)^{n-1}(P_1Q_0 - P_0Q_1) \\ &= (-1)^{n-1}[(d_1d_0 + 1)1 - d_0d_1] \\ &= (-1)^{n-1}. \quad \blacklozenge \end{aligned}$$

Corollary 3.25.1 Let $(a, m) = 1$. Then the congruence $ax \equiv b \pmod{m}$ has the sole solution $x \equiv (-1)^N P_{N-1} b \pmod{m}$, where P_{N-1} is determined from $P_N/Q_N = m/a$.

Proof By Theorem 3.24, the rational number m/a is representable by a finite simple continued fraction $[d_0; d_1, d_2, \ldots, d_N]$. By Theorem 3.23 we can write $m/a = P_N/Q_N$. In Theorem 3.25 let $n = N$ there and write the result as

$$|P_N Q_{N-1} + Q_N(-P_{N-1})| = |(-1)^{N-1}| = 1.$$

By Theorem 2.12 this says $(P_N, Q_N) = 1$. But since $(m, a) = 1$ and $m/a = P_N/Q_N$, then $m = P_N$ and $a = Q_N$. Hence, $mQ_{N-1} - aP_{N-1} = (-1)^{N-1}$.
In congruence notation, this last relation is

$$aP_{N-1} \equiv (-1)^N \pmod{m},$$

or upon combining this with $b(-1)^N \equiv b(-1)^N \pmod{m}$,

$$ab(-1)^N P_{N-1} \equiv (-1)^{2N} b \pmod{m}.$$

Since $(-1)^{2N} = 1$, then the comparison with $ax \equiv b \pmod{m}$ yields $x \equiv (-1)^N bP_{N-1} \pmod{m}$. By Theorem 3.22 this is the only solution since $(a, m) = 1$. ◆

Example 3.36 Solve the nontrivial congruence $113x \equiv 51 \pmod{173}$ that was mentioned after Example 3.33.

The continued fraction expansion of $m/a = 173/113$ is found to be

$$\frac{173}{113} = [1; 1, 1, 7, 1, 1, 3], \qquad N = 6.$$

The sequence $\{P_n\}$ is then $P_0 = 1$, $P_1 = 1(1) + 1 = 2$, $P_2 = 1(2) + 1 = 3$, $P_3 = 7(3) + 2 = 23$, $P_4 = 1(23) + 3 = 26$, $P_5 = 1(26) + 23 = 49$, $P_6 = 3(49) + 26 = 173$. Hence, the solution is

$$\begin{aligned} x &\equiv (-1)^6(51)(49) \pmod{173} \\ &\equiv 77 \pmod{173}. \end{aligned}$$ ◆

The use of continued fractions to solve linear Diophantine equations, or what amounts to the same thing, linear congruences, dates back to the Indian mathematician **Aryabhata** (b. 476 A.D.) (Kline, 1972). His work was an advance beyond that of **Diophantus** since Aryabhata sought all integral solutions of an equation $ax + by = c$, whereas Diophantus was content with just one solution.

Problems

3.116. How many incongruent solutions are there of each of the following congruences? Deduce them.

(a) $4x \equiv 5 \pmod{8}$.
(b) $11x \equiv 3 \pmod{20}$.
(c) $15x \equiv 10 \pmod{35}$.
(d) $-9x \equiv 9 \pmod{18}$.
(e) $26x \equiv -30 \pmod{117}$.
(f) $64x \equiv 1 \pmod{211}$.

3.117. Work out the continued fraction expansions of each of the following rational numbers:

(a) 1.414.
(b) 103993/33102.
(c) 0.73205.
(d) 2.718281828.
(e) 11/20.
(f) 2/1717.

3.118. For part (b) of Problem 3.117 verify the content of Theorem 3.23.

3.119. Show that $[a_0; a_1, a_2, \ldots, a_{n-1}, a_n]$ and $[a_0; a_1, a_2, \ldots, a_{n-1}, a_n - 1, 1]$ represent the same number. Hence the continued fraction representation of a rational number is not unique.

3.120. Show that if P_k/Q_k is the kth-order convergent of a simple continued fraction, then $(P_k, Q_k) = 1$.

3.121. Show that if P_{k-1}/Q_{k-1}, P_k/Q_k are successive convergents of a simple continued fraction, then

$$\frac{P_k}{Q_k} = \frac{P_{k-1}}{Q_{k-1}} + \frac{(-1)^{k-1}}{Q_{k-1}Q_k}.$$

3.122. Write a computer program to convert a finite, simple continued fraction into a positive rational number, and then apply it to the following examples:

(a) $[2; 4, 4, 4, 4, 4, 4, 4]$.
(b) $[3; 7, 15, 1, 25, 1, 7, 4]$.

3.123. Consider the sequence of convergents $\{P_n/Q_n\}$ of a simple continued fraction.

(a) Prove that

$$\frac{P_n}{Q_n} - \frac{P_{n-2}}{Q_{n-2}} = \frac{(-1)^n d_n}{Q_{n-2}Q_n}.$$

[Hint: Proceed in a manner similar to Theorem 3.25, but with $P_nQ_{n-2} - P_{n-2}Q_n$.]

(b) Prove that the even convergents $\{P_{2k}/Q_{2k}\}$ are a strictly increasing sequence, and the odd convergents $\{P_{2k+1}/Q_{2k+1}\}$ are a strictly decreasing sequence.

3.12 Cryptography and the RSA Method*

We conclude this chapter with a very interesting application of the mathematics of congruences. Cryptography, or the science of "secret writing," has been in existence for well over 2000 years. What originally had been used primarily by the military and governments has now become important for businesses and individuals who wish to safeguard professional and personal electronic data.

We start by going back to the time of **Julius Caesar**, who used a simple method of **enciphering**, or **encrypting**, to communicate with his generals during the Gallic Wars (58-50 B.C.). Caesar would replace each letter in the original form of the message, known as the **plaintext**, by the letter a fixed distance d (called the **shift constant**) later in the alphabet. This resultant message, which appears as gibberish to all who read it, is what is known as the **ciphertext**.

Example 3.37 Caesar wanted to send the message ATTACK NOW, disguised with a shift constant $d = 5$. The ciphertext message received by his generals was FYYFHP STB. ◆

In order to more fully disguise the preceding message, there would be no gaps, or blank spaces, between the words in the ciphertext. This would prohibit one from determining the lengths of the words.

The Caesar cipherment can be quantified, and hence computerized, by associating different numbers with different letters. In particular, if we make the correspondences $A \longleftrightarrow 00$, $B \longleftrightarrow 01$, $C \longleftrightarrow 02$, . . . , $Y \longleftrightarrow 24$, $Z \longleftrightarrow 25$, then the plaintext message ATTACK NOW is represented by the string of numbers

$$001919000210131422.$$

If P and C denote the numerical value of each plaintext and ciphertext letter, respectively, then the previous encryption can be denoted

$$C \equiv P + 5 \pmod{26}$$

or equivalently

$$P \equiv C - 5 \pmod{26}.$$

Example 3.38 With a shift constant of $d = 17$, decipher the message whose ciphertext is 2421020205.

* Sections marked with an asterisk can be skipped without impairing the continuity of the text.

The congruence $P \equiv C - 17 \pmod{26}$ gives the five values for P as

$$P = 24 - 17 \equiv 7 \pmod{26}$$

$$P = 21 - 17 \equiv 4 \pmod{26}$$

$$P = 2 - 17 = -15 \equiv 11 \pmod{26}$$

$$P = 2 - 17 \equiv 11 \pmod{26}$$

$$P = 5 - 17 \equiv 14 \pmod{26}$$

so the plaintext, in numerical form, is 0704111114, which corresponds to the message HELLO. ◆

Clearly the Caesar encryption method is relatively simple, although we could alter the scheme and make it more complex by changing the correspondence between the alphabet letters and their associated numbers. Actually, this correspondence, together with the shift constant, constitute the real "keys" to the encryption. Knowledge of these two pieces of information allows one to **decode**, **decipher**, or **decrypt** any such message.

Another variation of the Caesar encryption method would be to introduce a **multiplier constant** m, and then define C by

$$C \equiv mP + d \pmod{26}.$$

This more general Caesar transformation maps P to C in much the same way that a linear function maps x to y.

Example 3.39 We wish to encipher ATTACK NOW with a shift constant $d = 2$, a multiplier constant $m = 3$, and the correspondence $A \longleftrightarrow 00$, $B \longleftrightarrow 01, \ldots, Z \longleftrightarrow 25$.

letter	A	T	C	K	N	0	W
P	00	19	02	10	13	14	22
$3P + 2 \bmod 26$	02	07	08	06	15	18	16

Thus, the ciphertext in numerical form is

$$02 \quad 07 \quad 07 \quad 02 \quad 08 \quad 06 \quad 15 \quad 18 \quad 16,$$

while in alphanumeric form it reads CHHCIGPSQ. ◆

To reverse the process and decipher the message, we solve the congruence for P, and get

$$P \equiv m^{-1}(C - d) \pmod{26}$$

where m^{-1} is the integer (if it exists) that represents the inverse of m, and hence it is the solution to

$$mx \equiv 1 \pmod{26}.$$

This congruence has a solution when $(m, 26) = 1$ (Thorem 3.22), and this implies that $mP + d$ runs through a complete residue system modulo 26 (Theorem 3.14). Thus, C can assume all integral values in the interval [0, 25]. Note that if $m = 2$ and $d = 4$, for example, then $(m, 26) \neq 1$ and both $P = 0$ and $P = 13$ map to $C = 4$. This means that the letters A and N both map to the letter E. This is an unpleasant situation since it would be impossible to recover the exact plaintext letter when the cipher letter is E.

Example 3.40 Suppose we wish to decipher the message

$$001711011403$$

that had been enciphered with $d = 1$, $m = 3$.

Solving $3x \equiv 1 \pmod{26}$ gives $x = 9$, so the deciphering transformation is $P \equiv 9(C - 1) \pmod{26}$. So now the plaintext is 171412001318, which translates as ROMANS. ◆

It may prove desirable to include characters other than just letters. Say, for instance, we would like to include a blank space that can separate words. One way to accommodate this is to assign the number 00 to the blank and readjust the remaining letters A ⟷ 01, B ⟷ 02, . . . , Z ⟷ 26. Because there are 27 characters in use, the Caesar cipher now takes the form $C \equiv mP + d \pmod{27}$, and we should require $(m, 27) = 1$.

The most undesirable feature of the Caesar ciphers is the 1-to-1 correspondence between letters and numbers. This makes it easy for cryptanalysts to break the code because the numbers that occur most often probably correspond to the most frequently used letters, such as E or T. With minimal trial and error the message can then be deciphered.

This 1-to-1 correspondence can be eliminated by matching up blocks of letters with strings of digits. Consider, for instance, blocks of length two. Each separate block of two plaintext numbers P_1, P_2 gets mapped to the string $C_1 C_2$ where

$$C_1 \equiv aP_1 + bP_2 \pmod{26}$$

$$C_2 \equiv cP_1 + dP_2 \pmod{26}$$

and a, b, c, d are four given integers. This pair of congruences is equivalent to the single matrix congruence

$$\begin{bmatrix} C_1 \\ C_2 \end{bmatrix} \equiv \begin{bmatrix} a & b \\ c & d \end{bmatrix} \begin{bmatrix} p_1 \\ p_2 \end{bmatrix} \pmod{26}.$$

Consequently, if we want to retrieve P_1, P_2, given C_1, C_2, we have to invert the matrix and obtain

$$\begin{bmatrix} P_1 \\ P_2 \end{bmatrix} \equiv e^{-1} \begin{bmatrix} d & -b \\ -c & a \end{bmatrix} \begin{bmatrix} C_1 \\ C_2 \end{bmatrix} \pmod{26},$$

where e^{-1} is the integer solution to $(ad - bc)x \equiv 1 \pmod{26}$.

Example 3.41 The message JULIUS, represented by 092011082018, is to be enciphered with the "keys" $a = 2$, $b = 1$, $c = 1$, $d = 3$. Starting then with $P_1 = 9$, $P_2 = 20$, we find $C_1 = 12$ and $C_2 = 17$. Next, $P_1 = 11$ and $P_2 = 8$ gives $C_1 = 4$, $C_2 = 9$, while $P_1 = 20$, $P_2 = 18$ yield $C_1 = 64$, $C_2 = 22$. So the ciphertext is 121704090622, or MREJGW. It is of interest to note here that the plaintext message contains two identical letters (U), but the letters of the ciphertext are distinct. ◆

In matrix notation, the enciphering of the previous example would be written

$$\begin{bmatrix} C_1 \\ C_2 \end{bmatrix} \equiv \begin{bmatrix} 2 & 1 \\ 1 & 3 \end{bmatrix} \begin{bmatrix} P_1 \\ P_2 \end{bmatrix} \pmod{26}$$

and so the deciphering would be

$$\begin{bmatrix} P_1 \\ P_2 \end{bmatrix} \equiv e^{-1} \begin{bmatrix} 3 & -1 \\ -1 & 2 \end{bmatrix} \begin{bmatrix} C_1 \\ C_2 \end{bmatrix} \pmod{26}$$

where e^{-1} is the integer solution to $5x \equiv 1 \pmod{26}$, or $x \equiv 21 \pmod{26}$. Hence,

$$\begin{bmatrix} P_1 \\ P_2 \end{bmatrix} \equiv \begin{bmatrix} 63 & -21 \\ -21 & 42 \end{bmatrix} \begin{bmatrix} C_1 \\ C_2 \end{bmatrix} \equiv \begin{bmatrix} 11 & 5 \\ 5 & 16 \end{bmatrix} \begin{bmatrix} C_1 \\ C_2 \end{bmatrix} \pmod{26},$$

so the pair of congruences

$$P_1 \equiv 11C_1 + 5C_2 \pmod{26}$$

$$P_2 \equiv 5C_1 + 16C_2 \pmod{26}$$

serves to decipher the message. In particular, if $C_1 = 12$ and $C_2 = 17$, then $P_1 = 11(12) + 5(17) = 217 \equiv 9 \pmod{26}$ and $P_2 = 5(12) + 16(17) \equiv 20 \pmod{26}$, and P_1, P_2 correspond to the first two letters in JULIUS.

If it is not necessary to express the ciphertext with letters, then the string of digits would be sufficient. In this case, reduction modulo 26 can essentially be eliminated. Instead we could just write the ciphertext as

$$\begin{bmatrix} C_1 \\ C_2 \end{bmatrix} = \begin{bmatrix} a & b \\ c & d \end{bmatrix} \begin{bmatrix} P_1 \\ P_2 \end{bmatrix}$$

and the only stipulation here is that $ad \neq bc$ so that the 2×2 matrix can be inverted.

Example 3.42 The matrix $\begin{bmatrix} 2 & 1 \\ 4 & 3 \end{bmatrix}$ is the "key" that was used to encipher an unknown message. The ciphertext, which allows for blank spaces, is represented by

$$51 \quad 117 \quad 31 \quad 67 \quad 12 \quad 36 \quad 40 \quad 102 \quad 29 \quad 77.$$

To decipher, we compute the inverse of the matrix and multiply by the ciphertext,

$$(1/2)\begin{bmatrix} 3 & -1 \\ -4 & 2 \end{bmatrix}\begin{bmatrix} 51 & 31 & 12 & 40 & 29 \\ 117 & 67 & 36 & 102 & 77 \end{bmatrix} = \begin{bmatrix} 18 & 13 & 0 & 9 & 5 \\ 15 & 5 & 12 & 22 & 19 \end{bmatrix}$$

which gives the plaintext

$$18 \quad 15 \quad 13 \quad 5 \quad 0 \quad 12 \quad 9 \quad 22 \quad 5 \quad 19$$

and an original message of ROME LIVES. ◆

All of these encryption schemes share several significant characteristics. First, given enough time, they can all be broken (i.e., the keys can be deduced) with the aid of today's powerful computers. Second, the key needs to be transmitted to the recipient, and therein lies the risk of its being confiscated.

During World War II the German and Japanese military each made use of a cipher machine called the Enigma. German aggression against the British was hindered when the latter were finally able to crack the code for the Enigma that the German military used. This happened because the British had been fortunate enough to obtain one of the Enigma machines, and had a group of scientists study it. The group, led by the mathematician–computer scientist **Alan Turing** (1913–1954), studied the machine for over a year before they determined how it worked. An historical account of this appears in the film and book *The Codebreakers* (Kahn, 1967).

Major advances in the complexity of encryption schemes began around 1977, thanks in part to three mathematicians at the Massachusetts Institute of Technology (**Ronald Rivest**, **Adi Shamir**, and **Leonard Adleman**) and to an important result of Euler's (Theorem 3.21). The method developed by Rivest, Shamir, and Adleman, known as the RSA method, employs a simple algorithm to form the ciphertext because it just requires numbers to be multiplied together, an easy task for a computer. The secret to deciphering, though, lies in factoring, and this is the real rub because factoring is *very* difficult and time-consuming!

Because of the importance of the method, MIT actually patented the formulas. Shortly thereafter, the university licensed the patent to a newly formed company, RSA Data Security. Located in Redwood City, California, the company can be found on the Internet at www.rsa.com, where one can read, among other things, about the latest-breaking news in cryptography. The company provides encryption security for its clients, which include such well-known names as Apple, Microsoft, Motorola, Lotus, and Visa International.

Around 1976 two scientists at Stanford University, **Martin Hellman** and **Whitfield Diffie**, discovered and published a different system of cryptography that shares some intricate properties of the "keys" structure of the RSA method. Stanford, like MIT, patented the method and then licensed it out to a young company, Cylink Corporation, located in Sunnyvale, California, which happens to be just a short distance down the road from Redwood City, on the outskirts of San Jose. Cylink and RSA Data Security are major rivals in the modern business of providing electronic security.

Let's discuss now how the RSA encryption method works. Two large, odd and distinct primes p, q are selected, and we set $n = pq$. These primes should each have well over 100 digits if the code is to be virtually "unbreakable." Recalling the Euler phi-function, we now choose an odd integer k that is relatively prime to $\phi(n) = (p-1)(q-1)$. Finally, to convert a message from plaintext to ciphertext, we set up the congruence

$$C \equiv \mathrm{M}^k \pmod{n},$$

where M represents a typical block of plaintext of some constant length, with $\mathrm{M} < n$.

Example 3.43 Suppose we wish to encipher HI CLASS with $p = 73$, $q = 89$, and $k = 5$. These data are allowable since 73 and 89 are primes, and $(5, 6336) = 1$. Making use of the blank space in the message, the plaintext now reads

$$08 \quad 09 \quad 00 \quad 03 \quad 12 \quad 01 \quad 19 \quad 19.$$

If we let M be blocks of length 2, the ciphertext is found by

$$C \equiv \mathrm{M}^5 \pmod{6497}.$$

Hence, the string of values for C is

$$8^5 \pmod{6497},\ 9^5 \pmod{6497},\ \ldots,\ 19^5 \pmod{6497}$$

or

$$283 \quad 576 \quad 000 \quad 243 \quad 1946 \quad 001 \quad 742 \quad 742.$$

If we regroup the plaintext into blocks of length 3, and add a blank at the right end, then the string of values for C is

$$80^5 \pmod{6497},\ 900^5 \pmod{6497},\ \ldots,\ 900^5 \pmod{6497}$$

or

$$5565 \quad 587 \quad 3369 \quad 1643 \quad 4346 \quad 587.$$

As long as we know that each of the preceding integers is at most four digits long, then the ciphertext could be written as a continuous string of digits

$$5565058733691643434 60587. \quad \blacklozenge$$

The real beauty of the RSA method comes in the deciphering. Starting with the congruence

$$C \equiv \mathrm{M}^k \pmod{n}$$

what needs to be done is to compute the kth root of both sides, so that we can retrieve M by

$$\mathrm{M} \equiv C^{1/k} \pmod{n}.$$

How we do this is quite remarkable, for it involves elementary number theory.

Because $(k, \phi(n)) = 1$ there exists a solution to

$$kx \equiv 1 \pmod{\phi(n)},$$

and we can choose this solution, call it j, to lie in the interval $[1, \phi(n))$. This means $kj = 1 + z\phi(n)$ for some integer z. So now, if each side of the congruence

$$C \equiv \mathrm{M}^k \pmod{n}$$

is raised to the jth power, we get

$$\begin{aligned} C^j \equiv \mathrm{M}^{kj} \equiv \mathrm{M}^{1+z\phi(n)} &\equiv \mathrm{M} \cdot (\mathrm{M}^{\phi(n)})^z \\ &\equiv \mathrm{M} \cdot 1^z \\ &\equiv \mathrm{M} \pmod{n} \end{aligned}$$

where we note that $\mathrm{M}^{\phi(n)} \equiv 1 \pmod{n}$ is a direct application of Euler's Theorem. We have therefore retrieved our plaintext message M, so raising C to the jth power acts like a kth root. We do need to check, though, whether $(\mathrm{M}, n) = 1$ so that Euler's Theorem can be applied. Since $\mathrm{M} < n$ it is only a question of whether $(\mathrm{M}, p) = 1$ and $(\mathrm{M}, q) = 1$. These conditions unfortunately, may not always be true, but will frequently be so.

Example 3.44 Let us decipher the first letter in the ciphertext from the previous example. To find the critical value j, we need to solve $5x \equiv 1 \pmod{6336}$. Applying the Euclidean Algorithm as in Example 2.19, we obtain $x = -1267$, so then $j = -1267 + 6336 = 5069$. With the value $C = 283$ we then get

$$\mathrm{M} \equiv 283^{5069} \equiv 8 \pmod{6497},$$

which is the desired value, and this corresponds to the letter H. Clearly this illustrates one important reason for being able to efficiently reduce powers modulo m. ◆

The RSA encryption method is commonly known as a public-key cryptosystem. The reason for this is that an individual user (who expects to receive important messages, and wants the information secure) will publish, or make available to the public, the two integral values n, k that are necessary for enciphering. These two values are known as the **public keys**. The user can then decipher the message because he or she knows the other pertinent information, namely p, q, and hence, $\phi(n)$, j. These latter constitute the **private keys**. The system is as secure as the difficulty involved in factoring n, because if one knows $n = pq$, then one knows $\phi(n)$, and hence j. The ability to perform this factoring is thus a function of the power and speed of the available machines, and the algorithms used for factoring. Currently, it is generally agreed that to make the system as secure as possible, each of p and q should have about the same number of digits (over 100 each), and that $p - 1$ and $q - 1$ have large prime factors and share only small factors, if any at all.

There are other cryptosystems besides RSA's public key system. Some popular systems are **DES** (the federal Data Encryption Standard), **RC-4** (also devised by RSA Data Security), Ralph Merkle's **Khufu**, Bruce Schneier's **Blowfish**, and the classified U.S. government scheme **Skipjack**. Most likely, no matter what system you use, you'll soon need a better one.

Problems

3.124. Decipher the message CVUUBKBUUH that had been encrypted with a Caesar cipher with multiplier constant = 11 and shift constant = 3. Assume that blank spaces are ignored.

3.125. Decipher the message (blanks are assigned 00)

84 49 43 28 51 28 40 20 34 21 81 48 75 50

that had been coded with the matrix

$$A = \begin{bmatrix} 3 & 2 \\ 2 & 1 \end{bmatrix}.$$

3.126. (a) Encipher PAUL ERDOS using the congruences (blanks $\longleftrightarrow$ 00)

$$\begin{cases} C_1 \equiv 2P_1 + 3P_2 \pmod{27} \\ C_2 \equiv 5P_1 + 2P_2 \pmod{27}. \end{cases}$$

(b) Check your answer to part (a) by deciphering it, thereby, obtain the original message.

3.127. (a) Express NUMBER into blocks M of three digits each, and then encipher with an RSA cipher using $p = 31$, $q = 59$, and $k = 7$.

(b) When you decipher the message in part (a), what is the value of the recovery exponent j?

3.128. Your public keys are $n = 391$, $k = 5$ and your private keys are $p = 17$, $q = 23$. You received the message

$$242 \quad 388 \quad 199 \quad 243 \quad 388.$$

What do you think it says?

3.129. If you are using the RSA encryption scheme, and the enemy obtains the value of $\phi(n)$, then your code will be broken instantly. Show that

(a) $p + q = n + 1 - \phi(n)$.
(b) If $p > q$, then $p - q = \sqrt{(p + q)^2 - 4n}$.
(c) Find the two primes p, q if $n = 50{,}439{,}919$ and $\phi(n) = 50{,}425{,}696$.

RESEARCH PROBLEMS

1. Consider the congruence $ax^2 \equiv b \pmod{m}$. When does this have solutions x? Investigate this question by looking at several combinations of a, b, m. See if it is possible to reformulate Theorem 3.22 so as to apply to the congruence $ax^2 \equiv b \pmod{m}$.

2. Lists of pseudoprimes to various bases would be very interesting for further study. Investigate methods by which you could calculate on the computer the first 100 or so odd pseudoprimes to a given base a; consider cases where a is prime. Use your lists to look at questions such as these:

(a) Are there any twin psp(a)'s, that is, pairs n, $n + 2$ of pseudoprimes to the base a?
(b) Are there any psp(a)'s of the form $n = p(p + 2)$, where p, $p + 2$ are primes?
(c) Are there any psp(2)'s of the form $p - 1$, where p is a prime?

3. Table I(6) in Appendix II gives a short list of Carmichael numbers. It is of interest to look at their arithmetic structure. Investigate this on the computer by factoring the entries in the table. If possible, try to prove any pattern that you observe.

References

Brillhart, J., "Derrick Henry Lehmer," *Acta Arithmetica*, **62**, 207–220 (1992). A wonderful synopsis of the life and work of this American giant of number theory.

Bühler, W.K., *Gauss: A Biographical Study*, Springer, New York, 1981. A highly recommended treatment of this immortal mathematician by the late **Walter Bühler** (1944–1986); the book contains more actual mathematics in it than did older biographies.

Dence, T.P., "The Digital Root Function," *Two-Year College Mathematics Readings*, Mathematical Association of America, Washington, DC, 1982, pp. 96–103. More material than in our Problem 3.22 on this simple but interesting function can be found here.

Fletcher, C.R., "Fermat's Theorem," *Historia Math.*, **16**, 149–153 (1989). The author gives a slightly different interpretation of how **Fermat** discovered his theorem.

Golomb, S.W., "Combinatorial Proof of Fermat's 'Little' Theorem," *Amer. Math. Monthly*, **63**, 718 (1956). Who says that manipulatives have no place in advanced mathematics?

Granville, A., "Primality Testing and Carmichael Numbers," *Notices Amer. Math. Soc.*, **39**, 696–700 (1992). Exciting account of contemporary research that solved an 80-year-old problem.

Kahn, D., *The Codebreakers*, Macmillan, New York, 1967. An interesting account of cryptography during World War II. Excerpts of this have appeared as a television documentary presented by NOVA.

Kline, M., *Mathematical Thought from Ancient to Modern Times*, Oxford University Press, New York, 1972, pp. 184–188. Summary of Hindu arithmetic and algebra from 200–1200 A.D., which time period includes **Aryabhata**, **Brahmagupta**, and **Bhaskara**.

Kolata, G., "Perfect Shuffles and Their Relation to Math," *Science*, **216**, 505–506 (1982). Find out how card shuffling is an entry into some advanced mathematics.

Mahoney, M.S., *The Mathematical Career of Pierre de Fermat*, 2nd ed., Princeton University Press, Princeton, NJ, 1994. A detailed look at the work of **Fermat** in algebra, number theory, analytic geometry, and early calculus.

McDaniel, W.L., "Some Pseudoprimes and Related Numbers Having Special Forms," *Math. Comp.*, **53**, 407–409 (1989). Note that since $n = 2^{465,794} - 2$ is even (and hence composite), we must use the definition that n is a pseudoprime if it divides $2^n - 2$.

Olds, C.D., *Continued Fractions*, Mathematical Association of America, Washington, DC, 1963. Wonderful little (150-page) book that you can read from cover to cover; keep it under your pillow.

Pinch, R.G.E., "Some Primality Testing Algorithms," *Notices Amer. Math. Soc.*, **40**, 1203–1210 (1993). Focuses on where some popular primality testing algorithms break down.

Pomerance, C., Selfridge, J.L., and Wagstaff, Jr., S.S., "The Pseudoprimes to 25×10^9," *Math. Comp.*, **35**, 1003–1026 (1980). Packed with great stuff to challenge you; the source of our Problem 3.107.

Rotkiewicz, A., *Pseudoprime Numbers and Their Generalizations*, University of Novi Sad, Warsaw, 1972. This book is quite scarce, but is filled with interesting theorems, many of them due to the author himself.

Sierpiński, W., *A Selection of Problems in the Theory of Numbers*, Macmillan, New York, 1964, p. 52. The source of our Problem 3.97, but have a look at several of the other problems in this superb little book.

Silverman, J.H., *A Friendly Introduction to Number Theory*, Prentice–Hall, Englewood Cliffs, NJ, 1997, pp. 92–97. A brief discussion of the Method of Successive Squaring.

Chapter 4
Polynomial Congruences

4.1 Introduction to Polynomial Congruences

We saw in Theorem 3.22 the full statement of when a linear congruence in one unknown is solvable and, if solvable, of how many incongruent solutions it will possess. That statement contrasts with an analogous statement for the solution of a linear equation in one unknown: if $a, b \in \mathbf{Q}$ and $a \neq 0$, then $ax = b$ has exactly one solution in $\mathbf{Q}$.

The expression $\Sigma_{k=0}^{n} c_k x^k$, $c_n \neq 0$, is termed a **polynomial of degree *n***[1] in the indeterminate x. In this chapter we look at polynomial congruences of degree greater than 1. Table 4.1 shows some interesting examples.

Obviously, the relation between the algebraic degree and the number of incongruent solutions is complex for congruences. We shall develop in this chapter only a small fraction of the theory pertaining to the solution of polynomial congruences (Apostol, 1976).

DEF. Let $f(x) = \Sigma_{k=0}^{n} c_k x^k$ have integral coefficients. Then x_0 is said to be a **root** of the congruence $f(x) \equiv 0 \pmod{m}$ if $f(x_0) \equiv 0 \pmod{m}$. Congruent roots are equivalent, so if $f(x) \equiv 0 \pmod{m}$ has r roots, this means that there are r incongruent numbers $\{x_1, x_2, \ldots, x_r\}$, each of which is a root of $f(x) \equiv 0 \pmod{m}$.

[1] Nonzero, constant polynomials have degree 0; the degree of the zero polynomial is left undefined.

Table 4.1 Solutions of Polynomial Congruences of the Form $f(x) \equiv 0 \pmod{m}$

$f(x)$	m	Incongruent Solutions	No. of Solutions in $\mathbf{Q}$ of $f(x) = 0$	No. of Solutions in $\mathscr{C}$ of $f(x) = 0$
$x^4 - 1$	5	1, 2, 3, 4	2	4
$x^3 - 4$	4	0, 2	0	3
$x^2 + x$	6	0, 2, 3, 5	2	2

Two polynomials, $f(x) = \Sigma_{k=0}^{n} c_k x^k$ and $g(x) = \Sigma_{k=0}^{m} b_k x^k$, are said to be identical if $m = n$ and $c_k = b_k$ for $k = 0, 1, 2, \ldots, n$. We make a similar definition for congruences.

DEF. Two polynomials, $f(x) = \Sigma_{k=0}^{n} c_k x^k$ and $g(x) = \Sigma_{k=0}^{n} b_k x^k$, are said to be **identically congruent** modulo m if $c_k \equiv b_k \pmod{m}$ for $k = 0, 1, 2, \ldots, n$.

If two polynomials are identically congruent modulo m, then we can write $f(x) \equiv g(x) \pmod{m}$. Thus, $3x^2 + 2x + 1 \equiv x^2 - 1 \pmod{2}$ is an identical congruence but $x^5 \equiv x \pmod{5}$ is not, even though this congruence is true for all $x \in \mathbf{Z}$ (why?). The reader will now realize that a second usage of the symbol $\equiv$ has been quietly introduced. On the one hand, when discussing the roots of a congruence $f(x) \equiv 0 \pmod{m}$, we naturally intend a congruence relation between numbers (certain values of $f(x)$ and 0). On the other hand, when no assertion about roots is intended or when there is no explicit mention of numbers, then $f(x) \equiv g(x) \pmod{m}$ means an identical congruence. Thus, under these conditions, if $f(x) \equiv 0 \pmod{m}$ is written, then it is meant that each coefficient in the polynomial representation of $f(x)$ is congruent to 0 modulo m.[2]

Example 4.1 $3x^2 - x - 5 \equiv -x^2 - x - 1 \pmod{4}$, because $3 \equiv -1 \pmod{4}$, $-1 \equiv -1 \pmod{4}$, and $-5 \equiv -1 \pmod{4}$. Similarly, $12x^3 - 18 \equiv 0 \pmod{6}$. ◆

[2] The equality sign often does double duty too. In "solve $x^3 + x + 1 = 0$" or in "solve $x^2 + 2x = 5x - 7$" we intend an *equality of numbers* among certain values of the left-hand and right-hand sides. But in "consider $x^2 - 1 = (x + 1)(x - 1)$," finding a root is not intended; the use of "=" here is to indicate an *identity*.

In continuing here the theme of the parallelism between congruence and equality, we prove as our first theorem a result that extends the Division Algorithm (Theorem 1.3) to other rings besides $\mathbf{Z}$. The theorem is suggested by the following sort of elementary operation from college algebra.

Example 4.2 Let $f(x) = -2x^3 + x + 6$ and $g(x) = x^2 + 2x + 2$. Then long division of $f(x)$ by $g(x)$ gives

$$
\begin{array}{r|l}
 & -2x \quad + 4 \\
\hline
x^2 + 2x + 2 & -2x^3 + 0x^2 + x + 6 \\
 & -2x^3 - 4x^2 - 4x \\
\hline
 & \qquad 4x^2 + 5x + 6 \\
 & \qquad 4x^2 + 8x + 8 \\
\hline
 & \qquad\qquad -3x - 2
\end{array}
$$

So we have, uniquely, $f(x) = g(x)(-2x + 4) + (-3x - 2)$. ◆

Theorem 4.1 (Division Algorithm for $\mathbf{R}[x]$). Let $\mathbf{R}$ be a commutative ring with identity and let $\mathbf{R}[x]$ be the set of all polynomials in x with coefficients in $\mathbf{R}$. Let $f(x),\ g(x) \in \mathbf{R}[x]$ be of degrees n, m, respectively,

$$
f(x) = \sum_{k=0}^{n} a_k x^k \quad (a_n \neq 0) \qquad \text{and} \qquad g(x) = \sum_{k=0}^{m} b_k x^k, \quad (b_m \neq 0,\ m > 0)
$$

with b_m a unit in $\mathbf{R}$. Then there are unique elements $q(x),\ r(x) \in \mathbf{R}[x]$ such that $f(x) = g(x)q(x) + r(x)$, and either $r(x)$ is identically 0 or degree of $r(x) < m$.

Proof Let polynomials $f(x)$, $g(x)$ be given, as described in the theorem. It may happen that there is some polynomial $h(x)$ such that $f(x) = g(x)h(x)$. This polynomial is unique, because if $h'(x) \neq h(x)$ existed and $f(x) = g(x)h'(x)$, then subtraction of the two equations would give $0 = g(x)[h(x) - h'(x)]$. A contradiction is now reached if one considers the degree on both sides. Hence, the required $q(x)$ of the theorem is the unique $h(x)$, and the required $r(x)$ is identically 0.

If no $h(x)$ exists such that $f(x) = g(x)h(x)$, then consider the set

$$
\mathbf{S} = \{f(x) - g(x)h(x) : h(x) \in \mathbf{R}[x]\}.
$$

Every element in $\mathbf{S}$ has degree 0 or greater. By Theorem 1.2 there is at least one element in $\mathbf{S}$ of minimal degree. Let $r(x)$ be one such element; assume that its degree is $N \geq m$, where m is the degree of $g(x) = \Sigma_{k=0}^{m} b_k x^k$. Our object is to reach a contradiction by discovering another element in $\mathbf{S}$ whose degree is less than that of $r(x)$.

Write $r(x) = \Sigma_{k=0}^{N} c_k x^k$; because $r(x) \in \mathbf{S}$, we can also express it as $r(x) = f(x) - g(x)h(x)$, for some $h(x) \in \mathbf{R}[x]$. Now consider the polynomial $P(x) = r(x) - (c_N/b_m)x^{N-m}g(x)$, this being possible because b_m is a unit in $\mathbf{R}$. The polynomial $P(x)$ resides in $\mathbf{S}$ because it can be reexpressed as

$$\begin{aligned} P(x) &= f(x) - g(x)h(x) - \left(\frac{c_N}{b_m}\right) x^{N-m} g(x) \\ &= f(x) - g(x)\left[h(x) + \left(\frac{c_N}{b_m}\right) x^{N-m}\right]. \end{aligned}$$

However, $P(x) = r(x) - (c_N/b_m)x^{N-m}g(x)$ is now of degree $N - 1$ at most, because the leading term of $r(x)$ is $c_N x^N$ and the leading term of $(c_N/b_m)x^{N-m}g(x)$ is $(c_N/b_m)x^{N-m}(b_m x^m) = c_N x^N$. The two leading terms cancel in the expression for $r(x)$. Thus, we have a contradiction, and so degree $r(x) \geq m$ is false.

It follows that $h(x)$ is the $q(x)$ of the theorem and $r(x)$ is the desired remainder in the theorem provided we can show that they are unique. To show this, assume there is another pair of polynomials $q'(x)$, $r'(x)$ such that $f(x) = g(x)q'(x) + r'(x)$. We leave it to you to show that $q'(x) = q(x)$, $r'(x) = r(x)$ (Problem 4.5). ◆

Notice that Theorem 4.1 applies to a fairly broad class of rings $\mathbf{R}$, namely, commutative ones that have some units. The price we pay for this broadness is that one can only consider division by polynomials whose leading coefficient is one of the units. In the ring $\mathbf{R} = \mathbf{Z}$, for example, this means the units are either 1 or -1. Of course, if $\mathbf{R}$ is actually a field $\mathbf{F}$, then one can consider division by polynomials whose leading coefficient is any nonzero element of $\mathbf{F}$, because any such element is a unit.

Example 4.3 Let $\mathbf{R} = \mathbf{Z}_6$, $f(x) = 2x^4 + 3x^3 + 5x + 1$, $g(x) = 5x^2 + 4x + 2$. Modular division gives $f(x) = g(x)(4x^2 + x) + (3x + 1)$, uniquely (verify!). ◆

Corollary 4.1.1 Let $\mathbf{R}$ be a commutative ring with identity. An element $a \in \mathbf{R}$ is a zero of $f(x) \in \mathbf{R}[x]$ iff $(x - a)$ is a factor of $f(x)$ in $\mathbf{R}[x]$.[3]

Proof $(\rightarrow)$ By Theorem 4.1 there are unique elements $q(x)$, $r(x) \in \mathbf{R}[x]$ such that

$$f(x) = (x - a)q(x) + r(x),$$

[3] As a minor point of terminology, we say that equations have **roots**, but polynomials (or functions, more generally) have **zeros**.

and either the degree of $r(x)$ is less than that of $x - a$, or $r(x)$ is identically 0. Hence, $r(x)$ is a constant K.

$$f(x) = (x - a)q(x) + K$$

If a is a zero of $f(x)$, then $f(a) = 0$ and the preceding equation becomes $0 = (a - a)q(a) + K$, or $K = 0$, so $(x - a) | f(x)$.

($\leftarrow$) If $x - a$ is a factor of $f(x)$ in $\mathbf{R}[x]$, then we have

$$f(x) = (x - a)q(x),$$

for some $q(x) \in \mathbf{R}[x]$. Evaluating both sides at $x = a$, we obtain $f(a) = (a - a)q(a) = 0$, so a is a zero of $f(x)$. ◆

The corollary is a generalization of the **Factor Theorem** (Problem 4.9), due originally to **René Descartes** (1596–1650). A special case of Corollary 4.1.1 is where the commutative ring with identity, **R**, is a complete residue system modulo m. We have immediately Corollary 4.1.2.

Corollary 4.1.2 $f(a) \equiv 0 \pmod{m}$ iff $x - a$ divides $f(x)$ modulo m, that is, iff there is an element $q(x) \in \mathbf{R}[x]$ such that $(x - a)q(x)$ is identically congruent to $f(x)$ modulo m.

Example 4.4 Let $f(x) = 2x^2 + 3x + 3$ and $m = 6$. We find that $f(3) \equiv 0 \pmod{6}$. Hence, from Corollary 4.1.2 we have that $x - 3$ divides $2x^2 + 3x + 3$ (modulo 6). In fact, $2x^2 + 3x + 3 \equiv (x - 3)(2x + 3) \pmod{6}$. Note that $2x^2 + 3x + 3$ does not factor in **Z**. ◆

Problems

4.1. For each of the following choices of $f(x)$ and m, complete a row of entries in an extension of Table 4.1.

(a) $f(x) = x^3 + x^2, m = 6$.
(b) $f(x) = 2x^2 - 4x + 7, m = 8$.
(c) $f(x) = x^5 - 5x^4 + 10x^3 - 10x^2 + 5x - 1, m = 13$.

4.2. In the hypothesis of Theorem 4.1, why do we need to stipulate that b_m is a unit of **R**, that is, why is it insufficient to simply say $b_m \neq 0$?

4.3. Suppose that we attempt to redo Example 4.3, but with $g(x) = 2x^2 + 4x + 2$. After some work we are unable to find a unique $q(x)$, $r(x)$ such that $f(x) = g(x)q(x) + r(x)$ and degree $r(x) < 2$. Is this a violation of Theorem 4.1? Explain.

4.4. Find the q(x), r(x) of Theorem 4.1 if $\mathbf{R} = \mathbf{Z}$ and

(a) $f(x) = 2x^5 - x^4 + 3x^3 - 5$, $g(x) = x^3 + 10x^2 - 10x - 4$.
(b) $f(x) = x^7 + x^6 + x^5 + x^4 + x^3 + x^2 + x + 1$, $g(x) = -x^3 + 2$.
(c) $f(x) = 2x^6 + x + 1$, $g(x) = x^4 + 1$.
(d) $f(x) = -x^{15} + 3x^{10} - 3x^5 + 1$, $g(x) = x^5 - 1$.

4.5. Prove the uniqueness part of Theorem 4.1.

4.6. Let $\mathbf{R}[x]$ be the set of polynomials in the indeterminate x with coefficients in the ring $\mathbf{R}$. Show that $\langle \mathbf{R}[x], +\rangle$ is an Abelian group.

4.7. Let $\mathbf{R}[x]$ be as in Problem 4.6. Show that the distributive law for multiplication over addition holds for elements in $\mathbf{R}[x]$.

4.8. Again, let $\mathbf{R}[x]$ be as in Problem 4.6. Accept that the associative law for multiplication holds in $\mathbf{R}[x]$ (this is straightforward but messy). Then is $\mathbf{R}[x]$ itself a ring? If $\mathbf{R}$ is commutative, is $\mathbf{R}[x]$ also a commutative ring? If $\mathbf{R}$ has the unity 1, what is the unity for $\mathbf{R}[x]$?

4.9. What does the Factor Theorem of Descartes say? Apply it to the following cases:

(a) $x - 3$ is a factor of $2x^3 - 3x - 45$.
(b) $-1/2$ is a zero of $6x^6 + 3x^5 - 4x - 2$.
(c) $7x$ is a factor of $14x^5 - 21x^3 - 7x$.

4.10. Let $F(x) = 2x^2 - x + 5$.

(a) Show that by adding (or subtracting) some multiple of 6 to F(x), the congruence $F(x) \equiv 0 \pmod 6$ becomes factorable in $\mathbf{Z}$.
(b) Apply Corollary 4.1.2 to the result in (a) and obtain one root of the congruence.
(c) Are there any other roots of the congruence?

4.11. Consider the f(x) in Example 4.4, but let $m = 2^n$, $n \geq 2$. Prove that $f(x) \equiv 0 \pmod{2^n}$ has exactly one incongruent solution.

4.2 A Theorem of Lagrange

Algebraic results for polynomial congruences become somewhat analogous to those for polynomial equations if the moduli are restricted to primes. The basic reason for this is that the ring $\mathbf{Z}_p$ is a field if p is a prime (Theorem 3.12), so some of the properties of polynomials defined over $\mathscr{C}$ are mimicked by those of polynomial congruences to prime moduli. In particular, the principal result of this section is a kind of generalization of the Fundamental Theorem of Algebra to arbitrary fields **F**.

Theorem 4.2 (Lagrange). A nonzero polynomial $f(x) \in \mathbf{F}[x]$ of degree n can have at most n zeros in a field **F**.

Proof If $a_1 \in \mathbf{F}$ is a zero of $f(x)$, then Corollary 4.1.1 gives

$$f(x) = (x - a_1)q_1(x),$$

where the degree of $q_1(x)$ is $n - 1$. The corollary is applicable because any field **F** is automatically a commutative ring with identity (Fig. 3.2). A zero $a_2 \in \mathbf{F}$ of $q_1(x)$ yields the factorization

$$f(x) = (x - a_1)(x - a_2)q_2(x).$$

We continue in this way until we arrive at

$$f(x) = (x - a_1)(x - a_2) \ldots (x - a_t)q_t(x),$$

where $q_t(x)$ has no zeros in **F**. Clearly $t \leq n$ holds because $f(x)$ is of degree n. Let $b \in \mathbf{F}$ be arbitrary but distinct from $a_1, a_2, \ldots, a_t$, that is, $b - a_i \neq 0$ for $i = 1, 2, \ldots, t$. Then

$$f(b) = (b - a_1)(b - a_2) \ldots (b - a_t)q_t(b).$$

This is nonzero because $q_t(x)$ has no zeros in **F** and **F** has no divisors of 0. Therefore, $\{a_1, a_2, \ldots, a_t\}$ is the full set of zeros of $f(x)$ in **F**. ◆

Corollary 4.2.1 If the polynomial $f(x)$, of degree n, is not identically congruent to 0 modulo the prime p, then $f(x) \equiv 0 \pmod{p}$ has at most n roots.

Proof This follows immediately from Theorems 3.12 and 4.2. ◆

Example 4.5 Solve $2x^4 + 4x^3 + 3x^2 + 3x + 3 \equiv 0 \pmod{5}$.

Since $2 + 4 + 3 + 3 + 3 \equiv 0 \pmod{5}$, one root is $x \equiv 1 \pmod{5}$. Modular division by $x - 1$ as in Example 4.3 gives

$$2x^3 + x^2 + 4x + 2 \equiv 0 \pmod{5}.$$

Trial shows that a root of this is $x \equiv 2 \pmod{5}$. Hence, modular division by $x - 2$ now yields

$$2x^2 + 4 \equiv 0 \pmod{5}.$$

This is found to have no roots in $\mathbf{Z}_5$. Thus, the original congruence of degree 4 has just two roots in $\mathbf{Z}_5$. ◆

The requirement that p be a prime is necessary; look again at the third congruence in Table 4.1. Note also that Lagrange's result does not say how many roots in $\mathbf{F}$ a polynomial will have, or even if it will have any root at all in $\mathbf{F}$. For example, $f(x) = x^2 - 2 \in \mathbf{Q}[x]$ and $\mathbf{Q}$ is a field, but $f(x)$ has no root in $\mathbf{Q}$.

Corollary 4.2.2 Let p be a prime and suppose that $p > n$.

If $f(x) = \sum_{k=0}^{n} c_k x^k$ and if $f(x) \equiv 0 \pmod{p}$ has more than n roots, then $c_k \equiv 0 \pmod{p}$ for $k = 0, 1, 2, \ldots, n$.

Proof Suppose, to the contrary, that c_m $(0 \leq m \leq n)$ is the coefficient of largest index k that is not congruent to 0 modulo p. Define $g(x) = \sum_{k=0}^{m} c_k x^k$; this polynomial is not identically congruent to 0. All of the roots of $f(x) \equiv 0 \pmod{p}$ are roots of $g(x) \equiv 0 \pmod{p}$ because $c_{m+1}, c_{m+2}, \ldots, c_n$ are all congruent to 0 modulo p. So this means that $g(x) \equiv 0 \pmod{p}$ has more than n roots and thus more than m roots. This contradicts Corollary 4.2.1. Therefore, m does not exist, and every c_k is congruent to 0 modulo p. ◆

In the next section we will present a very interesting and somewhat significant application of Corollary 4.2.2. The following example illustrates the content of the corollary.

Example 4.6 Suppose that the following congruence is required to have fewer than three incongruent roots:

$$(4a + 1)x^2 + (a^3 + 2) \equiv 0 \pmod{3}.$$

What possible values could a have?

From Corollary 4.2.2 we require that at least one of $4a + 1$, $a^3 + 2$ be incongruent to 0 modulo 3. But $4a + 1 \not\equiv 0 \pmod{3}$ if $a \equiv 0, 1 \pmod{3}$, and $a^3 + 2 \not\equiv 0 \pmod{3}$ if $a \equiv 0, 2 \pmod{3}$. Hence, if a is any integer whatsoever, the given congruence will have 0, 1, or 2 incongruent roots modulo 3. ◆

Finally, a theorem that is very useful in the solution of polynomial congruences (to prime moduli) is the following easy consequence of Lagrange's theorem.

Theorem 4.3 Let p be a prime and suppose $f(x), g(x), h(x) \in \mathbf{Z}[x]$. If $f(x)$ is identically congruent to $g(x)h(x)$ modulo p, then any root of $f(x) \equiv 0 \pmod{p}$ is a root of $g(x) \equiv 0 \pmod{p}$ or of $h(x) \equiv 0 \pmod{p}$.

Proof The proof is by contrapositive. Suppose $f(x), g(x), h(x)$ are of algebraic degrees $n, m, n - m$, respectively. By Theorem 4.2 $g(x) \equiv 0 \pmod{p}$ has no more than m roots in $\mathbf{Z}_p$: $\{a_1, a_2, \ldots, a_k\}$, $k \leq m$. Similarly, $h(x) \equiv 0$

(mod p) has no more than $n - m$ roots in $\mathbf{Z}_p$: $\{b_1, b_2, \ldots, b_L\}$, $L \le m$. Let x_0 be incongruent to all of the a's and to all of the b's. Then $g(x_0) \not\equiv 0 \pmod{p}$ and $h(x_0) \not\equiv 0 \pmod{p}$. Since $\mathbf{Z}_p$ has no divisors of 0, then $g(x_0)h(x_0) \equiv f(x_0) \not\equiv 0 \pmod{p}$, so x_0 is not a root of $f(x_0) \equiv 0 \pmod{p}$. ◆

Example 4.7 Find the roots of $2x^2 + 7x + 7 \equiv 0 \pmod{11}$.

The congruence is equivalent to $2x^2 + 7x - 15 \equiv 0 \pmod{11}$. The left-hand side of this factors in $\mathbf{Z}$ to give $(2x - 3)(x + 5) \equiv 0 \pmod{11}$. The solution of $x + 5 \equiv 0 \pmod{11}$ is $x \equiv 6 \pmod{11}$. We can solve $2x \equiv 3 \pmod{11}$ either by inspection or by proceeding as in Example 3.32. Multiply both sides by 2^{10-1} and reduce modulo 11. The result is $x \equiv 2^9 \cdot 3 \pmod{11} \equiv 7 \pmod{11}$. Checking shows that both 6 and 7 satisfy the original congruence. ◆

Problems

4.12. Let $\mathbf{F} = \mathbf{Z}_5$; in each of the following cases find a cubic polynomial $f(x) \in \mathbf{F}[x]$ such that

(a) $f(x)$ has no zero in $\mathbf{F}$.
(b) $f(x)$ has one zero in $\mathbf{F}$.
(c) $f(x)$ has two zeros in $\mathbf{F}$.
(d) $f(x)$ has three zeros in $\mathbf{F}$.

4.13. Let p_1, p_2 be two odd primes, with $p_2 > p_1$, and let $f(x)$ be a cubic polynomial in $\mathbf{F}[x]$, where $\mathbf{F} = \mathbf{Z}_{p_1}$. Find an $f(x)$ and primes p_1, p_2 such that $f(x) \equiv 0 \pmod{p_2}$ has more roots than does $f(x) \equiv 0 \pmod{p_1}$.

4.14. In contrast to Problem 4.13, find a cubic polynomial $f(x)$ and primes p_1, p_2 such that $p_2 > p_1$, and $f(x) \equiv 0 \pmod{p_2}$ has fewer roots than does $f(x) \equiv 0 \pmod{p_1}$. This problem and the previous one show that there is no simple relationship between the number of roots of a polynomial of a given degree and the size of the field in which these roots are sought.

4.15. Let $f(x) = x^3 - 2$. How many roots does $f(x) \equiv 0 \pmod{5}$ have? How many roots does $5f(x) \equiv 0 \pmod{5}$ have? How do you explain this?

4.16. Find all roots of the following congruences:

(a) $x^2 + 7x \equiv 5 \pmod{17}$.
(b) $x^3 \equiv 1 \pmod{13}$.
(c) $51x \equiv 113 \pmod{173}$.
(d) $x^2 - x \equiv -3 \pmod{7}$.
(e) $x^3 + 11x \equiv 6(x^2 + 1) \pmod{23}$.
(f) $(x - 1)^2 \equiv 0 \pmod{5}$.
(g) $x^4 \equiv -1 \pmod{19}$.
(h) $5x \equiv 11 \pmod{18}$.

4.17. Let p be a prime and suppose that $d \mid (p - 1)$. Show that $x^d - 1 \equiv 0 \pmod{p}$ has d incongruent roots. [Hint: Use Fermat's Theorem.]

4.18. Let p be an odd prime. Show that $x^{(p-1)/2} + 1 \equiv 0 \pmod{p}$ has $(p - 1)/2$ incongruent roots.

4.19. Solve parts (b) and (e) of Problem 4.16 using Theorem 4.3.

4.20. Show with an example that Theorem 4.3 can fail if the modulus is not a prime p.

4.3 Wilson's Theorem

An interesting application of Corollary 4.2.2 to a particular choice of polynomial f(x) leads unexpectedly to a global result about primes. The following technical lemma is needed.

Lemma 4.4.1 If p is a prime and f(x) is defined by

$$\mathrm{f}(x) = (x - 1)(x - 2) \ldots (x - p + 1) - x^{p-1} + 1,$$

then f(x) is identically congruent to 0 modulo p.

Proof Let $\mathrm{A}(x) = (x - 1)(x - 2) \ldots (x - p + 1)$ and $\mathrm{B}(x) = x^{p-1} - 1$. In $\mathfrak{R}$, A(x) has zeros $1, 2, \ldots, p - 1$ (which all belong to $\mathbf{Z}_p$) by Corollary 4.1.2. Hence, $\mathrm{A}(x) \equiv 0 \pmod{p}$ holds, although not identically so.

By Corollary 3.19.2, B(x) has zeros $1, 2, \ldots, p - 1$ modulo p. But the term of highest degree in $\mathrm{f}(x) = \mathrm{A}(x) - \mathrm{B}(x)$ has degree $p - 2$. Corollary 4.2.2 then applies and we conclude that each coefficient in f(x) is congruent to 0 modulo p. ◆

Theorem 4.4 (Wilson's Theorem).[4] If p is a prime, then $(p - 1)! \equiv -1 \pmod{p}$.

Proof The result is trivial when $p = 2$. For $p > 2$, consider the function f(x) defined in Lemma 4.4.1. The constant term in the expansion of f(x) is given by $\mathrm{f}(0) = (-1)^{p-1}(p - 1)! + 1$. Hence, by the lemma we have $(-1)^{p-1}(p - 1)! + 1 \equiv 0 \pmod{p}$, and since $p - 1$ is even, this is equivalent to $(p - 1)! \equiv -1 \pmod{p}$. ◆

[4] **John Wilson** (1741–1793) was a senior wrangler in mathematics at Cambridge University. He conjectured the result named after him, but did not prove it. The result was actually known to mathematicians before Wilson's time, but it became known as Wilson's Theorem after Wilson's friend and teacher **Edward Waring** (1734–1798) published it in the latter's *Meditationes Algebraicae*. Proofs of Wilson's Theorem were given by **Lagrange** in 1771, by **Euler** in 1783, and by **Gauss** in 1801. Wilson, himself, left mathematics for the law, where he rose to a judgeship and to knighthood (Boyer and Merzbach, 1991).

It is not hard to show that the converse of Wilson's Theorem is also true (Problem 4.23). Hence, we have a necessary *and* sufficient condition for a positive integer to be a prime. Although this is interesting, it is not a very practical test for primality. Thus, in order to discover if 997 is a prime (it is!), one must examine 996!, an integer of more than 2550 digits.

Wilson's Theorem can be used to give a necessary and sufficient condition for $(p, p+2)$ to be twin primes (Dence and Dence, 1995). Since $(p-1)! = (p-1)(p-2)!$ and $p-1 \equiv -1 \pmod{p}$, it follows that $(-1)(p-2)! \equiv -1 \pmod{p}$ iff p is a prime. We continue reducing the $(p-2)!$ until we reach an arbitrary $(p-n)!$. These reductions give

$$(n-1)!(-1)^{n-1}(p-n)! \equiv -1 \pmod{p}, \qquad 1 \leq n < p.$$

Choose $n = (p+1)/2$; the preceding congruence then simplifies to

$$\left[\left(\frac{p-1}{2}\right)!\right]^2 \equiv \begin{cases} -1 \pmod{p} & \text{iff } p = 4k+1 \\ +1 \pmod{p} & \text{iff } p = 4k+3 \end{cases} \qquad (*)$$

Theorem 4.5 p is a $(4k+1)$-prime and $p+2$ is a $(4k+3)$-prime iff $2[\{[(p-1)/2]!\}^2 + 1] + 5p \equiv 0 \pmod{p(p+2)}$.

Proof Relation (*) gives immediately

$$\left[\left(\frac{p-1}{2}\right)!\right]^2 \equiv -1 \pmod{p}$$

and

$$\left[\left(\frac{p+1}{2}\right)!\right]^2 \equiv 1 \pmod{p+2}$$

iff $(p, p+2)$ is a twin-prime pair. The latter congruence is equivalent to

$$\left[\frac{p+1}{2}\left(\frac{p-1}{2}\right)!\right]^2 \equiv 1 \pmod{p+2},$$

or

$$(p^2+2p+1)\left[\left(\frac{p-1}{2}\right)!\right]^2 \equiv 4 \pmod{p+2}.$$

Reduction of p^2+2p+1 modulo $p+2$ gives $1 \pmod{p+2}$, so $(p, p+2)$ is a twin-prime pair iff $\{[(p-1)/2]!\}^2 \equiv -1 \pmod{p}$ and

$$\left[\left(\frac{p-1}{2}\right)!\right]^2 \equiv 4 \pmod{p+2}.$$

Thus, $\{[(p-1)/2]!\}^2 = 4 + b(p+2)$ for some $b \in \mathbf{N}$, and from (*) again we have $4 + b(p+2) = -1 + cp$, $c \in \mathbf{N}$. Hence, $2b = -5 + mp$ for some $m \in \mathbf{Z}$, and back substitution gives

$$\left[\left(\frac{p-1}{2}\right)!\right]^2 = 4 + \left(\frac{-5+mp}{2}\right)(p+2),$$

or

$$2\left[\left(\frac{p-1}{2}\right)!\right]^2 = -2 - 5p + mp(p+2).$$

In congruence notation this is equivalent to

$$2\left[\left\{\left(\frac{p-1}{2}\right)!\right\}^2 + 1\right] + 5p \equiv 0 \pmod{p(p+2)}. \quad ◆$$

Example 4.8 Let $p = 17$, a $(4k+1)$-prime whose twin is $p + 2 = 19$. Then

$$2\left[\left\{\left(\frac{17-1}{2}\right)!\right\}^2 + 1\right] + 5(17) = 3{,}251{,}404{,}887$$

$$= 17(19)(10{,}066{,}269).$$

On the other hand, $p = 11$ has the twin $p + 2 = 13$, but $2[\{(\frac{11-1}{2})!\}^2 + 1] + 5(11) = 28{,}857 \not\equiv 0 \pmod{143}$. ◆

Problems

4.21. Verify Lemma 4.4.1 for $p = 7$.

4.22. Show that 361 divides $19! + 19$, by applying Wilson's Theorem.

4.23. Prove the converse of Theorem 4.4. [Hint: Proof by contrapositive!]

4.24. **Gottfried Wilhelm Leibniz** (1646–1716) knew the theorem $(p-2)! \equiv 1 \pmod{p}$ iff p is a prime. Prove it. Then show that $2(p-3)! \equiv -1 \pmod{p}$ iff p is an odd prime.

4.25. Let p be an odd prime.

(a) Show that $(p-1)! = [1 \cdot 2 \cdot 3 \ldots (p-1)/2]\{p - (p-1)/2\} \cdot \{p - (p-3)/2\} \ldots (p-1)$.
(b) Let $q = (p-1)/2$. Prove that $(q!)^2 + (-1)^q \equiv 0 \pmod{p}$.

(c) Verify part (b) for $p = 11, 17$.
(d) Show that if $p \equiv 1 \pmod 4$, then $[(p-1)/2]!$ is a solution to the congruence $x^2 + 1 \equiv 0 \pmod p$.

4.26. Show how a generalization of Problem 4.24 can provide another proof of Problem 4.25(d).

4.27. A **Wilson prime** is a prime p that satisfies $(p-1)! + 1 \equiv 0 \pmod{p^2}$. Three were discovered by **C. E. Fröberg** in 1963; no others are known below 5×10^8 (Ribenboim, 1996; Crandall, Dilcher, and Pomerance, 1997). Write a computer program to discover two of these three scarce primes.

4.28. Let p be an odd prime, and let $q = (p-1)/2$.

(a) Prove that $2^2 4^2 6^2 \ldots (p-1)^2 \equiv (-1)^{q+1} \pmod p$.
(b) Then show that $1^2 3^2 5^2 \ldots (p-2)^2 \equiv (-1)^{q+1} \pmod p$.

4.29. Show that $(n+1)! \equiv 2(n-1)! \pmod{n+2}$.

4.30. Let $n > 2$ be odd. Prove that if

$$4[(n-1)! + 1] + n \equiv 0 \pmod{n(n+2)}$$

holds, then n and $n+2$ are twin primes. [Hint: Use Problem 4.29.]

4.31. Prove that the converse of the theorem in the preceding problem is also true. Thus, the two problems together constitute a necessary and sufficient condition for $(n, n+2)$ to be a pair of twin primes (Clement, 1949).

4.32. Prove the following companion to Theorem 4.5: p is a $(4k-1)$-prime and $p+2$ is a $(4k+1)$-prime iff

$$2\left[\left\{\left(\frac{p-1}{2}\right)!\right\}^2 - 1\right] - 5p \equiv 0 \pmod{p(p+2)}.$$

Illustrate this with $p = 11$ and $p = 19$.

4.4 The Lucas–Lehmer Test

There are no known easy tests that can be applied to integers, generally, in order to discover which are primes. Specialized classes of primes may be easier to examine. In Conjecture 3 in Section 1.8 we introduced **Mersenne primes**,[5] M_p,

[5] After the French Franciscan monk and amateur mathematician **Marin Mersenne** (1588–1648), a contemporary of **Fermat**. Mersenne was an important intermediary in communication among European mathematicians.

Table 4.2 Some Mersenne Primes, M_p

p	No. of Digits	p	No. of Digits
19	6	11,213	3,376
31	10	44,497	13,395
521	157	132,049	39,751
1279	386	216,091	65,050

primes of the form $2^p - 1$. They are needed, for example, in the construction of perfect numbers (Chapter 7). Table 4.2 gives a sampling of the 36 or so known Mersenne primes; see also Table I(7) in Appendix II. Supercomputers are now used to discover the largest ones.

The primality test to be described in this section is a test just for Mersenne primes. Our first property of Mersenne primes is easy to establish, and was known to **Fermat** (who called Mersenne primes the "radicals" of perfect numbers) (Mahoney, 1994).

Theorem 4.6 If M_n is a Mersenne prime, then n is prime.

Proof We argue by contrapositive. Let $n = rs$, $1 < r, s < n$. Then $M_n = 2^{rs} - 1 = (2^r - 1)(2^{rs-r} + 2^{rs-2r} + 2^{rs-3r} + \ldots + 2^r + 2^0)$. Since $r > 1$, then $2^r - 1 > 1$ and $2^r + 2^0 > 1$. Thus, both factors of M_n exceed 1, so M_n is composite. ◆

The converse of Theorem 4.6 is false. For example, when $n = 11$ (a prime), then $M_n = 2047 = 23 \cdot 89$. But even in cases like this, the divisors of M_n (including 1 and M_n itself) are restricted to certain forms (the result, Theorem 4.7, was also known to **Fermat**). We need the following lemma.

Lemma 4.7.1 Let m, n be integers satisfying $1 \leq m \leq n$. Then $(2^m - 1, 2^n - 1) = 2^{(m,n)} - 1$.

Proof By the Division Algorithm there are unique integers q, r such that $n = qm + r, 0 \leq r < m$. Then

$$\begin{aligned} 2^n - 1 &= 2^{qm+r} - 1 \\ &= (2^m - 1)(2^{qm+r-m} + 2^{qm+r-2m} + \ldots + 2^{m+r} + 2^r) + (2^r - 1). \end{aligned}$$

Clearly, $0 \leq 2^r - 1 < 2^m - 1$. If $2^r - 1$ is not 0, then write $m = q_1 r + r_1$, $0 \leq r_1 < r$, and perform a second division to give

$$\begin{aligned} 2^m - 1 &= 2^{q_1 r + r_1} - 1 \\ &= (2^r - 1)(2^{q_1 r + r_1 - r} + 2^{q_1 r + r_1 - 2r} + \ldots + 2^{r_1 + r} + 2^{r_1}) + (2^{r_1} - 1), \end{aligned}$$

where $0 \leq 2^{r_1} - 1 < 2^r - 1$. Continuing in this way, we generate a strictly decreasing sequence $\{2^{r_k} - 1\}$ that parallels the strictly decreasing sequence $\{r_k\}$. For some k we eventually obtain $r_k = 0$ and $r_{k-1} = (m, n)$ from Theorem 2.13. Correspondingly,

$$2^{r_k} - 1 = 2^0 - 1 = 0, \quad \text{so} \quad 2^{r_{k-1}} - 1 = (2^m - 1, 2^n - 1). \quad \blacklozenge$$

Theorem 4.7 Let p be an odd prime. Then any divisor of M_p is of the form $2kp + 1$, $k \in \mathbf{N}$.

Proof Let q be a prime that divides $M_p - 1$. From Fermat's Theorem (Corollary 3.19.2) we have $q \mid (2^{q-1} - 1)$. By definition of the greatest common divisor, it follows that $q \mid (2^p - 1, 2^{q-1} - 1)$, so $(2^p - 1, 2^{q-1} - 1) = (2^{(p,q-1)} - 1) > 1$ from Lemma 4.7.1. Hence, $2^{(p,q-1)} > 2$ or $(p, q-1) > 1$. This implies $(p, q-1) = p$, since p is a prime. So $p \mid (q-1)$, or

$$cp = q - 1$$

for some $c \in \mathbf{N}$. But $q - 1$ is even while p is odd; therefore c must be even and we can write $c = 2k$. Rearrangement then gives $q = 2kp + 1$. $\blacklozenge$

Example 4.9 Determine if M_{17} is a prime.

As 17 is a prime, $M_{17} = 131{,}071$ may be a prime. We need only investigate factors up to $\sqrt{M_{17}} \approx 361$. Of these factors, only ones of the form $34k + 1$ need be considered (Theorem 4.7). Thus, the members of the set

$$\mathbf{S} = \{35, 69, 103, 137, 171, 205, 239, 273, 307, 341\}$$

are divided in turn into M_{17}. None is found to divide 131,071, so M_{17} is a Mersenne prime. $\blacklozenge$

The trouble with the use of Theorem 4.7 to determine the primality of M_p is that for very large Mersenne numbers it requires too many operations. The number of divisions needed is of the order $\sqrt{M_p}/2p$, or about $(1/p)2^{(p-2)/2}$ (Problem 4.42). If p were about 10,000, this would be about 10^{1500} operations, which is prohibitive.

A shorter test is one based on work done by the French number theorist **Edouard Lucas** (1842–1891) and later improved by **D. H. Lehmer**. This test requires on the order of $p - 2$ squarings followed by $p - 2$ modular reductions, each of which requires one multiplication and one division. Thus, if p is about 10,000, we need to do about $3p$, or 3×10^4, multiplications and divisions, far less than 10^{1500}.

Our proof of the **Lucas–Lehmer test** is taken from (Bruce, 1993). We start by defining a sequence $\{S_n\}_{n=1}^{\infty}$ as follows:

$$S_n = \begin{cases} 4 & \text{if } n = 1 \\ (S_{n-1})^2 - 2 \pmod{M_p} & \text{if } n > 1, \end{cases}$$

where M_p is the Mersenne number whose primality is to be investigated.

Lemma 4.8.1 Let $w = 2 + \sqrt{3}$ and $\overline{w} = 2 - \sqrt{3}$. Then

$$S_n \equiv w^{2^{n-1}} + \overline{w}^{2^{n-1}} \pmod{M_p}.$$

Proof The result is trivial for $n = 1$. Use mathematical induction; the proof is left to the reader (Problem 4.46). ◆

Lemma 4.8.2 Let q be a prime and $\mathbf{R} = \{a + b\sqrt{3}\colon a, b \in \mathbf{Z}_q\}$, where $\mathbf{Z}_q = \{0, 1, 2, \ldots, q - 1\}$. Define addition and multiplication on $\mathbf{R}$ as follows:

$$(a + b\sqrt{3}) + (c + d\sqrt{3}) = (a + c) + (b + d)\sqrt{3}$$

$$(a + b\sqrt{3}) \cdot (c + d\sqrt{3}) = (ac + 3bd) + (ad + bc)\sqrt{3},$$

with $(a + c)$, $(b + d)$, $(ac + 3bd)$ and $(ad + bc)$ all being reduced modulo q. Then the set $\mathbf{R}^\times$ of units in $\mathbf{R}$ is a group under modular multiplication.

Proof $\mathbf{R}$ meets all of the requirements of a commutative ring with identity (Section 3.3); Problem 3.59 then applies. ◆

Lemma 4.8.3 Let $\mathbf{G}$ be a finite group and let $g \in \mathbf{G}$ be of order s. Suppose $g^d = \mathbf{I}$; then $s \mid d$.

Proof Suppose $s \nmid d$; then by the Division Algorithm there are integers q, r such that $d = sq + r$, $0 < r < s$. Hence, $g^d = (g^s)^q g^r = \mathbf{I}^q g^r = g^r = \mathbf{I}$; this contradicts s being the order of g, so $s \nmid d$ is false. ◆

Of course, $s \leq |\mathbf{G}|$ must also hold since g can be used to generate a subgroup $\mathbf{G}' \subset \mathbf{G}$. We are now ready to state the Lucas–Lehmer test (sufficiency part, only).

Theorem 4.8 (Lucas–Lehmer Test). Let $p > 2$ be a prime. If $M_p \mid S_{p-1}$, then M_p is a prime.

Proof If $S_{p-1} \equiv 0 \pmod{M_p}$, then $w^{2^{p-2}} + \overline{w}^{2^{p-2}} = cM_p$ for some $c \in \mathbf{N}$ from Lemma 4.8.1. Multiplication of this by $w^{2^{p-2}}$ gives (note: $w\overline{w} = 1$)

$$w^{2^{p-1}} = cM_p w^{2^{p-2}} - 1 \qquad (*)$$

and squaring,

$$w^{2^p} = [c\mathrm{M}_p w^{2^{p-2}} - 1]^2. \tag{**}$$

Now assume that M_p is not a prime; let $q > 2$ be a prime divisor of M_p, and let $\mathbf{Z}_q$ be as in Lemma 4.8.2. Equations (*) and (**) then reduce to

$$\begin{cases} w^{2^{p-1}} \equiv -1 \pmod{q} & (*)' \\ w^{2^p} \equiv 1 \pmod{q}. & (**)' \end{cases}$$

Let $\mathbf{R}$ be as in Lemma 4.8.2; then since $w \cdot w^{2^p-1} \equiv 1 \pmod{q}$, w must belong to the group $\mathbf{R}^\times$. By Lemma 4.8.3 the order of w divides 2^p, but by (*)′ it cannot be less than 2^p. Hence, the order of w is 2^p.

The order of $\mathbf{R}^\times$ is at most $q^2 - 1$, so $2^p \leq q^2 - 1$ by the remark after Lemma 4.8.3. On the other hand, if proper prime divisors of M_p do exist, we can certainly choose q so that $q^2 \leq \mathrm{M}_p$ or, equivalently, $q^2 - 1 \leq \mathrm{M}_p - 1 = (2^p - 1) - 1$. Combination with the first inequality in this paragraph gives $2^p \leq 2^p - 2$, an impossibility. It follows that M_p cannot be composite.
◆

Example 4.10 Determine if $\mathrm{M}_{13} = 8191$ is a prime.

We arrange the work conveniently in a table.

k	S_k	k	S_k	k	S_k
1	4	5	3953	9	1294
2	14	6	5970	10	3470
3	194	7	1857	11	128
4	4870	8	36	12	0

Since $S_{12} \equiv 0 \pmod{8191}$, we conclude from Theorem 4.8 that M_{13} is prime.
◆

Problems

4.33. What are the first eight values of the positive integer n for which M_n is composite? Give the prime factorizations of those eight Mersenne numbers.

4.34. Show that a Mersenne prime is a $(4k + 3)$-prime.

4.35. Illustrate Lemma 4.7.1 with

(a) $m = 6, n = 8$. (b) $m = 14, n = 15$. (c) $m = 8, n = 16$.

4.36. Let $[\mathrm{M}_n, \mathrm{M}_m]$ denote the least common multiple of the Mersenne numbers M_n, M_m. If m, n are coprime, then show that $[\mathrm{M}_n, \mathrm{M}_m] = \mathrm{M}_{n+m} - \mathrm{M}_n - \mathrm{M}_m$.

4.37. Illustrate Theorem 4.7 with (a) $p = 11$, (b) $p = 19$.

4.38. Prove that if p is an odd prime, then there exist an infinite number of Mersenne numbers M_n such that $p \mid \mathrm{M}_n$.

4.39. The Mersenne number M_n with $n = 11$ is the first instance of a counterexample to the converse of Theorem 4.6 where n is a $(4k + 3)$-prime. Find an analogous first instance of a counterexample where n is a $(4k + 1)$-prime, and give the factorization of M_n.

4.40. Recall from Problem 3.22 the **digital root function**, DR(N), of a positive integer N.

(a) Show that if M_p is a Mersenne prime, then $\mathrm{DR}(\mathrm{M}_p) = 1$ or 4 for $p > 5$. [Hint: p is of the form $6k + 1$ or the form $6k + 5$.]
(b) In Table 4.2, what is $\mathrm{DR}(\mathrm{M}_{216091})$?

4.41. Show that if p is an odd prime, n is composite, and $d = (\mathrm{M}_p, \mathrm{M}_n) > 1$, then there is an integer $k > 0$ such that $(2kp + 1) \mid d$. Illustrate this result with $p = 7, n = 21$.

4.42. Account for the statement that the number of divisions needed to determine if M_p is a Mersenne prime is of the order $p^{-1}\sqrt{2^{p-2}}$. At about what value of p does the Lucas–Lehmer Test become more efficient than Theorem 4.7 to determine the primality of M_p?

4.43. Write a computer program and apply the Lucas–Lehmer Test to M_p for (a) $p = 19$, (b) $p = 31$.

4.44. Let $2n + 1$ be a prime, $n \in \mathbf{N}$. Prove that $\mathrm{M}_n(\mathrm{M}_n + 2) \equiv 0 \pmod{2n + 1}$.

4.45. Let $\mathbf{S}$ be the set of all Mersenne primes, M_p. Prove that $\sum_{\mathrm{M}_p \in \mathbf{S}} \mathrm{M}_p^{-1/2}$ converges.

4.46. Prove Lemma 4.8.1.

4.47. Show that M_{2k} is not expressible in the form a^b, where $a, b \in \mathbf{N}$ and $a, b > 1$. Actually, the theorem is true for all Mersenne numbers, but the case of those with even index is easier to prove than those with odd index.

4.5 The Chinese Remainder Theorem

We return now to the solving of congruences. So far, we have dealt only with single congruences. In ancient times, however, systems of linear congruences arose in connection with calendar making, especially in China and India (Kang-sheng, 1988). Consider the following noncalendrical problem which appeared in the work *Sun Zi suanjing* ("Master Sun's Arithmetical Manual"), written sometime between 280 and 473 A.D. by **Sun Zi** (Needham, 1959).[6]

> There are certain things whose number is unknown. If we count them by threes, we have two left over; by fives, we have three left over; and by sevens, two are left over. How many things are there?

In modern notation, the problem is to find an x such that

$$\begin{cases} x \equiv 2 \pmod{3} \\ x \equiv 3 \pmod{5} \\ x \equiv 2 \pmod{7}. \end{cases}$$

Sun Zi's solution of the problem gave, as we will show,

$$x \equiv 140 + 63 + 30 = 233 \equiv 23 \pmod{105}.$$

Although Sun Zi was not in full possession of the underlying theory, it is clear that he understood the kernel of the problem. Hence, the basic theorem, to be given shortly, is called appropriately the **Chinese Remainder Theorem.**

Just as a system of two or more linear equations may be inconsistent and have no solution, so a system of two or more linear congruences may have no solution. For example, the system

$$\begin{cases} x \equiv 2 \pmod{3} \\ x \equiv 1 \pmod{6} \end{cases}$$

has no solution, even though each congruence separately has a solution. The problem arises because 3 and 6 are not relatively prime; change the 6 to a 5 or a 7, for example, and the system is solvable. The Chinese Remainder Theorem considers this point; the form of the theorem is suggested by the following sample calculation.

Example 4.11 Solve the system of linear congruences

$$\begin{cases} x \equiv 5 \pmod{7} \\ x \equiv 4 \pmod{5}. \end{cases}$$

[6] Written variously also as **Sun-Tzu** or **Sun-Tsu**. In certain works (e.g., Dickson) Sun Zi is placed in the first century A.D.

Multiply the first congruence by 5 to give $5x \equiv 25 \pmod{35}$; multiply the second congruence by 7 to give $7x \equiv 28 \pmod{35}$. Addition of the two resulting congruences is now legitimate; we obtain $12x \equiv 53 \pmod{35}$. The system has been reduced to a unique linear congruence with modulus $7 \cdot 5$. Observing that $12 \cdot 3 \equiv 1 \pmod{35}$, we multiply $12x \equiv 53 \pmod{35}$ by 3. The result is $x \equiv 159 \equiv 19 \pmod{35}$. Check: $19 \equiv 5 \pmod 7$ and $19 \equiv 4 \pmod 5$. ◆

Theorem 4.9 (The Chinese Remainder Theorem). Let positive moduli m_1, $m_2, \ldots, m_k$ be pairwise coprime, let $m = \Pi_{i=1}^{k} m_i$, and let the integers a_1, $a_2, \ldots, a_k$ be arbitrary. Then the system of congruences

$$\begin{cases} x \equiv a_1 \pmod{m_1} \\ x \equiv a_2 \pmod{m_2} \\ \quad\vdots \\ x \equiv a_k \pmod{m_k} \end{cases}$$

has a unique solution modulo m.

Proof Let $M_i = m/m_i$; since the m_i are pairwise coprime, then $(M_i, m_i) = 1$. By Theorem 3.22 there is a unique solution x_i modulo m_i of

$$M_i x \equiv 1 \pmod{m_i},$$

for each i. Now define

$$x_0 = \sum_{i=1}^{k} M_i x_i a_i.$$

Pick an arbitrary index j. When $i = j$, then $M_j x_j \equiv 1 \pmod{m_j}$, so $M_j x_j a_j \equiv a_j \pmod{m_j}$ by Theorem 3.2, part (2). When $i \neq j$, then $m_j | M_i$, so $M_i x_i a_i \equiv 0 \pmod{m_j}$. Adding up all the congruences modulo m_j, we obtain

$$x_0 \equiv a_j \pmod{m_j}.$$

Since the choice of j was arbitrary, analogous congruences hold for all the m_i.

Now suppose that y_0 is another solution of the system. Then $x_0 \equiv y_0 \pmod{m_i}$ for each i. No two of the moduli have a common prime factor, so by the Fundamental Theorem of Arithmetic all the prime factors of all the m_i's appear in the factorization of $x_0 - y_0$. This says $m | (x_0 - y_0)$, or $x_0 \equiv y_0 \pmod m$. But for $0 \leq x_0, y_0 < m$ we must have $x_0 = y_0$, so the solution is unique. ◆

Example 4.12 Solve Sun Zi's congruence problem.

Here, $m_1 = 3, m_2 = 5, m_3 = 7$, and $M_1 = 35, M_2 = 21, M_3 = 15$. We obtain the modular inverses of the M_i's as follows:

$$35x_1 \equiv 1 \pmod 3 \quad \to \quad x_1 \equiv 2 \pmod 3$$
$$21x_2 \equiv 1 \pmod 5 \quad \to \quad x_2 \equiv 1 \pmod 5$$
$$15x_3 \equiv 1 \pmod 7 \quad \to \quad x_3 \equiv 1 \pmod 7.$$

These calculations can be performed rapidly because the moduli are all small. For larger moduli, not necessarily primes, the techniques in Examples 3.32 and 3.36 would be useful.

Hence, the preceding results, together with Theorem 4.9, give

$$\begin{aligned} x_0 &= 35(2)(2) + 21(1)(3) + 15(1)(2) \\ &= 140 + 63 + 30 \\ &= 233 \\ &\equiv 23 \pmod{105}. \end{aligned}$$ ◆

Corollary 4.9.1 Let the positive moduli $m_1, m_2, \ldots, m_k$ be pairwise coprime, let the integers $a_1, a_2, \ldots, a_k$ be arbitrary, and let the integers $c_1, c_2, \ldots, c_k$ satisfy $(c_i, m_i) = 1$. Then the system of congruences

$$\begin{cases} c_1 x \equiv a_1 \pmod{m_1} \\ c_2 x \equiv a_2 \pmod{m_2} \\ \quad\vdots \\ c_k x \equiv a_k \pmod{m_k} \end{cases}$$

has a unique solution modulo $m = m_1 m_2 \ldots m_k$.

Proof By Theorem 3.22 each c_i has a multiplicative inverse c_i^{-1} modulo m_i. The original system of congruences is equivalent to

$$\begin{cases} x \equiv c_1^{-1} a_1 \pmod{m_1} \\ x \equiv c_2^{-1} a_2 \pmod{m_2} \\ \quad\vdots \\ x \equiv c_k^{-1} a_k \pmod{m_k}. \end{cases}$$

The Chinese Remainder Theorem now applies to this system. ◆

Example 4.13 Let $N > 0$ be divisible by 10. If 1/2 of the integers from 1 to N are counted off in threes, one remains. If 1/5 of the integers in $[1, N]$ are counted off in sevens, three remain. Finally, if all N integers are counted off in elevens, seven remain. What is the minimum possible value for N?

Let $N = 10 \cdot M$; then the problem consists in solving the system

$$\begin{cases} 5M \equiv 1 \pmod{3} \\ 2M \equiv 3 \pmod{7} \\ 10M \equiv 7 \pmod{11}. \end{cases}$$

Modulo 3, the inverse of 5 is 2, whence $2 \cdot 5M \equiv 2 \cdot 1 \pmod{3}$, or $M \equiv 2 \pmod{3}$. Modulo 7, the inverse of 2 is 4, whence $4 \cdot 2M \equiv 4 \cdot 3 \pmod{7}$, or $M \equiv 5 \pmod{7}$. Finally, modulo 11, the inverse of 10 is 10, whence $10 \cdot 10M \equiv 10 \cdot 7 \pmod{11}$, or $M \equiv 4 \pmod{11}$. The solution of the new system

$$\begin{cases} M \equiv 2 \pmod{3} \\ M \equiv 5 \pmod{7} \\ M \equiv 4 \pmod{11} \end{cases}$$

is, from Theorem 4.9 (verify details!),

$$\begin{aligned} M &= 77(2)(2) + 33(3)(5) + 21(10)(4) \\ &\equiv 1643 \pmod{231} \\ &\equiv 26 \pmod{231}. \end{aligned}$$

The minimum $N > 0$ is thus $N = 10M = 260$. ◆

Corollary 4.9.1 is easily implemented on a computer. For systems of more than three congruences, use of the computer is preferred. An alternative systematic procedure for solving systems of linear congruences, also easily implemented on a computer, is the following iterative method. We illustrate it by reworking Example 4.13.

STEP 1. Work from the top downward in the system:

$$\begin{cases} 5M \equiv 1 \pmod{3} \\ 2M \equiv 3 \pmod{7} \\ 10M \equiv 7 \pmod{11}. \end{cases}$$

Write the first congruence in equivalent algebraic form:

$$5M = 1 + 3t.$$

STEP 2. Substitute this algebraic form into the second congruence, adjusting by a scale factor if necessary:

$$\begin{aligned} 10M &\equiv 15 \pmod{7} \\ 2 + 6t &\equiv 15 \pmod{7} \\ 6t &\equiv 6 \pmod{7}. \end{aligned}$$

STEP 3. Solve for t: $t \equiv 1 \pmod 7$, or $t = 1 + 7v$.
STEP 4. Substitute this back into the expression for $5M$:

$$5M = 1 + 3(1 + 7v) = 4 + 21v.$$

STEP 5. Substitute this into the third congruence:

$$\begin{aligned} 10M &\equiv 7 \pmod{11} \\ 8 + 42v &\equiv 7 \pmod{11} \\ 42v &\equiv 10 \pmod{11} \\ 21v &\equiv 5 \pmod{11}. \end{aligned}$$

STEP 6. Solve for v: $v \equiv 6 \pmod{11}$, or $v = 6 + 11w$.
STEP 7. Substitute this back into the expression for $5M$ in Step 4:

$$5M = 4 + 21(6 + 11w) = 130 + 231w, \quad \text{or} \quad 5M \equiv 130 \pmod{231}.$$

STEP 8. Solve for M by inspection or by use of the methods in Example 3.32 or Corollary 3.25.1: $M \equiv 26 \pmod{231}$.

The procedure can also be used in systems with moduli that are not pairwise coprime.

Problems

4.48. Find the two smallest positive integers N, each of which leaves a remainder of 4 upon division by 9 and a remainder of 5 upon division by 10.

4.49. Find the least positive multiple of 7 that leaves a remainder of 1 when divided by 2, 3, 4, 5, or 6.

4.50. An integer is drawn at random from [1, 10000]. What is the probability that it leaves a remainder of 3 on division by 5 and a remainder of 5 on division by 7?

4.51. Solve the following systems of linear congruences, if possible:

(a) $\begin{cases} x \equiv 1 \pmod{M_3} \\ x \equiv 2 \pmod{M_5} \\ x \equiv 3 \pmod{M_7}. \end{cases}$

(b) $\begin{cases} x \equiv 1 \pmod 5 \\ 3x \equiv 6 \pmod{12} \\ x \equiv 4 \pmod 9 \\ x \equiv 0 \pmod{19}. \end{cases}$

(c) $\begin{cases} -2x \equiv 5 \pmod 7 \\ 19x \equiv 1 \pmod{23} \\ 4x \equiv 10 \pmod{15}. \end{cases}$

(d) $\begin{cases} x \equiv 4 \pmod 6 \\ x \equiv 5 \pmod 7 \\ x \equiv 4 \pmod{15}. \end{cases}$

(e) $\begin{cases} 2x \equiv 1 \pmod{11} \\ x \equiv 11 \pmod{24} \\ x \equiv 14 \pmod{30}. \end{cases}$ (f) $\begin{cases} 3x \equiv 5 \pmod{10} \\ 7x \equiv 9 \pmod{12} \\ 7x \equiv 9 \pmod{16}. \end{cases}$

4.52. Consider the system of linear congruences

$$\begin{cases} x \equiv b_1 \pmod{m_1} \\ x \equiv b_2 \pmod{m_2}, \end{cases}$$

and suppose that $d = (m_1, m_2) > 1$. Prove that a necessary and sufficient condition that the system has a solution is $d \mid (b_1 - b_2)$.

4.53. In Problem 4.52, if the system has a solution, show that it is unique modulo $[m_1, m_2]$, where [,] means least common multiple.

4.54. Prove the following result from elementary set theory, which is needed for Problem 4.55: Let f: $\mathbf{S}_1 \to \mathbf{S}_2$ be a map between finite sets, and suppose $|\mathbf{S}_1| = |\mathbf{S}_2|$. Then f is a bijection iff it is an injection.

4.55. Our proof of the Chinese Remainder Theorem is a constructive proof. An interesting, alternative, existence proof is given in Mozzochi (1967). Study this and write out the proof in full detail, making use of your work from Problem 4.54.

4.6 Polynomial Congruences with Prime-Power Moduli

The Chinese Remainder Theorem of the previous section leads directly into our next results. These deal with polynomial congruences with composite moduli.

Theorem 4.10 Let $\mathrm{f}(x) \in \mathbf{Z}[x]$ and let $m_1, m_2, \ldots, m_k$ be positive moduli that are pairwise coprime. Then the congruence

$$\mathrm{f}(x) \equiv 0 \left(\bmod \prod_{i=1}^{k} m_i \right)$$

has a root iff each congruence

$$\mathrm{f}(x) \equiv 0 \pmod{m_i}, \quad i = 1, 2, \ldots, k,$$

has a root.

Proof ($\rightarrow$) Let $m = \prod_{i=1}^{k} m_i$ and suppose a is a root of $\mathrm{f}(x) \equiv 0 \pmod{m}$, that is, $\mathrm{f}(a) \equiv 0 \pmod{m}$. Then Theorem 3.2, part (7), gives us immediately $\mathrm{f}(a) \equiv 0 \pmod{m_i}$ for each i. So a is a root of $\mathrm{f}(x) \equiv 0 \pmod{m_i}$.

(←) Suppose $a_1, a_2, \ldots, a_k$ are roots of $f(x) \equiv 0 \pmod{m_1}$, $f(x) \equiv 0 \pmod{m_2}, \ldots, f(x) \equiv 0 \pmod{m_k}$, respectively. By the Chinese Remainder Theorem the system of congruences

$$\begin{cases} x \equiv a_1 \pmod{m_1} \\ x \equiv a_2 \pmod{m_2} \\ \quad\vdots \\ x \equiv a_k \pmod{m_k} \end{cases}$$

has a unique solution $0 \leq a < m$ such that $a \equiv a_i \pmod{m_i}$ for each i. By Theorem 3.4 we obtain for each i,

$$f(a) \equiv f(a_i) \equiv 0 \pmod{m_i}.$$

But the m_i's are pairwise coprime and if each $m_i | f(a)$, then $m | f(a)$, so $f(a) \equiv 0 \pmod{m}$. This says that a is a root of $f(x) \equiv 0 \pmod{m}$. ◆

Corollary 4.10.1 The solution of a polynomial congruence

$$f(x) \equiv 0 \pmod{m}$$

where m is composite, is reducible to that of a system of polynomial congruences with prime-power moduli.

Proof Let $m = p_1^{\alpha_1} p_2^{\alpha_2} \ldots p_k^{\alpha_k}$, and in Theorem 4.10 take $m_1 = p_1^{\alpha_1}$, $m_2 = p_2^{\alpha_2}, \ldots, m_k = p_k^{\alpha_k}$. ◆

In view of the Corollary, the roots of $f(x) \equiv 0 \pmod{m}$ are synthesized from the roots of the congruences $f(x) \equiv 0 \pmod{p^\alpha}$, $\alpha \geq 2$. The case of $\alpha = 1$ is Corollary 4.2.1.

Suppose now that x_0 is a root of $f(x) \equiv 0 \pmod{p^\alpha}$ for some fixed α, where

$$f(x) = a_0 x^n + a_1 x^{n-1} + a_2 x^{x-2} + \ldots + a_n.$$

Consider the variable quantity $x_0 + zp^\alpha$, $z \in \mathbf{Z}$. Taylor's Theorem gives

$$f(x_0 + zp^\alpha) = f(x_0) + zp^\alpha f'(x_0) + \frac{z^2 p^{2\alpha}}{2!} f''(x_0) + \cdots + \frac{z^n p^{n\alpha}}{n!} f^{(n)}(x_0).$$

The series terminates because $f(x)$ is a polynomial. All of the coefficients in this equation are integers because for any k, $1 \leq k \leq n$, each coefficient in $f^{(k)}(x)$ is divisible by $k!$ (Problem 4.62).

Since $p^{\alpha+1}$ occurs in each term beyond the second in the preceding equation, we have

$$f(x_0 + zp^\alpha) =$$

$$f(x_0) + zp^\alpha f'(x_0) + p^{\alpha+1}\left[\frac{z^2 p^{\alpha-1}}{2!} f''(x_0) + \cdots + \frac{z^n p^{n\alpha-\alpha-1}}{n!} f^{(n)}(x_0)\right].$$

The expression $p^{\alpha+1}[\ldots]$ is clearly congruent to 0 modulo $p^{\alpha+1}$. Hence, $f(x_0 + zp^\alpha)$ is also congruent to 0 modulo $p^{\alpha+1}$ iff $f(x_0) + zp^\alpha f'(x_0) \equiv 0 \pmod{p^{\alpha+1}}$. From Theorem 3.3 this is equivalent to requiring

$$zf'(x_0) \equiv -\frac{f(x_0)}{p^\alpha} \pmod{p}.$$

There are now two cases. In the first, suppose $p \nmid f'(x_0)$. Application of Theorem 3.22 shows that there is a unique root z_0 modulo p obtained by solving the previous congruence. This means that $x_0 + z_0 p^\alpha$ is the unique root of $f(x) \equiv 0 \pmod{p^{\alpha+1}}$. One says that x_0 has been **lifted** uniquely from p^α to $p^{\alpha+1}$.

In the second case, suppose $p \mid f'(x_0)$, that is, $f'(x_0) \equiv 0 \pmod{p}$, or $p^\alpha f'(x_0) \equiv 0 \pmod{p^{\alpha+1}}$, and we have

$$f(x_0 + zp^\alpha) \equiv f(x_0) \pmod{p^{\alpha+1}}.$$

If $f(x_0) \not\equiv 0 \pmod{p^{\alpha+1}}$, then there can be no z that gives $f(x_0 + zp^\alpha) \equiv 0 \pmod{p^{\alpha+1}}$. But if $f(x_0) \equiv 0 \pmod{p^{\alpha+1}}$, then any $z \in \mathbf{Z}$ will work because $f(x_0)$ is *always* congruent to 0 modulo $p^{\alpha+1}$. In this case, there are exactly p incongruent roots corresponding to $0 \le z < p$ and to the root x_0 of $f(x) \equiv 0 \pmod{p^\alpha}$. We summarize the preceding discussion as Theorem 4.11.

Theorem 4.11 The number of roots of $f(x) \equiv 0 \pmod{p^{\alpha+1}}$ corresponding to a root x_0 of $f(x) \equiv 0 \pmod{p^\alpha}$ is

(1) One, if $p \nmid f'(x_0)$.
(2) None, if $p \mid f'(x_0)$ and x_0 is not a root of $f(x) \equiv 0 \pmod{p^{\alpha+1}}$.
(3) p, if $p \mid f'(x_0)$, and x_0 is a root of $f(x) \equiv 0 \pmod{p^{\alpha+1}}$.

Combination of Theorems 4.10 and 4.11 provides the means of obtaining the complete solution set of $f(x) \equiv 0 \pmod{m}$. Note that if $m = p_1^{\alpha_1} p_2^{\alpha_2} \ldots p_k^{\alpha_k}$ and if for one of the p_i's it is not possible to lift *any* of the roots of $f(x) \equiv 0 \pmod{p_i}$ all the way to $p_i^{\alpha_i}$, then $f(x) \equiv 0 \pmod{m}$ will have no root. The algorithm for solving $f(x) \equiv 0 \pmod{m}$ is diagrammed in Fig. 4.1.

Example 4.14 Solve $x^3 + 3x + 14 \equiv 0 \pmod{36}$.

Part I. Since $36 = 2^2 \cdot 3^2$, let $p_1 = 2$, $\alpha_1 = 2$, $p_2 = 3$, $\alpha_2 = 2$, and $f'(x) = 3x^2 + 3$. If $x^3 + 3x + 14 \equiv 0 \pmod{p_1}$ then $x_0 = 0, 1$. Hence, $f(x_0) = 14, 18$ and $f'(x_0) = 3, 6$. First consider $x_0 = 0$. Since $p_1 \nmid f'(0)$, there is one root of $x^3 + 3x + 14 \equiv 0 \pmod{p_1^2}$. The congruence $zf'(x_0) \equiv [-f(x_0)]/p^\alpha \pmod{p}$ becomes $3z_0 \equiv -14/p_1 \pmod{p_1} \equiv -7 \pmod{2}$, so $z_0 \equiv 1 \pmod{2}$. Then $x_0 + z_0 p_1 = 0 + 1 \cdot 2^1 = 2$ is the unique root of $f(x) \equiv 0 \pmod{p_1^2}$ corresponding to $x_0 = 0$.

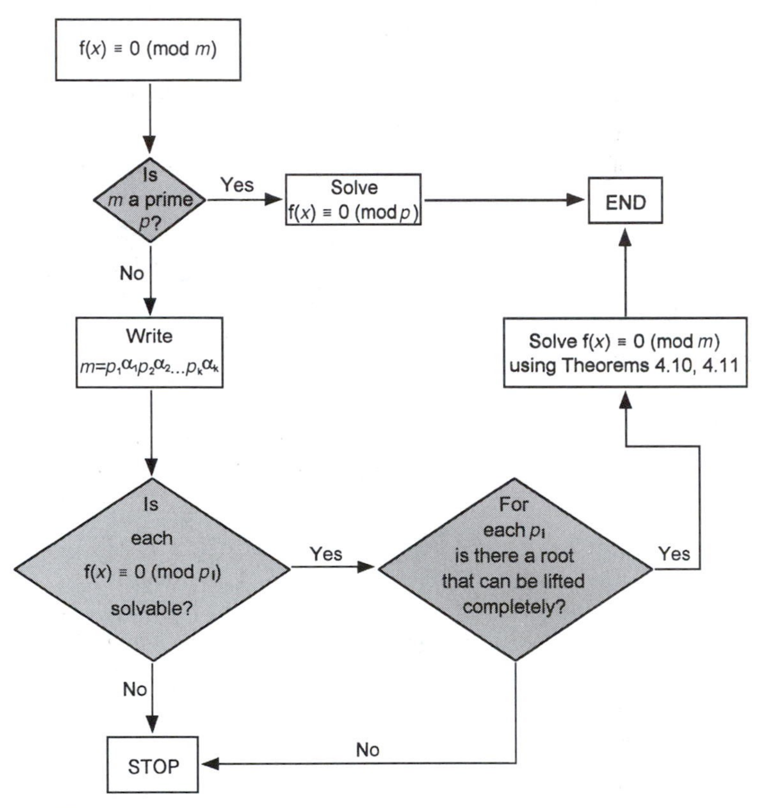

Figure 4.1 Solving the polynomial congruence $f(x) \equiv 0 \pmod{m}$

Next consider $x_0 = 1$. Since $p_1 | f'(1)$ but $f(1) \not\equiv 0 \pmod{p_1^2}$, then 1 does not lift to a root of $f(x) \equiv 0 \pmod{p_1^2}$.

Part II. If $x^3 + 3x + 14 \equiv 0 \pmod{p_2}$, then $x_0 = 1$ (again, by inspection). This leads to $f(x_0) = 18$ and $f'(x_0) = 6$. Since $p_2 | f'(1)$ and $f(1) \equiv 0 \pmod{p_2}$, then there are $p_2 = 3$ solutions to $f(x) \equiv 0 \pmod{p_2^2}$. These are given by

$$x_0 + 0 \cdot p_2^1 = 1, \quad x_0 + 1 \cdot p_2^1 = 4, \quad x_0 + 2 \cdot p_2^1 = 7.$$

Part III. The Chinese Remainder Theorem now gives

$$\begin{cases} x \equiv 2 \pmod{4} \\ x \equiv 1 \pmod{9} \end{cases} \qquad \begin{cases} x \equiv 2 \pmod{4} \\ x \equiv 4 \pmod{9} \end{cases} \qquad \begin{cases} x \equiv 2 \pmod{4} \\ x \equiv 7 \pmod{9} \end{cases}$$

so so so

$$x \equiv 10 \pmod{36} \qquad x \equiv 22 \pmod{36} \qquad x \equiv 34 \pmod{36}. \quad \blacklozenge$$

Problems

4.56. Solve $5x \equiv 1 \pmod{21}$ by using Theorem 4.10 together with a quick application of the Chinese Remainder Theorem. In view of the known divisibility rule for division of integers by 5, how could you also solve this congruence by inspection?

4.57. Let p be a prime and suppose $p \nmid a$. If x_0 is the unique root of $ax \equiv b \pmod{p}$, is it also a root of $ax \equiv b \pmod{p^2}$? Would $x_0 + p$ be a root of $ax \equiv b \pmod{p^2}$? Would px_0 be a root of $ax \equiv b \pmod{p^2}$? Answer these questions by looking at examples.

4.58. How can you tell by instant inspection that the polynomial congruence $2x^4 + 4x^3 - 2x + 3 \equiv 0 \pmod{6}$ has no solution? How can you tell with a minimum of labor that $x^2 + 1 \equiv 0 \pmod{24}$ has no solution?

4.59. Solve the following polynomial congruences, if possible:

(a) $2x^3 + x \equiv 0 \pmod{27}$. (b) $3x^2 + 10x + 3 \equiv 0 \pmod{20}$.
(c) $4x^4 - 3x^3 + 2x^2 - x + 1 \equiv 0 \pmod{2401}$.

4.60. Let us follow up on Problem 4.57. Again, let x_0 be the unique root of $ax \equiv b \pmod{p}$, and now consider $ax \equiv b \pmod{p^2}$.

(a) How do you know that $ax \equiv b \pmod{p^2}$ even has a root?
(b) There exists an integer k such that $ax_0 = b + kp$. How do you know this?
(c) Suppose, from part (a), that $ax \equiv b \pmod{p^2}$ has a root; denote it by x_1. Assume that $x_1 = x_0 + cp$, where the constant c is to be determined. Show that this leads to the congruence $ac \equiv -k \pmod{p}$.
(d) The congruence in part (c) has a unique solution c_0 modulo p. How do you know this?
(e) Hence, x_1 can be determined. Apply the steps just outlined to the solution of $11x \equiv 6 \pmod{169}$.

4.61. Write a computer program that is sufficiently general to solve the following congruences:

(a)
$$\begin{cases} x \equiv 7 \pmod{9} \\ 7x \equiv 1 \pmod{10} \\ 8x \equiv 3 \pmod{13} \\ 2x \equiv 21 \pmod{77} \\ x \equiv 16 \pmod{23}. \end{cases}$$

(b) $2x^5 + 11x^3 + x + 6 \equiv 0 \pmod{2025}$.

(c) $\begin{cases} 2x^3 + 7x^2 + 4x + 3 \equiv 0 \pmod{60} \\ x^4 + x^3 + 5 \equiv 0 \pmod{91}. \end{cases}$

4.62. Define **Pochhammer's symbol**[7] by

$$(m)_n = m(m + 1) \ldots (m + n - 1),$$

where $m, n \in \mathbf{N}$ and $(m)_0 = 1$.

(a) First show that $n! \mid (m)_n$. [Hint: Look at the connection between Pochhammer's symbol and binomial coefficients.]
(b) Use this result to show that each coefficient in $f^{(k)}(x)$ is divisible by $k!$, where $f(x) \in \mathbf{Z}[x]$ and is of degree n, with $1 \leq k \leq n$. This result was needed in the proof of Theorem 4.11.

4.63. The congruence $f(x) \equiv 0 \pmod{p}$ is defined to have a ***k*-fold root** a modulo p $(k > 1)$ iff $f(x) \equiv (x - a)^k g(x) \pmod{p}$ is an identical congruence and $g(x) \in \mathbf{Z}[x]$, $g(a) \not\equiv 0 \pmod{p}$. Less precisely, we say that a is a **multiple root** of $f(x) \equiv 0 \pmod{p}$. The following theorem is an analog of a well-known result from algebra: A root a is a multiple root of $f(x) \equiv 0 \pmod{p}$ iff $f'(a) \equiv 0 \pmod{p}$. Try to prove this on your own. If you need help, consult the short proof of it given in Grosswald (1984).

RESEARCH PROBLEMS

1. Consider the general quadratic congruence $\Sigma_{k=0}^{2} a_k x^k \equiv 0 \pmod{p}$, where $p > 2$ is a prime. Investigate, for various choices of the coefficients a_0, a_1, a_2, the number of roots of the congruences. Also investigate possible connections between the roots and the coefficients. Then extend your investigation to the general cubic congruence $\Sigma_{k=0}^{3} b_k x^k \equiv 0 \pmod{p}$. Formulate conjectures.

2. Let p be an odd prime, and consider the family of congruences $(p - 1)! \equiv \pm D^{(p-1)/2} \pmod{p}$, where D is variable, $1 \leq D < p$. Select several primes p, and investigate for each choice which values of D will satisfy the indicated congruence when the $+$ sign is chosen, and which will satisfy it when the $-$ sign is chosen. Formulate conjectures.

3. Investigate the polynomial congruence

$$(2n - 1)! \equiv (n - 1)!n! \pmod{n^3}$$

for different choices of n.

[7] After the German mathematician **Leo A. Pochhammer** (1841–1920).

References

Apostol, T.M., *Introduction to Analytic Number Theory*, Springer, New York, 1976, pp. 106–128. These pages roughly parallel our two chapters on congruences; Apostol's writing is always clear and to the point.

Boyer, C.B., and Merzbach, U.C., *A History of Mathematics*, 2nd ed., John Wiley & Sons, New York, 1991, pp. 456–458. A few summarizing paragraphs on the theory of numbers in the Age of Euler, including the note on **John Wilson**.

Bruce, J.W., "A Really Trivial Proof of the Lucas–Lehmer Test," *Amer. Math. Monthly*, **100**, 370–371 (1993). The source of our proof of Theorem 4.8. A proof of the necessity part of the Lucas–Lehmer test would make a good project for extra credit.

Clement, P.A., "Congruences for Sets of Primes," *Amer. Math. Monthly*, **56**, 23–25 (1949). The source of our Problems 4.30 and 4.31.

Crandall, R., Dilcher, K., and Pomerance, C., "A Search for Wieferich and Wilson Primes," *Math. Comp.*, **66**, 433–449 (1997). This computer search uncovered no new Wilson primes beyond the three that **Fröberg** discovered.

Dence, J.B., and Dence, T.P., "A Necessary and Sufficient Condition for Twin Primes," *Missouri J. Math. Sci.*, **7**, 129–131 (1995). The source of our presentation of Theorem 4.5 and of our Problem 4.32.

Grosswald, E., *Topics from the Theory of Numbers*, 2nd ed., Birkhäuser, Boston, 1984, pp. 55–61. Nice section on congruences of higher order.

Kangsheng, S., "Historical Development of the Chinese Remainder Theorem," *Arch. Hist. Exact. Sci.*, **38**, 285–305 (1988). The writing is spotty in places, but the exposition is interesting. Unfortunately, the author does not give a clear indication of when **Sun Zi** lived.

Mahoney, M.S., *The Mathematical Career of Pierre de Fermat: 1601–1665*, 2nd ed., Princeton University Press, Princeton, NJ, 1994, pp. 288–302. An interesting, detailed account (as far as can be reconstructed) of the work by **Fermat** on aliquot parts (proper divisors).

Mozzochi, C.J., "A Simple Proof of the Chinese Remainder Theorem," *Amer. Math. Monthly*, **74**, 998 (1967). The source of our Problem 4.55, a nice exercise for extra credit.

Needham, J., *Science and Civilization in China*, Vol. 3, "Mathematics and the Sciences of the Heavens and the Earth," Cambridge University Press, Cambridge, England, 1959, pp. 1–168. A massive scholarly work by one of the foremost authorities on China in antiquity. **L.E. Dickson** claims that **Nicomachus of Gerasa** (ca. 100 A.D.) in Judea give the same treatment of systems of linear congruences as did **Sun Zi**. Other authors have perpetuated this statement. **Needham** points out that this is false and arose from a copying error of Nicomachus's work made in the West during the Middle Ages. Interestingly, later Chinese mathematicians after Sun Zi considered systems of congruences where the moduli are not pairwise coprime (pp. 119–122).

Ribenboim, P., *The New Book of Prime Number Records*, Springer, New York, 1996, pp. 346–350. A terse summary of what is known about Wilson primes. The author points out that the search for them could be aided by the identity

$$(p-1)! \equiv (-1)^{(p-1)/2}4^{p-1}\left[\left(\frac{p-1}{2}\right)!\right]^2 \pmod{p^2}.$$

Chapter 5
Primitive Roots

5.1 Order of an Integer

In this chapter (and the next) we look at various aspects of the family of congruences $x^k \equiv A \pmod{n}$, where the integer $k \geq 2$ and n may be prime or composite. We begin with a definition that is analogous to the definition of the order of an element of a group (see Problem 2.41).

DEF. Suppose $1 \leq x < n$ and $(x, n) = 1$. If k is the least positive integer such that $x^k \equiv 1 \pmod{n}$, then k is said to be the **order** of $x \pmod{n}$. Alternative (older) terminology is that x is said to **belong to the exponent** k modulo n.

That k must exist follows from Euler's Theorem (Theorem 3.21), that is, for any positive integer x coprime to n, $x^{\phi(n)} \equiv 1 \pmod{n}$ automatically holds. There may be positive integers, s, less than $\phi(n)$ and such that $x^s \equiv 1 \pmod{n}$ also holds. By the Well-Ordering Property, the set of all such integers has a smallest member k.

Example 5.1 Let $n = 9, x = 4$; then $4^2 \equiv 7 \pmod{9}$, but $4^3 \equiv 1 \pmod{9}$. Thus, the order of $4 \pmod{9}$ is 3. ◆

Given a positive integer x and a modulus n, the question arises as to how many different choices of the exponent k would one have to check in order to determine the order of $x \pmod{n}$. Euler's Theorem tells us that we need only examine choices of k in the interval $[1, \phi(n)]$. The following short theorem narrows the search even further.

Theorem 5.1 Let the order of $x \pmod{n}$ be k. If $x^m \equiv 1 \pmod{n}$, then $k|m$.

Proof The proof is identical to the proof of Lemma 4.8.3. ◆

Example 5.2 Let $n = 35$, $x = 3$; since $\phi(35) = \phi(5)\phi(7) = 4 \cdot 6 = 24$, we check only the divisors of 24 as possible exponents k. The set of divisors is $\{1, 2, 3, 4, 6, 8, 12, 24\}$. We now find that $3^1 \not\equiv 1 \pmod{35}$, $3^2 \not\equiv 1 \pmod{35}$, $3^3 \not\equiv 1 \pmod{35}$, $3^4 \not\equiv 1 \pmod{35}$, $3^6 \not\equiv 1 \pmod{35}$, $3^8 \not\equiv 1 \pmod{35}$, but $3^{12} \equiv 1 \pmod{35}$. Thus, the order of 3 (mod 35) is 12. ◆

It is plausible to look for a relationship between the order of $x \pmod{n}$ and the order of powers of $x \pmod{n}$. For example, from Example 5.2 we have that the order of 3^2 (mod 35) is 6 (because $3^{12} = (3^2)^6$), and that the order of 3^3 (mod 35) is 4 (because $3^{12} = (3^3)^4$).

What if we were interested in the order of 3^5 (mod 35)? In this case $5 \nmid 12$, so the reasoning is not so quick as in the two previous cases. What do you think? The general relationship is given shortly, which we state as a lemma since it is needed subsequently for the most important result in this section.

Lemma 5.2.1 Let $(x, n) = 1$ and k be the order of $x \pmod{n}$. Then for $h \in \mathbf{N}$, one has:

$$\text{order of } x^h \pmod{n} = k/(h, k).$$

Proof By definition the order of $x^h \pmod{n}$ is the smallest positive integer k' such that

$$x^{hk'} \equiv 1 \pmod{n}.$$

Since $x^0 \equiv x^k \equiv x^{2k} \equiv \ldots \equiv 1 \pmod{n}$, we need to find the smallest positive integer k' such that $hk' \equiv 0 \pmod{k}$. From Theorem 3.3, this latter congruence is equivalent to $k' \equiv 0 \pmod{[k/(h, k)]}$. Clearly, the least positive solution of this is $k' = k/(h, k)$, and this is the lemma. ◆

Example 5.3 As in Example 5.2, let $n = 35$ and $x = 3$; we have $k = 12$. Suppose $h = 8$; then $k/(h, k) = 12/(8, 12) = 3$. That is, 3 is the smallest integer k' such that $(3^8)^{k'} \equiv 1 \pmod{35}$: the order of 3^8 (mod 35) is 3. ◆

When the modulus is a prime p, the order of an integer could conceivably be as large as $p - 1$, but no larger (why?). For this special case, we can use Lemma 5.2.1 to prove the concluding result in this section, a result that leads directly into the concept of primitive roots.

Theorem 5.2 Let p be a prime. If there is at least one integer of order $p - 1$ modulo p, then there are actually $\phi(p - 1)$ incongruent integers of that order, where ϕ is the Euler ϕ-function.

Proof The theorem is trivial for $p = 2, 3$, so assume that $p \geq 5$. Let x_0 be of order $p - 1$; then $(x_0^k)^{p-1} \equiv 1 \pmod{p}$ is true for $k = 1, 2, \ldots, p - 1$. Thus, the congruence

$$x^{p-1} - 1 \equiv 0 \pmod{p}$$

has $(p - 1)$ solutions, namely, $x_0, x_0^2, x_0^3, \ldots, x_0^{p-1}$. These solutions are incongruent modulo p because $p - 1$ is the order of x_0.

The theorem of Lagrange (Theorem 4.2) tells us that there are at most $p - 1$ solutions modulo p to the preceding congruence, so $x_0, x_0^2, x_0^3, \ldots, x_0^{p-1}$ must be the complete set of solutions. Therefore, each incongruent integer of order $p - 1$ is necessarily of the form x_0^k for some k satisfying $1 \leq k \leq p - 1$. By Lemma 5.2.1 this is true iff $(k, p - 1) = 1$, so by definition k is one of the $\phi(p - 1)$ integers (mod p) that are coprime to $p - 1$. ◆

Example 5.4 Let $p = 11$, so $\phi(p - 1) = 4$. By trial, 2 is found to be of order 10 (mod 11). By Theorem 5.2 there should be three other incongruent integers of order 10. Since $10/(3, 10) = 10$, we see from Lemma 5.2.1 that $2^3 = 8$ has order 10 (mod 11). Similarly, $2^7 \equiv 7 \pmod{11}$ and $2^9 \equiv 6 \pmod{11}$ are of order 10 (mod 11) because $10/(7, 10) = 10/(9, 10) = 10$. ◆

Problems

5.1. In each case find the order of $x \pmod{n}$:

(a) $x = 2, n = 7$.
(b) $x = 5, n = 11$.
(c) $x = 3, n = 23$.
(d) $x = 10, n = 33$.
(e) $x = 3, n = 19$.
(f) $x = 4, n = 17$.

5.2. Can an integer have order that is a prime? Does the answer to this depend on whether the modulus is itself prime or composite?

5.3. Refer to Problem 5.1. Now compute the order of $x^h \pmod{n}$:

(a) $x = 5, h = 3, n = 11$.
(b) $x = 3, h = 4, n = 19$.
(c) $x = 4, h = 2, n = 17$.

5.4. If the order of $x^2 \pmod{n}$ is r_1, the order of $x^3 \pmod{n}$ is r_2, can we conclude that $r_1 < r_2$? That $r_1 > r_2$?

5.5. The order of 9 to some unspecified modulus is 5. Does this information determine uniquely the order of 3 to that same modulus?

5.6. Show that if x has order 3 (mod p), then $x + 1$ has order 6 (mod p). [Hint: Factor $x^3 - 1$.]. Illustrate with a concrete example.

5.7. Consider the prime $p = 31$ and let N_{31} be the number of incongruent integers modulo p whose order k satisfies $2 \leq k < p - 1$. Show that $N_{31} = 21$.

5.8. Show that for any integer $n > 1$ one has $n | \phi(2^n - 1)$. Illustrate with a concrete example.

5.9. Suppose that $n - 1$ is the order of x modulo n. Prove that n must be a prime.

5.2 Some Exploratory Computations

In this section we shall do some suggestive calculations that will point the way toward material in the next two sections. Our object here is to investigate the orders of the elements in various reduced residue systems.

To begin, we let the symbol $\mathbf{Z}_p^\times$ denote the set of units in the field $\mathbf{Z}_p$. It is, of course, a reduced residue system modulo p, and from Corollary 3.13.2 is a multiplicative group of order $p - 1$. Corollary 3.19.1 tells us that the order of any $x \in \mathbf{Z}_p^\times$ must be a divisor of $p - 1$.

Let us now choose for study the primes $p = 3, 5, 7, 11$. For each of these we work out the order of every element in $\mathbf{Z}_p^\times$. We note that the simple reduced residue system $\mathbf{Z}_2^\times$ has only one element (the unity), and its order is 1.

Example 5.5 Do the calculations just suggested for the primes $p = 3, 5, 7, 11$. For each prime rank the elements in increasing order. ◆

Is there in each case an element x of order $p - 1$? If so, that element can generate the entire group because all of the powers x^k, $1 \leq k \leq p - 1$, are incongruent modulo p (Problem 5.12). Table 5.1 shows the results.

Table 5.1 Orders of Elements in $\mathbf{Z}_p^\times$

p	Element (Order)
3	1(1); 2(2)
5	1(1); 4(2); 2(4), 3(4)
7	1(1); 6(2); 2(3), 4(3); 3(6), 5(6)
11	1(1); 10(2); 3(5), 4(5), 5(5), 9(5); 2(10), 6(10), 7(10), 8(10)

We see that in each case there is an element $x \in \mathbf{Z}_p^\times$ that can generate all of $\mathbf{Z}_p^\times$. We surmise (and will reexamine later) Theorem 5.3.

Theorem 5.3 If p is a prime, then $\mathbf{Z}_p^\times$ is a cyclic group of order $p - 1$.

Now let us consider some rings $\mathbf{Z}_n$, where n is composite. By extension, the notation $\mathbf{Z}_n^\times$ shall mean the set of all units in $\mathbf{Z}_n$. It is also a multiplicative group, but not one of order $n - 1$. It is not generally true that the order of every $x \in \mathbf{Z}_n^\times$ is a divisor of $n - 1$. However, $\mathbf{Z}_n^\times$ is a reduced residue system modulo n and thus contains $\phi(n)$ elements. Hence, the order of each $x \in \mathbf{Z}_n^\times$ must be a divisor of $\phi(n)$.

Let us choose for study $n = 8, 9, 10, 12, 14, 15, 21, 25$. For each of these we work out the order of every element in $\mathbf{Z}_n^\times$.

Example 5.6 Do the calculations just suggested for the integers $n = 8, 9, 10, 12, 14, 15, 21, 25$. For each n rank the elements in increasing order. ◆

Is there in each case an element x of order $\phi(n)$? Table 5.2 shows the results; we see that the reduced residue systems for $n = 9, 10, 14, 25$ have an element x of order $\phi(n)$, but those systems for $n = 8, 12, 15, 21$ do not.

In extending (the alleged) Theorem 5.3, we see that there are some composite integers n such that $\mathbf{Z}_n^\times$ is a cyclic group of order $\phi(n)$. We gather together in Table 5.3 all of the cases of moduli m that we know for which the reduced residue systems are cyclic multiplicative groups.

Table 5.2 Orders of Elements in $\mathbf{Z}_n^\times$

n	$\phi(n)$	Element (Order)	Number of Generators of $\mathbf{Z}_n^\times$
8	4	1(1); 3(2), 5(2), 7(2)	—
9	6	1(1); 8(2); 4(3), 7(3); 2(6), 5(6)	2
10	4	1(1); 9(2); 3(4), 7(4)	2
12	4	1(1); 5(2), 7(2), 11(2)	—
14	6	1(1); 13(2); 9(3), 11(3); 3(6), 5(6)	2
15	8	1(1); 4(2), 11(2), 14(2); 2(4), 7(4), 8(4), 13(4)	—
21	12	1(1); 8(2), 20(2); 4(3), 16(3); 13(4); 2(6), 5(6), 10(6), 11(6), 17(6), 19(6)	—
25	20	1(1); 24(2); 7(4), 18(4); 6(5), 11(5), 16(5), 21(5); 4(10), 9(10), 14(10), 19(10); 2(20), 3(20), 8(20), 12(20), 13(20), 17(20), 22(20), 23(20)	8

Table 5.3 Moduli m for Which $\mathbf{Z}_m^\times$ is Cyclic

m	Order of $\mathbf{Z}_m^\times$	m	Order of $\mathbf{Z}_m^\times$
2	1	3^2	6
3	2	$2 \cdot 5$	4
2^2	2	11	10
5	4	$2 \cdot 7$	6
7	6	5^2	20

At this point can you make any conjectures based on the data in Tables 5.1–5.3? If necessary, extend the tables by adding a few more entries.

Problems

5.10. Add the prime $p = 13$ to Table 5.1 and work out the orders of all the elements in $\mathbf{Z}_{13}^\times$. Is $\mathbf{Z}_{13}^\times$ cyclic?

5.11. Refer to the previous problem. For each $\mathbf{Z}_p^\times$ in your Table 5.1, determine how many generators $\mathbf{Z}_p^\times$ has. Is Theorem 5.2 upheld?

5.12. Suppose that $x \in \mathbf{Z}_p^\times$ has order $p - 1$. Prove that the integers x^1, x^2, $x^3, \ldots, x^{p-1}$ are incongruent modulo p.

5.13. Show explicitly that 11 generates the entire multiplicative group $\mathbf{Z}_{17}^\times$. Show that $\mathbf{Z}_{20}^\times$, the set of units in $\mathbf{Z}_{20}$, is not cyclic.

5.14. Add to Table 5.2 the cases $n = 24, 27$. Then, if possible, extend Table 5.3. Do you see now the relationship for determining the number of generators in $\mathbf{Z}_n^\times$ for composite n?

5.15. In this problem we define the Cartesian product group

$$\langle \mathbf{G}, \cdot \rangle = \langle \mathbf{Z}_4 \times \mathbf{Z}_3, \cdot \rangle \quad \text{by} \quad \mathbf{G} = \{g\colon g = (x, y), x \in \mathbf{Z}_4, y \in \mathbf{Z}_3\},$$

where $\mathbf{Z}_4$, $\mathbf{Z}_3$, are the additive groups modulo 4, 3, respectively, (x, y) means the ordered pair of integers x and y, and $g \cdot g' = (x, y) \cdot (x', y') = (x + x' \pmod 4, y + y' \pmod 3))$.

(a) Work out the group multiplication table for **G**.
(b) Sort the elements of **G** according to their orders. Is **G** cyclic?
(c) Does **G** conform to Problem 2.43?

5.16. Let the set $\mathbf{Z}_p^{\times 2}$ be defined as follows:

$$\mathbf{Z}_p^{\times 2} = \{x^2\colon x \in \mathbf{Z}_p^\times\}.$$

(a) Show that $\mathbf{Z}_p^{\times 2}$ is a subgroup of $\mathbf{Z}_p^{\times}$.
(b) If $|\mathbf{Z}_p^{\times 2}|$ denotes the order of $\mathbf{Z}_p^{\times 2}$, show that $|\mathbf{Z}_p^{\times}|/|\mathbf{Z}_p^{\times 2}| = 2$ when $p \geq 3$.
(c) Work out the elements of $\mathbf{Z}_{17}^{\times 2}$. How many of the generators of $\mathbf{Z}_{17}^{\times}$ are found outside of $\mathbf{Z}_{17}^{\times 2}$?

5.3 Primitive Roots

Theorem 5.2 is the inspiration for the definition of a primitive root. We will see in the next section that primitive roots are somewhat analogous to logarithmic bases.

DEF. A positive integer $g \pmod{p}$ whose order is $p - 1$ is called a **primitive root** of p. More generally, a positive integer g whose order $\pmod{n}$ is $\phi(n)$ is called a primitive root of n.

An integer n may have several primitive roots or it may have none (Problem 5.18). For $n > 2$, a primitive root (if it exists) of n must exceed 1. Also note that a primitive root g must satisfy $(g, n) = 1$. We define $n = 1$ to have the primitive root $g = 1$.

Example 5.7 From Example 5.4 or Table 5.1 we see that $p = 11$ has four primitive roots. On the other hand, let $n = 8$; the integers (mod 8) coprime to it are 1, 3, 5, 7. Checking only 3, 5, 7, we find $3^2 \equiv 1 \pmod{8}$, $5^2 \equiv 1 \pmod{8}$, and $7^2 \equiv 1 \pmod{8}$, so 8 has no primitive roots (see Table 5.2). ◆

Table 5.4 gives a brief listing of primitive roots $\{g\}$ of both primes and composite integers. A larger selection of data is given in Table D in Appendix II. A still larger selection is in (Osborne, 1961). You are invited to scrutinize Table 5.4 and see if you can formulate any conjectures.

Theorem 5.4 Let g be a primitive root of n. Then

(i) $g^0, g^1, g^2, \ldots, g^{\phi(n)-1}$ are incongruent $\pmod{n}$.
(ii) g^h is also a primitive root of n iff $(h, \phi(n)) = 1$.

Proof (i) Suppose, to the contrary, that $g^i \equiv g^j \pmod{n}$ holds for some i, j satisfying $0 \leq j < i < \phi(n)$. Then $g^{i-j} \equiv 1 \pmod{n}$, since $(g, n) = 1$ holds; this contradicts the hypothesis that the order of g is $\phi(n)$. Hence, $\{1, g,$

Table 5.4 Primitive Roots

n	$\phi(n)$	$\phi\{\phi(n)\}$	g
3	2	1	2
4	2	1	3
5	4	2	2, 3
6	2	1	5
7	6	2	3, 5
9	6	2	2, 5
10	4	2	3, 7
11	10	4	2, 6, 7, 8
13	12	4	2, 6, 7, 11
14	6	2	3, 5
17	16	8	3, 5, 6, 7, 10, 11, 12, 14
19	18	6	2, 3, 10, 13, 14, 15
26	12	4	7, 11, 15, 19
31	30	8	3, 11, 12, 13, 17, 21, 22, 24

$g^2, \ldots, g^{\phi(n)-1}\}$ is a set of $\phi(n)$ incongruent integers modulo n, and is thus a reduced residue system modulo n.

(ii) In Lemma 5.2.1 set $k = \phi(n)$. Then we have immediately: order of $g^h \pmod{n} = [\phi(n)]/(h, \phi(n)) = \phi(n)$ iff $(h, \phi(n)) = 1$. This says g^h is then, and only then, a primitive root of n. ◆

Part (ii) of Theorem 5.4 is especially significant because it allows us to find easily all of the other (incongruent) primitive roots of n if we know one primitive root. We've already done this once, in Example 5.4.

Example 5.8 One primitive root of $n = 14$ is $g = 3$. Find the others. Since $\phi(14) = 6$, we use those values of h such that $(h, 6) = 1$; hence, $h = 1, 5$. We find $3^h = 3^5 = 243 \equiv 5 \pmod{14}$, so 5 is the only other primitive root of 14. ◆

That there are only two incongruent primitive roots of $n = 14$ is assured by Corollary 5.4.1.

Corollary 5.4.1 If n has a primitive root g, then it has exactly $\phi\{\phi(n)\}$ incongruent primitive roots.

The corollary is a generalization of Theorem 5.2. Various examples for composite n are given in Table 5.4. The corollary effectively answers Problem 5.14.

Problems

5.17. Find (and justify) two primitive roots of each of the following moduli: (a) $n = 22$. (b) $n = 23$. (c) $n = 37$. (d) $n = 41$. (e) $n = 74$.

5.18. By trial show that 35 has no primitive roots.

5.19. From Table 5.4 we see that 3 is a primitive root of 14. Show that the first six powers of 3 yield a reduced residue system of 14, but that the first six powers of 11 or 13 do not yield a reduced residue system of 14.

5.20. Let $\mathbf{G}$ be the set of incongruent primitive roots of n, and define $\mathbf{S} = \{1\} \cup \mathbf{G}$. Can $\mathbf{S}$ ever be a multiplicative group?

5.21. Assume that the prime $p > 2$ has an odd primitive root g. Prove that $2p$ must have a primitive root. In fact, how many primitive roots should $2p$ possess?

5.22. Let $p > 2$ be a prime and g be a primitive root of p. Show that

$$\prod_{i=1}^{p-1} g^i \equiv -1 \pmod{p}.$$

5.23. Suppose $(x, n) = 1$. Then if the set $\{x, x^2, \ldots, x^{\phi(n)}\}$ forms a reduced residue system modulo n, show that the integer x is a primitive root (mod n).

5.24. The smallest modulus n that has 10 as its smallest primitive root is larger than 100. Write a computer program to uncover this smallest modulus n.

5.25. Write a computer program that will identify all of the primitive roots of a prime p and all of the primitive roots of $2p$ (Problem 5.21). Run the program with a few choices of p, and in each case compare the two sets of primitive roots. Formulate a conjecture.

5.26. Let $p > 9$ be a prime. Examination of Table D in Appendix II suggests that 9 is never a primitive root of p. Prove why this is so. [Hint: Make use of Euler's Theorem (Theorem 3.21).]

5.27. You are given that 2 is a primitive root of 29 and 5 is a primitive root of 73. From this deduce an integer that is a primitive root simultaneously of 29 and 73.

5.28. In this problem let p be an odd prime greater than 3.

(a) If $1 \le x < p - 1$ satisfies $(x, p - 1) = 1$, then show that $(p - 1 - x, p - 1) = 1$.

(b) Let S_p be the sum of all the members of the least positive reduced residue system modulo $p - 1$. Then show that $(p - 1) | S_p$.

(c) Let $g_1, g_2, \ldots, g_{\phi(p-1)}$ be the primitive roots of p. Choosing g_i arbitrarily, show that each primitive root is a power of g_i (mod p), where the exponent is coprime to $p - 1$. Illustrate this with the case $p = 11$.

(d) Finally, using parts (b) and (c), prove that $\Pi_{i=1}^{\phi(p-1)} g_i \equiv 1 \pmod p$. Illustrate this with the case of $p = 19$.

5.4 Indices

Since the powers of a primitive root g constitute a reduced residue system, then corresponding to any integer A coprime to the modulus n there exists an a such that

$$A \equiv g^a \pmod n, \qquad 0 \leq a < \phi(n).$$

We call a the **index** of A (mod n) and write $a = \text{ind}_g A \pmod{\phi(n)}$ or, when the modulus $\phi(n)$ is understood, as just $a = \text{ind}_g A$.

Example 5.9 From Example 5.4 a primitive root of $n = 11$ is $g = 2$. As seen there, $2^7 \equiv 7 \pmod{11}$, so $\text{ind}_2 7 \pmod{10} = 7$. ◆

It is an interesting fact (first pointed out by **Gauss** in 1801) that the function ind_g obeys a calculus similar to that of the logarithmic function to the base g. The following theorem shows how far one can go with the analogy.

Theorem 5.5 Let g be a primitive root of n, and suppose that $(A, n) = (B, n) = 1$. Then

(i) $\text{Ind}_g 1 = 0$, $\text{Ind}_g g = 1$,
(ii) $\text{Ind}_g AB \equiv \text{ind}_g A + \text{ind}_g B \pmod{\phi(n)}$,
(iii) $\text{Ind}_g A^k \equiv k \cdot \text{ind}_g A \pmod{\phi(n)}$,
(iv) (the "base interchange formula") If $g_1 \equiv g^b \pmod n$ is another primitive root of n, then $\text{ind}_g A \equiv (\text{ind}_g g_1)(\text{ind} g_1 A) \pmod{\phi(n)}$.

Proof Part (i) is immediate from the definition of the index. We prove part (iv) and leave (ii) and (iii) as exercises. Let $a = \text{ind}_g A$, $a' = \text{ind}_{g_1} A$, and $b = \text{ind}_g g_1$. Then $A \equiv g^a \pmod n$ and $A \equiv g_1^{a'} \pmod n$, so $g^a \equiv g_1^{a'} \pmod n$. Using Theorem 3.2, part (6), we have

$$(g^b)^a \equiv g_1^{a'b} \pmod n.$$

But $g^b \equiv g_1 \pmod n$, so $g_1^a \equiv g_1^{a'b} \pmod n$ and this is true iff $a \equiv a'b \pmod{\phi(n)}$. ◆

Table 5.5 Indices of Selected Integers A Modulo Selected Primes p and to Selected Bases g (Primitive Roots)*

p	3	5		7		11		13		17		19		23	
A \ g	2	2	3	3	5	2	6	2	6	3	5	2	3	5	7
1	0	0	0	0	0	0	0	0	0	0	0	0	0	0	0
2	1	1	3	2	4	1	9	1	5	14	6	1	7	2	14
3		3	1	1	5	8	2	4	8	1	13	13	1	16	2
4		2	2	4	2	2	8	2	10	12	12	2	14	4	6
5				5	1	4	6	9	9	5	1	16	4	1	7
6				3	3	9	1	5	1	15	3	14	8	18	16
7						7	3	11	7	11	15	6	6	19	1
8						3	7	3	3	10	2	3	3	6	20
9						6	4	8	4	2	10	8	2	10	4
10						5	5	10	2	3	7	17	11	3	21
11								7	11	7	11	12	12	9	19
12								6	6	13	9	15	15	20	8
13										4	4	5	17	14	10
14										9	5	7	13	21	15
15										6	14	11	4	17	9
16										8	8	4	10	8	12
17												10	16	7	5
18												9	9	12	18
19														15	17
20														5	13
21														13	3
22														11	11

*Taken from the University of Oklahoma Mathematical Tables Project (1962).

In Table 5.5 we give a brief selection of indices. A larger collection appears in Table E in Appendix II. Note that as with logarithms, the index of an integer A depends upon which primitive root (base) g is chosen (Problem 5.32). Conventional tables usually employ the smallest primitive roots.

Indices are useful in solving congruences. The following examples are illustrative.

Example 5.10 Solve $3x \equiv 11 \pmod{23}$.

The relations in Theorem 5.5 do not depend upon the choice of g, provided the same g is used throughout a given congruence. For $p = 23$, let us arbitrarily use $g = 7$. Then the problem is equivalent to

$$\text{ind}_7 3 + \text{ind}_7 x \equiv \text{ind}_7 11 \pmod{22}$$

or

$$2 + \text{ind}_7 x \equiv 19 \pmod{22}.$$

Hence, $\text{ind}_7 x \equiv 17 \pmod{22}$, so reading downward we find $x \equiv 19 \pmod{23}$. To check: $3 \cdot 19 - 11 = 46 = 2 \cdot 23$. ◆

Example 5.11 Solve $x^3 \equiv 5 \pmod{13}$.

Choose $g = 2$; the congruence is equivalent to

$$3 \ \text{ind}_2 x \equiv \text{ind}_2 5 \pmod{12}.$$

From Theorem 3.22 this is solvable for $\text{ind}_2 x$ iff $\text{ind}_2 5$ is divisible by $d = (3, 12) = 3$, in which case there are $d = 3$ incongruent solutions modulo 12. We obtain $\text{ind}_2 x \equiv 3 \pmod 4$, or $\text{ind}_2 x = 3, 7, 11$. From the table we read downward $x \equiv 8, 7, 11 \pmod{13}$. You may check these results. ◆

We now have five techniques for solving the linear congruence $ax \equiv b \pmod n$:

1. By inspection.
2. By the use of Euler's Theorem (Example 3.32).
3. By the use of continued fractions (Corollary 3.25.1).
4. By the use of the Chinese Remainder Theorem (Corollary 4.10.1).
5. By the use of indices.

Although the last method is the most rapid (provided that the necessary indices are available), it has the narrowest scope of applicability. The modulus n must possess a primitive root. We face the issue of when this is true in the next section.

Problems

5.29. Prove parts (ii) and (iii) of Theorem 5.5.

5.30. Let $p = 17$; compute the indices of $A = 1, 2, 3, \ldots, 16$ with the base $g = 6$ using the "base interchange formula" of Theorem 5.5.

5.31. In Table 5.5 notice that for $p = 13$ one has $\text{ind}_2 A + \text{ind}_6 A \equiv 0 \pmod 6$ for all A in the interval $[1, 12]$. Why is this?

5.32. Rework Examples 5.10 and 5.11 using different bases.

5.33. Solve the following congruences using indices:

(a) $41x \equiv 5 \pmod{43}$. (b) $11x + 5 \equiv 0 \pmod{17}$.

(c) $5x^2 \equiv 1 \pmod{11}$.
(d) $2(3x+4)^2 \equiv 9 \pmod{23}$.
(e) $x^4 + x^2 + 1 \equiv 4 \pmod{43}$.
(f) $4^x \equiv 29 \pmod{53}$.

5.34. Find a primitive root of $n = 25$ and use it to prepare a table of indices for A, $1 \leq A < 25$.

5.35. Use your table in Problem 5.34 to solve

(a) $3 \cdot 2^t \equiv 7 \pmod{25}$.
(b) $2^t + 2^{3t} \equiv 0 \pmod{25}$.
(c) $24 \cdot t^t \equiv 7 \pmod{25}$.

5.36. Find an infinite family of positive integers h such that the units and tens digits of $3 \cdot 2^h$ are the same as the units and tens digit of 6^{2h}.

5.37. Let $p > 2$ be a prime and g be one of its primitive roots. Let $\{a_1, a_2, \ldots, a_{p-1}\}$ be the set of indices (mod p) of the integers $1, 2, \ldots, p-1$. Prove that $p-2$ divides $\Sigma_{i=1}^{p-1} a_i$. Illustrate this result with $p = 37$.

5.38. Solve this system of congruences:

$$\begin{cases} 3x^3 \equiv 5y \pmod{19} \\ 2x \equiv -15y^4 \pmod{19}. \end{cases}$$

5.5 The Existence of Primitive Roots

We now have to address the question of which integers n have primitive roots. This section supplies first the necessary condition for this, and then the sufficient condition (Hua, 1982). The results of this section were first established by **Gauss** in 1801.

5.5.1 Necessity

This part of the proof considers the congruence $g^k \equiv 1 \pmod{n}$ and determines what structure n must have if the minimum k is to be $\phi(n)$. To get started, we set out two elementary lemmas: one dealing with the least common multiple of divisors of n (needed because n may be composite), and one dealing with the Euler ϕ-function (needed because of the special requirement just noted for k). The lemma on the least common multiple is motivated by the following calculations.

Example 5.12 Let $n_1 = 3$, $n_2 = 4$, $n_3 = 6$, and $n = n_1 n_2 n_3 = 72$. Then $[n_1, n_2, n_3] = 12$; by rough analogy to Theorem 2.15, we attempt to express $[n_1, n_2, n_3]$ as a quotient n/D. Clearly, $D = 6$, but $6 \neq (n_1, n_2, n_3)$ as one might have wished. Just what is the significance of the 6?

Or consider $n_1 = 8$, $n_2 = 10$, $n_3 = 12$, $n = 960$. We compute $[n_1, n_2, n_3] = 120 = 960/D$; hence, $D = 8 > (n_1, n_2, n_3) = 2$. Again, what is the significance of the 8? Before reading on, see if you can make any reasonable guesses as to what is happening. Here is still one more example: $n_1 = 2$, $n_2 = 9$, $n_3 = 15$, and $n_1 n_2 n_3 / [n_1, n_2, n_3] = 270/90 = 3$. In general, it appears that $(\Pi_{i=1}^{r} n_i)/[n_1, n_2, \ldots, n_r]$ does not necessarily equal one of the n_i's, and is greater than $(n_1, n_2, \ldots, n_r)$. ◆

Lemma 5.6.1 Let $n = \Pi_{i=1}^{r} n_i$, where the positive integers n_i are not necessarily pairwise coprime. Then $[n_1, n_2, \ldots, n_r] = n/D$, where D is the greatest common divisor of all the products of the n_i's taken $r - 1$ at a time.

Proof The lemma is true for $r = 2$ by Theorem 2.15. Use mathematical induction (Problem 5.42) to show that it is true for an arbitrary integer $r \geq 2$. ◆

In the special case where the n_i's are pairwise coprime, then $D = 1$ and the lemma reduces to the expected result $[n_1, n_2, \ldots, n_r] = \Pi_{i=1}^{r} n_i = n$.

Example 5.13 Let $n_1 = 5$, $n_2 = 9$, $n_3 = 15$, $n_4 = 21$; then $n = 14{,}175$. By inspection, $[5, 9, 15, 21] = 315$. We compute

$$
\begin{aligned}
D &= (n_1 n_2 n_3,\ n_1 n_2 n_4,\ n_1 n_3 n_4,\ n_2 n_3 n_4) \\
&= (675,\ 945,\ 1575,\ 2835) \\
&= 45.
\end{aligned}
$$

Finally, we observe that $14{,}175/45 = 315$. ◆

Lemma 5.6.2 Let $n = \Pi_{i=1}^{r} p_i^{\alpha_i}$, where the p_i are odd primes and $r \geq 2$. Then

$$[\phi(p_1^{\alpha_1}),\ \phi(p_2^{\alpha_2}),\ \ldots,\ \phi(p_r^{\alpha_r})] < \phi(n).$$

Proof The integers $1, 2, \ldots, p_i - 1$ are coprime to p_i. For $p_i^{\alpha_i} = (p_i^{\alpha_i - 1})p_i$ there are $p_i^{\alpha_i - 1}$ similar sets of integers less than and coprime to $p_i^{\alpha_i}$. Thus, $\phi(p_i^{\alpha_i}) = (p_i^{\alpha_i - 1})(p_i - 1)$, which is necessarily even.

Let D be the greatest common divisor of all of the products of the $\phi(p_i^{\alpha_i})$'s taken $r-1$ at a time. Since each $\phi(p_i^{\alpha_i})$ is even, then $D \geq 2$ and from Lemma 5.6.1 we have

$$\begin{aligned}[\phi(p_1^{\alpha_1}), \phi(p_2^{\alpha_2}), \ldots, \phi(p_r^{\alpha_r})] &= D^{-1} \prod_{i=1}^{r} \phi(p_i^{\alpha_i}) \\ &= \frac{\phi(n)}{D} \\ &\leq \frac{\phi(n)}{2} \\ &< \phi(n).\end{aligned}$$ ◆

Example 5.14 In Example 5.13 write $n = 3^4 \cdot 5^2 \cdot 7^1$ and compute $\phi(3^4) = 54$, $\phi(5^2) = 20$, $\phi(7^1) = 6$, $\phi(n) = 6480$. Then $[54, 20, 6] = 540 < 6480$. ◆

Theorem 5.6 If $n > 1$ has a primitive root, then n must be of the form 2^k, p^k, or $2p^k$, where p is some odd prime.

Proof Let n have the factorization $n = \Pi_{i=1}^{r} p_i^{\alpha_i}$, with $p_1 < p_2 < \ldots < p_r$. By Euler's Theorem we have

$$a^{\phi(p_i^{\alpha_i})} \equiv 1 \pmod{p_i^{\alpha_i}}$$

for any $a \not\equiv 0 \pmod{p_i}$. Similar congruences for all the p_i's in n can be simultaneously expressed by $a^L \equiv 1 \pmod n$, where L is the least common multiple of all of the $\phi(p_i^{\alpha_i})$'s. If n has a primitive root g, then we must have $g^{\phi(n)} \equiv 1 \pmod n$ and $L < \phi(n)$ is prohibited. But from Lemma 5.6.2 we see that for $p_1 \geq 3$ and $r \geq 2$, $L < \phi(n)$ always holds. Thus, n cannot contain two or more distinct odd primes, that is, if n is odd, it must be of the form p^k. If n is even ($p_1 = 2$), then $L = \phi(n)$ implies $r \leq 2$ and

$$[\phi(2^{\alpha_1}), \phi(p^{\alpha_2})] = \phi(2^{\alpha_1})\phi(p^{\alpha_2})$$

only when $\alpha_2 = 0$, or $\alpha_2 > 0$ and $\alpha_1 = 1$, so n can also have the form 2^k or $2p^k$. ◆

Theorem 5.7 If $n = 2^k$ has a primitive root, then $k \leq 2$.

Proof We argue by contrapositive. The congruence

$$a^{2^{k-2}} \equiv 1 \pmod{2^k}$$

holds for all odd $a > 0$ and $k = 3$ (see Example 3.3). Assume it holds for $k = m > 3$, so that $a^{2^{m-2}} = 1 + 2^m c$ for some $c \in \mathbf{N}$. Squaring gives

$$a^{2^{m-1}} = 1 + 2^{m+1}(c + 2^{m-1}c^2)$$

or, equivalently, $a^{2^{m-1}} \equiv 1 \pmod{2^{m+1}}$. Therefore, by the Principle of Finite Induction the initial congruence holds for all integers $k \geq 3$. Since $2^{k-2} < 2^{k-1} = \phi(2^k)$, it follows that if $n = 2^k$ has a primitive root, then $k \geq 3$ is prohibited. ◆

5.5.2 Sufficiency

In this part of the proof we show that if an integer n has a certain form, then it must have a primitive root. In the trivial cases of the moduli $n = 2^0$ and 2^1, a primitive root is $g = 1$; when $n = 2^2$, a primitive root is $g = 3$.

Next, we want to show that any odd prime p has a primitive root. A standard proof of this uses a neat property of the Euler ϕ-function that we shall prove in Chapter 7. Alternatively, we present here a proof that uses only the concept of order of an integer. The strategy of the proof is to show that if p does not have a primitive root, that is, if every element in $\mathbf{Z}_p^\times$ is assumed to be of order less than $p - 1$, then this will not account for the existence of all of the elements in $\mathbf{Z}_p^\times$. We proceed in stages (Chahal, 1988).

Lemma 5.8.1 Let the Abelian group $\mathbf{G}$ have elements x, y of order m and n, respectively, and suppose that $(m, n) = 1$. Then $\mathbf{G}$ contains an element of order $[m, n]$.

Proof Since $x^m = y^n = \mathbf{I}$ and $\mathbf{G}$ is Abelian, then $(xy)^{mn} = \mathbf{I}$. Let h be the order of xy; by Lemma 4.8.3 we have $h|mn$. Write $h = h_1 h_2$, where $h_1|m$ and $h_2|n$. Then

$$[(xy)^h]^{m/h_1} = \mathbf{I} = (x^m)^{h_2} \cdot y^{mh_2} = y^{mh_2},$$

so by Lemma 4.8.3, again, $n|mh_2$. Euclid's Lemma (Corollary 2.12.3) then gives us $n|h_2$. This, together with $h_2|n$, implies that $h_2 = n$.

Similarly, $[(xy)^h]^{n/h_2} = \mathbf{I}$ leads to $h_1 = m$, so the order of xy is $h = h_1 h_2 = mn$. ◆

To prepare us for the case where the m, n of Lemma 5.8.1 are not relatively prime, we require the following minor result.

Lemma 5.8.2 Let $(m, n) = d > 1$. Then d can be written as a product $d = d_1 d_2$ in such a way that $(m/d_1, n/d_2) = 1$.

Proof Let $m = \Pi_{i=1}^{r} p_i^{\alpha_i}$ and $n = \Pi_{i=1}^{r} p_i^{\beta_i}$. Define d_1 by the formula $d_1 = \Pi_{\alpha_i \leq \beta_i} p_i^{\alpha_i}$, where the product is taken over just those primes p_i for which $\alpha_i \leq \beta_i$. Similarly, define $d_2 = \Pi_{\beta_i < \alpha_i} p_i^{\beta_i}$. The product $d_1 d_2$ contains all of, and only those, primes that are common to m, n. Each such prime p_i in $d_1 d_2$ occurs to a positive power that is min $\{\alpha_i, \beta_i\}$. Hence, $d_1 d_2 = d$. Further, we see that m/d_1 contains only those primes p_i such that $\alpha_i > \beta_i$, and n/d_2 contains only those primes p_i such that $\alpha_i \leq \beta_i$. The two sets of primes are thus disjoint, so $(m/d_1, n/d_2) = 1$. ◆

Example 5.15 Let $m = 3^4 \cdot 5^2 \cdot 7 \cdot 11^2 \cdot 13$ and $n = 3^5 \cdot 5 \cdot 11 \cdot 13$. Then we have $d_1 = 3^4 \cdot 13$ and $d_2 = 5 \cdot 11$, so $d = 3^4 \cdot 5 \cdot 11 \cdot 13$. Further, $m/d_1 = 5^2 \cdot 7 \cdot 11^2$ and $n/d = 3 \cdot 13$, and we see that $(5^2 \cdot 7 \cdot 11^2, 3 \cdot 13) = 1$. ◆

Lemma 5.8.3 Let the Abelian group **G** have elements x, y of orders m, n, respectively, and suppose that $(m, n) = d > 1$. Then **G** contains an element of order $[m, n]$.

Proof By Lemma 5.8.2 we can write $d = d_1 d_2$, where $(m/d_1, n/d_2) = 1$. Now consider the element $x^{d_1} \in \mathbf{G}$. It has order m/d_1 because $(x^{d_1})^{m/d_1} = x^m = \mathbf{I}$ and m is the order of x. Similarly, the element $y^{d_2} \in \mathbf{G}$ has order n/d_2. Hence, from Lemma 5.8.1 the order of $x^{d_1}y^{d_2}$ is $(m/d_1)(n/d_2) = mn/d = [m, n]$, from Theorem 2.15. ◆

Lemma 5.8.1 is, of course, merely a special case of Lemma 5.8.3 in the instance where $[m, n] = mn$.

Example 5.16 In $\mathbf{Z}_{22}^{\times}$ the order of 3 (mod 22) is 5 and the order of 21 (mod 22) is 2 (verify!). Hence, by Lemma 5.8.1, $3 \cdot 21 \equiv 19$ (mod 22) is of order 10 (mod 22) (verify!). ◆

Example 5.17 In $\mathbf{Z}_{35}^{\times}$ the order of 9 (mod 35) is 6 and the order of 22 (mod 35) is 4 (verify!). Hence, by Lemma 5.8.3, $9 \cdot 22 \equiv 23$ (mod 35) is of order $[4, 6]$, or 12 (mod 35) (verify!). ◆

We now return to Theorem 5.3, reworded here as Theorem 5.8 in the language of primitive roots, and proved by contradiction.

Theorem 5.8 Any odd prime p has a primitive root.

Proof Assume that p does not have a primitive root, that is, the order of every $x \in \mathbf{Z}_p^\times$ is less than $p - 1$. As $\mathbf{Z}_p^\times$ is finite, there is a maximum value, K, of the orders of all $x \in \mathbf{Z}_p^\times$; Corollary 3.19.1 tells us that $K \leq (p - 1)/2$. The order h of each $x \in \mathbf{Z}_p^\times$ must divide K, because if $h \nmid K$ were true for some $x \in \mathbf{Z}_p^\times$, then by Lemma 5.8.3 there would also be an element in $\mathbf{Z}_p^\times$ of order $[h, K]$, which is clearly larger than K and is thus a contradiction.

It follows that $x^K \equiv 1 \pmod{p}$ for every $x \in \mathbf{Z}_p^\times$ because K is a multiple of the order of each x. But by the theorem of Lagrange (Theorem 4.2) there can be no more than K incongruent roots of $x^K \equiv 1 \pmod{p}$, fewer than the $p - 1$ roots just deduced. The contradiction implies that $K \leq (p - 1)/2$ is false, so that $K = p - 1$ and at least one $x \in \mathbf{Z}_p^\times$ has order $p - 1$. This says that p has a primitive root. ◆

Example 5.18 The integer $p = 61$ has a primitive root. By trial we find that 2 is one such primitive root. ◆

Theorem 5.9 If $n = p^k$, where p is an odd prime and $k \in \mathbf{N}$, then n has a primitive root.

Proof Theorem 5.8 has established that an odd prime p has a primitive root g. There are now two cases to consider.

CASE 1. Suppose $g^{p-1} \not\equiv 1 \pmod{p^2}$. Assume, more generally, that $g^{\phi(p^{k-1})} \not\equiv 1 \pmod{p^k}$. When $k = 2$ this reduces to $g^{p-1} \not\equiv 1 \pmod{p^2}$. From Euler's Theorem we get $g^{\phi(p^{k-1})} \equiv 1 \pmod{p^{k-1}}$, or $g^{\phi(p^{k-1})} = 1 + cp^{k-1}$, and $p \nmid c$ because of the incongruence. Now raise both sides of this equality to the pth power; making use of the identity $p\phi(p^{k-1}) = \phi(p^k)$ (Problem 5.44), we obtain

$$g^{p\phi(p^{k-1})} = g^{\phi(p^k)} = [1 + cp^{k-1}]^p$$

$$= 1 + cp^k + \frac{c^2p(p-1)}{2}p^{2k-2} + \text{higher terms}.$$

Since $pp^{2k-2} = p^{2k-1} \geq p^{k+1}$ because $k \geq 2$, then all terms from the third one on are divisible by p^{k+1}. Thus, we can write

$$g^{\phi(p^k)} \equiv 1 + cp^k \pmod{p^{k+1}},$$

and so by the Principle of Finite Induction

$$g^{\phi(p^k)} \not\equiv 1 \pmod{p^{k+1}}, \qquad (*)$$

holds for all $k \in \mathbf{N}$.

Now choose any k and let the order of g (mod p^k) be h: $g^h \equiv 1 \pmod{p^k}$. Then obviously $g^h \equiv 1 \pmod{p}$ too, so $(p-1)|h$ from Theorem 5.1. Write

$$h = a(p-1)$$

for some $a \in \mathbf{N}$. However, $h|\phi(p^k)$ also from Theorem 5.1, and combination with the preceding equation gives $a(p-1)|p^{k-1}(p-1)$. Hence, $a|p^{k-1}$ and so $a = p^\alpha$ for some $\alpha \le k-1$. Assume $\alpha \le k-2$; then $h = p^\alpha(p-1)$ and $h|p^{k-2}(p-1)$, that is, $h|\phi(p^{k-1})$. As h is the order of g (mod p^k), then $g^{\phi(p^{k-1})} \equiv 1 \pmod{p^k}$. This contradicts (*), so $\alpha = k-1$ and $h = p^{k-1}(p-1) = \phi(p^k)$. This says g is a primitive root of p^k. As k was arbitrary, then g is a primitive root for all powers of p.

CASE 2. Suppose $g^{p-1} \equiv 1 \pmod{p^2}$. Define $g_1 = g + p$; then g_1 is a primitive root modulo p. Raising g_1 to the $(p-1)$st power, we obtain

$$g_1^{p-1} = (g+p)^{p-1} \equiv 1 + p(p-1)g^{p-2} + \text{higher terms in } p,$$

and so $g_1^{p-1} \equiv 1 - pg^{p-2} \pmod{p^2}$. Thus, $g_1^{p-1} \not\equiv 1 \pmod{p^2}$ and we return to Case 1. Repetition of the steps there leads to the conclusion that g_1 is a primitive root for all powers of p. By either case we have identified a primitive root of p^k. ◆

Example 5.19 The following are some integers that have primitive roots:

$$25,\ 49,\ 343,\ 729,\ 1331,\ 10201. \quad ◆$$

To decide whether integers such as 50, 98, 686, 1458, 2662, and 20,402 have primitive roots, we need the remaining theorem of this section.

Theorem 5.10 If $n = 2p^k$, $k \in \mathbf{N}$, then n has a primitive root.

Proof Let g be a primitive root of p^k: $g^{\phi(p^k)} \equiv 1 \pmod{p^k}$. Since $\phi(2p^k) = \phi(2)\phi(p^k) = \phi(p^k)$; then $g^{\phi(2p^k)} \equiv 1 \pmod{p^k}$. Suppose g is odd; then $2|(g^{\phi(2p^k)} - 1)$, so $g^{\phi(2p^k)} \equiv 1 \pmod{2p^k}$. If h is the order of g (mod $2p^k$), then $h \le \phi(2p^k)$. Assume $h < \phi(2p^k) = \phi(p^k)$; then $g^h \equiv 1 \pmod{2p^k}$, and so also $g^h \equiv 1 \pmod{p^k}$. This is a contradiction since the order of g (mod p^k) is $\phi(p^k)$. Therefore, $h = \phi(2p^k)$, and g is a primitive root of $n = 2p^k$.

On the other hand, suppose g is an even primitive root of p^k. Define $g_1 = g + p^k$; this is now odd and is also a primitive root of p^k. The argument now proceeds as above and we conclude that g_1 is a primitive root of $2p^k$. ◆

Example 5.20 Let $p = 11$; from Table 5.4, $g = 6$ is a primitive root of p. But $6^{11-1} \not\equiv 1 \pmod{121}$, so g is a primitive root of 121 from Case 1 of Theorem 5.9. As g is even, then $g_1 = g + p^2 = 6 + 121 = 127$ is a primitive root of $n = 2p^2 = 242$, from Theorem 5.10. ◆

Possession of a primitive root is clearly a very specialized property of an integer and has algebraic significance. We arrive at the following beautiful statement, suggested by the calculations in Section 5.2, and confirmed by the results in the present section.

Theorem 5.11 A reduced residue system $\mathbf{S}'_n$ modulo n is a cyclic multiplicative group iff n is $1, 2, 4, p^a$ or $2p^a$, where p is an odd prime and $a \in \mathbf{N}$.

Example 5.21 The reduced residue system $\mathbf{S}'_{250}$ is a cyclic multiplicative group, but the reduced residue system $\mathbf{S}'_{150}$ is not. ◆

Problems

5.39. Which of these integers have primitive roots: 139, 203, 722, 4913? Which of these reduced residue systems are cyclic multiplicative groups: $\mathbf{S}'_{52}$, $\mathbf{S}'_{83}$, $\mathbf{S}'_{529}$, $\mathbf{S}'_{579}$?

5.40. Compute $M = [2, 4, 6, 8, 10]$ by using Lemma 5.6.1. How do you know that $16 \nmid M$?

5.41. Illustrate Lemma 5.6.2 with $n = 7{,}014{,}007$.

5.42. Prove Lemma 5.6.1 along the lines suggested in the text.

5.43. In the group $\mathbf{Z}_{61}^{\times}$ the element 3 is of order 10 and the element 11 is of order 4. Discover an element of order 20. Discover an element of order 5. How many elements of order 60 are in $\mathbf{Z}_{61}^{\times}$?

5.44. If p is an odd prime, prove that $p\phi(p^{k-1}) = \phi(p^k)$, where the integer $k > 1$. Does the result hold if $p = 2$?

5.45. Prove that 10 is a primitive root of 7^k for all $k \in \mathbf{N}$. Similarly, prove that 2 is a primitive root of 5^k for all $k \in \mathbf{N}$.

5.46. Show that no integer n can have exactly 26 primitive roots.

5.47. The set $\{4, 5, 6, 7\}$ is a set of four consecutive integers, each of which has a primitive root. Show that there is no larger such set.

5.48. A conjecture says that if p is a $(4k + 1)$-prime, then the primitive roots of p can be arranged in pairs, each of whose sum is p. Explore this conjecture on the computer with several choices of p. Then explore similarly the primitive roots of $2p$, where p is again a $(4k + 1)$-prime.

5.49. Write out the proof of Theorem 5.11.

5.50. Suppose g is a primitive root of p^2, where p is an odd prime. Show that it is also a primitive root of p^3. Illustrate this with some examples.

5.51. Let p be an odd prime and let the integer A satisfy $2 \leq A < p$. Prove that if the congruence $x^2 \equiv A \pmod{p}$ has a solution $x \equiv x_0 \pmod{p}$, then A cannot be a primitive root of p.

RESEARCH PROBLEMS

1. Let p be an odd prime and S_p the sum of the primitive roots of p. Choose several primes, including some from beyond Table D in Appendix II, and investigate the residues A obtained for the family of congruences

$$S_p \equiv A \pmod{p}.$$

See if you can classify the primes into groups according to the value of A obtained.

2. A working conjecture is that the smallest primitive root of a prime p does not exceed $1 + \lfloor\sqrt{p}\rfloor$. Investigate this on the computer, and determine the first prime that has 14 as its smallest primitive root.

3. Investigate what connection there might be between those primes $p > 7$ for which the decimal expansion of $1/p$ has period length $p - 1$ and the primitive roots of those primes. Also, let $N(x)$ denote the number of primes $p \leq x$ for which the period length of the decimal expansion of $1/p$ is $p - 1$, and let $\pi(x)$ denote the number of all primes not exceeding x. Investigate the ratio $N(x)/\pi(x)$.

References

Chahal, J.S., *Topics in Number Theory*, Plenum Press, New York, 1988, p. 31. The inspiration for our sequence of Lemmas 5.8.1–5.8.3.

Hua, L.-K., *Introduction to Number Theory*, Springer–Verlag, Berlin, 1982, pp. 49–50. The source of our development in Section 5.5. This fine book has much to offer to a student who wishes to pursue number theory in more depth.

Osborn, R., *Tables of All Primitive Roots of Odd Primes Less Than 1000*, University of Texas Press, Austin, 1961. A very useful reference as a motivational source of conjectures.

A Table of Indices and Power Residues for All Primes and Prime Powers Below 2000, University of Oklahoma Mathematical Tables Project, W.W. Norton & Co., New York, 1962. This monograph contains much more data than our Table E in Appendix II.

Chapter 6
Residues

6.1 Quadratic Residues and Nonresidues

We saw in Section 4.6 that solving a polynomial congruence in one variable, $f(x) \equiv 0 \pmod{m}$, reduces to solving polynomial congruences with prime moduli plus a system of linear congruences (using the Chinese Remainder Theorem). We have already studied some techniques for solving linear congruences with prime moduli.

The general second-order (quadratic) congruence with an odd prime modulus p is $ax^2 + bx + c \equiv 0 \pmod{p}$, $p \nmid a$. If we write this as $4a(ax^2 + bx + c) \equiv 0 \pmod{p}$ or equivalently as

$$(2ax + b)^2 + (4ac - b^2) \equiv 0 \pmod{p},$$

then this is equivalent to the system of congruences

$$\begin{cases} y^2 \equiv A \pmod{p}, \\ 2ax + b \equiv y \pmod{p}, \end{cases}$$

where $A \equiv b^2 - 4ac \pmod{p}$. Since $p \nmid a$, the congruence $2ax + b \equiv y \pmod{p}$ has a unique solution $x \equiv x_0 \pmod{p}$, so the solution of the original quadratic congruence reduces to that of solving $y^2 \equiv A \pmod{p}$.

In the first part of this chapter we study in detail $y^2 \equiv A \pmod{p}$. Later, we consider various aspects of $y^n \equiv A \pmod{m}$, where $n > 2$ and m may be composite.

Example 6.1 Solve $2x^2 + 3x + 5 \equiv 0 \pmod{7}$.

Multiply by $4a = 8$ to give $16x^2 + 24x + 40 \equiv 0 \pmod{7}$, or $(4x + 3)^2 + (40 - 9) \equiv 0 \pmod{7}$. Let $y \equiv 4x + 3 \pmod{7}$; this gives, upon substitution, $y^2 + 31 \equiv 0 \pmod{7}$, or $y^2 \equiv 4 \pmod{7}$. This is easily solved by inspection.

$$\begin{aligned} y^2 &\equiv 4 \pmod{7} & \qquad y^2 &\equiv 4 \pmod{7} \\ y &\equiv 2 \pmod{7} & y &\equiv 5 \pmod{7} \\ 4x + 3 &\equiv 2 \pmod{7} & 4x + 3 &\equiv 5 \pmod{7} \\ 4x &\equiv 6 \pmod{7} & 4x &\equiv 2 \pmod{7} \\ 2x &\equiv 3 \pmod{7} & 2x &\equiv 1 \pmod{7} \\ x &\equiv 5 \pmod{7} & x &\equiv 4 \pmod{7}. \end{aligned}$$ ◆

In the example just worked, the 4 in $y^2 \equiv 4 \pmod{7}$ is considered important enough to be given a special name.

DEF. Let p be a prime and A an integer such that $p \nmid A$. If the congruence $x^2 \equiv A \pmod{p}$ has a solution, then A is called a **quadratic residue** modulo p. If no solution exists, then A is termed a **quadratic nonresidue** modulo p.

Since p is prime, Theorem 4.2 tells us that $x^2 \equiv A \pmod{p}$ has at most two solutions. Further, if x_0 is a solution modulo p of $x^2 \equiv A \pmod{p}$, then so is $x_1 = -x_0 = p - x_0$, and therefore $x^2 \equiv A \pmod{p}$ has either two solutions or no solutions modulo p, provided $p > 2$.

Example 6.2 Let $p = 13$, $A = 9$. An immediate solution of $x^2 \equiv A \pmod{p}$ is $x_0 \equiv 3 \pmod{13}$. Hence, $p - x_0 = 13 - 3 = 10$, so $x_1 \equiv 10 \pmod{13}$ is the other solution. Indeed, $10^2 \equiv 9 \pmod{13}$. ◆

Example 6.3 Let $p = 11$, $A = 6$. We square the integers 1, 2, 3, 4, 5 and reduce modulo 11:

$$1^2 \equiv 1, \quad 2^2 \equiv 4, \quad 3^2 \equiv 9, \quad 4^2 \equiv 5, \quad 5^2 \equiv 3.$$

Hence, 6 is a quadratic nonresidue modulo 11. We do not need to check the integers 6, 7, 8, 9, 10 (why?). ◆

Of course, a prime p always has some quadratic residues; 1 is a quadratic residue of any prime. So are all the perfect squares less than p. We denote a complete set of incongruent quadratic residues of a prime p by $\mathbf{A}_p^{(2)}$. Notice in Example 6.3 that $|\mathbf{A}_{11}^{(2)}| = (p - 1)/2$. This is a general relationship.

Theorem 6.1 Let p be an odd prime. Then half of the integers A that satisfy $0 < A < p$ are quadratic residues modulo p and the rest are quadratic nonresidues modulo p.

Proof Let x_i be a representative of one of the $p - 1$ residue classes of p. Then $x_i \not\equiv p - x_i \pmod{p}$, but $x_i^2 \equiv (p - x_i)^2 \pmod{p}$. This shows that $|\mathbf{A}_p^{(2)}| \le (p - 1)/2$.

On the other hand, let x_i, x_j satisfy $1 \le x_i, x_j \le (p - 1)/2$, $x_i \ne x_j$. If $x_i^2 - x_j^2 \equiv 0 \pmod{p}$ held, then either $p \mid (x_i - x_j)$ or $p \mid (x_i + x_j)$ would be true. Both are impossible since $0 < (x_i \pm x_j) < p$. So $x_i^2 \not\equiv x_j^2 \pmod{p}$, and $|\mathbf{A}_p^{(2)}| \ge (p - 1)/2$. We conclude that $|\mathbf{A}_p^{(2)}| = (p - 1)/2$. ◆

If a quadratic residue or nonresidue of a prime p lies in the interval $[1, p - 1]$, it is termed **least positive**. Table 6.1 gives a listing for the first few primes; a more extensive compilation is in Table F in Appendix II. The distribution of quadratic residues and nonresidues is intriguing, but is not completely understood (Rose, 1994).

We stress again, as a purely computational matter (see Example 6.3), that when the modulus is an odd prime p, $\mathbf{A}_p^{(2)}$ can be obtained by squaring and reducing modulo p the first $(p - 1)/2$ positive integers.

We are now going to work out two results concerning quadratic residues: one of an arithmetic nature and one of an algebraic nature. It is observed from Table 6.1 that for $p = 13$ the sum of the elements of $\mathbf{A}_{13}^{(2)}$ is 39 and $13 \mid 39$. Similarly, for $p = 19$ we see that the sum of the elements of $\mathbf{A}_{19}^{(2)}$ is 76 and $19 \mid 76$. Of course, these two divisibility results hold whether or not $\mathbf{A}_p^{(2)}$ consists of the least positive residues. The following theorem seems plausible.

Table 6.1 Quadratic Residues and Nonresidues Modulo Primes

p	Least Positive Residues	Least Positive Nonresidues
3	1	2
5	1, 4	2, 3
7	1, 2, 4	3, 5, 6
11	1, 3, 4, 5, 9	2, 6, 7, 8, 10
13	1, 3, 4, 9, 10, 12	2, 5, 6, 7, 8, 11
17	1, 2, 4, 8, 9, 13, 15, 16	3, 5, 6, 7, 10, 11, 12, 14
19	1, 4, 5, 6, 7, 9, 11, 16, 17	2, 3, 8, 10, 12, 13, 14, 15, 18

Theorem 6.2 Let $T_p^{(2)}$ denote the sum of the members of $\mathbf{A}_p^{(2)}$. Then $p \mid T_p^{(2)}$, provided $p > 3$.

Proof In view of Theorem 6.1, and especially of the discussion following it, we must have

$$T_p^{(2)} \equiv \sum_{i=1}^{(p-1)/2} i^2 \pmod{p}.$$

From elementary algebra, the sum is $[(p-1)/2][1+(p-1)/2]p/6$ (Problem 6.5). Hence, after simplification, we obtain

$$T_p^{(2)} \equiv \frac{p(p^2-1)}{24} \pmod{p},$$

and $T_p^{(2)} \equiv 0 \pmod{p}$ iff $(p^2-1)/24$ is integral. But $8 \mid (p^2-1)$ from Example 3.3, and $3 \mid (p^2-1)$ for any prime $p > 3$ from Problem 3.12. As $(3, 8) = 1$, we conclude from the Fundamental Theorem of Arithmetic that $24 \mid (p^2-1)$, and we are done. ◆

Look once again, for example, at $p = 11$ in Table 6.1. We see that $3 \cdot 5 = 15 \equiv 4 \pmod{11}$ and $4 \in \mathbf{A}_{11}^{(2)}$; $5 \cdot 5 = 25 \equiv 3 \pmod{11}$ and $3 \in \mathbf{A}_{11}^{(2)}$; $3 \cdot 9 = 27 \equiv 5 \pmod{11}$ and $5 \in \mathbf{A}_{11}^{(2)}$. Modular multiplication appears to be a binary operation on $\mathbf{A}_{11}^{(2)}$. We know from Theorem 5.3 that $\mathbf{Z}_{11}^{\times}$ is a group and $\mathbf{A}_{11}^{(2)}$ is a subset of this group. So maybe $\mathbf{A}_{11}^{(2)}$ is itself a group.

Theorem 6.3 For any prime p, $\mathbf{A}_p^{(2)}$ is a group under multiplication modulo p.

Proof Let $A_1, A_2 \in \mathbf{A}_p^{(2)}$; then there are integers x, y such that $x^2 \equiv A_1 \pmod{p}$ and $y^2 \equiv A_2 \pmod{p}$. Multiplication of the congruences gives $(x \cdot y)^2 \equiv A_1A_2 \pmod{p}$, so $A_1A_2 \in \mathbf{A}_p^{(2)}$. This shows that modular multiplication is a binary operation on $\mathbf{A}_p^{(2)}$.

We also have $1 \in \mathbf{A}_p^{(2)}$ because $1^2 \equiv 1 \pmod{p}$. Further, the elements of $\mathbf{A}_p^{(2)}$ obey the associative law because they are real numbers. It remains to show that $\mathbf{A}_p^{(2)}$ contains all of its own inverses. Let $A_i \in \mathbf{A}_p^{(2)}$; then $(A_i, p) = 1$ implies that $A_i x \equiv 1 \pmod{p}$ has a solution, say $x \equiv x_0 \pmod{p}$. Hence, $A_i x_0^2 \equiv x_0 \pmod{p}$, but since $A_i \in \mathbf{A}_p^{(2)}$, there is an integer x_i such that $A_i \equiv x_i^2 \pmod{p}$. This gives $(x_i \cdot x_0)^2 \equiv x_0 \pmod{p}$, so $x_0 \in \mathbf{A}_p^{(2)}$ and we have thus met all of the requirements for a group. ◆

Example 6.4 The multiplication table for $\mathbf{A}_{11}^{(2)}$ is that of a group.

$\cdot$	1	3	4	5	9
1	1	3	4	5	9
3	3	9	1	4	5
4	4	1	5	9	3
5	5	4	9	3	1
9	9	5	3	1	4

◆

Problems

6.1. Use the method shown at the start of this section and solve the following congruences completely:

(a) $x^2 + 5x + 7 \equiv 0 \pmod{13}$.
(b) $-3x^2 - x + 9 \equiv 0 \pmod{31}$.
(c) $11x^2 + 20x + 1 \equiv 0 \pmod{47}$.
(d) $2x^3 + 6x \equiv 0 \pmod{97}$.

Also show how to solve (b) and (c) by factoring.

6.2. Find all the elements of $\mathbf{A}_p^{(2)}$ corresponding to

(a) $p = 29$. (b) $p = 37$.

Also, verify Theorem 6.2 for both parts.

6.3. In view of your answers to Problem 6.2, state which of the following congruences have solutions:

(a) $x^2 \equiv 60 \pmod{29}$. (b) $x^2 \equiv -211 \pmod{37}$.
(c) $x^2 \equiv 288 \pmod{37}$. (d) $x^2 \equiv 602 \pmod{29}$.

6.4. Let p be a prime. The statement "A is a quadratic nonresidue of p" emphasizes what A IS NOT. We could make this statement more positive and emphasize what A IS by saying "A is a nonquadratic residue of p."

(a) From Table 6.1 we have that 13 is a quadratic nonresidue of $p = 19$. Find a positive integer n such that $x^n \equiv 13 \pmod{19}$ has a solution. This will justify our calling 13 a nonquadratic residue (here, an nth-power residue).
(b) If $A \not\equiv 0 \pmod{p}$, can we always find an integer n such that A is an nth-power residue of p?

6.5. In the proof of Theorem 6.2 we used the theorem that

$$\sum_{i=1}^{n} i^2 = \frac{n(n+1)(2n+1)}{6}.$$

Prove this.

6.6. Select four $(4k+3)$-primes beyond those given in Table 6.1. For each prime compute the number of quadratic residues in the interval $[1, (p-1)/2]$ and the number of quadratic nonresidues in this same interval. Formulate a conjecture (Moser, 1951).

6.7. This problem is an unimaginative extension of Theorem 6.2 to second powers of the members of $\mathbf{A}_p^{(2)}$.

(a) Prove that $\Sigma_{i=1}^{n} i^4 = n(n+1)(2n+1)(3n^2+3n-1)/30$.
(b) Next, show that the prime p divides $\Sigma_{i=1}^{(p-1)/2} i^4$ if $p > 5$.
(c) Define $T_{p,2}^{(2)} = \Sigma_i A_i^2$, where the summation is over all the members of $\mathbf{A}_p^{(2)}$. For example, $T_{11,2}^{(2)} = 1^2 + 3^2 + 4^2 + 5^2 + 9^2 = 132$. Prove that $p \mid T_{p,2}^{(2)}$.

6.8. Let $p \geq 7$ be a prime. It is asserted that corresponding to p there is a positive integer A, not exceeding 9, such that A, $A+1$ are both quadratic residues modulo p. Write a computer program and test this out on several challenging primes.

6.9. Let g be a primitive root of the odd prime p. Show that the even powers g^2, $g^4, \ldots, g^{p-1}$ are quadratic residues modulo p, and the odd powers $g^1, g^3, \ldots,$ g^{p-2} are quadratic nonresidues.

6.10. Let p be an odd prime, $\mathbf{S}_1$ be the set of least positive quadratic nonresidues modulo p, and $\mathbf{S}_2$ be the set of primitive roots of p.

(a) Prove that $\mathbf{S}_2 \subset \mathbf{S}_1$. This is basically Problem 5.51.
(b) Consider $p = 59$. What percentage of its quadratic nonresidues are primitive roots?
(c) Find another prime whose percentage of primitive roots is higher than the value calculated in (b).

6.11. Let p be an odd prime.

(a) What term should be applied to the description of the set of quadratic nonresidues of p?
(b) Investigate the groups of quadratic residues for $p = 11, 13, 17, 19, 23$. Are they cyclic?

6.12. (a) Prove that any subgroup of a cyclic group is itself a cyclic group.
(b) Then show that the observation made in Problem 6.11(b) holds for all primes p.

6.13. Select at least half a dozen odd primes; determine which primes p have $p - 1$ as one of their quadratic residues. Formulate a conjecture based on your results; test it by examining a few more primes.

6.14. Let p be a $(4k + 1)$-prime.

(a) How do we know that $-k$ is a quadratic residue of p?
(b) How do we know that $1, p - 1$ are the *only* elements of $\mathbf{A}_p^{(2)}$ that are their own multiplicative inverses, assuming the truth of your conjecture in Problem 6.13?

6.15. Let A be a quadratic residue of the $(4k + 3)$-prime p.

(a) What is the value of $A^{(p-1)/2} \pmod{p}$?
(b) In view of part (a), show that $p - 1$ is not a quadratic residue of p.
(c) Hence, how many elements of $\mathbf{A}_p^{(2)}$ for a $(4k + 3)$-prime are their own multiplicative inverses?

6.16. Let $P_p^{(2)}$ denote the product of all the elements of $\mathbf{A}_p^{(2)}$. Use the results of Problems 6.14 and 6.15 to show that for $p \geq 3$

$$\begin{cases} p \mid (P_p^{(2)} + 1) & \text{if } p = 4k + 1 \\ p \mid (P_p^{(2)} - 1) & \text{if } p = 4k + 3. \end{cases}$$

Illustrate each case with an example.

6.17. Let p be a $(4k + 3)$-prime and define $P_p^{(2)}$ as in Problem 6.16. What are the roots of the congruence $x^{(p-1)/2} - P_p^{(2)} \equiv 0 \pmod{p}$? Illustrate with a concrete case.

6.2 Legendre's Symbol

The French mathematician **Adrien-Marie Legendre** (1752–1833), in the first edition of his book *Essai sur la théorie des nombres* (1798), introduced the symbol that is now known as **Legendre's symbol,** $(\frac{A}{p})$:

$$\left(\frac{A}{p}\right) = \begin{cases} +1 & \text{if } A \text{ is a quadratic residue modulo } p > 2 \\ 0 & \text{if } p \mid A \\ -1 & \text{if } A \text{ is a quadratic nonresidue modulo } p > 2. \end{cases}$$

The symbol is not defined for $p = 2$. Note also that the definition is not restricted to the least positive quadratic residues and nonresidues. Shortly, we'll see that $(\frac{A}{p})$ bears a simple relationship to $(\frac{B}{p})$ if B is a least positive quadratic residue (or nonresidue) and $A \equiv B \pmod{p}$.

Example 6.5 From Example 6.3 we have $(\frac{6}{11}) = -1$. From Table 6.1 we see that $(\frac{16}{17}) = 1$. Also, $(\frac{30}{7}) = 1$ since $3^2 \equiv 30 \pmod{7}$. ◆

From Table 6.1 we have, for example, $10 \in \mathbf{A}_{13}^{(2)}$. Fermat's Theorem gives $10^{12} \equiv 1 \pmod{13}$, so $10^{12/2}$ is congruent modulo 13 to one of the two roots of $x^2 - 1 \equiv 0 \pmod{13}$. In fact, $10^6 \equiv 1 \pmod{13}$. On the other hand, $6 \not\in \mathbf{A}_{13}^{(2)}$ but $6^{12/2}$ must still be congruent modulo 13 to one of the two roots of $x^2 - 1 \equiv 0 \pmod{13}$. Here, $6^6 \equiv -1 \pmod{13}$. These examples suggest a test for quadratic residues.

Although he used neither the symbolism of congruences nor the Legendre symbol, **Euler** gave a criterion for a number A to be a quadratic residue. We require the following lemma (which appeared earlier as Problem 4.17).

Lemma 6.4.1 Let p be an odd prime. If $n \mid (p - 1)$, then the congruence $x^n \equiv 1 \pmod{p}$ has exactly n solutions modulo p.

Proof By Fermat's Theorem $x^{p-1} - 1 \equiv 0 \pmod{p}$ has $p - 1$ incongruent roots. Set $p - 1 = nk$, where $k \in \mathbf{N}$. Then $x^{p-1} - 1 = (x^n - 1) \cdot \mathrm{h}(x)$, where

$$\mathrm{h}(x) = x^{n(k-1)} + x^{n(k-2)} + \cdots + x^n + 1.$$

By Theorem 4.2 the congruences $x^n - 1 \equiv 0 \pmod{p}$ and $\mathrm{h}(x) \equiv 0 \pmod{p}$ can have no more than n, $n(k - 1)$ incongruent roots, respectively. If either of the congruences has fewer than these number of roots, then the total number of incongruent roots of $x^{p-1} - 1 \equiv 0 \pmod{p}$ would be fewer than $n + n(k - 1) = nk = p - 1$, a contradiction. Hence, $x^n - 1 \equiv 0 \pmod{p}$ has exactly n solutions. ◆

Theorem 6.4 (Euler's Criterion). Let p be an odd prime and $(A, p) = 1$. Then $(\frac{A}{p}) = 1$ iff

$$A^{(p-1)/2} \equiv 1 \pmod{p}.$$

Proof $(\rightarrow)$ If $(\frac{A}{p}) = 1$, there is a y, $0 < y < p$, such that $y^2 \equiv A \pmod{p}$. Theorem 3.2 then gives

$$(y^2)^{(p-1)/2} \equiv A^{(p-1)/2} \equiv y^{p-1} \equiv 1 \pmod{p}$$

from Fermat's Theorem.

($\leftarrow$) Let $\mathbf{S}$ be the set of solutions $\{x_i\}$ modulo p to $x^{(p-1)/2} \equiv 1 \pmod{p}$. The sufficiency condition just proved implies that $\mathbf{A}_p^{(2)} \subset \mathbf{S}$. Since $[(p-1)/2] \mid (p-1)$, then Lemma 6.4.1 says that $|\mathbf{S}| = (p-1)/2$. But as $|\mathbf{A}_p^{(2)}| = (p-1)/2$ also, it follows that $\mathbf{A}_p^{(2)} = \mathbf{S}$, so x_i must be one of the A's. That is, $A^{(p-1)/2} \equiv 1 \pmod{p}$ implies $\left(\frac{A}{p}\right) = 1$. ◆

Example 6.6 Let $p = 13$, $A = 9$. Then we have

$$9^{(13-1)/2} \equiv 9^6 \equiv (9^2)^3 \equiv 3^3 \equiv 1 \pmod{13},$$

so $\left(\frac{9}{13}\right) = 1$ (see Table 6.1). ◆

Corollary 6.4.1 If p is an odd prime and $(A, p) = 1$, then $\left(\frac{A}{p}\right) = -1$ iff

$$A^{(p-1)/2} \equiv -1 \pmod{p}.$$

Proof The proof is left to the reader. ◆

Example 6.7 Let $p = 59$, $A = 6$. Then we have

$$6^{(59-1)/2} \equiv 6^{29} \equiv (6^4)^7 \cdot 6 \equiv (-2)^7 \cdot 6 \equiv -768 \equiv -1 \pmod{59},$$

so $\left(\frac{6}{59}\right) = -1$. ◆

The next theorem gives the most fundamental properties of the Legendre symbol. These properties expedite the labor needed to decide whether $x^2 \equiv A \pmod{p}$ is solvable.

Theorem 6.5 Let p be an odd prime, $p \nmid A$ and $p \nmid B$. Then

(a) $\left(\frac{A^2}{p}\right) = 1$.
(b) $\left(\frac{1}{p}\right) = 1$.
(c) $\left(\frac{A}{p}\right) \equiv A^{(p-1)/2} \pmod{p}$.
(d) If $A \equiv B \pmod{p}$, then $\left(\frac{A}{p}\right) = \left(\frac{B}{p}\right)$.
(e) $\left(\frac{-1}{p}\right) = (-1)^{(p-1)/2} = \begin{cases} +1 & \text{if } p \equiv 1 \pmod{4} \\ -1 & \text{if } p \equiv 3 \pmod{4}. \end{cases}$
(f) $\left(\frac{A}{p}\right)\left(\frac{B}{p}\right) = \left(\frac{AB}{p}\right)$.

Proof (a) is immediate from the definition of a quadratic residue. When $A^2 = 1^2$, part (a) reduces to (b). Part (c) is merely a restatement of Theorem 6.4 and its corollary. For part (d), $A \equiv B \pmod{p}$ implies $A^{(p-1)/2} \equiv B^{(p-1)/2}$

(mod p), and then apply part (c). Part (e) follows from part (c) and is straightforward; it and part (f) are left as exercises for you (Problem 6.21). ◆

Example 6.8 To evaluate $(\frac{1000}{23})$, write this as $(\frac{2}{23})(\frac{5}{23})(\frac{10^2}{23})$ from Theorem 6.5, part (f). Then $(\frac{1000}{23}) = (\frac{2}{23})(\frac{5}{23})$ from part (a). Use of Theorem 6.5, part (c), finally gives

$$\left(\frac{2}{23}\right) \equiv 2^{11} \equiv 1 \pmod{23}$$

$$\left(\frac{5}{23}\right) \equiv 5^{11} \equiv 5(5^2)^5 \equiv -1 \pmod{23},$$

so $(\frac{1000}{23}) = -1$. The result tells us that the congruence $x^2 \equiv 1000 \pmod{23}$ has no solution. ◆

Part of the work in Example 6.8 could have been circumvented if we had known Theorem 6.6, which follows. A proof of this could be presented at this point, but would be somewhat lengthy. We shall delay the proof until Section 6.4, at which time it will follow more readily from Eisenstein's Lemma there.

Theorem 6.6 $(\frac{2}{p}) = (-1)^{(p^2-1)/8} = \begin{cases} +1 & \text{if } p \equiv \pm 1 \pmod 8 \\ -1 & \text{if } p \equiv \pm 3 \pmod 8. \end{cases}$

Example 6.9 Let $p = 103 = 8(13) - 1$; then $(\frac{2}{103}) = +1$ by Theorem 6.6. Hence, the congruence $x^2 \equiv 2 \pmod{103}$ has a solution; in fact, $x \equiv 38 \pmod{103}$ is a solution. Can you find another solution? ◆

Problems

6.18. Compute the following Legendre symbols:

(a) $(\frac{-3}{11})$. (b) $(\frac{6}{17})$. (c) $(\frac{163}{23})$.
(d) $(\frac{196}{23})$. (e) $(\frac{30}{31})$. (f) $(\frac{231}{233})$.

6.19. Write out the proof of Corollary 6.4.1.

6.20. In view of material in this section, why is Problem 6.13 now an easy theorem?

6.21. Write out the proofs of parts (e) and (f) of Theorem 6.5.

6.22. Write a computer program that will employ Theorems 6.5 and 6.6 to evaluate a general Legendre symbol $(\frac{A}{p})$. Apply the program to these examples:

(a) $(\frac{280}{19})$.
(b) $(\frac{-63}{61})$.
(c) $(\frac{-1875}{101})$.
(d) $(\frac{10}{1117})$.

6.23. We have given no explicit algorithms or methods of solution of $x^2 \equiv A \pmod{p}$ for cases where $(\frac{A}{p}) = +1$. However,

(a) if $p = 4k + 3$ and $(\frac{A}{p}) = +1$, show that $x \equiv A^{(p+1)/4} \pmod{p}$ is a solution of $x^2 \equiv A \pmod{p}$.
(b) if $p = 8k + 5$ and $(\frac{A}{p}) = +1$, show that one of $x \equiv A^{(p+3)/8} \pmod{p}$, $x \equiv A^{(p+3)/8}\,[(p-1)/2]! \pmod{p}$ is a solution of $x^2 \equiv A \pmod{p}$.
(c) Illustrate parts (a) and (b) with examples.

6.24. For which primes p is $(\frac{-2}{p})$ equal to $+1$? When does $(\frac{-2}{p}) = -1$?

6.25. Let $p = 8191$. What is the value of

$$\sum_{A=1}^{p-1} \left(\frac{A}{p}\right)?$$

[Hint: This is not really a lengthy computation.]

6.26. Suppose $p > 5$. Then prove that

(a) Either one or all three of 2, 5, 10 are quadratic residues modulo p.
(b) There are two consecutive quadratic residues modulo p.
(c) There is a prime p that has all three of 2, 5, 10 as quadratic residues.

6.27. Let p be a $(4k + 1)$-prime.

(a) Show that $(\frac{A}{p}) = (\frac{p-A}{p})$. Does this hold if p is a $(4k + 3)$-prime?
(b) Then prove $\sum_{A=1}^{p-1} A\,(\frac{A}{p}) = 0$. Illustrate this numerically for $p = 17$.

6.28. Let $p = 107$. Show that $\sum_{k=0}^{p} (\frac{2^k}{p}) = 0$.

6.29. Here is another way to view Legendre's symbol. Given an odd prime p, define the **standard permutation** $\pi_0(p)$ of the least positive residues modulo p to be the listing of them in ascending order. Thus, $\pi_0(7) = \{1, 2, 3, 4, 5, 6\}$. The **sign** of any other permutation $\pi_j(p)$ of the least positive residues modulo p is $\operatorname{sgn} \pi_j(p) = (-1)^r$, where r is the number of pair interchanges needed to transform $\pi_j(p)$ into $\pi_0(p)$. Thus, $\operatorname{sgn}\{2, 1, 4, 3, 5, 6\} = (-1)^2 = +1$ because two interchanges are needed to transform $\{2, 1, 4, 3, 5, 6\}$ into $\pi_0(7) = \{1, 2, 3, 4, 5, 6\}$. Now let A be chosen from $1, 2, \ldots, p - 1$ and form the product $A \cdot \pi_0(p)$, then reduce modulo p to give some $\pi_j(p)$. Determine $\operatorname{sgn} \pi_j(p)$ and compare with $\left(\frac{A}{p}\right)$. Prepare a table for various choices of p and of A. Formulate a conjecture.

6.3 Jacobi's Symbol

Legendre's symbol $\left(\frac{A}{p}\right)$ is defined only when p is an odd prime. The symbolism is therefore useless as a guide to tell us whether the congruence $x^2 \equiv 6 \pmod{35}$, for example, has a solution. This congruence has a solution iff the system of congruences

$$\begin{cases} x^2 \equiv 6 \pmod{5} \\ x^2 \equiv 6 \pmod{7} \end{cases}$$

has a solution, that is, iff $\left(\frac{6}{5}\right) = \left(\frac{6}{7}\right) = 1$.

The German mathematician **Carl G.J. Jacobi** (1804–1851) introduced the following generalization of Legendre's symbol, now known as **Jacobi's symbol**, $(A \mid n)$. Let n be an odd positive integer with factorization

$$n = \prod_{i=1}^{m} p_i^{\alpha_i}.$$

Then we define

$$(A|n) = \begin{cases} \displaystyle\prod_{i=1}^{m} \left(\frac{A}{p_i}\right)^{\alpha_i} & \text{if } (A, n) = 1,\ n > 1 \\ 0 & \text{if } (A, n) > 1 \\ 1 & \text{if } n = 1. \end{cases}$$

When n is an odd prime, Legendre's symbol and Jacobi's symbol coincide.

Example 6.10 Does the congruence $x^2 \equiv 6 \pmod{35}$ have a solution? From Theorem 6.5 we compute

$$(6 \mid 35) = \left(\frac{6}{5}\right)\left(\frac{6}{7}\right) = \left(\frac{1}{5}\right)\left(\frac{-1}{7}\right) = 1 \cdot (-1)^{(7-1)/2} = -1.$$

We conclude that $x^2 \equiv 6 \pmod{35}$ has no solution. ◆

On the other hand, if n is odd and composite and $(A \mid n) = +1$, this does not necessarily mean that $x^2 \equiv A \pmod{n}$ has a solution (Problem 6.30). Can you see why? Look at the case of $x^2 \equiv 5 \pmod{21}$.

Many of the properties listed in Theorem 6.5 have their analogs for the Jacobi symbol. We prove some of them here.

Theorem 6.7 If n, n_1, n_2 are odd and positive, then

(a) $(A^2 \mid n) = 1$, if $(A, n) = 1$.
(b) $(1 \mid n) = 1$.
(c) if $A \equiv B \pmod{n}$, then $(A \mid n) = (B \mid n)$.
(d) $(A \mid n)(B \mid n) = (AB \mid n)$.
(e) $(A \mid n_1)(A \mid n_2) = (A \mid n_1 n_2)$.
(f) $(2 \mid n) = (-1)^{(n^2-1)/8}$.

Proof Parts (a) and (b) are immediate from the definition. Part (c) follows in a manner similar to that of part (d) of Theorem 6.5. Part (d) follows from the definition of the Jacobi symbol and the corresponding property for the Legendre symbol. Part (e) also follows from the definition. In part (f), note that for any positive odd n, $(n^2 - 1)/8$ is integral (see Example 3.3). Suppose $n = ab$, where a, b are odd primes. Then we have

$$\frac{n^2 - 1}{8} = \frac{a^2b^2 - 1}{8} = \frac{a^2 - 1}{8} + \frac{b^2 - 1}{8} + \frac{(a^2 - 1)(b^2 - 1)}{8},$$

and the last term is necessarily even. Hence,

$$\begin{aligned}
(-1)^{(n^2-1)/8} &= (-1)^{(a^2-1)/8}(-1)^{(b^2-1)/8} \\
&= \left(\frac{2}{a}\right)\left(\frac{2}{b}\right) && \text{from Theorem 6.6} \\
&= (2 \mid a)(2 \mid b) \\
&= (2 \mid n) && \text{from part (e).}
\end{aligned}$$

This process can clearly be extended by induction to odd n with more than two prime factors (Problem 6.36). ◆

Example 6.11

$$\begin{aligned}
(112 \mid 153) &= (112 \mid 9)(112 \mid 17) \\
&= (112 \mid 3)^2(112 \mid 17) \\
&= (112 \mid 17) \\
&= \left(\frac{10}{17}\right) \\
&= \left(\frac{2}{17}\right)\left(\frac{5}{17}\right) \\
&= \left(\frac{5}{17}\right) \\
&= -1 && \text{from Table 6.1.}
\end{aligned}$$

◆

Besides being a useful aid for deciding if a congruence of the form $x^2 \equiv A \pmod{n}$ has a solution or not, Jacobi symbols have other uses as well. One such use is in the study of certain sequences of integers. We defined the sequence of **Fibonacci numbers** in Problem 1.20. A long-standing question concerning these numbers had been how many of them are perfect squares. The Jacobi symbol was a useful tool in the solution of this problem (Cohn, 1964), which we sketch now.

The associated sequence of **Lucas numbers**, $\{L_n\}_{n=1}^{\infty}$, is defined by $L_0 = 2$, $L_1 = 1$, and

$$L_{n+1} = L_n + L_{n-1}, \qquad n \geq 1$$

(more on sequences like this in Section 11.3). The following pair of identities can be easily established (Problem 6.38):

1. $2F_{m+n} = F_m L_n + F_n L_m$.
2. $L_{2m} = L_m^2 - 2(-1)^m$.

We shall look at Fibonacci and Lucas numbers with certain values for their index n. First, $L_2 = 3$, $L_4 = 7$, $L_8 = 47$; assume, in general, that $L_n \equiv 3 \pmod{4}$, where $n = 2^r$. Then by relation (2) we have

$$\begin{aligned} L_{2n} = L_{2^{r+1}} &= L_n^2 - 2(-1)^{2^r} \\ &\equiv 9 - 2 \pmod{4} \\ &\equiv 3 \pmod{4}. \end{aligned}$$

Hence, by induction $L_n \equiv 3 \pmod{4}$ for $n = 2^r$ and all $r \geq 1$, so L_n is odd.

Next, we have from relation (1) for $n = 2^r$, $r \geq 1$,

$$\begin{aligned} 2F_{k+2n} &= F_k L_{2n} + F_{2n} L_k \\ &= F_k(L_n^2 - 2) + (F_n L_n) L_k \qquad \text{from (1), (2)} \\ &\equiv -2F_k \pmod{L_n}. \end{aligned}$$

But L_n is odd for $n = 2^r$, so the congruence reduces to

$$F_{k+2n} \equiv -F_k \pmod{L_n}, \tag{*}$$

for all $k \in \mathbf{N}$.

One more preliminary result is that if $n = 2^r$, $r \geq 1$, then $(-1 \mid L_n) = -1$. For this choice of n, L_n is an odd integer of the form $4k + 3$. Hence, in the factorization of L_n there appears an odd number of $(4k + 3)$-prime factors, each of which contributes a factor of -1 to $(-1 \mid L_n)$ according to Theorem 6.5, part (e).

We can now prove Theorem 6.8.

Theorem 6.8 F_n cannot be a perfect square if $n = 12m + q$, where $m \geq 1$ and $q = -1, 1$, or 2.

Proof Let $m = 2^{r-1}K$, where K is odd and $r \geq 1$. Then

$$\begin{aligned} n &= 12(2^{r-1}K) + q \\ &= 6kK + q, \qquad \text{where } k = 2^r \text{ and } r \geq 1. \end{aligned}$$

By (*) we have

$$\begin{aligned} F_n &= F_{6kK+q} \\ &\equiv -F_{(3K-1)2k+q} \pmod{L_k} \\ &\equiv (-1)^{3K} F_q \pmod{L_k} \end{aligned}$$

after $3K - 1$ more applications of (*). Since K is odd, then

$$F_n \equiv -F_q \pmod{L_k}.$$

The Fibonacci numbers can be extended into negative values of the index by $F_{-q} = (-1)^{q-1}F_q$. In this way, the desired recurrence relation $F_{-q} = F_{-q-1} + F_{-q-2}$ holds. Then for $q = -1, 1, 2$ we have $F_{-1} = F_1 = F_2 = 1$, so

$$F_n \equiv -1 \pmod{L_k}$$

or, equivalently, $(F_n \mid L_k) = (-1 \mid L_k) = -1$. Thus, F_n is a quadratic nonresidue modulo L_k and cannot be a perfect square. ◆

Further work with the Jacobi symbol results in elimination of the other choices (Problem 6.39) of q in F_{12m+q}. The final result is that a Fibonacci number F_n is a perfect square for positive n only when $n = 1, 2, 12$.

Problems

6.30. Explain why $(A \mid n) = -1$ *definitely implies* that $x^2 \equiv A \pmod{n}$ has no solution, but $(A \mid n) = +1$ only *suggests* that $x^2 \equiv A \pmod{n}$ has a solution. Assume n is composite.

6.31. Compute the Jacobi symbols:

(a) $(9 \mid 77)$. (b) $(58 \mid 87)$. (c) $(11 \mid 45)$.
(d) $(2 \mid 125)$. (e) $(1007 \mid 57)$. (f) $(2000 \mid 101)$.

6.32. Prove parts (c) and (d) of Theorem 6.7.

6.33. Suppose $(A \mid n) = 0$. What can we conclude about the solvability of $x^2 \equiv A \pmod{n}$?

6.34. Determine in each case whether the indicated congruence has a solution:

(a) $x^2 \equiv 10 \pmod{21}$.
(b) $x^2 \equiv 16 \pmod{105}$.
(c) $2x^2 \equiv 24 \pmod{50}$.
(d) $3x^2 - 27 \equiv 0 \pmod{147}$.
(e) $x^2 \equiv 7 \pmod{6!}$.

6.35. Find all integers n such that $(n \mid 15) = 1$, and yet $x^2 \equiv n \pmod{15}$ fails to have a solution.

6.36. The proof of part (f) of Theorem 6.7 was rather restricted. Extend the result to any odd $n \in \mathbf{N}$.

6.37. Show that the Fibonacci and Lucas numbers are given by the following formulas, referred to as **Binet formulas**:

$$\begin{cases} F_n = \dfrac{\alpha^n - \beta^n}{\alpha - \beta} \\ L_n = \alpha^n + \beta^n, \end{cases} \qquad (n \in \mathbf{N})$$

where $\alpha = (1 + \sqrt{5})/2$ and $\beta = (1 - \sqrt{5})/2$. Illustrate the formulas by calculating F_{10}, L_{10} from them.

6.38. Prove identities (1) and (2) in this section.

6.39. Prove that if F_n is a square, then $n \equiv 0, 1, 2, 6$, or $11 \pmod{12}$. [Hint: Use identity (1) and the fact that any square is congruent to 0, 1 or 4 (mod 8).]

6.4 Gauss's Law of Quadratic Reciprocity

Suppose the odd prime p is a quadratic residue modulo the odd prime q. Then there exists an $x \in \mathbf{Z}_q$ such that

$$x^2 \equiv p \pmod{q}.$$

Can it happen simultaneously that q is a quadratic residue modulo p, that is, does there exist a $y \in \mathbf{Z}_p$ such that

$$y^2 \equiv q \pmod{p}?$$

The examples of $(\frac{5}{11}) = (\frac{11}{5}) = +1$ and $(\frac{11}{47}) = -(\frac{47}{11}) = -1$ show that it happens some of the time, but not always.

The answer to the preceding question is given by the celebrated **Law of Quadratic Reciprocity**. Both **Euler** and **Legendre** knew of the law but lacked complete proofs. The law was independently discovered by **Gauss** at the age of 18 and first proved by him at the age of 19.[1] During his lifetime **Gauss** produced a total of eight proofs of the law. His motivation in doing this was to discover a proof that could be generalized to yield reciprocity theorems for higher-order residues. Such theorems exist and are now an important part of the theory of algebraic numbers. In fact, more than 150 proofs of the Law of Quadratic Reciprocity are now claimed to exist. Many employ topics that are beyond the scope of this book.

We present here the appealing proof of the law that was given by **Gotthold Eisenstein**[2] in 1844 (Laubenbacher and Pengelley, 1994a). It is a beautiful adaptation of Gauss's third proof.

Before beginning the proof, which requires one preliminary lemma (Eisenstein's Lemma), let us first state the law and see an application of it. Let p, q be distinct odd primes; then according to the law,

$$\left(\frac{p}{q}\right)\left(\frac{q}{p}\right) = (-1)^{(p-1)(q-1)/4}$$

holds. The law clears up completely the matter of when $(\frac{p}{q})$ and $(\frac{q}{p})$ are equal. If at least one of p, q is a $(4k+1)$-prime, then at least one of $(p-1)/2$, $(q-1)/2$ is even, so $(-1)^{(p-1)(q-1)/4} = +1$. This implies that the two Legendre symbols are either both $+1$ or both -1. Only when p, q are both $(4k+3)$-primes is the right-hand side -1, so that the two Legendre symbols are of opposite sign.

The idea behind Eisenstein's proof is to produce formulas for $(\frac{p}{q})$, $(\frac{q}{p})$, and to simply multiply them. The formulas are obtained in the following lemma by making a connection with Euler's Criterion in the form of part (c) of Theorem 6.5.

Lemma 6.9.1 (Eisenstein's Lemma). Let p, q be distinct odd primes and let $\mathbf{A} = \{a_i\} = \{2, 4, 6, \ldots, p-1\}$. Then

$$\left(\frac{q}{p}\right) = (-1)^{\sum_i \lfloor qa_i/p \rfloor}.$$

Proof Define the sequence of $(p-1)/2$ integers $\{r_i\}$ by $qa_i \equiv r_i \pmod{p}$, $0 < r_i < p$. If r_i is even, then $(-1)^{r_i} r_i$ is in $\mathbf{A}$, so $(-1)^{r_i} r_i \equiv a_j \pmod{p}$ for some $a_j \in \mathbf{A}$. If r_i is odd, then $p + (-1)^{r_i} r_i$ is in $\mathbf{A}$, so $p + (-1)^{r_i} r_i \equiv (-1)^{r_i} r_i \equiv a_j \pmod{p}$.

No two of the $(-1)^{r_i} r_i$ can be congruent modulo p since if

$$(-1)^{qa_i} qa_i \equiv (-1)^{qa_k} qa_k \pmod{p}$$

[1] To be precise, on April 8, 1796 the young **Gauss** recorded a cryptic statement in the famous Gauss diary of discoveries, which has been interpreted as signaling the completion of the first proof (Bühler, 1981).

[2] **Ferdinand Gotthold Max Eisenstein** (1823–1852), like **Abel** before him and **Ramanujan** after him, was of poor health. He enjoyed the highest admiration by **Gauss**, but a promising career was cut tragically short by tuberculosis (Gillispie, 1971).

held, then as a_i, a_k are even and $(q, p) = 1$, we would obtain $a_i \equiv a_k \pmod p$. This is impossible because $0 < |a_i - a_k| < p$. Hence, the integers $\{(-1)^{r_i} r_i\}$ are congruent modulo p in some order to the integers $\{a_i\}$, and we have

$$\prod_{i=1}^{(p-1)/2} (a_i) \equiv \prod_{i=1}^{(p-1)/2} (-1)^{r_i} r_i \pmod p$$

as well as (by definition)

$$\prod_{i=1}^{(p-1)/2} (qa_i) \equiv \prod_{i=1}^{(p-1)/2} r_i \pmod p.$$

Multiplication of the first congruence by $q^{(p-1)/2}$ and comparison with the second leads to

$$\prod_{i=1}^{(p-1)/2} r_i \equiv q^{(p-1)/2} (-1)^{\Sigma_i r_i} \prod_{i=1}^{(p-1)/2} r_i \pmod p,$$

which becomes $q^{(p-1)/2} \equiv (-1)^{\Sigma_i r_i} \pmod p$. Euler's Criterion (Theorem 6.5, part (c)) transforms this into

$$\left(\frac{q}{p}\right) = (-1)^{\Sigma_i r_i}.$$

For each a_i we have by simple division[3]

$$qa_i = p \left\lfloor \frac{qa_i}{p} \right\rfloor + r_i,$$

so $\Sigma_i r_i = \Sigma_i (qa_i) - \Sigma_i p \lfloor qa_i/p \rfloor$. Since the a_i's are even and $-p \equiv 1 \pmod 2$, then

$$\sum_i r_i \equiv \sum_i \left\lfloor \frac{qa_i}{p} \right\rfloor \pmod 2.$$

Using this in the expression for (q/p), we arrive at

$$\left(\frac{q}{p}\right) = (-1)^{\Sigma_i \lfloor qa_i/p \rfloor}. \quad ◆$$

Example 6.12 Let $p = 7$, $q = 11$; then $\mathbf{A} = \{2, 4, 6\}$. From $qa_i \equiv r_i \pmod p$, we obtain $r_1 = 1$, $r_2 = 2$, $r_3 = 3$. We observe, in agreement with certain key relations in the preceding proof, that

$$\prod_{i=1}^{(p-1)/2} (a_i) = 2 \cdot 4 \cdot 6 \equiv \prod_{i=1}^{(p-1)/2} (-1)^{r_i} r_i = (-1)(2)(-3) = 6 \pmod 7,$$

[3] Recall that $\lfloor qa_i/p \rfloor$ means the greatest integer not exceeding qa_i/p.

and that

$$\left(\frac{11}{7}\right) = 1 = (-1)^{\Sigma_i r_i} = (-1)^6,$$

and finally that $\Sigma_i \lfloor qa_i/p \rfloor = \lfloor 22/7 \rfloor + \lfloor 44/7 \rfloor + \lfloor 66/7 \rfloor = 18$, so

$$\left(\frac{q}{p}\right) = (-1)^{18} = 1. \quad \blacklozenge$$

Eisenstein's Lemma is a criterion for the solvability of $x^2 \equiv q \pmod p$. If the lemma is to be used for just that purpose, it is preferable to employ it in the form

$$\left(\frac{q}{p}\right) = (-1)^{\Sigma_i r_i}.$$

Even in this form it is not distinctly superior to Euler's Criterion.

However, our main interest in Eisenstein's Lemma is in its role as a stepping stone to Gauss's Law of Quadratic Reciprocity. For this purpose we use the lemma in the form

$$\left(\frac{q}{p}\right) = (-1)^{\Sigma_i \lfloor qa_i/p \rfloor}$$

because the exponent $\Sigma_i \lfloor qa_i/p \rfloor$ has a nice geometric interpretation (Fig. 6.1).

Unfortunately, there does not appear to be an easy formula for the quantity $\Sigma_i \lfloor qa_i/p \rfloor$ except in the case of $q = 2$ (Problem 6.42). This special case is not covered by Gauss's law.

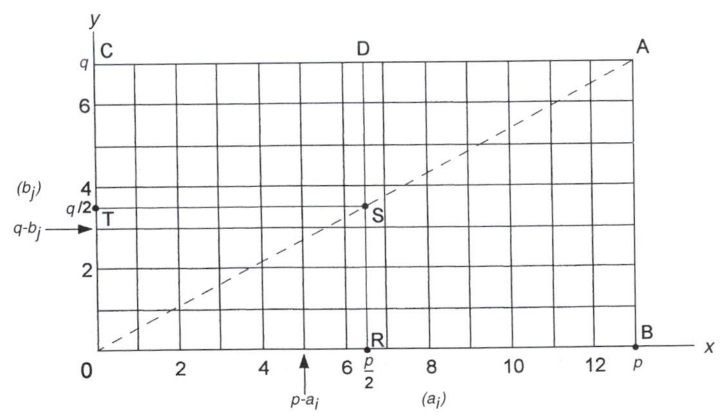

Figure 6.1 Diagram for Eisenstein's proof of Gauss's Law of Quadratic Reciprocity

Theorem 6.9 (Gauss's Law of Quadratic Reciprocity). If p, q are distinct odd primes, then

$$\left(\frac{p}{q}\right)\left(\frac{q}{p}\right) = (-1)^{(p-1)(q-1)/4}.$$

Proof Refer to Figure 6.1 for illustration. The quantity $\Sigma_i \lfloor qa_i/p \rfloor$ is the number of lattice points lying below the line OA, but not on the x-axis, and located in the even-numbered columns ($x = 2, 4, \ldots, p - 1$). This follows because each term $\lfloor qa_i/p \rfloor$ is the number of points with abscissa $x = a_i$ that lie below OA and above the x-axis.

Let $B = \{b_i\} = \{2, 4, 6, \ldots, q - 1\}$; by symmetry, Eisenstein's Lemma gives

$$\left(\frac{p}{q}\right) = (-1)^{\Sigma_j \lfloor pb_j/q \rfloor}$$

The quantity $\Sigma_j \lfloor pb_j/q \rfloor$ is the number of lattice points lying to the left of line OA, but not on the y-axis, and located in the even-numbered rows ($y = 2, 4, \ldots, q - 1$). Hence,

$$\left(\frac{p}{q}\right)\left(\frac{q}{p}\right) = (-1)^{\Sigma_j \lfloor pb_j/q \rfloor + \Sigma_i \lfloor qa_i/p \rfloor}.$$

We now want to do a count of all of the pertinent points in OBAC; this count is the exponent in the preceding formula. No lattice points can lie on line OA because $(p, q) = 1$. Let $a_i > p/2$ be an arbitrary even abscissa. Since $q - 1$ is even, the number of lattice points with abscissa a_i and lying below OA has the same parity as the number of lattice points with abscissa a_i that lie above OA:

$$\left\lfloor \frac{qa_i}{p} \right\rfloor \equiv q - 1 - \left\lfloor \frac{qa_i}{p} \right\rfloor \pmod 2.$$

This latter number, in turn, is equal to the number of lattice points with abscissa $p - a_i$ and lying in the interior of triangle ORS:

$$q - 1 - \left\lfloor \frac{qa_i}{p} \right\rfloor = \left\lfloor \frac{q(p - a_i)}{p} \right\rfloor. \qquad (*)$$

This equation is geometrically plausible. To see it algebraically, let $qa_i/p = n + \epsilon$, where n is an integer and $0 < \epsilon < 1$. Then $q - 1 - \lfloor qa_i/p \rfloor = q - 1 - n$, while on the right side of (*)

$$\left\lfloor q - \left(\frac{qa_i}{p}\right) \right\rfloor = \lfloor q - n - \epsilon \rfloor = \lfloor (q - 1 - n) + (1 - \epsilon) \rfloor = q - 1 - n.$$

We thus have a one-to-one correspondence between points with even abscissa in the interior of triangle ADS and those with odd abscissa in the interior of triangle ORS. As a result we obtain

$$\sum_i \left\lfloor \frac{qa_i}{p} \right\rfloor \equiv N_1 \pmod{2},$$

where N_1 is the total number of points in the interior of triangle ORS. Consequently,

$$\left(\frac{q}{p}\right) = (-1)^{N_1}.$$

Similar reasoning as we just presented shows that if $b_j > q/2$ is an arbitrary even ordinate, then the number of lattice points with this ordinate to the right of OA is equal to the number of lattice points with ordinate $q - b_j$ that lie in the interior of triangle OTS. Consequently, Eisenstein's Lemma leads us to

$$\left(\frac{p}{q}\right) = (-1)^{N_2},$$

where N_2 is the total number of points in the interior of triangle OTS. But, $N_1 + N_2 = [(p-1)/2][(q-1)/2]$, so we obtain finally

$$\left(\frac{p}{q}\right)\left(\frac{q}{p}\right) = (-1)^{(p-1)(q-1)/4}. \quad \blacklozenge$$

To illustrate, recall that in Problem 6.18(c) you found $(\frac{163}{23}) = 1$. Denoting $p = 163$ and $q = 23$, we obtain for $(\frac{23}{163})$

$$1 \cdot \left(\frac{23}{163}\right) = (-1)^{(162)(22)/4} = (-1)^{81(11)} = -1.$$

This says that $x^2 \equiv 23 \pmod{163}$ has no solution.

Gauss's Law allows us to more quickly evaluate complicated Jacobi (and Legendre) symbols. The following example is illustrative.

Example 6.13 Does $x^2 \equiv 58 \pmod{329}$ have a solution? We have

$$\begin{aligned}(58 \mid 329) &= \left(\frac{58}{7}\right)\left(\frac{58}{47}\right) && \text{by definition}\\ &= \left(\frac{2}{7}\right)\left(\frac{11}{47}\right) && \text{from Theorem 6.5}\\ &= \left(\frac{11}{47}\right) && \text{from Theorem 6.6.}\end{aligned}$$

Since $(47 - 1)(11 - 1)/4 = 115$, then from Theorem 6.9 we have $(\frac{11}{47}) = -(\frac{47}{11}) = -(\frac{3}{11})$, so $(58 \mid 329) = -(\frac{3}{11}) = -1$. The negative sign indicates that $x^2 \equiv 58 \pmod{329}$ has no solution. ◆

Notice in the preceding example that we have again used Theorem 6.6; we still owe a proof of that theorem. Eisenstein's beautiful proof of Gauss's Law began with Lemma 6.9.1. That lemma can also be used to prove Theorem 6.6 because the proof of Eisenstein's Lemma does not exclude the possibility that $q = 2$.

Theorem 6.6[4]

$$\left(\frac{2}{p}\right) = (-1)^{(p^2-1)/8} = \begin{cases} +1 & \text{if } p \equiv \pm 1 \pmod 8) \\ -1 & \text{if } p \equiv \pm 3 \pmod 8). \end{cases}$$

Proof From the proof of Lemma 6.9.1 we have for $q = 2$

$$\left(\frac{2}{p}\right) = (-1)^{\Sigma_i r_i}.$$

Next, we note that if r_i is a remainder, so is $p + 2 - r_i$ because

$$p + 2 - r_i \equiv p + 2 - 2a_i \equiv 2(1 - a_i) \pmod p$$

and $2(1 - a_i)$ is clearly congruent to some element of **A**. But r_i and $p + 2 - r_i$ are distinct, since otherwise $2r_i = p + 2$, an impossibility.

Hence, if $p = 4k + 1$, the remainders occur as $(1/2)[(p - 1)/2]$ complementary pairs whose sums are $r_i + (p + 2 - r_i) = p + 2$:

$$\sum_{i=1}^{(p-1)/2} r_i = \frac{(p+2)(p-1)}{4} = \frac{p^2 + p - 2}{4} \qquad (p = 4k + 1).$$

Since

$$\frac{p^2 + p - 2}{4} - \frac{p^2 - 1}{8} = \frac{p^2 + 2p - 3}{8} = 2k(k + 1)$$

and is even, it follows that

$$\sum_{i=1}^{(p-1)/2} r_i \equiv \frac{p^2 - 1}{8} \pmod 2.$$

[4] Sometimes referred to in the literature as the *Ergänzungssatz*, or "Complementary Theorem" (after **Gauss**), because it was an offshoot of Gauss's Lemma. We do not present Gauss's Lemma here (Problem 6.52), opting instead for the more convenient lemma due to **Eisenstein** (Laubenbacher and Pengelley, 1994b).

On the other hand, if $p = 4k + 3$, the remainder $r_i = 1$ is unpaired (this remainder arises from $a_i = (p + 1)/2$). The other $(p - 3)/2$ r_i's are paired as we just showed. We obtain

$$\sum_{i=1}^{(p-1)/2} r_i = \frac{(p + 2)(p - 3)}{4} + 1 = \frac{p^2 - p - 2}{4} \qquad (p = 4k + 3),$$

so

$$\frac{p^2 - p - 2}{4} - \frac{p^2 - 1}{8} = \frac{p^2 - 2p - 3}{8} = 2k(k + 1),$$

and the preceding congruence holds again. Thus, for both families of primes

$$\left(\frac{2}{p}\right) = (-1)^{(p^2-1)/8},$$

and the cases $p = 8k \pm 1, 8k \pm 3$ follow trivially because a $(4k + 1)$-prime is automatically either an $(8k + 1)$-prime or an $(8k - 3)$-prime, and a $(4k + 3)$-prime is automatically either an $(8k - 1)$-prime or an $(8k + 3)$-prime. ◆

Example 6.14 Consider the family of primes of the form $p = 3 \cdot 2^n + 5$. When is $x^2 \equiv 2 \pmod{p}$ solvable?

When $n = 1$, then $p = 11 = 8 \cdot 1 + 3$, so $x^2 \equiv 2 \pmod{11}$ has no solution from Theorem 6.6. When $n = 2$, then $p = 17 = 8 \cdot 2 + 3$, so $x^2 \equiv 2 \pmod{17}$ is solvable. Finally, for $n \geq 3$ we have $3 \cdot 2^n + 5 \equiv 0 - 3 \pmod{8}$, so $x^2 \equiv 2 \pmod{p}$ is not solvable. Thus, $x^2 \equiv 2 \pmod{3 \cdot 2^n + 5}$ is solvable only for $n = 2$; indeed, the solutions of $x^2 \equiv 2 \pmod{17}$ are $x \equiv 6 \pmod{17}$ and $x \equiv 11 \pmod{17}$. ◆

Can the Law of Quadratic Reciprocity be extended to pairs of Jacobi symbols, $(P \mid Q)$ and $(Q \mid P)$? Look at these examples.

Example 6.15 Let $P = 21, Q = 25$.

$$(21 \mid 25) = \left(\frac{21}{5}\right)^2 = 1, \qquad (25 \mid 21) = \left(\frac{25}{3}\right)\left(\frac{25}{7}\right) = \left(\frac{5}{3}\right)^2\left(\frac{5}{7}\right)^2 = 1.$$

So $(21 \mid 25)(25 \mid 21) = 1 = (-1)^{(P-1)(Q-1)/4} = (-1)^{120}$.

On the other hand, let $P = 39$, $Q = 55$. Then

$$(39 \mid 55) = \left(\frac{39}{5}\right)\left(\frac{39}{11}\right) \qquad (55 \mid 39) = \left(\frac{55}{3}\right)\left(\frac{55}{13}\right)$$
$$= \left(\frac{4}{5}\right)\left(\frac{6}{11}\right) \qquad = \left(\frac{1}{3}\right)\left(\frac{3}{13}\right)$$
$$= -1, \qquad = 1.$$

So $(39 \mid 55)(55 \mid 39) = -1 = (-1)^{(P-1)(Q-1)/4} = (-1)^{513}$. ◆

The general result, given in the next theorem, provides an alternative sequence for evaluating complicated Jacobi symbols (Problem 6.49).

Theorem 6.10 If P, Q are positive odd integers and $(P, Q) = 1$, then

$$(P \mid Q)(Q \mid P) = (-1)^{(P-1)(Q-1)/4}.$$

Proof Let $P = \Pi_{i=1}^{n} p_i$, $Q = \Pi_{j=1}^{m} q_j$ be factorizations into primes, not necessarily distinct. Then from Theorem 6.7

$$(P \mid Q)(Q \mid P) = \prod_{i=1}^{n}\prod_{j=1}^{m}\left(\frac{p_i}{q_j}\right)\left(\frac{q_j}{p_i}\right) = (-1)^s, \qquad (P, Q) = 1$$

for some $s \in \mathbf{N}$. We now apply the Law of Quadratic Reciprocity to each pair of Legendre symbols,

$$\left(\frac{p_i}{q_j}\right)\left(\frac{q_j}{p_i}\right) = (-1)^{(p_i-1)(q_j-1)/4}$$

This gives $s = \Sigma_{i=1}^{n} (p_i - 1)/2\ \Sigma_{j=1}^{m} (q_j - 1)/2$. What we'd like to show is that the summation over the p_i's has the same parity as $(P - 1)/2$, and analogously for the summation over the q_j's.

Now write

$$P = [1 + (p_1 - 1)][1 + (p_2 - 1)]\ldots[1 + (p_n - 1)]$$
$$= 1 + \sum_{i=1}^{n}(p_i - 1) + \sum_{i \neq j}(p_i - 1)(p_j - 1) + \sum_{i \neq j \neq k}(p_i - 1)(p_j - 1)(p_k - 1)$$
$$+ \ldots + \prod_{i=1}^{n}(p_i - 1).$$

Each $(p_i - 1)$ is even, so each term in the second and in all succeeding summations is divisible by 4. Hence,

$$P - 1 \equiv \sum_{i=1}^{n} (p_i - 1) \pmod 4$$

or

$$\frac{(P - 1)}{2} \equiv \sum_{i=1}^{n} \frac{(p_i - 1)}{2} \pmod 2.$$

A similar analysis is carried out for Q, and we obtain

$$\frac{(Q - 1)}{2} \equiv \sum_{j=1}^{n} \frac{(q_j - 1)}{2} \pmod 2.$$

It follows that $s \equiv [(P - 1)/2][(Q - 1)/2] \pmod 2$ and the theorem follows. ◆

Example 6.16 Evaluate $(29 \mid 329)$.

Since $(329 - 1)(29 - 1)/4 = 2296$, from Theorem 6.10 we have $(29 \mid 329) = (329 \mid 29) = (\frac{10}{29}) = (\frac{2}{29})(\frac{5}{29}) = -(\frac{5}{29})$. Now apply the Reciprocity Law again to $(\frac{5}{29})$. As $(29 - 1)(5 - 1)/4 = 28$, then

$$(29 \mid 329) = -\left(\frac{5}{29}\right) = -\left(\frac{29}{5}\right) = -\left(\frac{2^2}{5}\right) = -1. \quad ◆$$

By now we can see the great usefulness of Gauss's Law of Quadratic Reciprocity for the evaluation of Legendre and Jacobi symbols. We can also use the law to solve problems of an inverse nature (Problems 6.47 and 6.48).

Example 6.17 Find all primes $p > 3$ such that $(\frac{-3}{p}) = +1$. By Theorem 6.5(f) we have $(\frac{-1}{p})(\frac{3}{p}) = (\frac{-3}{p}) = +1$, and from part (e) of that theorem $(\frac{-1}{p}) = (-1)^{(p-1)/2}$. Hence, we desire all primes p such that $(\frac{3}{p}) = (-1)^{(p-1)/2}$. Now apply Theorem 6.9:

$$\left(\frac{3}{p}\right)\left(\frac{p}{3}\right) = (-1)^{(p-1)(3-1)/4} = (-1)^{(p-1)/2}.$$

It follows that $(\frac{p}{3}) = +1$. All primes $p > 3$ are of the form $3k + 1$ or $3k + 2$. If $p = 3k + 1$, then

$$\left(\frac{p}{3}\right) = \left(\frac{3k + 1}{3}\right) = \left(\frac{1}{3}\right) = +1,$$

and if $p = 3k + 2$, then

$$\left(\frac{p}{3}\right) = \left(\frac{3k + 2}{3}\right) = \left(\frac{2}{3}\right) = -1.$$

Hence, the answer to the original question is the set $\{7, 13, 19, 31, \ldots\}$ of $(3k + 1)$-primes or, equivalently, the set of $(6k + 1)$-primes. ◆

Problems

6.40. Using only Eisenstein's Lemma, determine in each case whether the indicated congruence has a solution:

(a) $x^2 \equiv 23 \pmod{31}$.
(b) $x^2 \equiv 5 \pmod{37}$.
(c) $x^2 \equiv 11 \pmod{23}$.
(d) $x^2 \equiv 31 \pmod{23}$.
(e) $x^2 \equiv 37 \pmod{7}$.
(f) $x^2 \equiv 97 \pmod{83}$.

6.41. Consider the sets $\{r_i\}$ of remainders referred to in the proof of Eisenstein's Lemma.

(a) Work out the set $\{r_i\}$ for each of these choices of prime pairs, $\{p, q\}$:
  (i) $p = 11, q = 7$.
  (ii) $p = 11, q = 13$.
  (iii) $p = 17, q = 11$.
  (iv) $p = 13, q = 23$.

(b) Does $\{r_i\}$ have the structure of a multiplicative group in any of the four preceding cases?

(c) 1 appears to be a common member of the sets of remainders. Show, however, that $1 \in \{r_i\}$ iff $q^{p-2} \equiv Q \pmod{p}$ is even $(0 < Q < p)$.

6.42. Show that if $q = 2$ and the prime $p = 4k + 1$, then the quantity $\Sigma_i \lfloor qa_i/p \rfloor$ in Eisenstein's Lemma is $(p - 1)/4$. If the prime $p = 4k + 3$, however, show that $\Sigma_i \lfloor qa_i/p \rfloor = (p + 1)/4$.

6.43. Use a geometric argument like that in connection with Figure 6.1 to rederive the first result in Problem 6.42.

6.44. Evaluate the following Legendre and Jacobi symbols:

(a) $(\frac{379}{761})$.
(b) $(68 \mid 111)$.
(c) $(1009 \mid 2307)$.
(d) $(180 \mid 3773)$.
(e) $(\frac{5503}{7331})$.
(f) $(2^{22} + 1 \mid 2^{11} + 1)$.

6.45. Select a $(4k + 1)$-prime p and a $(4k + 3)$-prime p and show in each case, in accordance with the proof of Theorem 6.6, that for $q = 2$ one has $\Sigma_{i=1}^{(p-1)/2} r_i \equiv \frac{p^2 - 1}{8} \pmod{2}$.

6.46. Prove that if $p = 4q + 1$ is a prime and q is a prime, then 2 is a primitive root of p. Corroborate this using data from Table D in Appendix II. [Hint: Theorem 6.6 is useful.]

6.47. Find all primes $p > 5$ such that $(\frac{5}{p}) = +1$. Illustrate with some examples.

6.48. Find all primes $p \geq 3$ such that $(\frac{-5}{p}) = +1$. Illustrate with some examples.

6.49. Use Theorem 6.10 as far as possible and decide which of the following congruences have solutions:

(a) $x^2 \equiv 27 \pmod{35}$.
(b) $x^2 \equiv 117 \pmod{119}$.
(c) $x^2 \equiv 91 \pmod{783}$.

6.50. For those congruences in Problem 6.49 that have solutions, find these solutions.

6.51. In considering the congruence $x^2 \equiv 15 \pmod{35}$, a student writes $(15 \mid 35)(35 \mid 15) = (-1)^{14(34)/4} = -1$. Then $(35 \mid 15)$ is reduced as follows:

$$(35 \mid 15) = (5 \mid 15) = \left(\frac{1}{3}\right) = 1,$$

so $(15 \mid 35) = -1$ and the congruence should have no solution. A simple computation, however, shows that $x \equiv 15 \pmod{35}$ is a solution. What has gone wrong?

6.52. **Gauss's Lemma** says: Let p be an odd prime and suppose $p \nmid q$. Let μ be the number of least positive residues modulo p of $q, 2q, 3q, \ldots, (p - 1)q/2$ that exceed $p/2$. Then

$$\left(\frac{q}{p}\right) = (-1)^{\mu}.$$

Apply Gauss's Lemma to the congruences in Problem 6.40. Is either Gauss's Lemma or Eisenstein's Lemma superior to the other as an alternative to Euler's Criterion?

6.53. Show that the Diophantine equation $-x^2 - x - 10 = 95y$ has no solutions (x, y) in integers.

6.54. Show that the Diophantine equation $2x^2 + x + 310 = 97y$ has an infinite number of integral solutions, but in none of these solutions are x, y *both* negative.

6.5 Cubic and Quartic Residues

By analogy to the definition of quadratic residue given in Section 6.1, we make the following more general definition:

> **DEF.** Suppose $p \nmid A$, where p is a prime. If the congruence $x^k \equiv A \pmod{p}$ has (has not) a solution, then A is called a ***k*****th-power residue** (**non-residue**) modulo p.

For example, since $2^5 \equiv 32 \equiv 10 \pmod{11}$, then 10 is a fifth-power residue modulo 11. The integer 1 is a kth-power residue modulo any prime for any $k \in \mathbf{N}$. We shall first prove a theorem for general k, and then for concreteness specialize thereafter to cubic residues ($k = 3$) and quartic residues ($k = 4$).

Theorem 6.11 The congruence $x^k \equiv A \pmod{p}$ has a solution iff $A^{(p-1)/d} \equiv 1 \pmod{p}$, where $d = (k, p - 1)$. If the congruence has a solution, then it actually has d incongruent solutions modulo p.

Proof By Theorem 5.8 the prime p has a primitive root g. Then $x^k \equiv A \pmod{p}$ holds iff $k \cdot \operatorname{ind}_g x \equiv \operatorname{ind}_g A \pmod{p - 1}$. Let $d = (k, p - 1)$. Then from Theorem 3.22 the congruence is solvable for $\operatorname{ind}_g x$ iff $d \mid \operatorname{ind}_g A$, and in the event that it is solvable, there are then d incongruent solutions modulo $p - 1$. This implies that there are then d incongruent values of x modulo p. Further, $d \mid \operatorname{ind}_g A$ iff $nd = \operatorname{ind}_g A$ for some $n \in \mathbf{N}$, or equivalently, $(p - 1)nd = (p - 1) \operatorname{ind}_g A$. Transforming this back to congruence notation, we obtain

$$(g^{p-1})^n \equiv 1^n \equiv A^{(p-1)/d} \pmod{p}. \qquad \blacklozenge$$

The theorem is a generalization of Euler's Criterion (Theorem 6.4), where for quadratic congruences $d = (k, p - 1) = (2, p - 1) = 2$. One difference, however, is that when A is a quadratic nonresidue modulo p, then

$$A^{(p-1)/d} \equiv -1 \pmod{p}$$

(Corollary 6.4.1), but when A is a kth-power nonresidue, then $A^{(p-1)/d}$ need not be congruent to -1 modulo p (Problem 6.57).

Another difference arises when we consider the number of kth-power residues modulo a given prime p. Suppose $k = 3$; then when $p = 5, 7, 11, 13, 17, 19$, and 23, the numbers of cubic residues modulo these primes are found to be 4, 2, 10, 4, 16, 6, and 22. A pattern is evident.

Theorem 6.12 The number of cubic residues modulo p is $(p - 1)/3$ or $p - 1$, depending on whether p is of the form $3j + 1$ or $3j + 2$, respectively.

Proof If $p = 3j + 1$, then $d = (k, p - 1) = (3, 3j) = 3$, so from Theorem 6.11, A is a cubic residue modulo p iff $A^{(p-1)/3} - 1 \equiv 0 \pmod{p}$. We recall now Lemma 6.4.1; since $(p - 1)/3 \mid (p - 1)$, then $x^{(p-1)/3} - 1 \equiv 0 \pmod{p}$ has exactly $(p - 1)/3$ incongruent roots. Accordingly, there are $(p - 1)/3$ cubic residues modulo p and $2(p - 1)/3$ cubic nonresidues modulo p.

If $p = 3j + 2$, then $d = (k, 3j + 1) = (3, 3j + 1) = 1$ and A is a cubic residue modulo p iff $A^{p-1} \equiv 1 \pmod{p}$. This holds for all integers A in $[1, p - 1]$ by Fermat's Theorem, so there are $p - 1$ cubic residues modulo p and no cubic nonresidues. ◆

Example 6.18 Theorem 6.11 shows that 2, 3, 4 are cubic nonresidues modulo $p = 13$, but 5 is a cubic residue because $5^{(13-1)/3} \equiv 1 \pmod{13}$. Can you find the other three residues that Theorem 6.12 says must exist? ◆

To further illustrate Theorem 6.12, we present in Table 6.2 the least positive cubic residues of the first few primes. This table is to be compared with Table 6.1. More data appear in Table F in Appendix II.

We think that Problems 5.51 and 6.10(a), though elementary in nature, brought out an interesting relationship between the primitive roots of a prime and the quadratic nonresidues. A similar relationship exists between the primitive roots and the cubic nonresidues, provided there are any of the latter (Problem 6.64). But now Theorem 6.12 has told us exactly when this happens, so we have Corollary 6.12.1.

Table 6.2 Some Cubic Residues and Nonresidues Modulo Primes

p	Least Positive Residues	Least Positive Nonresidues
3	1, 2	none
5	1, 2, 3, 4	none
7	1, 6	2, 3, 4, 5
11	1, 2, 3, 4, 5, 6, 7, 8, 9, 10	none
13	1, 5, 8, 12	2, 3, 4, 6, 7, 9, 10, 11
17	1, 2, 3, 4, 5, 6, 7, 8, 9, 10, 11, 12, 13, 14, 15, 16	none
19	1, 7, 8, 11, 12, 18	2, 3, 4, 5, 6, 9, 10, 13, 14, 15, 16, 17

Corollary 6.12.1 Let p be a $(3j + 1)$-prime, $j > 0$. Then the primitive roots of p are to be sought among the members of the intersection of the sets of all least positive quadratic nonresidues modulo p and cubic nonresidues modulo p.

Example 6.19 The least positive quadratic nonresidues modulo $p = 13$ are 2, 5, 6, 7, 8, and 11 (Table 6.1). The least positive cubic nonresidues modulo p are (from Table 6.2) 2, 3, 4, 6, 7, 9, 10, and 11. Hence, the primitive roots of p are a subset of $\{2, 6, 7, 11\}$. Since $\phi\{\phi(13)\} = 4$, this set is precisely the set of primitive roots of 13. ◆

Turning now to quartic residues, we make the observation that every quartic residue A modulo a prime p is automatically a quadratic residue, since if $x^4 \equiv A \pmod{p}$ has a solution, then $y^2 \equiv A \pmod{p}$ also holds, where $y \equiv x^2 \pmod{p}$. Thus, the set of all least positive quartic residues modulo p is a subset of the set of all least positive quadratic residues modulo p, and we may find all members of the former by squaring members of the latter.

Theorem 6.13 The number of quartic residues modulo p is $(p - 1)/4$ or $(p - 1)/2$, depending on whether p is of the form $4j + 1$ or $4j + 3$, respectively.

Proof If $p = 4j + 1$, then $d = (k, p - 1) = (4, 4j) = 4$, so from Theorem 6.11, A is a quadratic residue modulo p iff $A^{(p-1)/4} \equiv 1 \pmod{p}$. By Lemma 6.4.1 the congruence $x^{(p-1)/4} \equiv 1 \pmod{p}$ has $(p - 1)/4$ solutions, so there are $(p - 1)/4$ quartic residues modulo p.

If $p = 4j + 3$, then $d = (4, 4j + 2) = 2$, so Theorem 6.11 reduces to Euler's Criterion and the set of quartic residues modulo p is identical to the set of quadratic residues modulo p, in view of comments made prior to this theorem. ◆

Example 6.20 Let $p = 17 = 4(4) + 1$; then by Theorem 6.13 there are just four quartic residues modulo p. The quadratic residues modulo p are 1, 2, 4, 8, 9, 13, 15, and 16 (Table 6.1). Squaring these and reducing modulo 17, we obtain the set $\{1, 4, 13, 16\}$. These are all the quartic residues modulo 17. ◆

Table 6.3 gives the least positive quartic residues and nonresidues modulo some primes. Note, as just mentioned, that the sets of quartic residues are subsets of the analogous sets of quadratic residues.

In Theorem 6.3 it was proved that the set of quadratic residues modulo a prime p is a multiplicative group. The reasoning used there extends immediately to kth-power residues modulo a prime (Problem 6.65). Then let $\mathbf{A}_p^{(k)}$ denote the group of least positive kth-power residues $\{A_i\}$ modulo the prime p. When $p > 2$ it is

Table 6.3 Some Quartic Residues and Nonresidues Modulo Primes

p	Least Positive Residues	Least Positive Nonresidues
3	1	2
5	1	2, 3, 4
7	1, 2, 4	3, 5, 6
11	1, 3, 4, 5, 9	2, 6, 7, 8, 10
13	1, 3, 9	2, 4, 5, 6, 7, 8, 10, 11, 12
17	1, 4, 13, 16	2, 3, 5, 6, 7, 8, 9, 10, 11, 12, 14, 15
19	1, 4, 5, 6, 7, 9, 11, 16, 17	2, 3, 8, 10, 12, 13, 14, 15, 18

easy to show that $|\mathbf{A}_p^{(3)}|$ is always even (Problem 6.63). For $\mathbf{A}_p^{(4)}$, the group of least positive quartic residues modulo p, the situation is different (Dence and Dence, 1995):

Theorem 6.14 $\mathbf{A}_p^{(4)}$ is of even order iff $p = 8j + 1$.

Proof ($\leftarrow$) Suppose $p = 8j + 1$. Then $d = (k, p - 1) = (4, 8j) = 4$, so $(p - 1)/d = 2j$. Thus, $|\mathbf{A}_p^{(4)}| = 2j$ is even.

($\rightarrow$) Suppose $|\mathbf{A}_p^{(4)}|$ is even. When $p = 4k + 1$, then $|\mathbf{A}_p^{(4)}| = (p - 1)/4 = k$, and this is even iff $k = 2j$, so $p = 8j + 1$. When $p = 4k + 3$, then $|\mathbf{A}_p^{(4)}| = (p - 1)/2 = 2k + 1$, and this is never even. Hence, when $|\mathbf{A}_p^{(4)}|$ is even, p can only be of the form $8j + 1$. ◆

So far we have been considering only prime moduli. Moduli that are powers of 2 represent another special case that leads to many results with recognizable patterns (Dence and Dence, 1996). The following theorem is to be contrasted with Theorem 6.12.

Theorem 6.15 If $k \geq 3$ is odd, then every odd integer A satisfying $1 \leq A < 2^n$, $n \in \mathbf{N}$, is a kth-power residue modulo 2^n.

Proof Let $\{x_1, x_2, \ldots, x_n\}$ be the odd integers $\{1, 3, 5, \ldots, 2^n - 1\}$. Corresponding to each x_i we compute A_i, the least positive kth-power residue modulo 2^n. If $x_i \neq x_j$, then we must have $A_i \neq A_j$. For if the contrary held, then it would follow that

$$x_i^k - x_j^k \equiv 0 \pmod{2^n}$$

or $(x_i - x_j)(x_i^{k-1} + x_i^{k-2}x_j + \cdots + x_j^{k-1}) \equiv 0 \pmod{2^n}$. The binomial factor is divisible by at most $n - 1$ powers of 2 since $x_i, x_j < 2^n$. The polynomial factor contains k terms, each of which is odd, so the polynomial itself is odd and therefore not divisible by 2. The congruence is an impossibility, and the A_i's are merely a permutation of all of the x_i's. ◆

Table 6.4 Cubic Residues Modulo 32

A_i	x_i	A_i	x_i
1	1	17	17
3	27	19	11
5	29	21	13
7	23	23	7
9	25	25	9
11	19	27	3
13	21	29	5
15	15	31	31

Example 6.21 The congruence $x^3 \equiv A \pmod{32}$ is solvable for $A = 1, 3, 5, \ldots, 31$. That is, 32 has 16 incongruent cubic residues. ◆

This last example is intriguing because corresponding to each residue A_i there is but a single incongruent solution, $x_i^3 \equiv A_i \pmod{2^5}$. These are listed in Table 6.4. Closer examination of the table shows that in each case of $x_i^3 \equiv A_i \pmod{2^5}$ one also has $A_i^3 \equiv x_i \pmod{2^5}$. The A_i's and the x_i's exhibit a kind of interchangeability. This phenomenon holds for moduli 2^n when $n = 1, 2, 3, 4, 5$, but not for $n \geq 6$ (Problem 6.72).

In contrast to the abundance of cubic residues of 2^n, quartic residues are scarce. For example, there are only two incongruent quartic residues modulo 2^5.

Theorem 6.16 The integer A, $1 \leq A \leq 2^n - 1$, is a quartic residue modulo 2^n iff $A = 16k + 1$, for k a nonnegative integer.

Proof The theorem is trivial for $A = 1$ $(k = 0)$ for any $n \in \mathbf{N}$.

($\rightarrow$) Suppose A_i is a quartic residue $\pmod{2^n}$ and let $x_i = 2j + 1$ be a solution of $x^4 \equiv A_i \pmod{2^n}$. Then $x_i^4 = (2j + 1)^4 = 8j(j + 1)(2j^2 + 2j + 1) + 1 = 16m + 1$, $m > 0$, because either j or $j + 1$ is even. Hence, we have

$$16m + 1 \equiv A_i \pmod{2^n},$$

and as we may assume $n \geq 4$ (because the lower cases can be disposed of individually), then A_i itself is of the form $16k + 1$.

($\leftarrow$) The theorem can be shown directly to hold in this direction for the special cases of $n \leq 7$, so we shall assume that $n \geq 8$. The idea now is to show that for a given n and from a given solution to $x^4 \equiv 16k + 1 \pmod{2^n}$, one can construct a solution to $x^4 \equiv 16(k + 1) + 1 \pmod{2^n}$. The necessity direction of the theorem will follow from the Principle of Finite Induction.

We already know there is a solution to $x^4 \equiv 16k + 1 \pmod{2^n}$ when $k = 0$; assume that x_i is a solution when $k = k_i$. Now consider

$$x^4 \equiv 16(k_i + 1) + 1 \pmod{2^n}$$

and set $x = 3x_i + B \cdot 2^{n-3}$, where B is to be determined. Substitute the expression for x into the congruence and obtain

$$81x_i^4 + 108x_i^3 \cdot B \cdot 2^{n-3} + 54x_i^2 \cdot B^2 \cdot 2^{2n-6} + 12x_i \cdot B^3 \cdot 2^{3n-9} + B^4 \cdot 2^{4n-12} \equiv 16(k_i + 1) + 1 \pmod{2^n}.$$

Since $n \geq 8$, then $2n - 6 > n$, $3n - 9 > n$, and $4n - 12 > n$; also, $x_i^4 \equiv 16k_i + 1 \pmod{2^n}$. The preceding congruence then reduces to

$$27x_i^3 \cdot B \cdot 2^{n-1} \equiv -64 - 1280k_i \pmod{2^n}.$$

Let $z = B \cdot 2^{n-7}$, and divide both sides of the congruence by 2^6 (use Theorem 3.3) to give

$$(27x_i^3)z \equiv -1 - 20k_i \pmod{2^{n-6}}.$$

This has a unique solution $z \equiv z_i \pmod{2^{n-6}}$ because $(27x_i^3, 2^{n-6}) = 1$. It follows that x can be computed from $x = 3x_i + B \cdot 2^{n-3} = 3x_i + 16z_i$, and thus every integer of the form $16k + 1$, $k \geq 0$, is a quartic residue of 2^n. ◆

Example 6.22 A complete set of incongruent quartic residues of $2^8 = 256$ is $\{1, 17, 33, 49, \ldots, 241\}$. In the notation of the proof of Theorem 6.16, we have $x_1 = 1$ when $k_1 = 0$ and $n = 8$. Now solve $(27x_1^3)z \equiv -1 - 20k_1 \pmod{2^{n-6}}$, or $27z \equiv -1 \pmod{16}$, to give $z \equiv 13 \pmod{16}$. This yields $x_2 \equiv 3x_1 + 16z_1 \pmod{256}$, or $x_2 \equiv 211 \pmod{256}$. Indeed, $211^4 \equiv 17 \pmod{256}$. ◆

Problems

6.55. How many of the integers in [1, 100] are perfect cubes modulo 101? How many of the integers in [1, 150] are perfect cubes modulo 151?

6.56. How many cube roots, fourth roots, and fifth roots (all mod (11)) of 3 are present in $\mathbf{Z}_{11}$?

6.57. Find all the cubic nonresidues $\{A_i\}$ modulo $p = 31$. For each A_i find $A_i^{(p-1)/d}$ modulo p. In how many cases does $A_i^{(p-1)/d} \equiv -1 \pmod{p}$ hold?

6.58. Find all the quartic nonresidues $\{A_i\}$ modulo $p = 23$. For each A_i find $A_i^{(p-1)/d}$ modulo p. In how many cases does $A_i^{(p-1)/d} \equiv -1 \pmod{p}$ hold?

6.59. Explain the contrasting results of the two previous problems.

6.60. What is the order of the group of quintic residues modulo the odd prime p under modular multiplication?

6.61. Consider $p = 13$. We know that the $\phi\{\phi(13)\} = 4$ primitive roots of p can be found among the nine quartic nonresidues modulo p. By definition, the roots have order $\phi(p) = 12$. What are the orders of the remaining $9 - 4 = 5$ quartic nonresidues?

6.62. Which of these congruences has a solution?

(a) $2x^3 \equiv 12 \pmod{17}$.
(b) $x^4 \equiv 39 \pmod{53}$.
(c) $x^{12} \equiv 59 \pmod{67}$.

6.63. Show that if A is a cubic residue modulo the odd prime p, then so is $p - A$. How does this lead to the result that $p \mid T_p^{(3)}$, where $T_p^{(3)}$ denotes the sum of the members of $\mathbf{A}_p^{(3)}$? Compute easily $T_{97}^{(3)}$. How can we also conclude that $|\mathbf{A}_p^{(3)}|$ is even?

6.64. Let p be a $3j + 1$ prime, $j > 1$.

(a) Prove that all of the primitive roots of p are to be found among the cubic nonresidues of p.
(b) Use Corollary 6.12.1 to help you identify all of the primitive roots of $p = 37$.

6.65. Prove that for any prime p and any integer $k > 1$ the set $\mathbf{A}_p^{(k)}$ is a group, and that it is cyclic. Is $\mathbf{A}_n^{(3)}$ a cyclic group when $n = 26$? When $n = 36$?

6.66. Let $p = 8k + 1$ be a prime. Show that k is a cubic residue of p. [Hint: Use Problem 6.65.]

6.67. Let $P_p^{(3)}$ denote the product of all the elements of $\mathbf{A}_p^{(3)}$. Show that for all primes p one has $1 + P_p^{(3)} \equiv 0 \pmod{p}$.

6.68. Show that for all primes p one has

$$P_p^{(4)} \equiv \begin{cases} -1 & \pmod{p} \text{ if } p = 8j + 1 \\ +1 & \pmod{p} \text{ otherwise.} \end{cases}$$

6.69. Let the prime $p > 5$. Show that $p \mid T_p^{(4)}$, where $T_p^{(4)}$ is the sum of all the members of $\mathbf{A}_p^{(4)}$. [Hint: Use Theorem 6.13 and Problem 6.12(a).]

6.70. The integer 271 is a prime.

(a) Show that the set of distinct least positive residues obtained when powers of 10 are divided by 271 form a cyclic multiplicative group modulo 271 of order 5.

(b) How does the order of this group compare with the length of the repeat unit in the decimal expansion of $1/271$?

(c) Prove that if $271 \mid N$, then $\mathrm{DS}(N) \geq 5$, where $\mathrm{DS}(N)$ denotes the digital sum of the integer $N > 0$. This is an adaptation of a problem that was first proposed in (Cooper and Kennedy, 1994).

6.71. Recall the definition of a **repunit** from Problem 2.4. Let $p \neq 5$ be any odd prime. Prove that there exists a repunit x such that $p \mid x$.

6.72. Prove that $x_i^3 \equiv A_i \pmod{2^n}$ implies $A_i^3 \equiv x_i \pmod{2^n}$ only when $1 \leq n \leq 5$.

6.73. Prove that the integer $A > 0$ is a quadratic residue modulo 2^n iff $A = 8k + 1$, for k a nonnegative integer. Hence, also show that (a) the number of incongruent quadratic residues of 2^n is $(1/4)\phi(2^n)$ if $n \geq 3$, and (b) the number of incongruent quartic residues of 2^n is $(1/8)\phi(2^n)$ if $n \geq 4$.

6.74. Let p be a $(3n + 1)$-prime. It is known that unique numbers $L, M \in \mathbf{Z}$ exist such that

$$4p = L^2 + 27M^2,$$

and $L \equiv 1 \pmod 3$. A theorem says that 2 is a cubic residue modulo p iff L is even (Lehmer, 1951). Find, by computer, all $(3n + 1)$-primes less than 100 for which 2 is a cubic residue, and solve the associated congruences.

6.75. Another theorem says that 3 is a cubic residue modulo $p = 3n + 1$ iff $3 \mid M$ in the representation of the previous problem. Find, by computer, all $(3n + 1)$-primes less than 100 for which 3 is a cubic residue, and solve the associated congruences.

6.76. Let p be an $(8n + 1)$-prime. It is known that unique integers a, b exist such that

$$p = a^2 + b^2,$$

and $a \equiv 1 \pmod 4$. A theorem says that 2 is a quartic residue modulo p iff $b \equiv 0 \pmod 8$ (Lehmer, 1958). Find, by computer, all $(8n + 1)$-primes less than 200 for which 2 is a quartic residue, and give a solution to the associated congruences.

6.77. Consider the congruence $x^m - 1 \equiv 0 \pmod p$, and suppose that $m \mid (p - 1)$. Let $\{\alpha_i\}$ be the set of roots in $\mathbf{Z}_p$ of this congruence.

(a) Show that $\Sigma_{i<j}\, \alpha_i \alpha_j \equiv 0 \pmod p$.

(b) Let $\mathbf{A}_p^{(3)}$ be the set of cubic residues modulo p and let $T_{p,2}^{(3)}$ be the sum of the squares of all the members of $\mathbf{A}_p^{(3)}$. Use part (a) to prove that $p \mid T_{p,2}^{(3)}$ for $p = 5$ and all primes $p > 7$. Why do we exclude $p = 7$?

6.78. Let p, L, M be as in Problem 6.74. A theorem says that the sign of M can be chosen so that

$$\frac{L + 9M}{L - 9M}, \frac{L - 9M}{L + 9M}$$

become integers in $[2, p - 1]$ by addition of suitable multiples of p to numerators and denominators (Williams, 1975). The two preceding integers are then the other two cube roots of 1 besides 1 itself.

(a) Explain **Euler's Criterion for Cubic Residues**:

$$A^{(p-1)/3} \equiv \begin{cases} 1 \pmod{p} & \text{if } A \text{ is a cubic residue} \\ \dfrac{L + 9M}{L - 9M} \pmod{p} & \text{if } A \text{ is a cubic nonresidue.} \end{cases}$$

(b) Confirm Euler's Criterion for $p = 7$ and all A in $[1, 6]$.
(c) Find the three cube roots of 1 in $\mathbf{Z}_{61}$.

6.6 A Theorem on kth-Power Residues of an Arbitrary Modulus

It would be nice to have a criterion more general than Euler's Criterion that can be applied to congruences of the form

$$x^k \equiv A \pmod{m},$$

where the modulus m is arbitrary. We present an abbreviated discussion of this, omitting many details. Our final result here is Theorem 6.19 (Dence and Dence, 1997).

Suppose first that $m = p^\alpha$, where p is an odd prime. In Theorem 6.11 replace p there by p^α and replace $p - 1$ by $\phi(p^\alpha)$; then the proof carries through unchanged because p^α has a primitive root g (Theorem 5.9). Essentially the following theorem appeared in the *Disquisitiones Arithmeticae* (1801) of **Gauss**:

Theorem 6.17 Let p be an odd prime and $\alpha,\ k \in \mathbf{N}$. Then A is a kth-power residue of p^α iff

$$A^{\phi(p^\alpha)/d} \equiv 1 \pmod{p^\alpha}$$

where $d = (k,\ \phi(p^\alpha))$.

Example 6.23 Consider $x^8 \equiv A \pmod{125}$. We have $d = (8, \phi(5^3)) = 4$. Let $A = 11$ be an alleged octic residue of 125. Then

$$11^{100/4} = (11^5)^5 \equiv 51^5 \pmod{125}$$
$$\equiv 1 \pmod{125},$$

so 11 is indeed an octic residue of 125. In fact, $47^8 \equiv 11 \pmod{125}$. ◆

Turning now to moduli that are powers of 2, we recall from Theorem 6.15 that if $k \geq 3$ is odd, then all of the odd integers in $[1, 2^r - 1]$ are least positive kth-power residues of 2^r. The data in Table 6.5 suggest a very different behavior when $k \geq 2$ is itself a power of 2 ($k = 2^d$).

The cases of $d = 1, 2$ correspond, respectively, to Problem 6.73 and Theorem 6.16. We notice from the table that $A = 1$ is the only (2^dth)-power residue when $d \geq r - 2$. The general result is given by the following theorem:

Theorem 6.18 Let $k = 2^d$, $d \in \mathbf{N}$. Then A is a least positive kth-power residue of 2^r, $r \geq d + 2$, iff $A = 1 + 2^{d+2}\sigma$, where $\sigma \in \{0, 1, 2, \ldots, 2^{r-d-2} - 1\}$. If $r < d + 2$, only $A = 1$ is possible.

Example 6.24 A complete set of incongruent 16th-power residues of 512 is $\{1, 65, 129, 193, 257, 321, 385, 449\}$. ◆

What if k is even but is not a power of 2? Table 6.6 gives some numerical results in this direction. Compare the sets of residues here with those in Table 6.5. The

Table 6.5 Least Positive (2^dth)-Power Residues of 2^r

d	$r = 4$	$r = 5$	$r = 6$	$r = 7$
1	1, 9	1, 9, 17, 25	1, 9, 17, 25, . . . , 57	1, 9, 17, 25, . . . , 121
2	1	1, 17	1, 17, 33, 49	1, 17, 33, 49, . . . , 113
3	1	1	1, 33	1, 33, 65, 97
4	1	1	1	1, 65
5	1	1	1	1

Table 6.6 (kth)-Power Residues of 2^r, Where $k = 2^d n$, $(2, n) = 1$

d	n	$r = 5$	$r = 6$	$r = 7$
2	3	1, 17	1, 17, 33, 49	1, 17, 33, 49, . . . , 113
3	3	1	1, 33	1, 33, 65, 97
3	5	1	1, 33	1, 33, 65, 97
4	3	1	1	1, 65
4	15	1	1	1, 65

following corollary answers the question just posed and confirms what is indicated in Table 6.6.

Corollary 6.18.1 Let $k = 2^d n$, where $n \geq 3$ is odd. Then the kth-power residues of 2^r are the (2^dth)-power residues of 2^r.

Proof The congruence $x^k \equiv A \pmod{2^r}$ is simultaneously an nth-power congruence and a 2^dth-power congruence. Since $(2, n) = 1$, then the least positive $(2^d n\text{th})$-power residues of 2^r are in the intersection of the set $\mathbf{S}_1$ of least positive nth-power residues of 2^r and the set $\mathbf{S}_2$ of least positive (2^dth)-power residues of 2^r. But $\mathbf{S}_1$ is all the odd integers in $[1, 2^r - 1]$ by Theorem 6.15, and $\mathbf{S}_2$ is given by Theorem 6.18 if $r \geq d + 2$, or by the singleton set $\{1\}$ if $r < d + 2$. In either case, $\mathbf{S}_1 \cap \mathbf{S}_2 = \mathbf{S}_2$ and this is the corollary. ◆

Example 6.25 A complete set of incongruent (166th)-power residues of 256 is identical to the set of octic residues of 256 (Corollary 6.18.1), and is $\{1, 33, 65, 97, 129, 161, 193, 225\}$ from Theorem 6.18. ◆

We are now ready to assemble the various pieces and set up a generalization of Euler's Criterion. The general congruence of interest is

$$x^k \equiv A \pmod{m},$$

where k satisfies $2 \leq k \leq \phi(m)$. The restriction at the upper end is imposed because of Euler's Theorem. We write the modulus in the canonical form

$$m = 2^r \prod_i p_i^{\alpha_i},$$

where the p_i's are distinct odd primes.

Theorem 6.19 (Generalized Euler Criterion). Let $k = 2^d n$, $(2, n) = 1$, and let m be as just defined. Then A is a kth-power residue of m iff

1. $A^{\phi(p_i^{\alpha_i})/d_i} \equiv 1 \pmod{p_i^{\alpha_i}}$, $d_i = (k, \phi(p_i^{\alpha_i}))$ for each i

and

$$2.\ A \equiv \begin{cases} 1 + 2^{d+2}\sigma \pmod{2^r},\ 0 \leq \sigma \leq 2^{r-d-2} - 1 & \text{if } r \geq d+2 \\ 1 \pmod{2^r}, & \text{if } 0 < r < d+2, \end{cases}$$

when $d, r > 0$.

Proof By Theorem 4.10, for the congruence $x^k - A \equiv 0 \pmod{m}$, A is a kth-power residue of m iff

$$x^k - A \equiv 0 \pmod{p_i^{\alpha_i}}$$

has a solution for each i and iff

$$x^k - A \equiv 0 \pmod{2^r}$$

has a solution ($r > 0$). For the family of congruences involving the odd-prime powers as moduli, A is a residue iff

$$A^{\phi(p_i^{\alpha_i})/d_i} \equiv 1 \pmod{p_i^{\alpha_i}},$$

where $d_i = (k, \phi(p_i^{\alpha_i}))$, according to Theorem 6.17.

If $d = 0$, the congruence with modulus 2^r is solvable iff A is odd. However, when $d > 0$ then this congruence has these kth-power residues modulo 2^r:

$$\begin{cases} \{1 + 2^{d+2}\sigma\colon \sigma = 0, 1, 2, \ldots, 2^{r-d-2} - 1\} & \text{if } r \geq d + 2 \\ \{1\} & \text{if } 0 < r < d + 2, \end{cases}$$

according to Theorem 6.18 and its corollary (if $n \geq 3$ is odd). ◆

Example 6.26 Consider $x^{40} \equiv A \pmod{448}$. Here, $m = 2^r p_1 = 2^6 \cdot 7$ and $k = 2^d n = 2^3 \cdot 5$, so $r - d - 2 = 1$. Then $\phi(p_1) = 6$ and $d_1 = (k, \phi(p_1)) = (40, 6) = 2$, so Theorem 6.19 gives as criteria for A

$$\begin{cases} A^{6/2} \equiv 1 \pmod{7} \\ A \equiv 1 \text{ or } 33 \pmod{64}. \end{cases}$$

The cubic congruence yields $A \equiv 1$, 2, or 4 (mod 7). The allowed A's for the original congruence can now be found by repeated use of the Chinese Remainder Theorem (Theorem 4.9). For example, the unique solution (modulo 448) of the system

$$\begin{cases} A \equiv 2 \pmod{7} \\ A \equiv 33 \pmod{64} \end{cases}$$

is $A = 289$. Indeed, one can find as a solution $5^{40} \equiv 289 \pmod{448}$. ◆

Problems

6.79. In each of the following cases determine whether the indicated integer A is a kth-power residue of p^α:

(a) $A = 10, k = 3, p^\alpha = 7^2$.
(b) $A = 46, k = 4, p^\alpha = 5^3$.
(c) $A = 119, k = 6, p^\alpha = 11^2$.
(d) $A = 5, k = 12, p^\alpha = 3^5$.

6.80. In Table 6.5 why is $A = 1$ the only (2^dth)-power residue of 2^r when $d > r - 2$?

6.81. Table 6.5 also shows that $A = 1$ is the only (2^dth)-power residue of 2^r when $d = r - 2$. Prove this special case by using induction on r.

6.82. Write a computer program that can use Theorem 6.17 to find the cubic residues of a modulus of the form p^α. Use it to find $\mathbf{S}_1$, the group of least positive cubic residues of 7^2, and $\mathbf{S}_2$, the group of least positive cubic residues of 7^3. What connection do you see between $\mathbf{S}_1$ and $\mathbf{S}_2$?

6.83. Following up on Problem 6.82, we let p be an odd prime, and we denote the least positive kth-power residues of p^α ($2 \le k < \phi(p^\alpha)$) by $\{A_i\}_{i=1}^m$. Show that the elements of the following set $\mathbf{S}$ are least positive kth-power residues of $p^{\alpha+1}$:

$$S = \{A_i + ap^\alpha : a = 0, 1, 2, \ldots, p - 1, i = 1, 2, \ldots, m\}.$$

[Hint: Use Theorem 6.17.]

6.84. Show that the solution $x \equiv 5 \pmod{448}$ given in Example 6.26 can be obtained by solving the system of congruences

$$\begin{cases} x^4 \equiv 2 \pmod{7} \\ x^8 \equiv 33 \pmod{64}. \end{cases}$$

Does this system have only one incongruent solution?

6.85. The octic congruence itself in Problem 6.84 has 16 incongruent solutions.

(a) Find all of them. One way to get started is to make use of Problem 6.73.
(b) Use the result of (a) to find two other solutions modulo 448 (besides $x \equiv 5 \pmod{448}$) less than 64 to the pair of congruences in Problem 6.84.

6.86. Find the complete set of incongruent least positive residues A for the congruence of Example 6.26. Do they form a multiplicative group? If so, is it cyclic?

6.87. Work out the complete set of least positive incongruent 12th-power residues of 232.

6.7 Other Reciprocity Laws*

We mentioned earlier that there are so-called reciprocity laws for kth-order power residues ($k > 2$). The Law of Cubic Reciprocity ($k = 3$) was discussed by **Jacobi**, and the Law of Quartic Reciprocity ($k = 4$) was first guessed by **Gauss** as early as 1828 (Collison, 1977). The first published proofs of these two laws, however, were given in 1844 by the brilliant **F.G.M. Eisenstein**. Still higher order

* Sections marked with an asterisk can be skipped without impairing the continuity of the text.

reciprocity laws were sought, and these efforts contributed to the development of algebraic number theory into the 20th century.

The most modern developments in reciprocity laws are beyond the scope of this book, although some recent work on so-called rational reciprocity laws has borne results not greatly different in form from Gauss's Law of Quadratic Reciprocity (Lehmer, 1978). In this chapter's concluding section, however, we shall be content to just sketch out the main ideas leading up to a statement of the Law of Cubic Reciprocity. We shall prove little here, but the ideas discussed will give you some of the flavor of algebraic number theory. More on this topic will be presented in Chapter 9.

Our object is to first obtain a symbol, $(\frac{\alpha}{p})_3$, analogous to the Legendre symbol, that assumes different values depending on whether the congruence

$$x^3 \equiv \alpha \pmod{p}$$

does or does not have a solution. Then the next step is to seek results for the new symbol that are analogs of those in Theorem 6.5. Finally, with this background in place, we look for a relationship between $(\frac{\alpha}{p})_3$ and $(\frac{p}{\alpha})_3$, an analog of Theorem 6.9.

Here's our first crucial observation. Look at Table 6.2. You will notice several instances where the product of two cubic nonresidues is another cubic nonresidue. Suppose $(\frac{\alpha}{p})_3$ is defined to be positive if α is a cubic residue to the prime p. Then if we wish to maintain an analogy to Theorem 6.5, part (f), for the cubic residue symbol,

$$\left(\frac{\alpha}{p}\right)_3\left(\frac{\beta}{p}\right)_3 = \left(\frac{\alpha\beta}{p}\right)_3,$$

we immediately encounter a problem if α, β, $\alpha\beta$ are all cubic nonresidues. The three symbols cannot all be negative. A way out is to permit the symbols to have complex values.

Next, we have Fermat's Theorem,

$$a^{p-1} \equiv 1 \pmod{p},$$

if $p \nmid a$ and p is an odd prime. Hence, we have $a^{(p-1)/2} \equiv \pm 1 \pmod{p}$. We might expect the cubic case to read

$$\alpha^{(p-1)/3} \equiv 1,\ \omega,\ \text{or}\ \omega^2 \pmod{p},$$

where 1, $\omega = (-1 + i\sqrt{3})/2$, and ω^2 are the three cube roots of unity. But this congruence cannot be correct. The exponent $(p - 1)/3$ is not usually an integer. Further, in such cases, the left-hand side is not generally congruent to 1, ω, or ω^2. For example, if $p = 5$ and $\alpha = 2$, then it does not happen that 5 divides either $2^{4/3} - 1$, $2^{4/3} - \omega$, or $2^{4/3} - \omega^2$. In fact, what does divisibility even mean in this case?

The previous observations have brought us face to face with the central issue: The number system **Z** is not the appropriate number system if a cubic reciprocity law and a calculus of cubic residue symbols are to be developed. Although the complex number $\omega = (-1 + i\sqrt{3})/2$ has entered the picture dubiously, it is our only lead, so we shall pursue it.

Let $\mathbf{Z}[\omega]$ represent the set of all numbers α of the form $\alpha = a + b\omega$, $a, b \in \mathbf{Z}$. Addition and multiplication in this set can be defined as one would define them in high school algebra. It is easy to see that these operations are binary operations. Various other elementary properties are easily ascertained, and we can conclude that $\mathbf{Z}[\omega]$ is a ring (Section 3.3). Further, multiplication in $\mathbf{Z}[\omega]$ is readily seen to be commutative, and with a bit more work one can show that $\mathbf{Z}[\omega]$ has no divisors of 0. Hence, $\mathbf{Z}[\omega]$ is really an integral domain (Problem 6.88). The integral domain **Z** is a subdomain of it, so we hope to find our desired theorems in $\mathbf{Z}[\omega]$.

Much preliminary work needs to be done. Because we are now dealing with complex numbers, we need some measure of the size of an element in $\mathbf{Z}[\omega]$. We define the norm of $\alpha \in \mathbf{Z}[\omega]$.

DEF. Let $\alpha = a + b\omega$, where $a, b \in \mathbf{Z}$ and $\omega = (-1 + i\sqrt{3})/2$. Then the **norm** of α, written $N\alpha$, is defined as

$$N\alpha = \alpha\bar{\alpha} = (a + b\omega)(a + b\bar{\omega}),$$

where $\bar{\omega}$ means the complex conjugate of ω.

It follows that $N\alpha = a^2 - ab + b^2$, that $N\alpha = 0$ iff $\alpha = 0$, and that $N\alpha \geq 0$ for all $\alpha \in \mathbf{Z}[\omega]$ (Problem 6.89).

Example 6.27 Let $\alpha = 5 - 3\omega$. Then $\omega + \bar{\omega} = -1$ and $\omega\bar{\omega} = 1$, so

$$\begin{aligned} N\alpha &= (5 - 3\omega)(5 - 3\bar{\omega}) \\ &= 25 - 15(\omega + \bar{\omega}) + 9\omega\bar{\omega} \\ &= 25 - 15(-1) + 9(1) \\ &= 49 \\ &= 5^2 - (5)(-3) + (-3)^2. \quad ◆ \end{aligned}$$

The integral domain $\mathbf{Z}[\omega]$ is not a field, but it does have some units $(1, -1$, for example). Recall that a unit is an element $\alpha \in \mathbf{Z}[\omega]$ which has a multiplicative inverse $\alpha^{-1} \in \mathbf{Z}[\omega]$. The following theorem is very useful in this regard.

Theorem 6.20 $\alpha \in \mathbf{Z}[\omega]$ is a unit iff $N\alpha = 1$.

The theorem allows us to determine completely the set of units in $\mathbf{Z}[\omega]$.

Example 6.28 If $\alpha = a + b\omega$ is a unit, then $a^2 - ab + b^2 = 1$, so from the quadratic formula

$$a = \frac{b \pm \sqrt{4 - 3b^2}}{2}.$$

As $a, b \in \mathbf{Z}$, the only possibilities for the radicand are $4 - 3b^2 = 1, 4$. Hence, $b = \pm 1, 0$ and $a = \pm 1, 0$, and we obtain these six units in $\mathbf{Z}[\omega]$: $1 + \omega, -1 - \omega, \omega, -\omega, 1, -1$. Note that since $\omega^3 - 1 = (1 + \omega)(\omega^2 + \omega + 1) = 0$, then $\omega^2 = -1 - \omega$, so the six units may also be listed as $\pm\omega^2, \pm\omega, \pm 1$. ◆

If $\alpha, \beta \in \mathbf{Z}[\omega]$ and $\alpha = \beta\epsilon_i$, where ϵ_i is a unit, then α, β are said to be **associates**. Every nonzero $\alpha \in \mathbf{Z}[\omega]$ has six associates. We need the concept of associates in order to give a definition of primes in $\mathbf{Z}[\omega]$. If $\alpha, \beta \in \mathbf{Z}[\omega]$ and $\alpha | \beta$, we mean that there is a $\gamma \in \mathbf{Z}[\omega]$ such that $\alpha\gamma = \beta$. Both α and γ are termed **divisors** of β. We let the symbol π denote a prime in $\mathbf{Z}[\omega]$.

DEF. A **prime** π in $\mathbf{Z}[\omega]$ is any nonzero element, not a unit, that has only its six associates and the six units as divisors.

All other elements in $\mathbf{Z}[\omega]$ besides 0, the six units, and all the primes are referred to as **composites**. Note how the definition of prime is an extension of the one given in Section 1.4. In $\mathbf{Z}$ one has uniqueness of factorization of composites into primes up to an arbitrary assignment of order of listing and of algebraic sign. A somewhat analogous statement holds for $\mathbf{Z}[\omega]$; this is an important theorem and is by no means obvious.

Theorem 6.21 If $\alpha \in \mathbf{Z}[\omega]$ is composite, then it is uniquely factorable into primes up to an arbitrary assignment of associates and an arbitrary order of listing of the primes.

In spite of Theorem 6.21, the identification of the primes in $\mathbf{Z}[\omega]$ is not as straightforward as it is in $\mathbf{Z}$. One difficulty is that a prime in $\mathbf{Z}$ is not necessarily a prime in $\mathbf{Z}[\omega]$ (Problem 6.93). For example, $13 = (4 + \omega)(3 - \omega)$, so 13 is clearly composite in $\mathbf{Z}[\omega]$. Can you tell at a glance if $4 + \omega$, $3 - \omega$ are primes? (They are!) How about $7 - \omega$ and $1 + 5\omega$? (They aren't!) The following theorem is therefore very useful.

Theorem 6.22 If $\pi = a + b\omega \in \mathbf{Z}[\omega]$ is such that $N\pi = p$, a prime in $\mathbf{Z}$, then π is a prime in $\mathbf{Z}[\omega]$. Conversely, if π is a prime in $\mathbf{Z}[\omega]$, then $N\pi = p$ or p^2, the latter case arising if π is associate to some prime $p \in \mathbf{Z}$.

This theorem is more useful in one direction than the other. In particular, it does not allow us to identify the real primes in $\mathbf{Z}[\omega]$. For example, 2 is a prime but 3 is not in $\mathbf{Z}[\omega]$ (Problem 6.93), even though in both cases their norms are of the form p^2.

Example 6.29 Let $\alpha = 7 - \omega$; then $N\alpha = 7^2 - (7)(-1) + (-1)^2 = 57 = 19 \cdot 3$. By Theorem 6.22, α is not a prime in $\mathbf{Z}[\omega]$. ◆

Example 6.30 Suppose π is a prime in $\mathbf{Z}[\omega]$ and $N\pi = 3$. Then $\pi = a + b\omega$ and $a^2 - ab + b^2 = 3$, or $a = (b \pm \sqrt{12 - 3b^2})/2$. Choose $b = 2$, $a = 1$ so that $\pi = 1 + 2\omega$.

Similarly, if $\pi' = c + d\omega$ and $N\pi' = 19$, then $c = (d \pm \sqrt{76 - 3d^2})/2$. Choose $d = -5$, $c = -3$, so that $\pi' = -3 - 5\omega$. Then observe that $\pi\pi' = (1 + 2\omega)(-3 - 5\omega) = 7 - \omega = \alpha$ from Example 6.29.

On the other hand, $\omega(1 + 2\omega) = \omega + 2\omega^2 = \omega + 2(-1 - \omega) = -2 - \omega$ is an associate of $1 + 2\omega$, and $\omega^{-1}(-3 - 5\omega) = (-1 - \omega)(-3 - 5\omega) = 3 + 8\omega + 5\omega^2 = -2 + 3\omega$ is an associate of $-3 - 5\omega$. Then we have $(-2 - \omega)(-2 + 3\omega) = 4 - 4\omega - 3\omega^2 = 7 - \omega$, in agreement with Theorem 6.21. ◆

The notion of congruence is as fruitful in $\mathbf{Z}[\omega]$ as it is in $\mathbf{Z}$. Thus, if $\alpha, \beta, \pi \in \mathbf{Z}[\omega]$, the symbolism $\alpha \equiv \beta \pmod{\pi}$ shall mean that $\pi \mid (\alpha - \beta)$ in $\mathbf{Z}[\omega]$. In $\mathbf{Z}$, congruence modulo a prime p partitions all of $\mathbf{Z}$ into p residue classes. The analogous statement for $\mathbf{Z}[\omega]$ is that a congruence modulo a prime π partitions all of $\mathbf{Z}[\omega]$ into $N\pi$ residue classes. In $\mathbf{Z}$ the set of p residue classes constitutes a field; in $\mathbf{Z}[\omega]$ the set of $N\pi$ residue classes is also a field. In $\mathbf{Z}$ the field contains a multiplicative group of order $p - 1$; in $\mathbf{Z}[\omega]$ the field there contains a multiplicative group of order $N\pi - 1$. This gives an analog of Fermat's Theorem,

$$\alpha^{N\pi - 1} \equiv 1 \pmod{\pi}$$

if $\pi \nmid \alpha$.

If $N\pi \neq 3$, then one can show that no two of $1, \omega, \omega^2$ belong to the same residue class. Thus, the set of three residue classes to which these elements belong constitutes a group of order 3, a subgroup of the multiplicative group of order $N\pi - 1$. According to Lagrange's Theorem (Theorem 3.19) we must then have $3 \mid (N\pi - 1)$, if $N\pi \neq 3$. Hence, we have Theorem 6.23 (Problem 6.95).

Theorem 6.23 If $\pi \in \mathbf{Z}[\omega]$ is a prime, $N\pi \neq 3$, and $\pi \nmid \alpha$, then $\alpha^{(N\pi - 1)/3}$ is congruent to one and only one of $1, \omega, \omega^2$ modulo π.

Fermat's Theorem tells us that $\pi \mid (\alpha^{(N\pi - 1)} - 1)$. Since the three cube roots of 1 are 1, ω, ω^2, then $\pi \mid (\alpha^{(N\pi-1)/3} - 1)(\alpha^{(N\pi-1)/3} - \omega)(\alpha^{(N\pi-1)/3} - \omega^2)$. Since π is a prime, it divides one of the three factors on the right (a deduction from Theorem 6.21). We can now give the definition of the **cubic residue symbol**.

DEF. Consider the congruence $x^3 \equiv \alpha \pmod{\pi}$, $N\pi \neq 3$. Then

$$\left(\frac{\alpha}{\pi}\right)_3 = \begin{cases} 1 & \text{iff } \alpha \text{ is a cubic residue} \\ 0 & \text{iff } \pi \mid \alpha \\ \omega \text{ or } \omega^2 & \text{iff } \alpha \text{ is a cubic nonresidue.} \end{cases}$$

Of course, in order to determine the value of $(\frac{\alpha}{\pi})_3$ without actually solving $x^3 \equiv \alpha \pmod{\pi}$, we need a criterion, the analog of Theorem 6.4 or Theorem 6.5(c).

Theorem 6.24 If $N\pi \neq 3$, where $\pi \in \mathbf{Z}[\omega]$ is a prime, then

$$\alpha^{(N\pi-1)/3} \equiv \left(\frac{\alpha}{\pi}\right)_3 \pmod{\pi}.$$

Example 6.31 Let $\pi = 3 + 2\omega$ and $\alpha = 1 + 2\omega$. Then $N\pi - 1 = 3^2 - (2)(3) + 2^2 - 1 = 6$, and so one has

$$\begin{aligned} \alpha^6 = (1 + 2\omega)^6 &= 1 + 12\omega + 60\omega^2 + 160\omega^3 + 240\omega^4 + 192\omega^5 + 64\omega^6 \\ &= -27, \end{aligned}$$

after employing $\omega^3 = 1$ and $\omega^2 = -1 - \omega$. We now find

$$\begin{aligned} \frac{-28}{3 + 2\omega} = \frac{-28(3 + 2\bar{\omega})}{(3 + 2\omega)(3 + 2\bar{\omega})} = \frac{-28(3 + 2\bar{\omega})}{7} &= -12 - 8\bar{\omega} \\ &= -4 + 8\omega. \end{aligned}$$

Thus, $\alpha^6 \equiv 1 \pmod{\pi}$ as Fermat's Theorem requires. ◆

Example 6.32 Let α, π be as in Example 6.31. Then $\alpha^{(N\pi-1)/3} = \alpha^2 = (1 + 2\omega)^2 = -3$. We now find

$$\frac{-3 - \omega^2}{3 + 2\omega} = \frac{(-3 - \omega^2)(3 + 2\bar{\omega})}{7} = \frac{(-2 + \omega)[(3 + 2(-1 - \omega)]}{7} = \omega.$$

Hence, $\alpha^2 \equiv \omega^2 \pmod{\pi}$, so the congruence $x^3 \equiv 1 + 2\omega \pmod{\pi}$ has no solution. ◆

It follows from Theorem 6.24 that

$$\left(\frac{\alpha\beta}{\pi}\right)_3 \equiv (\alpha\beta)^{(N\pi-1)/3} \equiv \alpha^{(N\pi-1)/3}\beta^{(N\pi-1)/3} \equiv \left(\frac{\alpha}{\pi}\right)_3\left(\frac{\beta}{\pi}\right)_3 \pmod{\pi}.$$

This is the analog of Theorem 6.5(f). The result

$$\left(\frac{\alpha}{\pi}\right)_3 = \left(\frac{\beta}{\pi}\right)_3$$

if $\alpha \equiv \beta \pmod{\pi}$, the analog of Theorem 6.5(d), follows equally easily.

We are now ready to lay out the statement of the Law of Cubic Reciprocity. The motivation for its form is derived from the following observation made on real primes. Theorem 6.12 has told us that if p is a $(3j + 2)$-prime, then there are $p - 1$ cubic residues modulo p. Therefore, if p, q are both $(3j + 2)$-primes,

$$\left(\frac{p}{q}\right)_3 = \left(\frac{q}{p}\right)_3,$$

since each equals 1. To generalize this so as to cover complex primes in $\mathbf{Z}[\omega]$, we make this definition.

DEF. If π is a prime in $\mathbf{Z}[\omega]$, then π is termed **primary** if $\pi \equiv 2 \pmod 3$. That is, if $\pi = a + b\omega$, then $a \equiv 2 \pmod 3$ and $3 \mid b$.

It can be proved that all of the $(3j + 2)$-primes in $\mathbf{Z}$ are also primes in $\mathbf{Z}[\omega]$, so these are automatically primary. None of the $(3j + 1)$-primes in $\mathbf{Z}$ are primes in $\mathbf{Z}[\omega]$. For $p = 3j + 1$ is composite in $\mathbf{Z}[\omega]$ iff $p = \alpha\beta$, iff $N\alpha = N\beta = p$. Let $\alpha = a + b\omega$; then $a^2 - ab + b^2 = p$, or

$$a = \frac{b \pm \sqrt{4p - 3b^2}}{2} \in \mathbf{N}.$$

This implies that $4p - 3b^2 = x^2$, for some nonzero integer x. We must have $3 \mid (x^2 - 4p)$, that is, $x^2 \equiv 4p \equiv 12j + 4 \equiv 1 \pmod 3$. This congruence is solvable, so nonzero integers a, b exist and p is not prime in $\mathbf{Z}[\omega]$.

Finally, let $\pi = a + b\omega$ be a complex prime in $\mathbf{Z}[\omega]$. We must have $N\pi \equiv 1 \pmod 3$. By examining the possible cases of $a \equiv 0$, 1 or 2 (mod 3) and $b \equiv 0$, 1 or 2 (mod 3), we can establish the general result (Problem 6.96).

Theorem 6.25 Let $\pi \in \mathbf{Z}[\omega]$ be any real or complex prime satisfying $N\pi \equiv 1 \pmod 3$. Then exactly one associate of π is primary.

Example 6.33 Let $\pi = 5 + 6\omega$; then $N\pi = 31$, so π is a complex prime in $\mathbf{Z}[\omega]$ (Theorem 6.22). Also, $N\pi \equiv 1 \pmod 3$, as it should be. The associates of π are $\pi_1 = \pi \times 1 = 5 + 6\omega$, $\pi_2 = \pi \times (-1) = -5 - 6\omega$, $\pi_3 = \pi \times \omega = -6 - \omega$, $\pi_4 = \pi \times (-\omega) = 6 + \omega$, $\pi_5 = \pi \times (1 + \omega) = -1 + 5\omega$, $\pi_6 = \pi \times (-1 - \omega) = 1 - 5\omega$. Only π_1 satisfies $\pi_1 \equiv 2 \pmod 3$, so π has only one primary associate, in agreement with Theorem 6.25. ◆

We now state Theorem 6.26.

Theorem 6.26 (The Law of Cubic Reciprocity). If π_1, π_2 are primary primes in $\mathbf{Z}[\omega]$, and $N\pi_1 \neq 3$, $N\pi_2 \neq 3$, $N\pi_1 \neq N\pi_2$ hold, then

$$\left(\frac{\pi_1}{\pi_2}\right)_3 = \left(\frac{\pi_2}{\pi_1}\right)_3.$$

The theorem applies to all real primes in $\mathbf{Z}[\omega]$ and to the unique primary associate of each complex prime. The following example illustrates the content of the theorem.

Example 6.34 Let $\pi_1 = 2$ and $\pi_2 = -1 + 3\omega$; both are easily verified to be primes and to be primary. By Theorem 6.24,

$$\left(\frac{\pi_1}{\pi_2}\right)_3 \equiv 2^{(13-1)/3} \equiv 16 \pmod{-1 + 3\omega}.$$

By trial, one finds that $(16 - \omega^2)/(-1 + 3\omega) = -5 - 4\omega$, so $(\frac{\pi_1}{\pi_2})_3 = \omega^2$.

Again, from Theorem 6.24,

$$\left(\frac{\pi_2}{\pi_1}\right)_3 \equiv (-1 + 3\omega)^{(4-1)/3} \equiv (-1 + 3\omega) \pmod 2.$$

By trial, one finds that $(-1 + 3\omega - \omega^2)/2 = 2\omega$, so $(\frac{\pi_2}{\pi_1})_3 = \omega^2$. Hence,

$$\left(\frac{2}{-1 + 3\omega}\right)_3 = \left(\frac{-1 + 3\omega}{2}\right)_3,$$

as required by Theorem 6.26, and incidentally, neither $x^3 \equiv 2 \pmod{-1 + 3\omega}$ nor $x^3 \equiv -1 + 3\omega \pmod 2$ has a solution in $\mathbf{Z}[\omega]$. ◆

If you have found the development in this section tantalizing, you are referred to (Ireland and Rosen, 1990) for many more details.

Problems

6.88. Without using the concept of norm, prove that if $(a + b\omega)(c + d\omega) = 0$, then either $a = b = 0$ or $c = d = 0$. This shows that $\mathbf{Z}[\omega]$ has no divisors of 0.

6.89. Prove, as stated in the text, that $N\alpha = 0$ iff $\alpha = 0$ and $N\alpha \geq 0$ for all $\alpha \in \mathbf{Z}[\omega]$.

6.90. You do not have available Theorem 6.20. Use elementary arguments to show that if $a, b \in \mathbf{Z}$ and $(a + b\omega)^{-1}$ is a member of $\mathbf{Z}[\omega]$, then $a^2 - ab + b^2 = 1$ follows.

6.91. Determine which of the following are primes in $\mathbf{Z}[\omega]$:

(a) $5 - 4\omega$. (b) $4 + 7\omega$. (c) $3 - \omega$.
(d) $2 + 15\omega$. (e) $1 + 5\omega$. (f) $13 + 14\omega$.

6.92. Suppose $\alpha, \beta, \gamma \in \mathbf{Z}[\omega]$ and $\alpha\beta = \gamma$. Show that $N\alpha N\beta = N\gamma$.

6.93. Regarding the integers 2, 3,

(a) Prove that 2 is a prime in $\mathbf{Z}[\omega]$.
(b) Prove that 3 is a composite in $\mathbf{Z}[\omega]$ by actually factoring it.

6.94. Factor into primes in $\mathbf{Z}[\omega]$:

(a) $10 + \omega$. (b) 13. (c) $3 - 5\omega$.
(d) $-1 - 8\omega$. (e) 19. (f) 31.

6.95. Given α, π in each case, find to which of $1, \omega, \omega^2$ the number $\alpha^{(N\pi - 1)/3}$ is congruent modulo π; that is, compute $(\frac{\alpha}{\pi})_3$:

(a) $\alpha = 1 + 2\omega, \pi = 1 + 3\omega$. (b) $\alpha = -\omega, \pi = 4 - 3\omega$.
(c) $\alpha = 3 - \omega, \pi = 5 + 2\omega$. (d) $\alpha = 5\omega, \pi = 4 + \omega$.

6.96. Define $\alpha = a + b\omega, \beta = c + d\omega$ to be in the same equivalence class if $a \equiv c \pmod 3$ and $b \equiv d \pmod 3$.

(a) How many equivalence classes of $\mathbf{Z}[\omega]$ are there?
(b) Show that the equivalence classes form an additive group by working out the group addition table.
(c) Find four subgroups of order 3.
(d) For which equivalence classes is $N\alpha \equiv 0 \pmod 3$?
(e) Show that the other equivalence classes form a multiplicative group by working out the group multiplication table.
(f) Hence, deduce that if $\alpha \in \mathbf{Z}[\omega]$ has $N\alpha \equiv 1 \pmod 3$, then there is just one associate of α that is primary.

6.97. Verify the Law of Cubic Reciprocity for the primary associates of the following pairs of primes:

(a) $\pi_1 = 5$. $\quad \pi_2 = 2 - \omega$.
(b) $\pi_1 = 3 - \omega$. $\quad \pi_2 = 2 + 5\omega$.

6.98. The Law of Quartic Reciprocity is stated in the context of the integral domain $\mathbf{Z}[i]$. An element in this domain has the form $\alpha = a + bi$, $i = \sqrt{-1}$, $a, b \in \mathbf{Z}$. The norm of such an element is $N\alpha = a^2 + b^2$. There are only four units in $\mathbf{Z}[i]$: $1, -1, i, -i$. Some examples of primes in $\mathbf{Z}[i]$ are 3, $1 + i$, 7, $3 + 2i$.

(a) Show that norms in $\mathbf{Z}[i]$ have the key characteristic of a norm: $N\alpha N\beta = N(\alpha\beta)$ if $\alpha, \beta \in \mathbf{Z}[i]$.
(b) A nonunit $\alpha \in \mathbf{Z}[i]$ is termed **primary** if for $\alpha = a + bi$ one has $b \equiv 0 \pmod 2$ and $a + b \equiv 1 \pmod 4$. Find the primary associates of the following: $-2 + i, 2 + 3i$.
(c) If $\pi \in \mathbf{Z}[i]$ is a prime and α is any integer in $\mathbf{Z}[i]$ not divisible by π, then the **quartic residue symbol** is defined by the congruence

$$\left(\frac{\alpha}{\pi}\right)_4 \equiv \alpha^{(N\pi - 1)/4} \pmod{\pi}.$$

Find $\left(\frac{-2 + i}{2 + 3i}\right)_4$; its value is one of the four units.

6.99. A form of the **Law of Quartic Reciprocity** says that if α, π are distinct primary primes in $\mathbf{Z}[i]$, then (Ireland and Rosen, 1990)

$$\left(\frac{\alpha}{\pi}\right)_4 = \left(\frac{\pi}{\alpha}\right)_4 (-1)^{(N\alpha - 1)(N\pi - 1)/16}.$$

Use the primary forms of the primes in Problem 6.98(b) and show that they conform to the Law of Quartic Reciprocity.

6.100. A **rational quartic reciprocity law** is a quartic reciprocity law for certain odd primes in $\mathbf{Z}$. One such law, due to **Ezra Brown** (Lehmer, 1978), says that if $p \equiv q \equiv 1 \pmod 4$ are such that $p = c^2 + qd^2$, where $c, d \in \mathbf{N}$, then

$$\left(\frac{p}{q}\right)_4 \left(\frac{q}{p}\right)_4 = \begin{cases} 1 & \text{if } q \equiv 1 \pmod 8 \\ (-1)^d & \text{if } q \equiv 5 \pmod 8). \end{cases}$$

The quartic residue symbol $\left(\frac{p}{q}\right)_4$ has the value $+1$ if the congruence $x^4 \equiv p \pmod q$ has a solution, and it has the value -1 otherwise. Show that the following pairs of primes conform to Brown's rational quartic reciprocity law, and find a solution to those quartic congruences $x^4 \equiv q \pmod{p}$ that are solvable.

(a) $q = 37, p = 149$.
(b) $q = 29, p = 277$.
(c) $q = 17, p = 593$.

RESEARCH PROBLEMS

1. Let $T^{(k)}_{n,m}$ denote the sum of the mth powers of the kth-power residues of n: $T^{(k)}_{n,m} = \Sigma_i A_i^m, A_i \in \mathbf{A}_n^{(k)}$. Theorem 6.2 has told us that $n \mid T^{(k)}_{n,m}$ when $m = 1$, $k = 2$, and $n = p$, a prime, and Problem 6.7 gives the result $n \mid T^{(k)}_{n,m}$ when $m = k = 2$ and $n = p$. Investigate the divisibility of $T^{(k)}_{n,m}$ by n when $m = 3, k = 2$ and when $m = k = 3$, n being arbitrary in both cases. Try to prove anything that you discover.

2. Let $n > 2$ and consider $x^2 \equiv -1 \pmod{n}$. Let $N(n)$ denote the number of incongruent solutions of the congruence. Investigate $N(n)$ for various choices of n, and see if you can discern a pattern.

3. A conjecture says that for any prime $p \geq 11$, there are cubic residues $A_i, A_j \in \mathbf{A}_p^{(3)}$ such that $A_j - A_i = 3$. Investigate this on the computer for several primes.

4. A **Sophie Germain prime** (after the French mathematician [1776–1831] of the same name) is a prime p such that $q = 2p + 1$ is also prime. Investigate, in the case where p is a Sophie Germain prime, (a) if $2 \cdot (-1)^{(p-1)/2}$ is a primitive root of q, (b) if q divides the Mersenne number $M_p = 2^p - 1$.

References

Bühler, W.K., *Gauss: A Biographical Study*, Springer, New York, 1981, pp. 32–33. The author points out the ambiguities in the interpretation of Gauss's many brief comments in his diary.

Cohn, J.H.E., "On Square Fibonacci Numbers," *J. London Math. Soc.*, **39**, 537–541 (1964). Interesting article shows the utility of the Jacobi symbol.

Collison, M.J., "The Origins of the Cubic and Biquadratic Reciprocity Laws," *Arch. Hist. Exact Sci.*, **17**, 63–69 (1977). The cubic and biquadratic (quartic) reciprocity laws were so highly esteemed in the 19th century that an acrimonious dispute about priority of their proofs arose between **Eisenstein** and **Jacobi**.

Cooper, C., and Kennedy, R.E., "Problem No. 74," *Missouri J. Math. Sci.*, **6**, 160 (1994). This journal problem was the inspiration for our Problem 6.70.

Dence, J.B., and Dence, T.P., "Cubic and Quartic Residues Modulo a Prime," *Missouri J. Math. Sci.*, **7**, 24–31 (1995). Our material for Sections 6.5 and 6.6 is drawn from this and the next two papers.

Dence, J.B., and Dence, T.P., "Residues. II. Congruences Modulo Powers of 2," *Missouri J. Math. Sci.*, **8**, 26–35 (1996). A continuation of work initiated in the previous reference.

Dence, J.B., and Dence, T.P., "Residues. III. Congruences to General Composite Moduli," *Missouri J. Math. Sci.*, **9**, 72–78 (1997). The concluding result in this paper is our Theorem 6.19.

Gillispie, C.C. (ed.), *Dictionary of Scientific Biography*, Vol. IV, "Ferdinand Gotthold Max Eisenstein," Charles Scribner's Sons, New York, 1971, pp. 340–343. Read about the unfortunate **Eisenstein** here.

Ireland, K., and Rosen, M., *A Classical Introduction to Modern Number Theory*, 2nd ed., Springer, New York, 1990, pp. 108–127. See these pages for a good discussion of the Law of Cubic Reciprocity and the Law of Quartic (Biquadratic) Reciprocity.

Laubenbacher, R.C., and Pengelley, D.J., "Eisenstein's Misunderstood Geometric Proof of the Quadratic Reciprocity Theorem," *Coll. Math. J.*, **25**, 29–34 (1994a). A broad comparison of Eisenstein's vs Gauss's (3rd) proof of the Law of Quadratic Reciprocity.

Laubenbacher, R.C., and Pengelley, D.J., "Gauss, Eisenstein, and the 'Third' Proof of the Quadratic Reciprocity Theorem: *Ein Kleines Schauspiel*," *Math. Intelligencer*, **16**, 67–72 (1994b). A slightly more detailed look at Eisenstein's proof than the previous reference. Includes an outline of what we have called Eisenstein's Lemma in Lemma 6.9.1.

Lehmer, E., "The Quintic Character of 2 and 3," *Duke Math. J.*, **18**, 11–18 (1951). The source of our material for Problems 6.74 and 6.75.

Lehmer, E., "Criteria for Cubic and Quartic Residuacity," *Mathematika*, **5**, 20–29 (1958). The source of our material for Problem 6.76; an interesting paper.

Lehmer, E., "Rational Reciprocity Laws," *Amer. Math. Monthly.*, **85**, 467–472 (1978). Our Problem 6.100 and much more inviting material is summarized in this paper.

Moser, L., "A Theorem on Quadratic Residues," *Proc. Amer. Math. Soc.*, **2**, 503–504 (1951). Our Problem 6.6 was motivated by the result in this paper; the first proof of the result actually goes back to **Dirichlet** (1839).

Rose, H.E., *A Course in Number Theory*, 2nd ed., Oxford University Press, Oxford, England, 1994, pp. 68–70. One known result is that the smallest quadratic nonresidue of the prime p is less than $p^{1/4+\epsilon}$ for any $\epsilon > 0$ and for sufficiently large p.

Williams, K.S., "On Euler's Criterion for Cubic Nonresidues," *Proc. Amer. Math. Soc.*, **49**, 277–283 (1975). Interesting paper from which our Problem 6.78 was taken.

Chapter 7
Multiplicative Functions

7.1 Some Common Multiplicative Functions

This chapter deals with a number of interesting functions that occur widely in number theory (McCarthy, 1986; Sivaramakrishnan, 1989). These functions are variously called **arithmetic functions** or **number-theoretic functions**. Their domain is **N** and their range lies in $\mathcal{R}$ or $\mathcal{C}$.

Of special interest to number theory is the following particular class of arithmetic functions.

> **DEF.** An arithmetic function $f(n)$ is called a **multiplicative function** if $(m, n) = 1$ implies $f(mn) = f(m)f(n)$.

This definition was previously given in Section 3.7.

Example 7.1 Let $f(n) = 1/n$, $n \in \mathbf{N}$. Then $f(n)$ is multiplicative since $f(mn) = 1/(mn) = (1/m)(1/n) = f(m)f(n)$. Note that here the same result clearly follows, whether or not the arguments are relatively prime. ◆

Example 7.2 Let $f(n) = \ln(n)$, $n \in \mathbf{N}$. Then $f(n)$ is not multiplicative since, for example, $\ln(1 \cdot 2) \neq (\ln 1)(\ln 2)$. ◆

The exemplar of all multiplicative functions is the Euler ϕ-function. Its multiplicative character was mentioned in Section 3.7. The next few pages survey some of the most prominent multiplicative functions, including the Euler ϕ-function. Throughout, *divisor* shall mean a positive integral factor.

7.1.1 $\tau(n)$: The Number-of-Divisors Function

The **number-of-divisors function** may be defined by

$$\tau(n) = \sum_{d \mid n} 1,$$

where the summation is over all divisors d of n. For example, $\tau(12) = 6$ because the divisors (six in number) of 12 are 1, 2, 3, 4, 6, 12. If $n = p^{\alpha}$, where p is a prime, then $\tau(n) = \alpha + 1$ because the divisors of p^{α} are $1, p, p^2, \ldots, p^{\alpha}$. Suppose n has more than one distinct prime in its factorization, $n = p_1^{\alpha_1} p_2^{\alpha_1}$. Then the divisors of n are constructed by making all possible multiplicative combinations of the powers of p_1 and the powers of p_2. From elementary combinatorial theory this yields $\tau(n) = (\alpha_1 + 1)(\alpha_2 + 1)$. The argument can be extended immediately to positive integers n of arbitrary factorization:

$$\tau(n) = \prod_{i=1}^{r} (\alpha_i + 1), \quad n = p_1^{\alpha_1} p_2^{\alpha_2} \cdots p_r^{\alpha_r}.$$

Example 7.3 If $m = 77 = 7 \cdot 11$ and $n = 225 = 3^2 \cdot 5^2$, then $\tau(m) = (1 + 1)(1 + 1) = 4$ and $\tau(n) = (2 + 1)(2 + 1) = 9$. Further, $\tau(mn) = \tau(3^2 \cdot 5^2 \cdot 7 \cdot 11) = (2 + 1)^2(1 + 1)^2 = 36 = \tau(m) \cdot \tau(n)$. The 36 divisors of $mn = 17{,}325$ are 1, 3, 5, 7, 9, 11, 15, 21, 25, 33, 35, 45, 55, 63, 75, 77, 99, 105, 165, 175, 225, 231, 275, 315, 385, 495, 525, 693, 825, 1155, 1575, 1925, 2475, 3465, 5775, 17,325. ◆

A brief table of values of $\tau(n)$ is given in Table G of Appendix II. The number-of-divisors function varies erratically with n (Fig. 7.1).

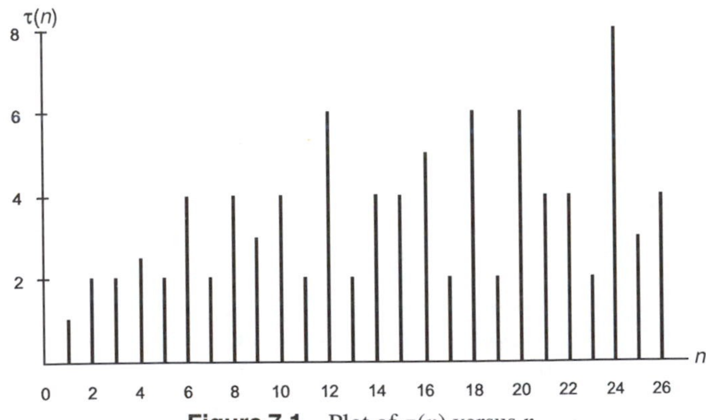

Figure 7.1 Plot of $\tau(n)$ versus n

7.1.2 $\sigma(n)$: The Sum-of-Divisors Function

We define the **sum-of-divisors function** by

$$\sigma(n) = \sum_{d|n} d,$$

where again the summation is over all of the divisors d of n. If $n = p_1 p_2$, then,

$$\sigma(n) = 1 + p_1 + p_2 + p_1 p_2 = (p_1 + 1)(p_2 + 1) = \frac{p_1^2 - 1}{p_1 - 1}\frac{p_2^2 - 1}{p_2 - 1}.$$

If $n = p^2$, then $\sigma(n) = 1 + p + p^2 = (p^3 - 1)/(p - 1)$. Finally, if $n = p_1^2 p_2$, then

$$\sigma(n) = 1 + p_1 + p_2 + p_1^2 + p_1 p_2 + p_1^2 p_2 = \frac{p_1^3 - 1}{p_1 - 1}\frac{p_2^2 - 1}{p_2 - 1}.$$

The trend is apparent, and by induction we have

$$\sigma(n) = \prod_{i=1}^{r} \frac{(p_i^{\alpha_i + 1} - 1)}{(p_i - 1)}, \quad n = p_1^{\alpha_1} p_2^{\alpha_2} \ldots p_r^{\alpha_r}.$$

Example 7.4 Let $n = 17{,}325 = 3^2 \cdot 5^2 \cdot 7 \cdot 11$. Then

$$\begin{aligned} \sigma(17{,}325) &= \left(\frac{3^3 - 1}{3 - 1}\right)\left(\frac{5^3 - 1}{5 - 1}\right)\left(\frac{7^2 - 1}{7 - 1}\right)\left(\frac{11^2 - 1}{11 - 1}\right) \\ &= 13(31)(8)(12) \\ &= 38{,}688. \end{aligned}$$

This agrees with the sum of the 36 divisors listed in Example 7.3. ◆

Values of $\sigma(n)$ are also gathered in Table G in Appendix II. Like $\tau(n)$, the sum-of-divisors function varies erratically with n, but here there is an upward bias in $\sigma(n)$ as n increases (Fig. 7.2). We will prove in Section 7.2 that $\tau(n)$, $\sigma(n)$ are multiplicative.

7.1.3 $\phi(n)$: Euler's Totient

Euler's totient,[1] $\phi(n)$, denotes the number of integers a such that $1 \leq a \leq n$ and $(a, n) = 1$; note, in particular, $\phi(1) = 1$. If p is a prime, then $\phi(p) = p - 1$.

[1] L. *totiens*, "so often." The function was introduced by **Euler**; the symbolism is due to **Gauss**, and the name totient is due to the English mathematician **J.J. Sylvester** (1814–1897).

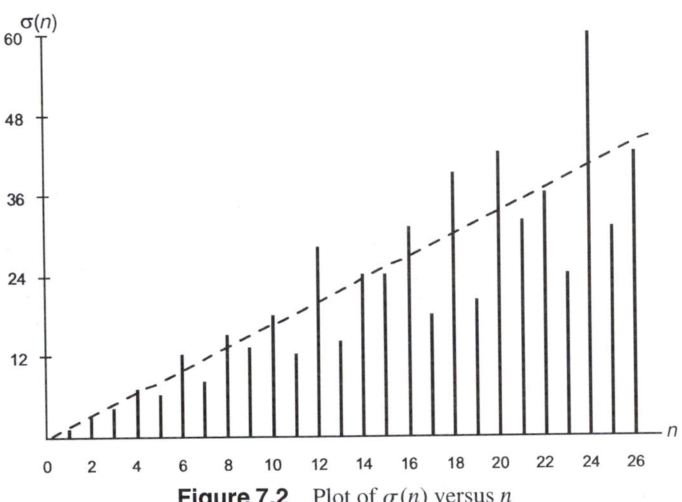

Figure 7.2 Plot of $\sigma(n)$ versus n

When $n = p^\alpha$, all of the positive integers not exceeding p^α are coprime to p^α except the $p^{\alpha-1}$ multiples of p $(1 \cdot p, 2 \cdot p, 3 \cdot p, \ldots, p^{\alpha-1} \cdot p)$. Hence, $\phi(p^\alpha) = p^\alpha - p^{\alpha-1} = p^{\alpha-1}(p - 1)$. Corollary 3.15.1 has already established that $\phi(n)$ is multiplicative, so we have

$$\phi(n) = \prod_{i=1}^{r} p_i^{\alpha_i - 1}(p_i - 1)$$

$$= n \prod_{i=1}^{r} (1 - p_i^{-1}), \quad n = p_1^{\alpha_1} p_2^{\alpha_2} \ldots p_r^{\alpha_r}.$$

Example 7.5 Let $n = 42 = 2 \cdot 3 \cdot 7$; $\phi(n) = 42(1 - 2^{-1})(1 - 3^{-1}) \cdot (1 - 7^{-1}) = 12$. Similarly, if $m = 55 = 5 \cdot 11$, then $\phi(55) = 55(1 - 5^{-1}) \cdot (1 - 11^{-1}) = 40$. Hence, $\phi(42 \cdot 55) = 12 \cdot 40 = 480$. From the formula directly, $\phi(2310)$ has the value

$$2310(1 - 2^{-1})(1 - 3^{-1})(1 - 5^{-1})(1 - 7^{-1})(1 - 11^{-1}),$$

which equals 480. ◆

Values of Euler's totient are also given in Table G of Appendix II, and the function is graphed in Fig. 7.3. Like $\sigma(n)$, the ϕ-function behaves erratically but has an upward bias.

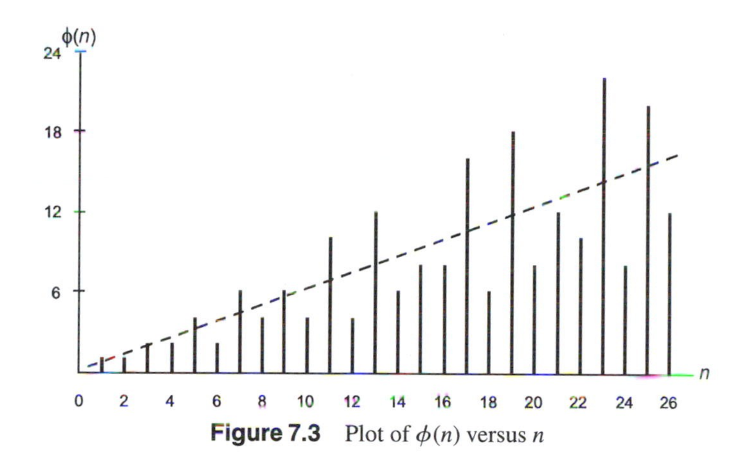

Figure 7.3 Plot of $\phi(n)$ versus n

7.1.4 $\mu(n)$: The Möbius Function

The **Möbius function** has only three possible values, and is defined by

$$\mu(n) = \begin{cases} 1 & n = 1 \\ (-1)^r & n = \prod_{i=1}^{r} p_i \\ 0 & \text{otherwise.} \end{cases}$$

The indicated extended product in the definition applies only when the p_i's are distinct (i.e., when n is squarefree). Thus, the frequency of occurrence of 0's for $\mu(n)$ is a measure of the frequency of occurrence of integers with square factors. Table 7.1 suggests that squarefree integers occur about 2/3 of the time (the limiting probability is $6/\pi^2$).

Clearly, if $n = \Pi_{i=1}^{r} p_i$ and $m = \Pi_{j=1}^{s} q_j$, where no prime p_i is a prime q_j, then mn factors as $r + s$ distinct primes. Since $(-1)^r(-1)^s = (-1)^{r+s}$, then $\mu(n) \cdot \mu(m) = \mu(nm)$. If $(m, n) = 1$ and either $\mu(m)$ or $\mu(n)$ is 0, then mn is not squarefree, and again $\mu(mn) = \mu(m) \cdot \mu(n) = 0$.

Table 7.1 Values of $\mu(n)$ and $\lambda(n)$

	1	2	3	4	5	6	7	8	9	10	11	12	13	14	15	16	17	18
$\mu(n)$	1	−1	−1	0	−1	1	−1	0	0	1	−1	0	−1	1	1	0	−1	0
$\lambda(n)$	1	−1	−1	1	−1	1	−1	−1	1	1	−1	−1	−1	1	1	1	−1	−1

7.1.5 $\lambda(n)$: The Liouville Function

Similar to the Möbius function is the **Liouville function**,[2] defined as

$$\lambda(n) = \begin{cases} 1 & \text{if } n = 1 \\ (-1)^{\sum_{i=1}^{r} \alpha_i} & \text{if } n = p_1^{\alpha_1} p_2^{\alpha_2} \ldots p_r^{\alpha_r}. \end{cases}$$

The multiplicative nature of $\lambda(n)$ is immediate from the definition. Some values of the Liouville function are also given in Table 7.1. Observe that whenever $\mu(n) \neq 0$, then $\lambda(n) = \mu(n)$ (why?).

7.1.6 u(n), $\epsilon(n)$: The Unit and the Identity Functions

The **unit function** u(n) and the **identity function** $\epsilon(n)$ are trivial but useful in manipulations:

$$\begin{array}{cc} \mathrm{u}(n) = 1 & \epsilon(n) = n \\ n \in \mathbf{N} & n \in \mathbf{N}. \end{array}$$

They are obviously multiplicative in character.

Number-theoretic functions such as the preceding multiplicative functions occur outside of number theory proper. For example, let a group **G** have order n. It has some cyclic subgroups ({**I**} is always one such subgroup). A theorem in the literature says that the number of cyclic subgroups of **G** is $\geq \tau(n)$, with equality holding iff **G** is itself cyclic (Richards, 1984). Thus, for $n = 4$ and $\tau(n) = 3$, we see from Fig. 2.1

$$\text{Cyclic subgroups of } \mathbf{V} = \{\mathbf{I}\}, \{\mathbf{I}, a\}, \{\mathbf{I}, b\}, \{\mathbf{I}, c\}$$

$$\text{Cyclic subgroups of } \mathbf{Z}_4 = \{\mathbf{I}\}, \{\mathbf{I}, B\}, \{\mathbf{I}, A, B, C\}.$$

The Euler ϕ-function is especially important in group theory. If **G** is a cyclic group of order n, there is an element $g \in \mathbf{G}$ of order n. Now consider the element g^m, $1 < m < n$, and let $d = (m, n)$. Since d is the largest divisor of both m and n, and since $m \times (n/d) = n \times (m/d)$, we see that the smallest power of g^m that yields a power of g^n is the (n/d)th:

$$(g^m)^{n/d} = (g^n)^{m/d} = 1.$$

Thus, g^m generates a cyclic subgroup of **G** of order n/d, so g^m can only generate all of **G** if $d = 1$. Hence, there are only as many generators of **G** as there are

[2] After the French mathematician **Joseph Liouville** (1809-1882). The Möbius function was studied by and is named after the German mathematician **August Ferdinand Möbius** (1790-1868), better known for his discovery of a surface with only one side.

positive integers less than but coprime to n, that is, $\phi(n)$. This result effectively answers Problem 5.14.

Look at the multiplication table for the cyclic group of order 6:

$(\cdot)$	**I**	a	b	c	d	e
I	**I**	a	b	c	d	e
a	a	b	c	d	e	**I**
b	b	c	d	e	**I**	a
c	c	d	e	**I**	a	b
d	d	e	**I**	a	b	c
e	e	**I**	a	b	c	d

It is cyclic because a is a generator ($a^0 = \mathbf{I}$, $a^1 = a$, $a^2 = b$, $a^3 = c$, $a^4 = d$, $a^5 = e$). There is only one other generator for the group (can you find it?), and this is consistent with $\phi(6) = 2$.

In fact, the multiplicative character of the Euler ϕ-function can be established group theoretically. Consider the cyclic group $\mathbf{Z}_{rr'}$; a result in group theory says that $\mathbf{Z}_{rr'}$ is isomorphic to the direct product $\mathbf{Z}_r \times \mathbf{Z}_{r'}$ iff $(r, r') = 1$ (Goldstein, 1973). No generator of $\mathbf{Z}_r$ can be a generator of $\mathbf{Z}_{r'}$, for if $g^r = g^{r'} = 1$ held ($r' > r$), then $g^{(r'-r)} = 1$ would follow, and this is a contradiction. Hence, the generators of $\mathbf{Z}_{rr'}$ are the pairs (g, g'), where g is a generator of $\mathbf{Z}_r$ and g' is a generator of $\mathbf{Z}_{r'}$. There are $\phi(r)\phi(r')$ such pairs, so

$$\phi(rr') = \phi(r)\phi(r').$$

Another appearance of $\phi(n)$ is the following. Let $\mathrm{N}(n)$ be the number of distinct (nonisomorphic) groups of order n. Some values of $\mathrm{N}(n)$ are tabulated below. A theorem in the literature (Murty and Murty, 1984) says that if n is squarefree (e.g., 5, 6, 10, 15), then $\mathrm{N}(n) \leq \phi(n)$. Thus, we see that $\mathrm{N}(6) = \phi(6)$, $\mathrm{N}(10) < \phi(10)$, and $\mathrm{N}(15) < \phi(15)$. There is no known theorem of a comparable nature for cases where n is not squarefree. A great many properties of the interesting ϕ-function are summarized in (Sivaramakrishnan, 1986).

n	$\mathrm{N}(n)$
3	1
4	2
5	1
6	2
8	5
9	2
10	2
15	1

Problems

7.1. (a) Is $\mathrm{f}(n) = \mathrm{e}^n, n \in \mathbf{N}$, multiplicative? How about $\mathrm{f}(n) = n^2$?

(b) An arithmetic function is defined by

$$\mathrm{f}(n) = \begin{cases} 1 & n = 1 \\ \sum_{p|n} 1 & n > 1, \end{cases}$$

where the summation is over all *prime* divisors of n. Is $\mathrm{f}(n)$ multiplicative?

7.2. (a) An arithmetic function is defined by

$$\mathrm{f}(n) = \begin{cases} 1 & n = 1 \\ \prod_{p|n} p & n > 1, \end{cases}$$

where the extended product is, as in Problem 7.1, over all *prime* divisors of n. Is $\mathrm{f}(n)$ multiplicative now?

(b) An arithmetic function is defined by

$$\mathrm{f}(n) = \begin{cases} 1 & n = 1 \\ \sum_{(m,n)=1} m & n > 1, \end{cases}$$

where the summation is over all positive integers m less than and coprime to n. Is $\mathrm{f}(n)$ multiplicative?

(c) Is Legendre's symbol (Section 6.2) a multiplicative function of its numerator if the domain of this is restricted to $\mathbf{N}$?

7.3. Equations involving arithmetic functions may or may not have solutions.

(a) Let $\mathrm{f}(n)$ be the function defined in Problem 7.2(b). Find three solutions to $\mathrm{f}(n) = 3n$.

(b) Again, with the $\mathrm{f}(n)$ of Problem 7.2(b), find the smallest integer $N > 0$ such that $\mathrm{f}(n) = nN$ has no solution in the interval $2 \le n \le 100$.

(c) Find the smallest solution to $\tau(n) = 4 + \mathrm{f}(n)$, where $\mathrm{f}(n)$ is the function defined in Problem 7.1(b).

7.4. Let $N > 1$ be an arbitrary positive integer. Prove that $\tau(n) = N$ has an infinite number of solutions. In view of this, find by elementary reasoning the smallest n when $N = 28$.

7.5. Show that $\tau(n^2) \equiv 1 \pmod 2$ for all $n \in \mathbf{N}$.

7.6. Show that $\sigma(n^2) \equiv 1 \pmod 2$ for all $n \in \mathbf{N}$.

7.7. Show that $\phi(n) \equiv 0 \pmod{2}$ for all $n > 2$.

7.8. Show that $\Pi_{d|n} d^2 = n^{\tau(n)}$ for all $n \in \mathbf{N}$.

7.9. Prove that the positive integer $p > 1$ is a prime iff $\sigma(p) = 1 + p$.

7.10. Prove that when p is a prime, then $\sigma(p) - \phi(p) = 2$. Is the converse true? Prove, or give a counterexample.

7.11. Evaluate: $\sigma(10^6)$, $\phi(10^6)$.

7.12. Write a computer program that will calculate $\tau(n)$ and $\sigma(n)$, given n.

7.13. Let m, n be arbitrary positive integers and suppose $m \mid n$. Then $n = km$ for some $k \in \mathbf{N}$. As ϕ is multiplicative, we have $\phi(n) = \phi(k)\phi(m)$, or $\phi(m) \mid \phi(n)$. Thus, the theorem $m \mid n \to \phi(m) \mid \phi(n)$ is proved.

(a) Criticize this argument.
(b) Repair the argument.

7.14. A **completely multiplicative function** $f(n)$ is an arithmetic function for which $f(mn) = f(m)f(n)$ for any pair m, n of positive integers. Prove that $\mu(n)$ is not completely multiplicative but that $\lambda(n)$ is completely multiplicative. Is Legendre's symbol a completely multiplicative function of its numerator (again, with domain restricted to $\mathbf{N}$)? How about Jacobi's symbol (Section 6.3)?

7.15. Prove that for any $n \in \mathbf{N}$, $\Sigma_{d|n} \lambda(d) = \begin{cases} 1 & \text{if } n \text{ is a square} \\ 0 & \text{otherwise.} \end{cases}$

7.16. Prove that for any $n \in \mathbf{N}$, $\Pi_{k=0}^{3} \mu(n + k) = 0$.

7.17. Find an $n \in \mathbf{N}$ such that $\Sigma_{k=0}^{2} \mu(n + k) = -3$.

7.18. (a) If $f(n)$, $g(n)$ are multiplicative functions, show that $h(n) = f(n) \cdot g(n)$ is also.

(b) Is the previous result true if we replace $f(n) \cdot g(n)$ by $f(n) + g(n)$?

7.19. Define the arithmetic function $F(n)$ by

$$F(n) = \sum_{d|n} [\epsilon(d)]^2.$$

Show that $F(n)$ is multiplicative. Illustrate this numerically with $m = 4$, $n = 15$.

7.20. Evaluate: $\Sigma_{n=1}^{\infty} \epsilon[\mu(n!)]$.

7.21. An arithmetic function f is called **additive** if for any $m, n \in \mathbf{N}$, $(m, n) = 1$, one has $f(m + n) = f(m) + f(n)$. Determine all arithmetic functions that are both additive and multiplicative. Are they also completely multiplicative?

7.22. Show that the sum of the kth-powers of the divisors of $n = p_1^{\alpha_1} p_2^{\alpha_2} \ldots p_r^{\alpha_r}$ is given by

$$\sigma_k = \prod_{i=1}^{r} \frac{p_i^{k(\alpha_i+1)} - 1}{p_i^k - 1}.$$

Illustrate this result numerically.

7.23. Let N(n) denote the number of fractions in lowest terms in the interval (0, 1) that have denominators not exceeding n. For example, when $n = 4$, the set of all such fractions is $\{1/2, 1/3, 2/3, 1/4, 3/4\}$, so N(4) = 5. Find a formula for N(n).

7.24. The odd divisors of 15 are 1, 3, 5, 15, and the sum of their reciprocals is $S_{15} = 1 + 1/3 + 1/5 + 1/15 = 1.60$. Let A_n be the average of the sum of the reciprocals of the odd divisors for all integers from 1 to n:

$$A_n = \left(\frac{1}{n}\right) \sum_{i=1}^{n} S_i.$$

The Austrian mathematician **Leopold Gegenbauer** (1849–1903) showed in 1885 that $\lim_{n\to\infty} A_n = \pi^2/8$. Choose a large n and then write a computer program that can estimate the value of π from Gegenbauer's result.

7.25. Let $n + 1$ be composite and suppose $\phi(n + 1) \mid n$. Show that $n + 1$ must be squarefree. How many distinct prime factors must $n + 1$ contain?

7.26. (a) Let $n = \prod_{j=1}^{r} p_j^{\alpha_j} > 1$ be odd; prove that $\phi(n) > \sqrt{n}$. Does this result extend to even n?

(b) It can be shown that for all integers $n > 1$

$$\frac{\sigma(n)\,\phi(n)}{n^2} < 1$$

(Annapurna, 1972). Hence, show that for all $n > 2$

$$1 + \frac{1}{n} \leq \frac{\sigma(n)}{n} < \sqrt{n}.$$

7.27. We denote by $\pi(n)$ the **product-of-divisors function**,

$$\pi(n) = \prod_{d \mid n} d.$$

From Problem 7.8 we have $\pi(n) = n^{\tau(n)/2}$.

(a) Show this if you did not do Problem 7.8 already.
(b) Is $\pi(n)$ multiplicative? Is it completely multiplicative (Problem 7.14)?

(c) A **triangular number** t is a number $t = x(x + 1)/2$, $x \in \mathbf{N}$. Prove that the number-theoretic equation

$$\pi(n) = p^{\alpha},$$

where p is a given prime, has a solution n iff α is triangular (Rahman, 1991).

7.28. Select six values of n and for each compute

$$\sum_{d|n} \epsilon\left(\frac{n}{d}\right) \mu(d),$$

where the summation is over all the divisors of n. Formulate a conjecture.

7.29. For $n = 10, 30, 50, 63$ compute the sums $\Sigma_{x\in \mathbf{S}'} (x - 1, n)$, where the x's are drawn from a reduced residue system $\mathbf{S}'$ modulo n. Compare your result in each case with the numbers $\phi(n)$, $\tau(n)$. Formulate a conjecture (this conjecture is known to be true (Menon, 1965)).

7.30. An arithmetic function $\mathrm{r}(n)$ is defined by the following generating function:

$$\sum_{n=1}^{\infty} \mathrm{r}(n)x^{n-1} = [(1 - x)(1 - x^2)(1 - x^3) \ldots]^{24}.$$

Values of $\mathrm{r}(n)$ may be called **Ramanujan numbers** because **Ramanujan** encountered them in his work on the representation of integers as sums of squares (Hardy, 1959).

(a) Determine the values of $\mathrm{r}(1)$ to $\mathrm{r}(7)$.
(b) It is known that $\mathrm{r}(n)$ is multiplicative. Illustrate this from data in part (a).

7.2 A General Theorem on Multiplicative Functions

Multiplicative functions enjoy a number of properties in common, as might be expected from the arithmetic character of their definition. The following is one such common property.

Theorem 7.1 If $\mathrm{f}(n)$ is a multiplicative function, other than the zero function, then $\mathrm{f}(1) = 1$.

Proof Since $\mathrm{f}(n)$ is multiplicative, then for any n

$$\mathrm{f}(n \cdot 1) = \mathrm{f}(n)\mathrm{f}(1).$$

Choose any $n > 1$ for which $\mathrm{f}(n) \neq 0$. This forces $\mathrm{f}(1) = 1$. ◆

This theorem makes it clear why Euler's totient and the Möbius and the Liouville functions must have a value of 1 at $n = 1$.

To motivate the next theorem, consider the following example, illustrated arbitrarily with the Euler totient function. Define $\Theta(n) = \Sigma_{d|n}\, \phi(d)$. Let $n = 4$, $m = 15$; then we obtain

$$\begin{aligned}\Theta(\text{n}) &= \phi(1) + \phi(2) + \phi(4)\\ &= 1 + 1 + 2\\ &= 4\end{aligned}$$

$$\begin{aligned}\Theta(\text{m}) &= \phi(1) + \phi(3) + \phi(5) + \phi(15)\\ &= 1 + 2 + 4 + 8\\ &= 15\end{aligned}$$

$$\begin{aligned}\Theta(\text{nm}) &= \phi(1) + \phi(2) + \phi(3) + \phi(4) + \phi(5) + \phi(6) + \phi(10)\\ &\quad + \phi(12) + \phi(15) + \phi(20) + \phi(30) + \phi(60)\\ &= 1 + 1 + 2 + 2 + 4 + 2 + 4 + 4 + 8 + 8 + 8 + 16\\ &= 60\end{aligned}$$

that is, $\Theta(4 \cdot 15) = \Theta(4)\ \Theta(15)$. This result suggests Theorem 7.2.

Theorem 7.2 If $f(n)$ is multiplicative, then so is $F(n) = \Sigma_{d|n}\, f(d)$, where the summation is over all of the divisors of n.

Proof Let $(m, n) = 1$; if $m = 1$, then $F(m) = 1$ and so $F(m)F(n) = F(n) = F(mn)$. Now assume $m, n > 1$ and that they have the prime factorizations

$$m = p_1^{\alpha_1} p_2^{\alpha_2} \ldots p_r^{\alpha_r}, \quad n = q_1^{\beta_1}\, q_2^{\beta_2} \ldots q_s^{\beta_s},$$

where no p_i is a q_j. Any divisor d_m of m and any divisor d_n of n have the factorizations

$$d_m = p_1^{a_1} p_2^{a_2} \ldots p_r^{a_r}, \quad d_n = q_1^{b_1} q_2^{b_2} \ldots q_s^{b_s},$$

where each a_i satisfies $0 \le a_i \le \alpha_i$ and each b_j satisfies $0 \le b_j \le \beta_j$. As d_m runs through all possible divisors of m and as d_n runs through all possible divisors of n, then $d_m d_n$ runs through all possible values of

$$d = d_m d_n = p_1^{a_1} p_2^{a_2} \ldots p_r^{a_r} q_1^{b_1} q_2^{b_2} \ldots q_s^{b_s},$$

which are all the possible divisors d of mn. Thus,

$$F(mn) = \sum_{d|mn} f(d) = \sum_{d_m|m} \sum_{d_n|n} f(d_m d_n).$$

Since no p_i is a q_j, then $(d_m, d_n) = 1$. By hypothesis $f(n)$ is multiplicative, so we can rewrite each term in the double summation as

$$\begin{aligned} F(mn) &= \sum_{d_m|m} \sum_{d_n|n} f(d_m)f(d_n) \\ &= \sum_{d_m|m} f(d_m) \sum_{d_n|n} f(d_n) \\ &= F(m)\ F(n). \end{aligned}$$

This says that F is multiplicative. ◆

Corollary 7.2.1 The function $\tau(n)$ is multiplicative.

Proof Let the multiplicative function $f(n)$ be $u(n)$, the unit function. Then for each divisor d of n we have $u(d) = 1$ and

$$F(n) = \sum_{d|n} u(d) = \sum_{d|n} 1 = \tau(n),$$

since the last summation records a single tally for each and every divisor of n. By Theorem 7.2, $\tau(n)$ must be multiplicative. ◆

In almost the same way we can prove Corollary 7.2.2.

Corollary 7.2.2 The function $\sigma(n)$ is multiplicative.

Example 7.6 Let $m = 24$, $n = 25$. Then $\sigma(m) = 60$ and $\sigma(n) = 31$. The divisors of $mn = 600$ are 1, 2, 3, 4, 5, 6, 8, 10, 12, 15, 20, 24, 25, 30, 40, 50, 60, 75, 100, 120, 150, 200, 300, 600, so $\sigma(mn) = 1860 = 60 \cdot 31$. ◆

The function $F(n)$ in Theorem 7.2 can be expressed in terms of the prime factors of n.

Theorem 7.3 If $n = p_1^{\alpha_1} p_2^{\alpha_2} \ldots p_r^{\alpha_r}$ and $f(n)$ is multiplicative, then

$$F(n) = \sum_{d|n} f(d) = \prod_{i=1}^{r} \{1 + f(p_i) + f(p_i^2) + \cdots + f(p_i^{\alpha_i})\}.$$

Proof The extended product is

$$\{1 + f(p_1) + f(p_1^2) + \cdots + f(p_1^{\alpha_1})\}\{1 + f(p_2) + f(p_2^2) + \cdots + f(p_2^{\alpha_2})\} \cdots \{1 + f(p_r) + f(p_r^2) + \cdots + f(p_r^{\alpha_r})\}.$$

Any term in the expansion is constructed by multiplying together as factors exactly one term from within each of the r pairs of braces. In this way, all possible combinations of powers of the p_i's appear as arguments of $f(n)$. That is, a typical term in the overall product is

$$f(p_1^{a_1} p_2^{a_2} \ldots p_r^{a_r}),$$

where $0 \leq a_i \leq \alpha_i$ for every i. Each such typical term appears only once in the expansion because the primes $p_1, p_2, \ldots, p_r$ are distinct. But the typical arguments $p_1^{a_1} p_2^{a_2} \ldots p_r^{a_r}$ are just the divisors d of n. Thus, the sum of all terms like the preceding one is the same as the sum of all terms of the form $f(d)$. ◆

Example 7.7 In Theorem 7.3 let $f(d) = 1/d^s$, $s \geq 2$. Then

$$\sum_{d|n} d^{-s} = \prod_{i=1}^{r} \{1 + p_i^{-s} + p_i^{-2s} + \cdots + p_i^{-\alpha_i s}\}.$$

What might we expect as $n \to \infty$? Naively, we let $r \to \infty$ and each $\alpha_i \to \infty$. The preceding equation then passes over into

$$\sum_{k=1}^{\infty} k^{-s} = \prod_{i=1}^{\infty} \{1 + p_i^{-s} + p_i^{-2s} + \cdots +\}$$

$$= \prod_{i=1}^{\infty} (1 - p_i^{-s})^{-1}$$

from calculus. Although this "derivation" of the final result is not valid, the result is correct. It is a famous formula due to the great manipulator **Euler**. ◆

Corollary 7.3.1 $\Sigma_{d|n} \mu(d) = \begin{cases} 1 & \text{if } n = 1 \\ 0 & \text{otherwise.} \end{cases}$

Proof The Möbius function is multiplicative. Hence, from Theorem 7.3 we have for $n = p_1^{\alpha_1} p_2^{\alpha_2} \ldots p_r^{\alpha_r} > 1$

$$\sum_{d|n} \mu(d) = \prod_{i=1}^{r} \{1 + \mu(p_i) + \mu(p_i^2) + \cdots + \mu(p_i^{\alpha_i})\}.$$

By definition, $\mu(p_i^k) = 0$ if $k \geq 2$ and -1 if $k = 1$. This gives for $n > 1$,

$$\sum_{d|n} \mu(d) = \prod_{i=1}^{r} \{1 + \mu(p_i)\} = (1 - 1)^r = 0.$$

When $n = 1$, then each $d = 1$ and $\Sigma_{d|n}\, \mu(d) = \mu(1) = 1$. ◆

Example 7.8 The divisors of $n = 12$ are 1, 2, 3, 4, 6, and 12. Hence,

$$\begin{aligned}\sum_{d|12} \mu(d) &= \mu(1) + \mu(2) + \mu(3) + \mu(2^2) + \mu(2 \cdot 3) + \mu(2^2 \cdot 3) \\ &= 1 + (-1) + (-1) + 0 + (+1) + 0 \\ &= 0.\end{aligned}$$ ◆

Problems

7.31. Prove Corollary 7.2.2.

7.32. If $f(n)$ is multiplicative, show that $F(n) = \Sigma_{d|n}\, [f(d)]^k$, $k \in \mathbf{N}$, is also multiplicative. Illustrate with the example: $f(n) = \phi(n)$, $n = 8$, $m = 9$, $k = 2$.

7.33. Consider the function $F(n) = \tau(n) - \Sigma_{d|n}\, \mu(d)\lambda(d)$. Is it a multiplicative function? What property of n does the function $F(n)$ describe?

7.34. Refer to Problem 7.14. If $f(n)$ is a completely multiplicative function, must $F(n) = \Sigma_{d|n}\, f(d)$ also be a completely multiplicative function? Prove or give a counterexample.

7.35. Consider the arithmetic function $F(n) = \Sigma_{d|n}\, \phi(d)$.

(a) Evaluate F(8), F(9), F(10).
(b) Prove the general theorem suggested by the results of part (a). [Hint: First evaluate $F(p_i^{\alpha_i})$; then apply Theorem 7.2.] This theorem is due to **Gauss** (1801), and it forms the basis for one of the popular proofs of the existence of primitive roots of a prime (see remarks on p. 183).

7.36. Define $F(n!) = \Sigma_{d|n!}\, \tau(d)$. It is conjectured that for $n > 1$, $F(n!) \equiv 0 \pmod 3$. Write a computer program and check this out for several n.

7.3 Highly Composite Integers*

In Section 2.1 we mentioned briefly the highly composite integers. These were extensively studied by **S. Ramanujan** in 1915. Our interest in them is due to their connection with the multiplicative function $\tau(n)$. We recall that n is a highly composite integer (HCI) if $\tau(n) > \tau(n')$ for all $n' < n$. We prove now some interesting facts about these numbers (Honsberger, 1985; Ratering, 1991).

Theorem 7.4 Let $n = p_1^{\alpha_1} p_2^{\alpha_2} \ldots p_r^{\alpha_r}$ be a highly composite integer. Then the primes $p_1, p_2, \ldots, p_r$ form a string of consecutive primes ($p_1 = 2, p_2 = 3, p_3 = 5$, etc.) as far as they go.

Proof Let the factorization of n be written in the order $p_1 < p_2 < \ldots < p_r$ and suppose that the jth natural prime P_j is missing from this string. That is, $P_j < p_j$ and $P_j \nmid n$. Now form the number n' by replacing $p_j^{\alpha_j}$ by $P_j^{\alpha_j}$. Then $n' < n$ and yet

$$\begin{aligned}\tau(n') &= (\alpha_1 + 1)(\alpha_2 + 1) \ldots (\alpha_j + 1) \ldots (\alpha_r + 1) \\ &= \tau(n).\end{aligned}$$

But if n is a highly composite integer, we must have $\tau(n') < \tau(n)$ for all $n' < n$. The contradiction shows that P_j cannot be absent from the string $p_1, p_2, \ldots, p_r$. ◆

Example 7.9 Refer to Table G in Appendix II for values of $\tau(n)$. We see that 60 is a highly composite integer since $\tau(60) = 12 > \tau(n')$ for any $n' < 60$. In accordance with Theorem 7.4, we have $60 = 2^2 \cdot 3^1 \cdot 5^1$. What is the next highly composite integer? ◆

Theorem 7.5 Let $n = p_1^{\alpha_1} p_2^{\alpha_2} \ldots p_r^{\alpha_r}$ be a highly composite integer. Then the exponents are nonincreasing: $\alpha_1 \geq \alpha_2 \geq \ldots \geq \alpha_r$.

Proof Suppose for some j, $1 \leq j < r$, one has $\alpha_j < \alpha_{j+1}$. What happens when these exponents are switched? The proof is left to the reader. ◆

Theorem 7.6 Let $n = p_1^{\alpha_1} p_2^{\alpha_2} \ldots p_r^{\alpha_r}$ be a highly composite integer. Then the final exponent $\alpha_r \leq 2$.

* Sections marked with an asterisk can be skipped without impairing the continuity of the text.

Proof Suppose, to the contrary, that $\alpha_r = 2 + a, a \geq 1$. Now consider the integer

$$n' = \left(\frac{p_{r+1}}{p_r^2}\right) n = p_1^{\alpha_1} p_2^{\alpha_2} \ldots p_r^a p_{r+1}.$$

Then $n' < n$ iff $p_{r+1} < p_r^2$. This holds for $p_r \geq 2$. A result known as **Bertrand's Postulate** (Section 1.5) declares that between the pair of integers m and $2m$ $(m > 1)$ there is at least one prime. Hence, between p_r and $2p_r$ there is a prime; clearly, p_{r+1} will do. That is, $p_{r+1} < 2p_r$, and since $2p_r \leq p_r^2$ for $p_r \geq 2$, we obtain $p_{r+1} < p_r^2$ for $p_r \geq 2$.

Since n is highly composite, then $\tau(n') < \tau(n)$, that is,

$$(\alpha_1 + 1)(\alpha_2 + 1) \ldots (a + 1)2 < (\alpha_1 + 1)(\alpha_2 + 1) \ldots (a + 2 + 1).$$

This reduces to $2a + 2 < a + 3$, or $a < 1$. But $a \geq 1$ by assumption. The contradiction implies $\alpha_r < 3$. ◆

We observe that the final exponent in the factorization of the highly composite integer $n = 60$ is a 1. Of course, Theorems 7.4–7.6 also apply to the special highly composite integers (Section 2.1).

Problems

7.37. Prove Theorem 7.5.

7.38. Let HCI $= p_1^{\alpha_1} p_2^{\alpha_2} \ldots p_r^{\alpha_r}$ be a highly composite integer.

(a) Prove that the largest HCI that has every $\alpha_i = 1$ in its factorization is 6.
(b) Prove that the largest HCI that has every $\alpha_i = 2$ in its factorization is 36.

7.39. Are there any highly composite integers greater than 60 whose factorization has the form HCI $= 2^2 \cdot 3^1 \cdot 5^1 \ldots p_r^1$?

7.40. Prove that there are infinitely many highly composite integers.

7.4 Perfect Numbers

Table 7.2 gives some data on the sum-of-divisors function, excerpted from Table G in Appendix II. Column 3 gives the deviation of $\sigma(n)$ from the number $2n$. Both positive and negative deviations occur; infrequently, $\sigma(n) = 2n$.

Table 7.2 Some Values of the Sum-of-Divisors Function

n	$\sigma(n)$	$\sigma(n) - 2n$
4	7	−1
6	12	0
9	13	−5
10	18	−2
22	36	−8
26	42	−10
28	56	0
32	63	−1
36	91	19
59	60	−58
100	217	17

To the Pythagoreans (550-350 B.C.) we owe the concepts of perfect number, abundant number, and deficient number.

DEF. A positive integer n is called a **perfect number** if the sum of the proper divisors of n is n (and, hence, $\sigma(n) = 2n$). If $\sigma(n) > 2n$ or if $\sigma(n) < 2n$, then n is termed **abundant** or **deficient**, respectively.

Thus, 6 and 28 are perfect, 36 and 100 are abundant, and 9 and 10 are deficient. All positive integers are either perfect, abundant, or deficient (Fig. 7.4). The reason for portraying the set of perfect numbers as a thin sector of **N** will be apparent after the proof of the next theorem.

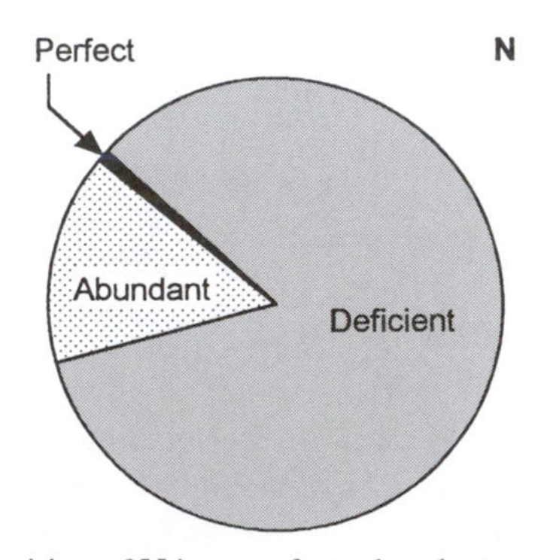

Figure 7.4 A partition of **N** into perfect, abundant, and deficient numbers

However, it is revealing that a simple count among the first 200 positive integers gives the following tallies of the three types of numbers:

Perfect numbers: 2 (1%)

Abundant numbers: 46 (23%)

Deficient numbers: 152 (76%).

Computer calculations show that as $N \to \infty$, the percentage of abundant numbers in the interval $[1, N]$ approaches a value that is bracketed by 24.4% on the low side and 29.1% on the high side (Wall, 1972). The precise limit is not yet known.

Sometime after the period of the Pythagoreans, **Euclid** in Book IX of his *Elements* gave a formula for computing even perfect numbers. Ever since, these fascinating numbers have intrigued amateur and professional mathematicians alike (Shoemaker, 1973; Brill and Stueben, 1993).

Incidentally, **Euclid** himself used the term *perfect*.

Theorem 7.7 Let p be a prime. Then $n = 2^{p-1} M_p$ is a perfect number if M_p is a Mersenne prime.

Proof The complete set of divisors of n is 1, the $p - 1$ positive powers of 2, and each of those p integers multiplied by $M_p = 2^p - 1$. Hence,

$$\sigma(n) = (1 + M_p) \sum_{k=0}^{p-1} 2^k = (1 + M_p)\left(\frac{1 - 2^p}{1 - 2}\right)$$

$$= 2^p(2^p - 1) = 2n. \quad \blacklozenge$$

Example 7.10 The next three Mersenne primes after M_3 are $M_5 = 31$, $M_7 = 127$, $M_{13} = 8191$. Therefore, three more even perfect numbers after 6, 28 are $2^4(31) = 496$, $2^6(127) = 8128$, and $2^{12}(8191) = 33{,}550{,}336$. ◆

In view of Conjecture 3 in Section 1.8, it is not known if there are an infinite number of even perfect numbers of the form $2^{p-1}M_p$, that is, the slice in Figure 7.4 may be an exaggeration.

Could some even perfect number have a form different from that just given? This question was answered centuries later by **Euler**; we give a contemporary proof (McDaniel, 1975).

Theorem 7.8 (Euler). If n is an even perfect number, then n is expressible as $2^{p-1}(2^p - 1)$ for some prime p.

Proof Let $n = 2^{k-1}m$, where $(m, 2) = 1$. We then have

$$2^k m = 2(2^{k-1}m) = 2n = \sigma(n) = \sigma(2^{k-1})\sigma(m),$$

because σ is multiplicative (Corollary 7.2.2) and $(2^{k-1}, m) = 1$. From the formula for σ given in Section 7.1, we have $\sigma(2^{k-1}) = 2^k - 1$, so

$$2^k m = (2^k - 1)\, \sigma(m).$$

Any prime divisor of $2^k - 1$ on the right-hand side is odd and must therefore divide m on the left-hand side. Let q be the smallest odd prime divisor of $2^k - 1$ and let x be the largest integer such that $q^x \mid m$. Then

$$1 = \frac{\sigma(n)}{2n} = \frac{(2^k - 1)\, \sigma(m)}{2^k m} \geq \frac{(2^k - 1)\, \sigma(q^x)}{2^k q^x}.$$

To see the inequality, set $m = q^x r$ and note that $(q^x, r) = 1$ because q is a prime and x is maximal. Then $\sigma(m)/m = \sigma(q^x)\sigma(r)/q^x r \geq \sigma(q^x)/q^x$ because $\sigma(r)/r \geq 1$ for any r (why?).

Further, we have

$$\frac{\sigma(q^x)}{q^x} = \frac{1 + q + \cdots + q^x}{q^x} \geq \frac{1 + q}{q} \text{ (why?)},$$

so we now have

$$1 \geq \frac{(2^k - 1)(1 + q)}{2^k q} = 1 + \frac{2^k - 1 - q}{2^k q}.$$

This is possible only if $2^k - 1 = q$ is a Mersenne prime ($k =$ a prime p).

If $x > 1$, then rearrangement of $2^k m = (2^k - 1)\sigma(m)$ gives, after substituting $2^k = 1 + q$,

$$\sigma(r) = \left[\frac{q^{x-1} + q^x}{1 + q + \cdots + q^x}\right] r < r.$$

This is impossible, so $x = 1$ and the equation reduces to $\sigma(r) = r$. The only solution of this is $r = 1$, and we obtain finally, $n = 2^{k-1}(2^k - 1)$ for k equal to some prime. ◆

By now you may be wondering why no odd perfect numbers have been exhibited. Oddly, none are known, despite a massive amount of work in this direction. On the other hand, there is no reason so far to rule out the existence of an odd perfect number. If one exists, it is known that it must exceed 10^{300} and have at least eight distinct prime factors (Ribenboim, 1996). By strictly elementary arguments, we can prove some facts about any alleged odd perfect number.

Theorem 7.9 If n is an odd perfect number, then in the factorization of n exactly one prime divisor must appear to an odd power.

Proof Let $n = p_1^{\alpha_1} p_2^{\alpha_2} \ldots p_r^{\alpha_r}$ be an odd perfect number. Then by definition one has

$$(1 + p_1 + p_1^2 + \cdots + p_1^{\alpha_1})(1 + p_2 + p_2^2 + \cdots + p_2^{\alpha_2}) \cdots (1 + p_r + p_r^2 + \cdots + p_r^{\alpha_r}) = 2n.$$

If α_i is even (odd), then $(1 + p_i + p_i^2 + \cdots + p_i^{\alpha_i})$ is odd (even). But as n is odd, $4 \nmid 2n$ and so only one factor on the left-hand side can be even. This means that one and only one α_i can be odd. ◆

We denote by π^α the prime power in n in which α is odd. This factor is called the **Euler factor**.

Theorem 7.10 For the Euler factor π^α of any odd perfect number, $\pi \equiv \alpha \equiv 1 \pmod 4$ holds.

Proof Let $\pi = 2k + 1$, $\alpha = 2h + 1$; then $\sigma(\pi^\alpha)$ is given by

$$1 + (2k + 1) + (2k + 1)^2 + (2k + 1)^3 + \cdots + (2k + 1)^{2h+1}.$$

All terms in the expansion of the preceding expression that are of quadratic or higher degree in $2k$ are divisible by 4 since $4 \mid 2^k$ for $k \geq 2$. If n is perfect, where $n = \pi^\alpha p_1^{\alpha_1} p_2^{\alpha_2} \ldots p_r^{\alpha_r}$, then we require $2 \mid \sigma(\pi^\alpha)$ but $4 \nmid \sigma(\pi^\alpha)$. Hence, the sum of the constants plus the terms linear in k must not be divisible by 4.

We find by easy computation:

Sum of constant terms: $S_1 = 1 + (2h + 1) = 2(h + 1)$

Sum of linear terms: $S_2 = 2k + 2(2k) + 3(2k) + \cdots + (2h + 1)(2k)$
$= 2k(h + 1)(2h + 1).$

Hence, $S_1 + S_2 = 2(h + 1)[k(2h + 1) + 1]$, and if this is not to be divisible by 4, then $h + 1$ must be odd and $k(2h + 1) + 1$ must be odd. These force $h = 2j$, $k = 2m$, whereupon $\pi = 4m + 1$, or $\pi \equiv 1 \pmod 4$, and $\alpha = 4j + 1$, or $\alpha \equiv 1 \pmod 4$. ◆

Several other facts are known about alleged odd perfect numbers, all in the form of restricted conditions that such numbers must obey. For example, if an odd integer $n < 10^{9118}$ is perfect, it must be divisible by the sixth power of some prime (McDaniel, 1971). If n is perfect but is not divisible by a sixth power of a prime, then the least prime factor of n is ≥ 101. An odd perfect number not divisible by 3 must have at least 11 distinct prime factors (Hagis, Jr., 1983).

Mathematicians love to change the rules of any game, with an eye either to discovering a generalization of some existing result or to uncovering some new result. For example, in one direction one can study properties of "integers" in a domain wider than **Z**; such a domain is **Z**$[i]$, the **Gaussian integers**, which are the set of numbers of the form $a + bi$, a, $b \in \mathbf{Z}$. In this domain, and in others, it is possible to formulate analogs of the Euclid–Euler theorems (Theorems 7.7 and 7.8) (McDaniel, 1974; McDaniel, 1990).

In another direction, one can continue to operate in **Z** but either expand or restrict the rules of the game at hand. As an example here, suppose that the only divisors of an odd integer n that can be considered are unitary divisors (Problem 7.56); then it is fairly easy to show that there can be no odd, unitary perfect numbers (Problem 7.57). In contrast, only five even, unitary perfect numbers are known explicitly, and it is not known if any others exist (Subbarao, 1970).

Problems

7.41. Determine which of the following integers are abundant, deficient, or perfect: 86, 405, 496, 8182, 13,824.

7.42. Mark the following statements as always true (T) or as false (F):

(a) The product of two deficient numbers is another deficient number.
(b) If x is an even perfect number, then either $x \equiv 0 \pmod 3$ or $x \equiv 1 \pmod 3$.
(c) If x is an abundant number, then $\tau(x) > 4$.
(d) If x is an abundant number, then $\sigma(x) < 3x$.

7.43. Write a computer program to find all integers $n < 500$ such that n is deficient but n^2 is abundant.

7.44. The first eight abundant integers are 12, 18, 20, 24, 30, 36, 40, and 42. All are even integers. Write a computer program to uncover the smallest odd abundant integer.

7.45. Show that if p is a prime and $k \in \mathbf{N}$, then p^k is deficient.

7.46. Prove that no even perfect number ends in a 0, 2, or 4.

7.47. In this and the next several problems, we look at a few of the many different special numbers that have been studied. Two positive integers m, n are said to be **amicable numbers** if $\sigma(m) = \sigma(n) = m + n$. Show that 79,750, 88,730 are amicable. The concept of amicable numbers originated with the Greeks of antiquity.

7.48. Both members of the smallest known pair of amicable numbers are less than 500. Write a computer program to discover this smallest pair.

7.49. Suppose $3 \cdot 2^{n-1} - 1$ and $3 \cdot 2^n - 1$ are both prime for some $n \in \mathbf{N}$, and define $M = 2^n(3 \cdot 2^{n-1} - 1)(3 \cdot 2^n - 1)$.

(a) Show that $\sigma(M) = 9 \cdot 2^{2n-1}(2^{n+1} - 1)$.

(b) Define $N = \sigma(M) - M$. Show that $\sigma(N) = \sigma(M)$ if $9 \cdot 2^{2n-1} - 1$ is prime. Then M, N are an amicable pair. This result was known to the Islamic mathematician **Thabit Ibn-Qurra** (836–901) of Baghdad.

(c) Find one other amicable pair besides the ones given in Problems 7.47 and 7.48.

It is known that Thabit's rule produces an amicable pair for only three values of n less than 2×10^4.

7.50. A **superperfect number** is a positive integer n satisfying $\sigma[\sigma(n)] = 2n$. Prove that n is an even superperfect number if $n = 2^k$ and $2^{k+1} - 1$ is prime. Find and verify two superperfect numbers larger than 16.

7.51. Abundant numbers have been generalized to **superabundant numbers** (Alaoglu and Erdös, 1944) defined as follows: A positive integer n is superabundant if $\sigma(n)/n > \sigma(k)/k$ for all $k \in \mathbf{N}$ satisfying $1 \le k < n$. Find the first 10 superabundant numbers.

7.52. Amicable numbers have been generalized as follows. We define the (non-multiplicative) arithmetic function s(n) by $s(n) = \sigma(n) - n$. Then the following sequence is defined:

$$s_k(n) = \begin{cases} n & k = 0 \\ s[s_{k-1}(n)] & k \ge 1. \end{cases}$$

For some k, $n \in \mathbf{N}$ the sequence may be periodic, that is, $s_k(n) = s_0(n)$. The set of integers $\{n, s_1(n), s_2(n), \ldots, s_{k-1}(n)\}$ is then termed a set of **sociable numbers** of order k. Amicable numbers are sociable numbers of order 2. In recent years, much computational effort has been expended in discovering sets of sociable numbers (Cohen, 1970; Flammenkamp, 1991). It is known that there are no sets of sociable numbers of order 3 whose smallest member is less than 5×10^7 (Bratley, Lunnon, and McKay, 1970). Also, the sequence $\{s_k\}$ might either converge, become periodic, or be unbounded. It is not known if there are any such unbounded sequences. Verify that the following is a set of sociable numbers of order 4:

$$1{,}264{,}460 = 2^2 \cdot 5 \cdot 17 \cdot 3719$$

$$1{,}547{,}860 = 2^2 \cdot 5 \cdot 193 \cdot 401$$

$$1{,}727{,}636 = 2^2 \cdot 521 \cdot 829$$

$$1{,}305{,}184 = 2^5 \cdot 40{,}787.$$

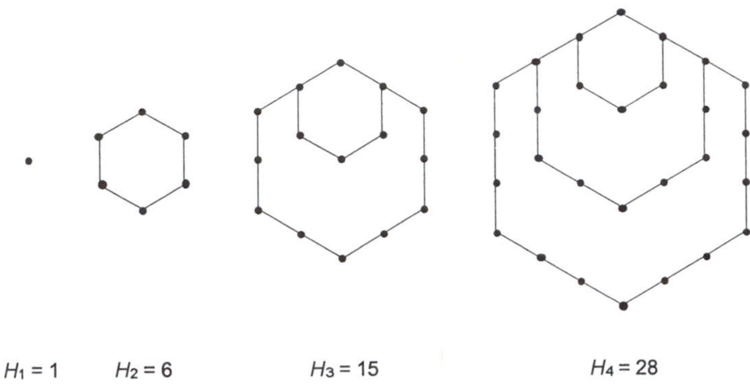

Figure 7.5 The first four hexagonal numbers

7.53. The **hexagonal numbers** $\{H_n\}_{n=1}^{\infty}$ are the integers whose geometric interpretations are illustrated by the following first four members of the sequence as shown in Fig. 7.5. We notice that two of the first four hexagonal numbers are perfect numbers. Prove, as did the Italian **Francesco Maurolico** (1494–1575), that every even perfect number is a hexagonal number.

7.54. The **abundancy index** of a positive integer n is the rational number

$$\mathrm{I}(n) = \frac{\sigma(n)}{n}.$$

When $\mathrm{I}(n)$ is an integer $r > 2$, then n is said to be **multiperfect of index *r*** (Laatsch, 1986). Let $k \in \mathbf{N}$; then prove that

$$\mathrm{I}(kn) > \mathrm{I}(n).$$

[Hint: Let the divisors of n be $\{1, d_1, d_2, \ldots, d_r, n\}$ and consider the set $\{1, k, kd_1, kd_2, \ldots, kd_r, kn\}$. How does it follow that there are infinitely many abundant numbers?]

7.55. The following result was stated in Alaoglu and Erdös (1944). Let $n = \prod_{i=1}^{r} p_i^{\alpha_i}$, $p_1 < p_2 < \cdots < p_r$, be a superabundant number (see Problem 7.51). Prove that $\alpha_1 \geq \alpha_2 \geq \alpha_3 \geq \cdots \geq \alpha_r$.

7.56. A **unitary divisor** of a positive integer n is a positive integer d such that $(d, n/d) = 1$. A **unitary abundant** (or **deficient**) **number** n is a positive integer such that $\sigma_u(n) > 2n$ (or $\sigma_u(n) < 2n$), where $\sigma_u(n)$ denotes the sum of the unitary divisors of n.

(a) Is $\sigma_u(n)$ multiplicative?
(b) Find the first unitarily abundant number.

7.57. In this problem we establish that there is no odd, **unitary perfect number**, that is, no integer n such that $\sigma_u(n) = 2n$. Let $n > 1$ be odd and write $n = \prod_{i=1}^{r} m_i$, $m_i = p_i^{\alpha_i}$, and the p_i's are distinct.

(a) Show that $\sigma_u(n) = \prod_{i=1}^{r} (1 + m_i)$.
(b) If n is odd and unitary perfect, why must $r > 1$ be true?
(c) Suppose, next, in the factorization of n there is some m_i (say, m_k), such that $m_k \equiv 3 \pmod 4$. How does $\sigma_u(n) \equiv 0 \pmod 4$ then follow? This, of course, is a contradiction (why?).
(d) On the other hand, if all m_i's are such that $m_i \equiv 1 \pmod 4$, then does $\sigma_u(n) \equiv 0 \pmod 4$ still follow?

7.5 More on Euler's Totient; Carmichael's Conjecture

An examination of a table of values of Euler's totient reveals many interesting facts. For $n > 2$, one has that $\phi(n)$ is always even (why?). But if we consider the number-theoretic equation

$$\phi(n) = x,$$

then not every choice of even integer x produces an equation that has a solution n. The smallest even x for which no solution exists is $x = 14$; the smallest integer x that is divisible by 4 and for which a solution fails to exist is $x = 68$. Further, in tables of ϕ-values up to $x = 2500$ (Carmichael, 1908; Glaisher, 1940), one finds no value of x for which only a single solution n exists. This is more interesting.

R.D. Carmichael made the following claim in 1907 (Carmichael, 1907).

CARMICHAEL'S CONJECTURE

For no x does the equation $\phi(n) = x$ have a unique solution or, equivalently, there exists no finite cyclic group that is characterized by its number of generators.

The equivalent algebraic statement follows from remarks made at the end of Section 7.1. Carmichael believed he had proved the assertion. Later, in his 1914 number theory text Carmichael submitted the proof as an exercise. Subsequently, the proof was discovered to have a flaw and the proposition reverted to the status of a conjecture. It remains a conjecture today; contemporary mathematicians believe the conjecture to be very deep. It is known that if there is an x such that $\phi(n) = x$ has a unique solution, then x must exceed $10^{10,000}$ (Wagon, 1986).

In this section we obtain some easy results for special cases of $\phi(n) = x$. The first theorem comes directly from (Carmichael, 1907).

Theorem 7.11 All the solutions of the equation $\phi(n) = 2m$, where $m > 1$ is odd, are of the form p^α or $2p^\alpha$, where p is an odd prime of the form $4k + 3$.

Proof First, $n \neq 4$ since $\phi(4) = 2$ and then $m = 1$, contrary to hypothesis. Next, suppose $n = p^a$, $a > 2$; then $\phi(n) = 2^{a-1}$ and $m = 2^{a-1}$, which is not odd. If $4 \mid n$, then write $n = 2^a q$, where $a \geq 2$, $q > 1$, and $(2, q) = 1$. This yields $\phi(n) = 2^{a-1}\ \phi(q)$, and since $\phi(q)$ is even we again obtain that m is even, contrary to hypothesis.

Suppose n contains at least two distinct odd primes in its factorization: $n = p_1^{\alpha_1} p_2^{\alpha_2} \ldots p_r^{\alpha_r}$, $r \geq 2$. Then

$$\phi(n) = \phi(p_1^{\alpha_1})\phi(p_2^{\alpha_2}) \ldots \phi(p_r^{\alpha_r}),$$

and again m is even. The only possibilities left are that n is a power of a single odd prime, or that n is twice a power of a single odd prime. These possibilities yield

$$\phi(p^\alpha) = p^{\alpha-1}(p - 1)$$

$$\text{and} \qquad \phi(2p^\alpha) = \phi(2)\ \phi(p^\alpha) = p^{\alpha-1}(p - 1),$$

and the hypothesis is satisfied if $2 \mid (p - 1)$ but $4 \nmid (p - 1)$. This holds iff $p = 4k + 3$. ◆

Example 7.11 Find all solutions of $\phi(n) = 18$.

We consider all the ways that 18 can be written as $p^{\alpha-1}(p - 1)$, where $p \mid 18$ if $\alpha > 1$. The only possibilities are $18 = 1 \cdot 18$ and $18 = 9 \cdot 2$, corresponding to $p = 19$, $\alpha = 1$ and $p = 3$, $\alpha = 3$. Then both p^α and $2p^\alpha$ are solutions; hence $n = 19^1 = 19$, $n = 2 \cdot 19^1 = 38$, $n = 3^3 = 27$, $n = 2 \cdot 3^3 = 54$. ◆

Note that in this Example we have obtained an even number of solutions. This is general (Klee, 1946).

Theorem 7.12 The number of distinct solutions of $\phi(n) = 2m$, where m is odd, is twice the number of ways in which m can be expressed in the form $q(2q + 1)^k$, where $2q + 1$ is prime and $k = 0$ or $k \in \mathbf{N}$.

Proof Suppose $\phi(n) = 2m$ has a nonempty solution set; by Theorem 7.11 one member of this set is $n = p^a = (4s + 3)^a$, for some prime p and some

$a \in \mathbf{N}$. Then $\phi(n) = p^a(1 - 1/p) = (4s + 3)^{a-1}(4s + 2) = 2m$; letting $q = 2s + 1$ and $k = a - 1$, we have $m = q(2q + 1)^k$.

Conversely, suppose m has the form $q(2q + 1)^k$, where $2q + 1$ is a prime and $k = 0$ or $k \in \mathbf{N}$. Then let $n_1 = (2q + 1)^{k+1}$ and $n_2 = 2(2q + 1)^{k+1}$. We find that $\phi(n_1) = \phi(n_2) = 2q(2q + 1)^k = 2m$. ◆

Example 7.12 How many distinct solutions are there to $\phi(n) = 70$? Here, $m = 35$ and we count the number of ways that 35 can be expressed in the form $q(2q + 1)^k$, where $2q + 1$ is a prime. Assume first that $k \geq 1$; then $35 = 5 \cdot 7$ implies that $2q + 1$ is either 5 or 7, so $q = 2$ or 3. But neither 2 nor 3 divides 35, so we must have $k = 0$. This forces q to be 35. As $2q + 1 = 71$ is a prime, then 35 can be expressed in the form $q(2q + 1)^k$ in only one way, namely, $35(71)^0$. By Theorem 7.12 there are just two solutions to $\phi(n) = 70$.

Corollary 7.12.1 If m has no divisor $d > 1$ such that $2d + 1$ is prime, then $\phi(n) = 2m$ has no solutions.

Corollary 7.12.2 If $m = p_1^{\alpha_1} p_2^{\alpha_2} \ldots p_r^{\alpha_r}$ and each prime is of the form $p_i = 3k_i + 1$, then $\phi(n) = 2m$ has no solutions.

Proof Any divisor $d > 1$ of m must be of the form $3k + 1$ (why?). But then $2d + 1 = 2(3k + 1) + 1 = 6k + 3$, and this is not prime. By Corollary 7.12.1, $\phi(n) = 2m$ then has no solutions. ◆

Example 7.13 The equation $\phi(n) = 182$ has no solutions since $182 = 2 \cdot 91 = 2[3 \cdot 2 + 1][3 \cdot 4 + 1]$. ◆

Corollary 7.12.3 If $m = p^2$, where $p > 3$ is a prime, then $\phi(n) = 2m$ has no solutions.

Proof The prime p must be of the form $3k + 1$ or $3k + 2$. If the former is the case, then $3 \mid (p - 1)$, and if the latter is the case, then $3 \mid (p + 1)$. Consequently, $2p^2 + 1 = 2(p - 1)(p + 1) + 3$ is divisible by 3, and so for the divisor $d = p^2$ of m, $2d + 1$ is composite.

The only other divisor of m larger than 1 is p itself. If $2p + 1$ is also composite, then Corollary 7.12.1 says that $\phi(n) = 2p^2$ has no solutions.

But if $2p + 1$ is prime, then $\phi(n) = 2p^2$ might have a solution. From Theorem 7.12, such a solution corresponds to $m = p(2p + 1)^k$, $k = 0$ or $k \in \mathbf{N}$. The equality $p^2 = p(2p + 1)^k$ holds only if $p = 1$, $k = 0$, and this contradicts the hypothesis that $p > 3$. It follows that there can be no solutions at all to $\phi(n) = 2p^2$. ◆

Problems

7.58. There are six solutions to $\phi(n) = 16$. Find them in a systematic manner (not by a hunt and search procedure). Explain your reasoning.

7.59. Find all solutions of $\phi(n) = x$ for the following choices of x. If no solutions exist, indicate so and tell why.

(a) $x = 162$.
(b) $x = 266$.
(c) $x = 146$.
(d) $x = 242$.
(e) $x = 102$.
(f) $x = 118$.

7.60. If x is given and n is an odd solution to $\phi(n) = x$, must there exist another solution to this equation?

7.61. How many distinct solutions are there to these equations?

(a) $\phi(n) = 1042$.
(b) $\phi(n) = 1870$.

7.62. Let the symbol $\phi^2(n)$ denote $\phi[\phi(n)]$. Find all solutions *of the form* $n = p^\alpha$, where p is a prime, of $\phi^2(n) = 4$. Are there solutions of any other form as well?

7.63. Does Corollary 7.12.3 still hold if p^2 there is replaced by p^3?

7.64. Find all n such that $\phi(pn) = p\phi(n)$, where p is an arbitrary fixed prime.

7.65. The following results were pointed out in (Moser, 1949).

(a) Prove that $\phi(n) = \phi(n + 2)$ is satisfied by $n = 2(2p - 1)$ if both p and $2p - 1$ are odd primes. Illustrate this with three numerical cases.
(b) Prove that $\phi(n) = \phi(n + 2)$ is satisfied by $n = 2^{2^a+1}$ if $2^{2^a} + 1$ is a Fermat prime. Illustrate this with the first four Fermat primes.

7.66. Suppose, in contrast to Carmichael's Conjecture, that there is some x such that $\phi(n) = x$ has a unique solution n_0. Let us show then that $1764 \mid n_0$ (Donnelly, 1973).

(a) Show that $2 \nmid n_0$ is false.
(b) Show that $2 \mid n_0$ but $2^2 \nmid n_0$ is false by looking at $\phi(2^{-1}n_0)$ and $\phi(n_0)$. Thus, $2^2 \mid n_0$.
(c) Show that $3 \nmid n_0$ is false by looking at $\phi(2^{-1} \cdot 3n_0)$ and $\phi(n_0)$.
(d) Show that $3 \mid n_0$ but $3^2 \nmid n_0$ is false by looking at $\phi(2^1 \cdot 3^{-1}n_0)$ and $\phi(n_0)$. Thus, $3^2 \mid n_0$, so $36 \mid n_0$.
(e) Show that $7 \nmid n_0$ is false by looking at $\phi(2^{-1} \cdot 3^{-1} \cdot 7n_0)$ and $\phi(n_0)$.
(f) Show that $7 \mid n_0$ but $7^2 \nmid n_0$ is false by looking at $\phi(2^1 \cdot 3^1 \cdot 7^{-1}n_0)$ and $\phi(n_0)$. Thus, $7^2 \mid n_0$, so $36(49) \mid n_0$.

7.67. In $\phi(n) = 2m$, m = odd, let $m = p_1 p_2$, where p_1, p_2 are distinct $(3k + 2)$-primes.

(a) Show that n cannot be prime.
(b) If, further, $p_2 > p_1$ and $p_2 \neq 2p_1 + 1$, then show that there is no solution of the form $n = p^a$, $a > 1$ (Theorem 7.11).

7.68. In $\phi(n) = 2m$, m = odd, let $m = p_1 p_2^2$, where p_1 is a $(3k + 1)$-prime and p_2 is a $(3k + 2)$-prime.

(a) Show that n cannot be prime.
(b) If, further, the condition $p_2 \neq [(p_1 - 1)/2]^{1/2}$ holds, then show that there is no solution of the form $n = p^a$, $a > 1$.

7.69. In $\phi(n) = 2m$, m = odd, let $m = p_1 p_2 p_2'$, where p_1 is a $(3k + 1)$-prime and p_2, p_2' are distinct $(3k + 2)$-primes.

(a) Show that n cannot be prime.
(b) Assume without loss of generality that $p_2 > p_2'$. If the condition $p_2 \neq 1 + 2p_1 p_2'$ holds, then show that there is no solution of the form $n = p^a$, $a > 1$.

7.70. Examination of tables of Euler's totient arranged according to increasing values of x, where $\phi(n) = x$, shows that certain even values of x are missing. In particular, in the interval [1, 1200] all but two of the numbers $x = 12k + 2$, $k = 1, 2, \ldots, 99$, are missing. Write a computer program that will identify the two numbers x for which $\phi(n) = x$ has solutions.

7.71. In connection with the previous problem, each of the remaining 97 values of x is absent because it meets one of the following five criteria (labeled as A, B, C, D, E):

A Corollary 7.12.2
B Corollary 7.12.3
C Problem 7.67
D Problem 7.68
E Problem 7.69.

For the 97 cases determine in each instance which is the applicable criterion. Which criterion predominates?

7.72. The phenomenon discussed in the two previous problems has been accounted for in a comprehensive manner (Dence and Dence, 1995). Consult the indicated reference, digest its contents, and write a short report on the results.

7.6 The Möbius Inversion Formula

Look again at Problem 7.28; if you worked through this, your conjecture should have been that

$$\sum_{d|n} \epsilon\left(\frac{n}{d}\right) \mu(d) = \phi(n).$$

We can show this in the following interesting way.

In the interval $[1, n]$ there are n integers. The set of integers coprime to n are all of those integers in [1, n] that are *not* multiples of the prime divisors of n. For each prime divisor p_i of n there are n/p_i such multiples; for example, in [1, 20] there are $20/5 = 4$ multiples of 5, namely, 5, 10, 15, 20. Hence, it would seem that

$$\phi(n) = n - \sum_i \left(\frac{n}{p_i}\right).$$

But this counts some multiples twice, namely, the multiples of $p_i\, p_j$, where $p_i \neq p_j$. So we need to add back the number of that kind of multiple:

$$\phi(n) = n - \sum_i \left(\frac{n}{p_i}\right) + \sum_{i<j} \left(\frac{n}{p_i\, p_j}\right).$$

This time we have added back too much because multiples involving three distinct primes have been counted twice. You can see the pattern now. If n contains r distinct prime divisors, then

$$\phi(n) = n\left[1 + \sum_{i=1}^{r} \frac{(-1)^1}{p_i} + \sum_{i<j} \frac{(-1)^2}{p_i p_j} + \sum_{i<j<k} \frac{(-1)^3}{p_i p_j p_k} + \cdots + \frac{(-1)^r}{\prod_{i=1}^{r} p_i}\right].$$

Let's try out this formula on a specific case.

Example 7.14 Compute $\phi(30)$.

We have $30 = 2 \cdot 3 \cdot 5$, so $r = 3$. The formula then gives

$$\begin{aligned}\phi(30) &= 30\left[1 + \left\{\frac{-1}{2} + \frac{-1}{3} + \frac{-1}{5}\right\} + \left\{\frac{1}{6} + \frac{1}{10} + \frac{1}{15}\right\} - \frac{1}{30}\right] \\ &= 30 - [15 + 10 + 6] + [5 + 3 + 2] - 1 \\ &= 8.\end{aligned}$$

As a check, we find $\phi(30) = \phi(2)\phi(3)\phi(5) = 1 \cdot 2 \cdot 4 = 8$. ◆

The formula just deduced for $\phi(n)$ is a special case of a very general statement in combinatorial mathematics known as the **Inclusion–Exclusion Principle**.

INCLUSION–EXCLUSION PRINCIPLE

If there are n objects, of which n_a have the property a, n_b have the property $b, \ldots, n_{ab}$ have both a and b, n_{abc} have $a, b, c, \ldots$, and so on, then the number of objects that have none of the properties $a, b, c, \ldots$, is given by

$$n - n_a - n_b - \cdots + n_{ab} + \cdots - n_{abc} - \cdots.$$

Noting the pattern of the numerators in the preceding formula for $\phi(n)$ and the fact that the denominators are squarefree, we see immediately from the definition of the Möbius function that

$$\phi(n) = n \sum_{d|n} \frac{\mu(d)}{d}.$$

But n/d is the value of $\epsilon(n/d)$, so the desired conjecture is verified.

It is not clear that anything more than just an unusual relationship has been obtained. However, look also at Problem 7.35(b) again. That problem (see also Problem 7.77) and Problem 7.28 together yield the pair of equations

$$\begin{cases} \epsilon(n) = \sum_{d|n} \phi(d) & \text{(A)} \\ \phi(n) = \sum_{d|n} \epsilon\left(\frac{n}{d}\right) \mu(d) & \text{(B)} \end{cases}$$

It looks like the Möbius function in some way inverts the relation between ϵ and ϕ in the first equation. A numerical example will help to illustrate this.

Example 7.15 Let $n = 6$; its divisors are 1, 2, 3, 6 and the corresponding values of Euler's totient are $\phi(1) = 1$, $\phi(2) = 1$, $\phi(3) = 2$, $\phi(6) = 2$. Hence, we have for equation (A),

$$\epsilon(6) = 1 + 1 + 2 + 2 = 6.$$

The values of the Möbius function are $\mu(1) = 1$, $\mu(2) = -1$, $\mu(3) = -1$, $\mu(6) = 1$. Hence, equation (B) is

$$\begin{aligned} \phi(6) &= [\epsilon(6)](1) + [\epsilon(3)](-1) + [\epsilon(2)](-1) + [\epsilon(1)](1) \\ &= 6(1) + 3(-1) + 2(-1) + 1(1) \\ &= 6 - 3 - 2 + 1 \\ &= 2. \quad \blacklozenge \end{aligned}$$

This seems like an amazing coincidence, but **Möbius** himself discovered in 1832 the general relationship.

Theorem 7.13 (The Möbius Inversion Formula). If f(n), g(n) are arithmetic functions and

$$\mathrm{g}(n) = \sum_{d|n} \mathrm{f}(d),$$

$$\text{then } \mathrm{f}(n) = \sum_{d|n} \mathrm{g}\left(\frac{n}{d}\right)\mu(d).$$

Proof We have directly from the hypothesis

$$\sum_{d|n} \mathrm{g}\left(\frac{n}{d}\right)\mu(d) = \sum_{d|n}\left[\sum_{d_i|(n/d)} \mathrm{f}(d_i)\right]\mu(d)$$

$$= \sum_{d_i|(n/d)} \mathrm{f}(d_i) \sum_{d|n} \mu(d)$$

upon reversing the order of the (finite) summations. Now observe that

$$\left\{d \mid n \text{ and } d_i \mid \left(\frac{n}{d}\right)\right\} \leftrightarrow dd_i \mid n \leftrightarrow \left\{d_i \mid n \text{ and } d \mid \left(\frac{n}{d_i}\right)\right\}.$$

This says that letting d_i range through each of the sets of divisors of n/d as d takes on the values of the divisors of n is the same as letting d_i range through the set of divisors of n as d takes on the values in each of the sets of divisors of the n/d_i's. That is, we can switch d, d_i in the indices of the summation signs. Consequently,

$$\sum_{d|n} \mathrm{g}\left(\frac{n}{d}\right)\mu(d) = \sum_{d_i|n} \mathrm{f}(d_i) \sum_{d|(n/d_i)} \mu(d).$$

By Corollary 7.3.1, the last summation is 0 except when $n/d_i = 1$, whereupon $\Sigma_{d_i|n}$ f(d_i) reduces to just a single term, namely, f(n). The final result is

$$\sum_{d|n} \mathrm{g}\left(\frac{n}{d}\right)\mu(d) = \mathrm{f}(n) \cdot 1. \quad \blacklozenge$$

Note that nowhere in the proof have we used the fact that f(n), g(n) are multiplicative. The theorem holds for arithmetic functions that are not multiplicative.

Example 7.16 Let f(n) denote the number of distinct prime divisors in n; consider $n = 6$. Then we have f(1) = 0, f(3) = 1, f(6) = 2. It follows that

$$\mathrm{g}(6) = \sum_{d|6} \mathrm{f}(d) = 0 + 1 + 1 + 2 = 4.$$

Additionally, we compute $g(3) = \Sigma_{d|3} f(d) = 0 + 1 = 1$, $g(2) = 1$, and $g(1) = 0$. Finally, we have

$$\begin{aligned}\sum_{d|6} g\left(\frac{6}{d}\right)\mu(d) &= 4 \cdot 1 + 1 \cdot (-1) + 1 \cdot (-1) + 0 \cdot 1 \\ &= 2 \\ &= f(6). \quad \blacklozenge\end{aligned}$$

Notice that the arithmetic function $f(n)$ in the preceding example is not multiplicative (for example, $f(2) \cdot f(3) \neq f(2 \cdot 3)$), and yet Theorem 7.13 is upheld. The general arithmetic nature of $f(n)$ in the statement of Theorem 7.13 allows us to prove the converse of Theorem 7.2.

Theorem 7.14 If $F(n) = \Sigma_{d|n} f(d)$ is a multiplicative function, then so is $f(n)$.

Proof Even though $f(n)$ is not known to be multiplicative *a priori*, we can write from Theorem 7.13

$$f(mn) = \sum_{\substack{d_i|m \\ d_j|n}} F\left(\frac{mn}{d_i d_j}\right) \mu(d_i d_j),$$

where $d_i d_j$ is any divisor of mn. Make the redefinitions $D_i = m/d_i$, $D_j = n/d_j$, so that

$$f(mn) = \sum_{\substack{D_i|m \\ D_j|n}} F(D_i D_j)\, \mu\left(\frac{mn}{D_i D_j}\right).$$

Now assume that $(m, n) = 1$; as F and μ are both multiplicative, then

$$\begin{aligned}f(mn) &= \sum_{\substack{D_i|m \\ D_j|n}} F(D_i)\, F(D_j)\, \mu\left(\frac{m}{D_i}\right) \mu\left(\frac{n}{D_j}\right) \\ &= \sum_{D_i|m} F(D_i)\, \mu\left(\frac{m}{D_i}\right) \sum_{D_j|n} F(D_j)\, \mu\left(\frac{n}{D_j}\right),\end{aligned}$$

upon grouping terms of like variables. But now Theorem 7.13 applies to each summation on the right-hand side. The result is

$$f(mn) = f(m)\, f(n),$$

and this says that $f(n)$ is indeed multiplicative. $\blacklozenge$

The Möbius function is really an unusual function. In fact, it is not too hard to show that if an arithmetic function $\mu^*(n)$ has the property described in Theorem 7.13, then $\mu^*(n)$ coincides with the Möbius function $\mu(n)$ (Satyanarayana, 1963). Also, we can use Theorem 7.13 to derive independently the formula for $\phi(p^\alpha)$ given in Section 7.1.3 (see Example 7.17, which follows).

In more recent times other inversion formulas besides that of Möbius have been developed. One, applied to functions of a continuous variable, was found useful in solving certain kinds of physics problems (Chen, 1990).

Example 7.17 The result of Problem 7.35(b), namely, $\Sigma_{d|n}\ \phi(d) = n$, can be rederived there without initially knowing the formulas for the $\phi(p_i^{\alpha_i})$'s. In Theorem 7.13 identify f(n) with $\phi(n)$ and g(n) with n. Then the theorem gives

$$\phi(p^\alpha) = \sum_{d|p^\alpha} \left(\frac{p^\alpha}{d}\right) \mu(d).$$

Since $\{d\} = \{1, p, p^2, \ldots, p^\alpha\}$ and $\mu(d)$ is nonzero only for $d = 1, p$, we obtain

$$\begin{aligned}\phi(p^\alpha) &= \left(\frac{p^\alpha}{1}\right) \mu(1) + \left(\frac{p^\alpha}{p}\right) \mu(p)\\ &= p^\alpha - p^{\alpha-1}.\end{aligned}$$ ◆

Problems

7.73. Use the Inclusion–Exclusion Principle to calculate

(a) $\phi(75)$. (b) $\phi(180)$. (c) $\phi(1155)$.

7.74. Use the Inclusion–Exclusion Principle to calculate

(a) The number of integers in [1, 100] that are simultaneously not multiples of 3 and not congruent to 2 modulo 4.

(b) The number of integers in [1, 200] that are simultaneously not squares and are not coprime to 200.

7.75. In Theorem 7.13 let f(n) be the **von Mangoldt function**, $\Lambda(n)$,

$$\Lambda(n) = \begin{cases} \ln p & \text{if } n = p^k \text{ for some } k \in \mathbf{N} \text{ and } p = \text{prime} \\ 0 & \text{otherwise.} \end{cases}$$

(a) Is $\Lambda(n)$ multiplicative?
(b) Compute g(1), g(3), g(9), g(27), where $g(n) = \Sigma_{d|n} \Lambda(d)$.
(c) Apply the Möbius Inversion Formula to the equation in (b) with $n = 27$.

7.76. In Theorem 7.13 let f(n) be $\sigma_p(n)$, the sum of the distinct prime divisors of n. For example, $\sigma_p(12) = 5$.

(a) Is $\sigma_p(n)$ multiplicative?
(b) Compute g(1), g(2), g(3), g(4), g(6), g(8), g(12), g(24), where $g(n) = \Sigma_{d|n} \sigma_p(d)$.
(c) Apply the Möbius Inversion Formula to the equation in (b) with $n = 24$.

7.77. Let n be a positive integer. Define the sets $\mathbf{C}_k$ by

$$\mathbf{C}_k = \{x : 1 \leq x \leq n, (x, n) = k\}.$$

(a) What choices of k yield nonempty $\mathbf{C}_k$?
(b) Prove that the $\mathbf{C}_k$'s are disjoint and that their union is all of $\mathbf{S} = \{1, 2, 3, \ldots, n\}$.
(c) In view of (b), how does this lead to the result of Problem 7.35(b)?

7.78. Prove the converse of the Möbius Inversion Theorem: If $f(n) = \Sigma_{d|n} g(n/d)\, \mu(d)$, then $g(n) = \Sigma_{d|n} f(d)$. [Hint: Start with $_{d|n}$ f(d)].

7.79. Refer to Problem 7.75.

(a) Then prove that for any $n \in \mathbf{N}$, one has $\ln n = \Sigma_{d|n} \Lambda(d)$, where the summation is over all divisors of n.
(b) Hence, show that

$$\Lambda(n) = -\sum_{d|n} \mu(d) \ln(d).$$

7.7 Convolution

In the Möbius Inversion Formula,

$$f(n) = \sum_{d|n} g\left(\frac{n}{d}\right) \mu(d)$$

the sum on the right involves a composition of the two functions g, μ in a manner that occurs frequently in number theory. Following a viewpoint advocated in 1915 by the Scottish-born American mathematician **Eric Temple Bell** (1883–1960), we regard this composition as a kind of multiplication.

DEF. If f(n), g(n) are arithmetic functions, the function h(n) defined by

$$h(n) = \sum_{d|n} f\left(\frac{n}{d}\right) g(d),$$

is called the **convolution** (or **Dirichlet product**[3]) of f, g. We indicate convolution by the notation $h(n) = (f * g)(n)$.

If (*) is to be regarded as a kind of multiplication, then we want to investigate what properties it has in common with ordinary multiplication.

Theorem 7.15 If f, g, h are arithmetic functions, then

(a) $f * g = g * f$.
(b) $(f * g) * h = f * (g * h)$.
(c) $f * (g + h) = (f * g) + (f * h)$.

Proof We prove part (b) here. The left-hand side is

$$[(f * g) * h](n) = \sum_{c|(n/d)} f\left(\frac{n}{dc}\right) g(c) \sum_{d|n} h(d).$$

Set $dc = D$; then

$$[(f * g) * h](n) = \sum_{D|n} f\left(\frac{n}{D}\right) \sum_{c|D} g(c)\, h\left(\frac{D}{c}\right)$$

$$= \sum_{D|n} f\left(\frac{n}{D}\right) \sum_{c|D} g\left(\frac{D}{c}\right) h(c)$$

from part (a). But the right-hand side is now $[f * (g * h)](n)$, so this proves part (b). ◆

[3] The term *Dirichlet product* is more precise since there are several different convolutions that are defined in number theory. In ordinary prose, the word *convolution* means a rolled up or coiled condition. In the present context, the sum $\Sigma_{d|n} f(n/d)g(d)$ is evaluated by letting the arguments of g(d) increase through all the divisors of n, while at the same time this same sequence of divisors is "folded back" or "rolled up" to form the decreasing sequence of arguments for f(n/d).

DEF. The **arithmetical unity function, I**(n), is the arithmetic function

$$\mathbf{I}(n) = \begin{cases} 1 & \text{if } n = 1 \\ 0 & \text{otherwise.} \end{cases}$$

The arithmetical unity function, not to be confused with u(n) in Section 7.1, takes its name from the following simple property.

Theorem 7.16 If f(n) is any arithmetic function, then $\mathbf{I} * \mathrm{f} = \mathrm{f} * \mathbf{I} = \mathrm{f}$.

Proof The proof is left to the reader. ◆

Note that Theorems 7.15 and 7.16 do not require the arithmetic functions to be multiplicative. However, being multiplicative is a strong feature of an arithmetic function, and we might expect this to have some effect on the convolution of two such functions.

Theorem 7.17 If f(n) and g(n) are multiplicative, then so is their convolution.

Proof Assume that $(m, n) = 1$ and let $\mathrm{h} = \mathrm{f} * \mathrm{g}$. Then we have

$$\mathrm{h}(mn) = \sum_{d \mid mn} \mathrm{f}\left(\frac{mn}{d}\right) \mathrm{g}(d).$$

From the Fundamental Theorem of Arithmetic it follows that every divisor d can be written as $d = bc$, where $(b, c) = 1$, $b \mid m$, and $c \mid n$. Hence, the equation can be rewritten as

$$\begin{aligned} \mathrm{h}(mn) &= \sum_{\substack{b \mid m \\ c \mid n}} \mathrm{f}\left(\frac{mn}{bc}\right) \mathrm{g}(bc) \\ &= \sum_{b \mid m} \mathrm{f}\left(\frac{m}{b}\right) \mathrm{g}(b) \sum_{c \mid n} \mathrm{f}\left(\frac{n}{c}\right) \mathrm{g}(c) \end{aligned}$$

because f, g are multiplicative and $(m/b, n/c) = 1$. But the right-hand side is just h(m)h(n), so we obtain $\mathrm{h}(mn) = \mathrm{h}(m)\mathrm{h}(n)$. ◆

Example 7.18 Let $\mathrm{f}(n) = \phi(n)$ and $\mathrm{g}(n) = \sigma(n)$. Choose arbitrarily the coprime integers $m = 6$, $n = 35$. We compute

$$\begin{aligned} h(6) = (\phi * \sigma)(6) &= \sum_{d|6} \phi\left(\frac{6}{d}\right) \sigma(d) \\ &= 2(1) + 2(3) + 1(4) + 1(12) \\ &= 24 \end{aligned}$$

$$\begin{aligned} h(35) = (\phi * \sigma)(35) &= \sum_{d|35} \phi\left(\frac{35}{d}\right) \sigma(d) \\ &= 24(1) + 6(6) + 4(8) + 1(48) \\ &= 140 \end{aligned}$$

$$\begin{aligned} h(210) = (\phi * \sigma)(210) &= \sum_{d|210} \phi\left(\frac{210}{d}\right) \sigma(d) \\ &= 48(1) + 48(3) + 24(4) + 12(6) \\ &\quad + 24(12) + 8(8) + 12(18) + 8(24) \\ &\quad + 6(24) + 4(32) + 6(72) + 2(48) \\ &\quad + 4(96) + 2(144) + 1(192) + 1(576) \\ &= 3360. \end{aligned}$$

Finally, $h(6)h(35) = 24(140) = 3360 = h(6 \cdot 35)$. ◆

It is easy to see that $\mathbf{I}(n)$ is itself a multiplicative function. Since Theorem 7.17 has just shown that f * g is multiplicative if f, g are multiplicative, the next logical step is to investigate if every multiplicative function f has an **inverse with respect to (*)**. That is, does there correspond to f a multiplicative function f^{-1} such that

$$(f * f^{-1})(n) = \mathbf{I}(n)$$

holds for all $n \in \mathbf{N}$?

Theorem 7.18 Any multiplicative function f has an inverse with respect to convolution.

Proof If f is to have an inverse, then at $n = 1$ one must have

$$\sum_{d|1} f\left(\frac{1}{d}\right) f^{-1}(d) = \mathbf{I}(1) = 1$$

or $f(1)f^{-1}(1) = 1$. By Theorem 7.1, $f(1) = 1$ and so $f^{-1}(1) = 1$.

We now proceed inductively. Assume that $f^{-1}(k)$ has been defined for all integers k satisfying $1 \leq k \leq n - 1$. Then at n we require

$$\sum_{d|n} f\left(\frac{n}{d}\right) f^{-1}(d) = \mathbf{I}(n) = 0.$$

This equation is solvable for $f^{-1}(n)$ because $f(1) \neq 0$. We obtain

$$\sum_{d<n} f\left(\frac{n}{d}\right) f^{-1}(d) + f(1)f^{-1}(n) = 0,$$

whereupon since $f(1) = 1$,

$$f^{-1}(n) = -\sum_{d<n} f\left(\frac{n}{d}\right) f^{-1}(d).$$

By the Principle of Finite Induction, $f^{-1}(n)$ is then defined for all $n \in \mathbf{N}$. ◆

Example 7.19 Compute $\mu^{-1}(4)$.

Since $\mu^{-1}(1) = 1$, then from the preceding we have

$$\begin{aligned}\mu^{-1}(2) &= -\sum_{d<2} \mu\left(\frac{2}{d}\right) \mu^{-1}(d)\\ &= -\mu(2)\,\mu^{-1}(1)\\ &= -(-1)(1)\\ &= 1\end{aligned}$$

and

$$\begin{aligned}\mu^{-1}(4) &= -\mu\left(\frac{4}{1}\right) \mu^{-1}(1) - \mu\left(\frac{4}{2}\right) \mu^{-1}(2)\\ &= -(0)(1) - (-1)(1)\\ &= 1.\end{aligned}$$ ◆

It is clear that if f is multiplicative, then $f^{-1}(1) = 1$ determines a unique value for $f^{-1}(2)$. In turn, these determine unique values for $f^{-1}(3), f^{-1}(4), \ldots$, so the inverse function $f^{-1}(n)$ is unique. We can also prove that if f is multiplicative, then so is f^{-1} (Section 7.8).

By the way, can you see that with this section we have led you back to algebra?

Example 7.20 A formula in the literature (Rabinowitz, Kennedy, and Cooper, 1997) reads

$$\sum_{\substack{k=1 \\ (k,n)=1}}^{n-1} k^3 = \frac{n^3}{4}\phi(n) + \frac{n^2}{4}\phi^{-1}(n),$$

where ϕ^{-1} is the Dirichlet inverse of ϕ. Verify this formula for $n = 15$.

In view of remarks just made about the multiplicativity of f^{-1} when f is multiplicative, we have

$$\phi^{-1}(15) = \phi^{-1}(3) \cdot \phi^{-1}(5).$$

From Theorem 7.18 we obtain

$$\phi^{-1}(3) = -\sum_{\substack{d|3 \\ d<3}} \phi\left(\frac{3}{d}\right)\phi^{-1}(d) = -\phi(3)\cdot\phi^{-1}(1) = -2$$

$$\phi^{-1}(5) = -\phi(5)\cdot\phi^{-1}(1) = -4,$$

so $\phi^{-1}(15) = 8$. Hence, we compute

$$\sum_{\substack{k=1 \\ (k,15)=1}}^{14} k^3 = \frac{15^3}{4}\phi(15) + \frac{15^2}{4}\phi^{-1}(15),$$

$$= \frac{3375}{4}(8) + \frac{225}{4}(8)$$

$$= 7200.$$

Indeed, $1^3 + 2^3 + 4^3 + 7^3 + 8^3 + 11^3 + 13^3 + 14^3 = 7200$. ◆

Problems

7.80. Prove parts (a) and (c) of Theorem 7.15.

7.81. Prove Theorem 7.16.

7.82. Compute $\mu^{-1}(n)$ for $n = 1, 2, 3, \ldots, 10$. Make a conjecture about the nature of μ^{-1}.

7.83. Compute the following convolutions, $(f * g)(n)$:

(a) $f(x) = \phi(x)$, $g(x) = \phi(x)$, $n = 24$.
(b) $f(x) = \mu(x)$, $g(x) = \mu(x)$, $n = 20$.
(c) $f(x) = \mu(x)$, $g(x) = \epsilon(x)$, $n = 50$.
(d) $f(x) = \mu(x)$, $g(x) = \lambda(x)$, $n = 35$.

7.84. Compute:

(a) $\phi^{-1}(6)$. (b) $\tau^{-1}(10)$. (c) $\lambda^{-1}(20)$.

7.85. Corollary 7.3.1 says that $\Sigma_{d|n}\, \mu(d) = \mathbf{I}(n)$, where $\mathbf{I}(n)$ is the arithmetical unity function. Now suppose that μ^* is some other arithmetic function such that

$$\sum_{d|n} \mu^*(d) = \mathbf{I}(n).$$

(a) Show that $\mu^*(\text{n})$ is multiplicative.
(b) Show that for any prime p one has $\mu^*(p) = -1$, $\mu^*(p^2) = 0$.
(c) Show that for any $\alpha \geq 2$ and any prime p one has $\mu^*(p^\alpha) = 0$.
(d) Then, is $\mu^*(n)$ identical to $\mu(n)$?

7.86. Let $f(n)$ be a completely multiplicative function (see Problem 7.14). Prove that $f^{-1}(n) = \mu(n)f(n)$. [Hint: Work directly with $(\mu f * f)(\text{n})$]. Tabulate values of $\lambda^{-1}(n)$ for $n = 1, 2, 3, \ldots, 10$, where $\lambda(n)$ is Liouville's function. Also, verify that $(\lambda * \lambda^{-1})(6) = \mathbf{I}(6)$.

7.87. This problem is a follow-up to the previous one. Let $f(n)$ be multiplicative.

(a) Prove that $f(n)$ is completely multiplicative iff for any prime p and any $\alpha \in \mathbf{N}$ one has $f(p^\alpha) = [f(p)]^\alpha$.
(b) Use part (a) to prove the converse of Problem 7.86: If $f(n)$ is multiplicative and $f^{-1}(n) = \mu(n)f(n)$, then $f(n)$ is completely multiplicative.

7.88. Another necessary and sufficient condition for a multiplicative function to be completely multiplicative is the following:

(a) Let $f(n)$ be completely multiplicative and let $g(n)$, $h(n)$ be arithmetic functions. Show that

$$f(n)(g * h)(n) = (fg * fh)(n).$$

(b) Conversely, suppose $f(n)$ is multiplicative and that for any arithmetic functions $g(n)$, $h(n)$ the previous equation holds. Show that $f(n)$ must then be completely multiplicative. [Hint: Specialize to $g(n) = \mu(n)$ and $h(n) = u(n)$, and make use of Problem 7.87(b).]

7.89. Refer to Problem 7.56 for the definition of a unitary divisor of an integer n. Let $\tau_u(n)$ be the arithmetic function that denotes the number of unitary divisors possessed by the integer n; for example, $\tau_u(4) = 2$.

(a) If $n = p_1^{\alpha_1} p_2^{\alpha_2} \ldots p_r^{\alpha_r}$, what is $\tau_u(n)$?
(b) Is $\tau_u(n)$ multiplicative?
(c) Can there be an n such that $\tau_u(n) = 6$? Can there be an $n < 200$ such that $\tau_u(n) = 16$?
(d) A **strictly highly unitary number** is a positive integer n such that $\tau_u(n) > \tau_u(k)$ for all k satisfying $1 \le k < n$. What are the first five strictly highly unitary numbers?
(e) We use the symbol $d \| n$ to indicate that d is a unitary divisor of n. The **unitary convolution** of two arithmetic functions f(n), g(n) is defined to be (Jager, 1961)

$$(\mathrm{f} \otimes \mathrm{g})(n) = \sum_{d \| n} \mathrm{f}\left(\frac{n}{d}\right) \mathrm{g}(d).$$

Is $\otimes$ commutative? Let $\mathrm{f}(n) = \tau_u(n)$ and $\mathrm{g}(n) = \phi(n)$, and compute $(\mathrm{f} \otimes \mathrm{g})(24)$.

7.90. If $n = \Pi_{i=1}^{r} p_i^{\alpha_i}$, develop a formula for $[\mu * (\mu * \mu)](n)$.

7.91. **D.H. Lehmer** (Lehmer, 1931) discussed a kind of convolution based on the least common multiple:

$$(\mathrm{f} ** \mathrm{g})(n) = \sum_{[d,d']=n} \mathrm{f}(d)\ \mathrm{g}(d'),$$

where the summation is taken over all pairs of positive integers d, d' such that their least common multiple is n.

(a) Evaluate $(\phi ** \phi)(n)$ for $n = 1$ through 12.
(b) Lehmer proved that if f(n), g(n) are multiplicative, then so is (f ** g)(n). Illustrate this with data from part (a).
(c) Find a formula for (f ** g)(p^a) for general f(n), g(n), if p^a is a power of a prime.

7.8 The Algebraic Structure of Arithmetic Functions

It would be surprising if there were not any underlying algebraic structure among the arithmetic functions (Cashwell and Everett, 1959; Carlitz, 1964; Wall, 1974).

In fact, the elements of structure are already in place. Let **A** be the set of all arithmetic functions, and let f, g, h $\in$ **A**. We can define addition pointwise:

$$(\mathrm{f} + \mathrm{g})(n) = \mathrm{f}(n) + \mathrm{g}(n).$$

Addition is clearly a binary operation on **A** and is associative and commutative. Any f $\in$ **A** has an additive inverse $(-\mathrm{f})(n) = -\mathrm{f}(n)$, and the zero function, $\theta(n) = 0$ for all n $\in$ **N**, is the additive identity. Hence, $\langle \mathbf{A}, + \rangle$ is an Abelian group.

Using the Dirichlet product (*) as a second binary operation on **A**, we have from Theorem 7.15 that (*) is associative and obeys the two distributive laws

$$\mathrm{f} * (\mathrm{g} + \mathrm{h}) = \mathrm{f} * \mathrm{g} + \mathrm{f} * \mathrm{h}$$

$$(\mathrm{f} + \mathrm{g}) * \mathrm{h} = \mathrm{f} * \mathrm{h} + \mathrm{g} * \mathrm{h}\ .$$

Hence, **A** is a commutative ring with identity (Theorem 7.16).

The question now is, what are the units of **A**? Theorem 7.18 looks like a result that points in the direction of the answer, but that theorem is not quite general enough. Reexamination of the proof of Theorem 7.18 shows that the analogous steps carry through if we replace "f(1) $=$ 1" by "f(1) $\neq$ 0." Thus, we can state

Theorem 7.19 Any arithmetic function f such that f(1) $\neq$ 0 has an inverse with respect to convolution.

Let the set of arithmetic functions in Theorem 7.19 be denoted $\mathbf{U}_A$. The elements of $\mathbf{U}_A$ are then the units of **A**. We have immediately the result in Theorem 7.20.

Theorem 7.20 The units of the ring **A** of arithmetic functions comprise an Abelian group under the group operation of convolution.

The group $\mathbf{U}_A$ includes the multiplicative functions because $\mathrm{f}(1) = 1 \neq 0$. Let the set of multiplicative functions be denoted by $\mathbf{A}_m$; trivially, $\mathbf{I}(n) \in \mathbf{A}_m$. We already know from Theorem 7.17 that if f, g $\in \mathbf{A}_m$, then so is f $*$ g, so convolution is a binary operation on $\mathbf{A}_m$. Consequently, $\mathbf{A}_m$ inherits associativity from $\mathbf{U}_A$. You can see where we are headed. We need only prove that if $\mathrm{f} \in \mathbf{A}_m$, then $\mathrm{f}^{-1} \in \mathbf{A}_m$, in order to establish that $\mathbf{A}_m$ is a subgroup of $\mathbf{U}_A$ (Fig. 7.6). Our proof is taken from (Berberian, 1992).

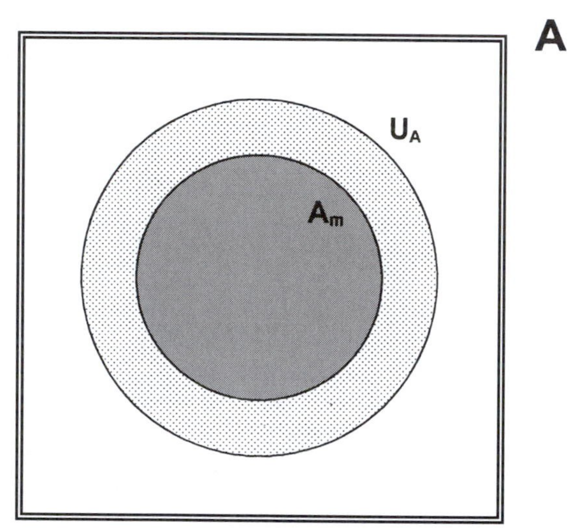

Figure 7.6 The group $\mathbf{A}_m$ of multiplicative functions is a subgroup of the group $\mathbf{U}_\mathrm{A}$ of units of the ring $\mathbf{A}$ of arithmetic functions.

Theorem 7.21 If f is multiplicative, then so is f^{-1}.

Proof Given $\mathrm{f} \in \mathbf{A}_m$, the idea is to concoct an arithmetic function g that is obviously multiplicative, and then show that $\mathrm{g} = \mathrm{f}^{-1}$. A clue as to how g might be constructed is given by Problem 2.38. The defined function is

$$\mathrm{g}(n) = \begin{cases} 1 & \text{if } n = 1 \\ \prod_{i=1}^{r} \mathrm{f}^{-1}(p_i^{\alpha_i}) & \text{if } n = \prod_{i=1}^{r} p_i^{\alpha_i}. \end{cases}$$

Now suppose that $(m, n) = 1, m, n > 1$, where $m = \Pi_{j=1}^{s} q_j^{\beta_j}$ and no prime p_i is identical to any prime q_j. Then $mn = \Pi_{i=1}^{r} \Pi_{j=1}^{s} p_i^{\alpha_i} q_j^{\beta_j}$, and from the definition of g we have

$$\begin{aligned} \mathrm{g}(mn) &= \prod_{i=1}^{r} \prod_{j=1}^{s} \mathrm{f}^{-1}(p_i^{\alpha_i})\, \mathrm{f}^{-1}(q_j^{\beta_j}) \\ &= \mathrm{g}(m)\, \mathrm{g}(n). \end{aligned}$$

This shows that g is multiplicative, that is, $\mathrm{g} \in \mathbf{A}_m$. By Theorem 7.17, $\mathrm{f} * \mathrm{g} \in \mathbf{A}_m$.

It remains to show that $(f * g)(n) = \mathbf{I}(n)$. First, $(f * g)(1) = f(1)g(1) = 1 = \mathbf{I}(1)$. Next, let $p_k^{\alpha_k}$ appear in the factorization of n. Then we obtain

$$\begin{aligned}(f * g)(p_k^{\alpha_k}) &= \sum_{i=0}^{\alpha_k} f(p_k^{\alpha_k - i})g(p_k^i) \\ &= \sum_{i=0}^{\alpha_k} f(p_k^{\alpha_k - i})\, f^{-1}(p_k^i) \\ &= (f * f^{-1})(p_k^{\alpha_k}) \\ &= \mathbf{I}(p_k^{\alpha_k}).\end{aligned}$$

By the multiplicativity of $f * g$ (Theorem 7.17), we have $(f * g)(n) = (g * f) \cdot (n) = \mathbf{I}(n)$. But this is the defining behavior of an inverse of f, and, since in a group an inverse is unique, then $g = f^{-1}$ and $f^{-1} \in \mathbf{A}_m$. ◆

Corollary 7.21.1 The multiplicative functions form a subgroup of the group of units of **A**.

Example 7.21 The function $\mu(n) \in \mathbf{A}_m$ generates an *infinite* subgroup of $\mathbf{A}_m$. When n is a prime p, for example, then

$$\mu^k(p) = \underbrace{[\mu * (\mu * (\mu * \ldots))](p)}_{k - 1 \text{ convolutions}} = -k,$$

that is, for no $k \in \mathbf{N}$ does $\mu^k(p) = \mathbf{I}(p)$. ◆

Problems

7.92. Verify the assertion made in Example 7.21.

7.93. Let **A** be the ring of arithmetic functions and let $\mathbf{B} \subset \mathbf{A}$ be the subset of arithmetic functions f such that $f(1) = 0$. Is **B** a subgroup of **A** under pointwise addition? Let $\mathbf{C} \subset \mathbf{A}$ be the subset of arithmetic functions f such that $f(1) = \pm 1$. Is **C** a subgroup of **A** under pointwise addition? Let $\mathbf{B}_m \subset \mathbf{A}_m$ be the subset of multiplicative functions f such that $f(2) = 0$. Is $\mathbf{B}_m$ a subgroup of $\mathbf{A}_m$ under pointwise addition?

7.94. If Fig. 7.6 is correct literally, then $\mathbf{U}_A - \mathbf{A}_m$ is nonempty. Give an example of an arithmetic function that is a unit of **A** but is not multiplicative.

7.95. Refer to Problem 7.89. Let **A** be the set of arithmetic functions, and let the binary operations of pointwise addition and unitary convolution be defined on **A**. Is $< \mathbf{A}, +, \otimes >$ a commutative ring with identity? Show that $\lambda^{-1}(2^n) = (-1)^{n+1}$, $n \in \mathbf{N}$, where the inverse is with respect to unitary convolution.

RESEARCH PROBLEMS

1. The pair of integers (a_0, a_1), $1 < a_0 \leq a_1$, is termed **ϕ-amicable with multiplier k** if $\phi(a_0) = \phi(a_1) = (a_0 + a_1)/k$ for some $k \in \mathbf{N}$. The pair is called **primitive** if GCD(a_0, a_1) is squarefree and GCD$(a_0, a_1, k) = 1$. Cohen and te Riele (1998) state that there are 27 primitive ϕ-amicable pairs whose larger member is less than 100,000. Investigate this claim on the computer. Are a_0, a_1 generally of the same parity?

2. The ring of polynomials $\mathbf{Z}_2[x]$ is analogous to the ring $\mathbf{Z}_3[x]$ mentioned in Research Problem 3 at the end of Chapter 1. Investigate possible definitions for arithmetic functions such as $\mu(\mathrm{f}(x))$, $\sigma(\mathrm{f}(x))$ that would make sense in $\mathbf{Z}_2[x]$. Are these functions multiplicative? How might you define $\phi(\mathrm{f}(x))$ in spite of the fact that there is no ordering of the elements in $\mathbf{Z}_2[x]$? You may assume that factorization is unique in $\mathbf{Z}_2[x]$, although it is not too hard to prove this by using polynomial analogs of theorems from Chapter 2 and without any of the machinery of ideals.

3. Investigate on the computer the conjecture that when $\phi(n) = \phi(n + 1)$, $n > 3$, then either n or $n + 1$ ends in a 5.

REFERENCES

Alaoglu, L., and Erdös, P., "On Highly Composite and Similar Numbers," *Trans. Amer. Math. Soc.*, **56**, 448–469 (1944). These authors tabulate the first 74 superabundant numbers. Our Problem 7.55 is Theorem 1 in this paper.

Annapurna, U., "Inequalities for $\sigma(n)$ and $\phi(n)$," *Math. Mag.*, **45**, 187–190 (1972). For all $n > 1$ except $n = 2, 3, 4, 6, 8$, and 12, the inequality $\sigma(n)/n\sqrt{n} < 6/\pi^2$ holds. The inequality that we actually used in Problem 7.26 was taken from this paper, but the result is probably very old. It appears as Theorem 329 in Hardy and Wright.

Berberian, S.K., "Number-Theoretic Functions via Convolution Rings," *Math. Mag.*, **65**, 75–90 (1992). A dynamite paper paralleling our Section 7.8, but in a somewhat broader setting. Strongly recommended.

Bratley, P., Lunnon, F., and McKay, J., "Amicable Numbers and Their Distribution," *Math. Comp.*, **24**, 431–432 (1970). A "crowd" is a sociable group of order 3; there are none below 5×10^7.

Brill, M.H., and Stueben, M., "A Magnificent Obsession: The Strange Story of Perfect (and Perfectly Useless) Numbers," *Quantum*, **3** (3), 18–23 (1993). Very enjoyable recreational article.

Carlitz, L., "Rings of Arithmetic Functions," *Pac. J. Math.*, **14**, 1165–1171 (1964). Duplicates our material somewhat; see also Ryden, R.W., "Groups of Arithmetic Functions under Dirichlet Convolution," *ibid.*, **44**, 355–360 (1973).

Carmichael, R.D., "On Euler's ϕ-Function," *Bull. Amer. Math. Soc.*, **13**, 241–243 (1907). This is the paper where Carmichael's Conjecture originated as a "theorem."

Carmichael, R.D., "A Table of the Values of m Corresponding to Given Values of $\phi(m)$," *Amer. J . Math.*, **30**, 394–400 (1908). The table runs from $\phi(m) = 1$ to 1000; there are nine errors in the table: $\phi(1785) = \phi(3570) = 768$, $\phi(2388$, not $2384) = 792$, $\phi(1043) = \phi(2086) = 888$, $\phi(1309) = \phi(2618) = 960$, $\phi(1467) = \phi(2934) = 972$.

Cashwell, E.D., and Everett, C.J., "The Ring of Number-Theoretic Functions," *Pac. J. Math.*, **9**, 975–985 (1959). Authors prove that the ring of number theoretic functions is a unique factorization domain.

Chen, N.-X., "Modified Möbius Inverse Formula and Its Applications in Physics," *Phys. Rev. Lett.*, **64**, 1193–1195 (1990). Let $\mathrm{A}(x)$, $\mathrm{B}(x)$ be continuous functions such that $\mathrm{A}(x) = \sum_{n=1}^{\infty} \mathrm{B}(x/n)$. If $|\mathrm{B}(x)| \leq cx^{1+\epsilon}$ for all $x > 0$ and some constants c, ϵ, then one has

$$\mathrm{B}(x) = \sum_{n=1}^{\infty} \mu(n)\, \mathrm{A}\left(\frac{x}{n}\right).$$

The author applies this to the heat capacity of solids. Still other inversion formulas appear in Ren, S.-Y., "New Möbius Inversion Formula," *Phys. Lett.*, **A164**, 1–5 (1992).

Cohen, H., "On Amicable and Sociable Numbers," *Math. Comp.*, **24**, 423–429 (1970). Display of several large amicable pairs and new sociable groups of order 10 or less.

Dence, J.B., and Dence, T.P., "A Surprise Regarding the Equation $\phi(x) = 2(6n + 1)$," *Coll. Math. J.*, **26**, 297–301 (1995). This paper defines explicitly those cases when the congruence in the title has a solution.

Donnelly, H., "On a Problem Concerning Euler's Phi-Function," *Amer. Math. Monthly*, **80**, 1029–1031 (1973). Interesting paper from which we extracted our Problem 7.66.

Flammenkamp, A., "New Sociable Numbers," *Math. Comp.*, **56**, 871–873 (1991). Display of new sociable groups of orders 4, 8, 9.

Glaisher, J.W.L., *Number-Divisor Tables*, Cambridge University Press, Cambridge, England, 1940. See Table II, where $\phi(n)$ runs up to 2500.

Goldstein, L.J., *Abstract Algebra: A First Course*, Prentice–Hall, Englewood Cliffs, NJ, 1973, pp. 258–268. Read about direct products of groups here.

Hagis, Jr., P., "Sketch of a Proof That an Odd Perfect Number Relatively Prime to 3 Has at Least Eleven Prime Factors," *Math. Comp.*, **40**, 399–404 (1983).

Hardy, G.H., *Ramanujan*, Chelsea Publishing Co., New York, 1959, pp. 155–156. Not much is known about the Ramanujan numbers (notated by Hardy as $\tau(n)$).

Honsberger, R., *Mathematical Gems III*, Mathematical Association of America, Washington, DC, 1985, pp. 193–200. A delightful, brief chapter entitled "An Introduction to Ramanujan's Highly Composite Numbers" by a skillful writer.

Jager, H., "The Unitary Analogs of Some Identities for Certain Arithmetical Functions," *Nederl. Akad. Wetensch., Proc. Ser. A*, **64**, 508–515 (1961). Accessible; overlaps our material somewhat.

Klee, V., "On the Equation $\phi(x) = 2m$," *Amer. Math. Monthly*, **53**, 327–328 (1946). The source of our Theorem 7.12 and its three corollaries.

Laatsch, R., "Measuring the Abundancy of Integers," *Math. Mag.*, **59**, 84–92 (1986). Very interesting paper; suitable as an outside reading assignment.

Lehmer, D.H., "A New Calculus of Numerical Functions," *Amer. J. Math.*, **53**, 843–854 (1931). The source of our Problem 7.91. This paper illustrates one of the many additional ways that convolution can be defined besides Dirichlet convolution.

McCarthy, P.J., *Introduction to Arithmetical Functions*, Springer, New York, 1986. If you think arithmetic functions are neat, dip into this book. Much of it will be accessible to you after you have finished our Chapter 7.

McDaniel, W.L., "On the Divisibility of an Odd Perfect Number by the Sixth Power of a Prime," *Math. Comp.*, **25**, 383–385 (1971). Any odd perfect number $< 10^{9118}$ is divisible by the sixth power of some prime.

McDaniel, W.L., "Perfect Gaussian Integers," *Acta Arith.*, **25**, 137–144 (1974). This paper characterizes the "even" perfect numbers (they are scarce) in $\mathbf{Z}[i]$.

McDaniel, W.L., "On the Proof That All Even Perfect Numbers Are of Euclid's Type," *Math. Mag.*, **48**, 107–108 (1975). The source of the proof of our Theorem 7.8.

McDaniel, W.L., "An Analogue in Certain Unique Factorization Domains of the Euclid–Euler Theorem on Perfect Numbers," *Intern. J. Math. & Math. Sci.*, **13**, 13–24 (1990). The author focuses mainly on $\mathbf{Z}[i]$ and $\mathbf{Z}[\omega]$, $\omega = (-1 + i\sqrt{3})/2$.

Menon, P.K., "On the Sum $\Sigma_a\ (a - 1, n), (a, n) = 1$," *J. Indian Math. Soc.*, **29** (3), 155–163 (1965). The proof of the theorem concerning the sum in the title is brief and you should be able to follow it.

Moser, L., "Some Equations Involving Euler's Totient Function," *Amer. Math. Monthly,* **56**, 22–23 (1949). The source of our Problem 7.65.

Murty, M.R., and Murty, V.K., "On the Number of Groups of a Given Order," *J. Number Theory*, **18**, 178–191 (1984). The source of $\mathrm{N}(n) \leq \phi(n)$ if n is squarefree; an advanced paper.

Rabinowitz, S., Kennedy, R.E., and Cooper, C., "Problem No. 95," *Missouri J. Math. Sci.*, **9**, 115–120 (1997) (solved by T.C. Leong, J.B. Dence, and others). The source of our Example 7.20 and of its interesting generalization.

Rahman, F., "A New Proof of the Product of Divisors Formula," *J. Nat. Sci. Math.*, **31**, 109–113 (1991). A proof of $\pi(n) = (\sqrt{n})^{\tau(n)}$.

Ratering, S., "An Interesting Subset of the Highly Composite Numbers," *Math. Mag.*, **64**, 343–346 (1991). Read this article and see that there are exactly six *special highly composite numbers*.

Ribenboim, P., *The New Book of Prime Number Records*, Springer, New York, 1996. See pp. 98–103 for an "Addendum on Perfect Numbers."

Richards, I.M., "A Remark on the Number of Cyclic Subgroups of a Finite Group," *Amer. Math. Monthly*, **91**, 571–572 (1984). This is a suitable outside reading assignment.

Satyanarayana, U.V., "On the Inversion Property of the Möbius μ-Function," *Math. Gaz.*, **47**, 38–42 (1963). The inversion property of Möbius's μ-function characterizes it uniquely.

Shoemaker, R.W., *Perfect Numbers*, National Council of Teachers of Mathematics, Reston, VA, 1973. A very elementary, slim (28-page) volume that would be usable even at the secondary level.

Sivaramakrishnan, R., "The Many Facets of Euler's Totient, I: A General Perspective," *Nieuw Arch. Wisk.* (4), **4**, 175–190 (1986). This is the first of three articles on the ϕ-function. It summarizes a wealth of results and has a good list of references. Parts II and III appeared in the same journal in 1990 and 1993, respectively.

Sivaramakrishnan, R., *Classical Theory of Arithmetic Functions*, Marcel Dekker, New York, 1989. This book collects a wealth of results concerning both the general theory and particular functions.

Subbarao, M.V., "Are There an Infinity of Unitary Perfect Numbers?" *Amer. Math. Monthly*, **77**, 389–390 (1970). An important lead paper into the field of unitary perfect numbers.

Wagon, S., "Carmichael's 'Empirical Theorem,'" *Math. Intelligencer*, **8**, 61–63 (1986). Up-to-date status summary on Carmichael's Conjecture.

Wall, C.R., "Density Bounds for the Sum-of-Divisors Function," in Gioia, A., and Goldsmith, D.L. (eds.), *The Theory of Arithmetic Functions*, Lecture Notes in Mathematics 251, Springer, Berlin, 1972, pp. 283–287.

Wall, C.R., *Selected Topics in Elementary Number Theory*, University of South Carolina Press, Columbia, 1974. A very engaging discussion of topics suitable for a second exposure to number theory. See the author's Chapters IV and V for material on arithmetic functions.

Part 2
Special Topics

Chapter 8
Representation Problems

8.1 The Equation $x^2 + y^2 = z^2$

A **representation problem** is a problem in which one seeks those numbers N that can be expressed according to a specified form. If N is stipulated to be an integer and if it is to be expressed by some specified arithmetic function of other integers, then the representation problem is an example of a **Diophantine equation** (Section 2.5).

The famous Diophantine equation known as **Fermat's Last Theorem**,

$$x^n + y^n = z^n, \qquad xyz \neq 0$$

has been shown by **A. Wiles** to have no solutions x, y, z if $n > 2$, as was conjectured by **Fermat** more than 350 years ago (Ribenboim, 1979). In this section we show that the preceding equation has solutions for $n = 2$ and we will find all of them.

First, we observe that if x, y, z is a solution (for $n = 2$) and $(x, y, z) = d > 1$, and we set $x = dx_1$, $y = dy_1$, $z = dz_1$, then substitution into the equation and division by d^2 throughout gives

$$x_1^2 + y_1^2 = z_1^2.$$

Hence, x_1, y_1, z_1 is also a solution and $(x_1, y_1, z_1) = 1$.

On the other hand, if x, y, z is any solution of the equation, then so is kx, ky, kz, where $k \in \mathbf{N}$. Hence, any solution with $(x, y, z) = 1$ can be used to generate an infinite number of other solutions. We call any solution x, y, z for which $(x, y, z) = 1$ a **primitive Pythagorean triple**, in view of its obvious connection to geometry (Fig. 8.1). It suffices to seek those primitive Pythagorean triples where $x > 0$, $y > 0$, $z > 0$ (referred to as **positive**, primitive Pythagorean triples).

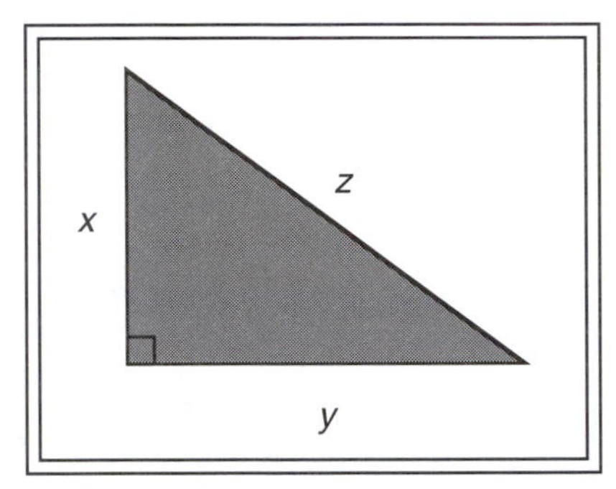

Figure 8.1 Pythagorean triples

Theorem 8.1 If x, y, z is a positive, primitive Pythagorean triple, then x, y are of opposite parity and z is odd.

Proof If x, y were both even, then z would also be even, and $(x, y, z) \geq 2$, a contradiction to the hypothesis. If x, y were both odd (and both greater than 1), then from Example 3.3

$$\begin{cases} x^2 \equiv 1 \pmod 8 \\ y^2 \equiv 1 \pmod 8, \end{cases}$$

so $z^2 \equiv 2 \pmod 8$. But 1, 4 are the only integers A in $[1, 7]$ for which $z^2 \equiv A \pmod 8$ is soluble. Hence, one of x, y is odd, the other is even, and the sum $x^2 + y^2$ is odd. ◆

Example 8.1 A positive, primitive Pythagorean triple is 5, 12, 13. We see that 5, 12 are of opposite parity and 13 is odd. ◆

Let us assume, then, that x is even and y is odd; set $x = 2x_1$. Then we have

$$z^2 - y^2 = (z - y)(z + y) = 4x_1^2.$$

The factors $z - y$, $z + y$ are both even because y, z are both odd (Theorem 8.1); set $z - y = 2s_1, z + y = 2t_1$. This gives $s_1 t_1 = x_1^2, 0 < s_1 < t_1$.

Theorem 8.2 s_1, t_1 are of opposite parity and $(s_1, t_1) = 1$.

Proof $(z - y) + (z + y) = 2z = 2(s_1 + t_1)$, so $s_1 + t_1 = z$, which is odd. Hence, one of s_1, t_1 is even and the other is odd.

Let $d = (s_1, t_1) = [\frac{1}{2}(z - y), \frac{1}{2}(z + y)]$. If the prime p divides d, then p must also divide $[\frac{1}{2}(z + y) + \frac{1}{2}(z - y)] = z$. Similarly, p must also divide $[\frac{1}{2}(z + y) - \frac{1}{2}(z - y)] = y$, and thus $(y, z) \geq p$. But x, y, z is a primitive Pythagorean triple, so $(x, y, z) = 1$. If a prime p divided y, z, it would also divide $x^2 = z^2 - y^2$, and hence $p \mid x$ would hold and $(x, y, z) \geq p$. This is a contradiction, so $(y, z) = 1$ and d itself can only be 1. ◆

Since $(s_1, t_1) = 1$ and $s_1 t_1 = x_1^2$, then s_1, t_1 must themselves be squares (why?); set $s_1 = s^2$ and $t_1 = t^2$. As s_1, t_1 are of opposite parity, so are s, t. We see, then, that every positive, primitive Pythagorean triple x, y, z implies the existence of integers $0 < s < t$ that are coprime, are of opposite parity, and satisfy $z - y = 2s^2, z + y = 2t^2, 0 < s < t$, or equivalently,

$$\begin{cases} z = t^2 + s^2 \\ y = t^2 - s^2 \\ x = 2st. \end{cases}$$

On the other hand, if s, t are any integers satisfying $0 < s < t$, are coprime, and are of opposite parity, then upon defining x, y, z as just shown, one finds $(2st)^2 + (t^2 - s^2)^2 = (t^2 + s^2)^2$. Equivalently, $x^2 + y^2 = z^2$, so x, y, z is a Pythagorean triple. We need to show that it is primitive.

Assume that $(x, y) > 1$; let p be a prime divisor of (x, y), that is, $p \mid (2st, t^2 - s^2)$. Since $t^2 - s^2$ is odd, p must be an odd prime, and so $p \mid st$ and $p \mid (t^2 - s^2)$. But $(s, t) = 1$, so either $p \mid s$ or $p \mid t$, but not both. This, however, contradicts $p \mid (t^2 - s^2)$; hence, $(x, y) > 1$ is false and $(x, y, z) = ((x, y), z) = (1, z) = 1$, so x, y, z is primitive. We have therefore proved Theorem 8.3.

Theorem 8.3 The integers x, y, z are a positive, primitive Pythagorean triple iff there are integers s, t such that $0 < s < t$, $(s, t) = 1$, s, t are of opposite parity, and $x = 2st$, $y = t^2 - s^2$, $z = t^2 + s^2$.

Example 8.2 Choose, arbitrarily, $s = 7$ and $t = 10$. These integers meet the conditions of Theorem 8.3. Hence, a primitive Pythagorean triple is $x = 140$, $y = 51$, $z = 149$. Indeed, $140^2 + 51^2 = 149^2$.

Conversely, 28, 45, 53 is a positive, primitive Pythagorean triple. We compute $t = [(y + z)/2]^{1/2} = \sqrt{49} = 7$, $s = [(z - y)/2]^{1/2} = \sqrt{4} = 2$; finally, $2st = 2(2)(7) = 28 = x$. ◆

Problems

8.1. Is there a positive Pythagorean triple x, y, z that satisfies $x + y + z = 60$? Is there a positive Pythagorean triple x, y, z that satisfies $2x - y = z$?

8.2. Determine all nonsimilar right triangles with integral sides and such that the hypotenuse is less than 60.

8.3. In the proof of Theorem 8.1 why is it legitimate to assume that both x, y exceed 1?

8.4. Identify all positive, primitive Pythagorean triples x, y, z that make the area of the triangle in Figure 8.1 less than 500.

8.5. Let x, y, z be a positive, primitive Pythagorean triple. Prove that $60 \mid xyz$. [Hint: What are the quartic residues of 5?]

8.6. By extension, a **positive, primitive Pythagorean quadruple** w, x, y, z is a set of positive integers such that $(w, x, y, z) = 1$ and $w^2 + x^2 + y^2 = z^2$. Find five such quadruples.

8.7. Suppose that w, x, y, z is a positive, primitive Pythagorean quadruple. Prove that two of w, x, y are even and the other is odd, and hence z is odd also.

8.2 The Two-Square Problem

Instead of representing a square of an integer as the sum of two other integral squares, as we did in the previous section, we can attempt more generally to represent any positive integer N as a sum of two squares (0^2 will now be permitted). It is easily observed that this is not always possible. Table 8.1 gives a sprinkling of results. Before reading on, you are invited to see if you can discover any pattern in the data of Table 8.1.

The so-called **Two-Square Problem** dates back at some level of conceptualization to **Diophantus**, but it appears to have been the Flemish mathematician **Albert Girard** (1590–1633), who first stated the necessary and sufficient condition for

$$x^2 + y^2 = N$$

to be solvable in integers. **Girard** supplied no proof; **Fermat** claimed to have one that was "irrefutable," but did not publish it. According to **Gauss**, it was **Euler** who gave the first acceptable proof of the Two-Square Problem.

We notice in the columns of Table 8.1 that both primes and composite integers occur. Also, both odd and even integers occur. The following lemma is easily established and is motivated by examples such as these:

$$65 = 1^2 + 8^2 \qquad 116 = 4^2 + 10^2$$
$$130 = 7^2 + 9^2 \qquad 232 = 6^2 + 14^2$$

Table 8.1 Some Integers N That Can and Cannot Be Written as Sums of Two Integral Squares

Can		Cannot	
$2 = 1^2 + 1^2$	$97 = 9^2 + 4^2$	3	95
$4 = 2^2 + 0^2$	$98 = 7^2 + 7^2$	6	102
$13 = 3^2 + 2^2$	$113 = 8^2 + 7^2$	7	115
$25 = 4^2 + 3^2$	$405 = 18^2 + 9^2$	23	460
$49 = 7^2 + 0^2$	$578 = 17^2 + 17^2$	54	607
$65 = 8^2 + 1^2$	$677 = 26^2 + 1^2$	63	686
$85 = 9^2 + 2^2$	$1573 = 33^2 + 22^2$	77	1900

Lemma 8.4.1 If N is expressible as the sum of two squares, then so is $2N$.

Proof Use factoring to show this. ◆

The lemma suggests that the presence or absence of powers of 2 in the factorization of N plays no role in determining if N is expressible as the sum of two squares. We are led to look at the odd primes in the factorization of N. It is simpler still to examine those N that are themselves prime. From Table 8.1 we see:

Can	13, 97, 113, 677
Cannot	3, 7, 23, 607.

One possibility is to look for distinct congruence classes that contain these two sets of numbers. Neither set is exclusively congruent to 1 or 2 modulo 3, but upon going to modulo 4 we discover that the first set is congruent to 1 modulo 4 and the second set is congruent to 3 modulo 4. This looks promising; further examples seem to confirm the pattern.

Next, we look at the composite numbers in Table 8.1. The *Can* column provides us with these examples:

$$\begin{array}{rclcrcl}
25 &=& 5^2 & \qquad & 49 &=& 7^2 \\
65 &=& 13^1 \cdot 5^1 & & 98 &=& 7^2 \cdot 2^1 \\
85 &=& 171 \cdot 5^1 & & 405 &=& 5^1 \cdot 3^4 \\
578 &=& 17^2 \cdot 2^1 & & 1573 &=& 13^1 \cdot 11^2 \\
& \text{(A)} & & & & \text{(B)} &
\end{array}$$

In column (A) all of the odd prime factors are $(4k + 1)$-primes, whereas in column (B) both $(4k + 1)$- and $(4k + 3)$-primes occur. However, we notice there that the $(4k + 3)$-primes appear to even powers.

From the *Cannot* column of Table 8.1 we obtain the following examples:

$$\begin{array}{rclcrcl}
6 &=& 3^1 \cdot 2^1 & \qquad & 95 &=& 19^1 \cdot 5^1 \\
54 &=& 3^3 \cdot 2^1 & & 102 &=& 17^1 \cdot 3^1 \cdot 2^1 \\
63 &=& 7^1 \cdot 3^2 & & 115 &=& 23^1 \cdot 5^1 \\
686 &=& 7^3 \cdot 2^1 & & 460 &=& 23^1 \cdot 5^1 \cdot 2^2 \\
& \text{(A)} & & & & \text{(B)} &
\end{array}$$

In column (A) all of the odd prime factors are $(4k + 3)$-primes, whereas in column (B) both $(4k + 1)$- and $(4k + 3)$-primes occur. However, we notice that in both columns one or more of the $(4k + 3)$-prime factors of each integer appear to an odd power.

Our ultimate goal in this section, based on the speculative ideas just formulated, is to solve the Two-Square Problem. We do this in stages. We need one more lemma, a result due to the Norwegian mathematician **Axel Thue** (1863–1922) (Nagell, 1951).

Lemma 8.4.2 (Thue's Remainder Theorem). Let $n > 1$ be a modulus and let $M = 1 + \lfloor\sqrt{n}\rfloor$. Then if $(a, n) = 1$, there exist two positive integers $x, y \leq M - 1$ such that $ay \equiv \pm x \pmod{n}$.

Proof Consider all integers of the form $ay + x$, where x, y belong to the complete residue system $\mathbf{Z}_M = \{0, 1, 2, \ldots, M - 1\}$. The total number of ordered pairs (x, y) is $M^2 > n$. Hence, at least two of the integers $ay + x$, although constructed differently, must have the same residue modulo n. If we write

$$ay_1 + x_1 \equiv ay_2 + x_2 \pmod{n},$$

then

$$a(y_1 - y_2) \equiv x_2 - x_1 \pmod{n}.$$

Since $y_1, y_2 \in \mathbf{Z}_M$, then $0 < |y_1 - y_2| \leq M - 1$; similarly, $0 < |x_2 - x_1| \leq M - 1$. The left-hand sides of the inequalities exclude 0, for if $0 = |y_1 - y_2|$, then $|x_2 - x_1|$ would also have to be 0. So then $y_1 = y_2$, $x_1 = x_2$, and consequently $ay_1 + x_1$ was not constructed differently from $ay_2 + x_2$, a contradiction. Finally, set $y = |y_1 - y_2|$ and $x = |x_2 - x_1|$, and the lemma follows. ◆

Example 8.3 Let $a = 9$, $n = 16$, so $M = 1 + \lfloor\sqrt{16}\rfloor = 5$. The M^2 integers $ay + x$ are:

$$\begin{array}{ccccc} 0 & 9 & 18 & 27 & \textcircled{36} \\ 1 & 10 & 19 & 28 & 37 \\ 2 & 11 & 20 & 29 & 38 \\ 3 & 12 & 21 & 30 & 39 \\ \textcircled{4} & 13 & 22 & 31 & 40. \end{array}$$

In this set there are many pairs of congruent integers modulo 16; 4, 36 is one such pair. ◆

Example 8.4 Consider $n = 10$, $a = 7$, so $(7, 10) = 1$. Then $M = 1 + \lfloor\sqrt{n}\rfloor = 4$, and we seek x, y such that $0 < x, y \leq 3$ and $7y \equiv \pm x \pmod{10}$. By inspection, we find that $x = 3$, $y = 1$ is a solution if we choose the negative sign. ◆

Theorem 8.4 If p is a $(4k + 1)$-prime, there exist $x, y \in \mathbf{N}$ such that $x^2 + y^2 = p$.

Proof By Theorem 6.5(e) we have $(\frac{-1}{p}) = 1$ if $p = 4k + 1$. Hence, a z_0 exists such that $z_0^2 \equiv -1 \pmod{p}$. As $(z_0, p) = 1$, then by Thue's Remainder Theorem there are integers x, y satisfying $0 < x, y < \sqrt{p}$ such that

$$yz_0 \equiv \pm x \pmod{p}$$

or

$$y^2 z_0^2 \equiv x^2 \pmod{p}.$$

Combination with the first congruence gives

$$y^2(-1) \equiv x^2 \pmod{p}$$

or, equivalently, $x^2 + y^2 = mp$ for some $m \in \mathbf{N}$. But as $x, y < \sqrt{p}$, only $m = 1$ is possible. ◆

Corollary 8.4.1 A composite integer N that contains only powers of 2 and $(4k + 1)$-primes in its factorization is expressible as the sum of two squares.

Proof It is sufficient to prove this for the special case $N = p_1 p_2$, where p_1, p_2 are $(4k + 1)$-primes, and then to argue for the general case using induction on the number of prime factors together with Lemma 8.4.1. But Theorem 8.4 gives us $p_1 = x^2 + y^2$ and $p_2 = w^2 + z^2$, where w, x, y, z are positive integers, and factoring shows that

$$\begin{aligned}(x^2 + y^2)(w^2 + z^2) &= (x^2w^2 + 2wxyz + y^2z^2) + (x^2z^2 - 2wxyz + y^2w^2) \\ &= (xw + yz)^2 + (xz - yw)^2.\end{aligned}$$

Thus, $p_1 p_2$ has now been expressed as the sum of two squares.[1] ◆

Notice that the corollary does not say that every composite integer N of the form $4k + 1$ or of the form a power of 2 times a $(4k + 1)$-integer is expressible as the sum of two squares. For example, when $N = 21 = 5 \cdot 4 + 1$ or when $N = 2 \cdot 21 = 42$, partitioning into the sums of two squares is not possible. What is important are the prime *factors* of N.

Now we turn to a consideration of $(4k + 3)$-primes. Suppose such a prime occurs to an even power, p^{2a}. Then obviously it can be expressed as $(p^a)^2 + 0^2$. By the factoring device used in the corollary, we see that any composite integer N that contains only powers of 2, powers of $(4k + 1)$-primes, and *even* powers of $(4k + 3)$-primes, is expressible as the sum of two squares. What if a $(4k + 3)$-prime p appears to an odd power in N?

Theorem 8.5 If N has a representation as a sum of two squares, then no $(4k + 3)$-prime p can appear to an odd power in the factorization of N.

[1] It is apparent that multiplication is an associative binary operation on the set of all integers expressible as sums of two squares. Such an algebraic structure, weaker than that of a group, is called a **semigroup**. If the semigroup contains an identity element, as in the present case, the algebraic structure is termed a **monoid**.

The preceding identity was crucial in Fermat's development of the theory of the decomposition of primes and their powers into squares. The identity dates back to the French mathematician **François Viète** (1540–1603), who presented it in his book *Prior Notes on Specious Logistic*, first published in Paris in 1631 (Mahoney, 1994).

Proof Assume p is a $(4k+3)$-prime and suppose $N = x^2 + y^2 = Kp^{2a+1}$, $(p, K) = 1$. Further, let $(x, y) = d, x/d = X, y/d = Y, N/d^2 = n$; we obtain

$$n = X^2 + Y^2, \ (X, \ Y) = 1.$$

If p^b is the highest power of p that divides d, then $p^{2b} \mid d^2$ and $p^{2a+1-2b} \mid n$. So $p \mid n$, but $p \nmid X$ and $p \nmid Y$, as $(X, \ Y) = 1$. The congruence

$$Xz \equiv Y \ (\text{mod } p)$$

thus has a solution, $z \equiv z_0 \ (\text{mod } p)$ according to Theorem 3.22. This gives

$$n \equiv X^2 + X^2 z_0^2 \equiv X^2(1 + z_0^2) \equiv 0 \ (\text{mod } p).$$

Since $X^2 \not\equiv 0 \ (\text{mod } p)$ because $p \nmid X$ and since $\mathbf{Z}_p$ is a field, then $1 + {z_0}^2 \equiv 0 \ (\text{mod } p)$. That is, -1 is a quadratic residue of p. But this contradicts Theorem 6.5 (e), so p could not have been a $(4k + 3)$-prime. ◆

Putting all the pieces together, we can now give the necessary and sufficient condition for the solvability of the Two-Square Problem.

Theorem 8.6 (Two-Square Problem). An integer $N > 0$ is expressible as the sum of two integral squares iff no $(4k + 3)$-prime in the factorization of N occurs to an odd power.

Example 8.5 The integer $N = 4840 = 11^2 \cdot 5^1 \cdot 2^3$ is expressible as the sum of two squares, $4840 = 66^2 + 22^2$, but the integer $N = 11^3 \cdot 5^1 = 6655$ is not expressible as the sum of two squares. ◆

The preceding theorem has numerous applications; here are two elementary ones.

Corollary 8.6.1 An odd perfect number, if one exists, has a representation as the sum of two squares of nonzero integers.

Proof From Theorems 7.9 and 7.10, an odd perfect number N, if it exists, has the form

$$N = q^2 \pi^\alpha,$$

where π is a prime and $\pi \equiv \alpha \equiv 1 \ (\text{mod } 4)$. Hence, any $(4k + 3)$-prime factor of N must occur in the factorization of q, and so appears to an even power. By Theorem 8.6, N can then be expressed as the sum of two squares; both are nonzero since α is odd. ◆

Corollary 8.6.2 A Mersenne prime does not have a representation as a sum of two squares.

Proof Let $M_p = 2^p - 1$ be a Mersenne prime; a necessary condition for this is that p be a prime. For $p \geq 2$, we obtain

$$M_p = (2^p - 4) + 3 = 4(2^{p-2} - 1) + 3 = 4k + 3, \qquad k \geq 0.$$

By Theorem 8.6 the prime M_p is then not expressible as a sum of two squares. ◆

A much more difficult question than that answered by Theorem 8.6 is the following: Given a positive integer N, estimate how many positive integers not exceeding N are representable as sums of two squares. This question was answered in 1908 by **Edmund Landau**. Let $B(N)$ be the number of positive integers not exceeding N that are expressible as sums of two squares. **Landau** showed that as $N \to \infty$

$$B(N) \sim K \cdot N \cdot (\ln N)^{-1/2}$$

$$K = \left\{ \frac{1}{2} \prod_{p=4k+3}^{\infty} \frac{1}{1 - p^{-2}} \right\}^{1/2},$$

where the infinite product runs over all $(4k + 3)$-primes. It is by no means obvious that K is finite. However, using nothing more than standard theorems from the calculus, it is possible to show that $K < 0.7670$ (Dence and Dence, 1992). A more accurate value is $K \approx 0.76422$ (Shiu, 1986). A sample calculation provides $B(10^5) = 24{,}028$ and $K(10^5)(\ln 10^5)^{-1/2} = 22{,}523$, an error of about 6%.

Problems

8.8. Suppose $N_1, N_2 > 0$ can each be written as the sum of two squares. Decide which of the following can also be so written:

(a) $N_1 - N_2$ (assume $N_1 > N_2$).
(b) $N_1 \cdot N_2$.
(c) N_1/N_2 (assume $N_2 \mid N_1$).

8.9. Prove Lemma 8.4.1. Illustrate it with two examples.

8.10. Let $N = p_1 p_2$, where p_1, p_2 are $(4k + 1)$-primes that may be either alike or distinct. Make a list of about 10 choices of N and for each N determine the number of distinct ways that it can be written as the sum of two squares. Formulate a conjecture.

8.11. Look at the proof of Corollary 8.4.1 again, and then express as the sum of two squares: (a) 6137. (b) 15,457.

8.12. Find a solution x, y to $ay \equiv \pm x \pmod{n}$, in accordance with Thue's Remainder Theorem, for each of the following:

(a) $a = 17, n = 10$.
(b) $a = 6, n = 13$.
(c) $a = 8, n = 35$.
(d) $a = 15, n = 8$.

8.13. Let A_1, A_2 be two quadratic residues of the $(4k + 3)$-prime p that satisfy $0 < A_1 < A_2 < p$. Prove that $A_1 + A_2 \equiv 0 \pmod{p}$ is impossible. Illustrate this result with $p = 23$.

8.14. Write a computer program that will count the number of positive integers not exceeding N that are expressible as the sums of two squares. Determine $B(10{,}000)$.

8.15. Present a simple argument that $K > 3/4$ in Landau's asymptotic formula. A theorem (Apostol, 1974) says that if each $a_n > 0$, then the infinite product $\Pi_{n=1}^{\infty} (1 + a_n)$ converges iff $\Sigma_{n=1}^{\infty} a_n$ converges. How does this imply that Landau's K is finite?

8.16. Consider the even perfect numbers. Are any of them representable as the sum of two squares? Consider the Fermat primes, F_n (Section 1.8). Are any of them representable as the sum of two squares?

8.17. Do a computer investigation of the Fibonacci numbers to see which ones are representable as the sum of two squares. If possible, formulate a conjecture.

8.3 The Four-Square Problem

At this stage it is logical to inquire about the conditions for solvability of

$$x^2 + y^2 + z^2 = N.$$

The necessary and sufficient conditions for the solvability of this Diophantine equation are simple to state but nontrivial to prove (Dickson, 1939a; Fraser, 1969; Mordell, 1969; Grosswald, 1985). The proof was first given by **Legendre** in 1798. We shall not give it, but instead will advance to the representation of an integer as a sum of *four* squares.

The **Four-Square Problem** was apparently recognized by **Diophantus**, but the problem was stated most explicitly (and without proof) by the Frenchman **Claude Gaspard de Bachet de Méziriac** (1581–1638), the individual who brought out in 1621 an annotated Greek and Latin translation of Diophantus's book *Arithmetica*. The first proof that

$$x^2 + y^2 + z^2 + w^2 = N$$

is solvable in integers for any positive integer N was given by **Lagrange** before 1770. We shall present here an exciting contemporary proof that is algebraic in nature (Small, 1982). As with the Two-Square Problem, the proof of the Four-Square Problem proceeds in stages.

Lemma 8.7.1 Let p be a prime and $\mathbf{Z}_p$ be the ring of integers modulo p. Then any $x \in \mathbf{Z}_p$ is a sum of two squares.

Proof The case $p = 2$ is immediate; hence we assume p is odd. Define the set $\mathbf{S}_1 = \{a^2 : a \in \mathbf{Z}_p\}$. From Theorem 6.1 there are $(p - 1)/2$ quadratic residues of p, so that together with $a = 0$ there are $(p + 1)/2$ elements in $\mathbf{S}_1$. For example, if $p = 5$, then modulo 5 we have $0^2 = 0$, $1^2 = 1$, $2^2 = 4$, $3^2 = 4$, $4^2 = 1$, so $\mathbf{S}_1$ is the set $\{0, 1, 4\}$.

Next, define the set $\mathbf{S}_2 = \{x - b^2 : b \in \mathbf{Z}_p\}$. Again, Theorem 6.1 allows us to say that together with $b = 0$ there are $(p + 1)/2$ elements in $\mathbf{S}_2$. For example, if $p = 5$ and $x = 3$, then $\mathbf{S}_2 = \{2, 3, 4\}$.

But $\mathbf{S}_1 \subset \mathbf{Z}_p$, $\mathbf{S}_2 \subset \mathbf{Z}_p$, and $\mathbf{Z}_p$ contains only p elements, whereas $(p + 1)/2 + (p + 1)/2 = p + 1$. Hence, $\mathbf{S}_1 \cap \mathbf{S}_2$ is nonempty, so there is some pair of integers a, $b \in \mathbf{Z}_p$ such that modulo p one has $a^2 = x - b^2$, or $a^2 + b^2 = x$. For example, if $p = 5$ and $x = 3$, then one finds $3 = 2^2 + 2^2$ (that is, $2^2 + 2^2 \equiv 3 \pmod 5$). ◆

Example 8.6 Let $x = 24$, $p = 13$. Then we find by trial that $24 \equiv 1^2 + 6^2 \pmod{13}$. ◆

Note that Lemma 8.7.1 does not contradict Theorem 8.6, for there the ring of interest was $\mathbf{Z}$, whereas in the lemma the ring is $\mathbf{Z}_p$. What if p in Lemma 8.7.1 is replaced by an arbitrary composite integer N? Then the lemma does not hold. For example, let $N = 8$; then modulo 8 we find that

$$\begin{array}{ll} 0^2 \equiv 0 & 4^2 \equiv 0 \\ 1^2 \equiv 1 & 5^2 \equiv 1 \\ 2^2 \equiv 4 & 6^2 \equiv 4 \\ 3^2 \equiv 1 & 7^2 \equiv 1. \end{array}$$

But now we see that it is impossible to express 3 modulo 8 as a sum of two squares. However, extension of the lemma can be salvaged if we restrict N to be squarefree, that is, $N = p_1 p_2 \ldots p_r$, where the prime p_i's are distinct.

Theorem 8.7 Let $\mathbf{Z}_N$ denote the ring of integers modulo N, where N is squarefree. Then every $x \in \mathbf{Z}_N$ is a sum of two squares.

Proof Suppose $N = p_1 p_2 \ldots p_r$, where the primes are distinct. By Lemma 8.7.1, any $x_0 \in \mathbf{Z}_N$ can be written as the sum of two squares in each $\mathbf{Z}_{p_i}$ (see Example 8.6):

$$\begin{cases} x_0 \equiv a_1^2 + b_1^2 \pmod{p_1} \\ x_0 \equiv a_2^2 + b_2^2 \pmod{p_2}. \end{cases}$$

We now apply the Chinese Remainder Theorem (Theorem 4.9) to the systems of congruences

$$\begin{cases} y \equiv a_1 \pmod{p_1} \\ y \equiv a_2 \pmod{p_2} \end{cases} \qquad \begin{cases} z \equiv b_1 \pmod{p_1} \\ z \equiv b_2 \pmod{p_2}. \end{cases}$$

These have unique solutions $y \equiv y_0 \pmod{p_1p_2}$, $z \equiv z_0 \pmod{p_1p_2}$. Hence, we obtain

$$\begin{cases} y_0^2 \equiv a_1^2 \pmod{p_1} \\ z_0^2 \equiv b_1^2 \pmod{p_1} \end{cases} \quad \rightarrow \quad y_0^2 + z_0^2 \equiv x_0 \pmod{p_1}$$

$$\begin{cases} y_0^2 \equiv a_2^2 \pmod{p_2} \\ z_0^2 \equiv b_2^2 \pmod{p_2} \end{cases} \quad \rightarrow \quad y_0^2 + z_0^2 \equiv x_0 \pmod{p_2}$$

and so $x_0 \equiv y_0^2 + z_0^2 \pmod{p_1p_2}$.

Since $(p_3, p_1p_2) = 1$, we can repeat the previous steps to obtain x_0 as a sum of two squares modulo $p_1p_2p_3$. By induction, we finally produce x_0 as a sum of two squares modulo $p_1p_2 \cdots p_r = N$. ◆

Example 8.7 Let $x_0 = 6$ and $N = 77 = 7 \cdot 11$. In conformity with Lemma 8.7.1, we find $x_0 \equiv 2^2 + 3^2 \pmod{7}$ and $x_0 \equiv 1^2 + 4^2 \pmod{11}$. The solution of

$$\begin{cases} y \equiv 2 \pmod{7} \\ y \equiv 1 \pmod{11} \end{cases}$$

is $y \equiv 23 \pmod{77}$. Similarly, the solution of

$$\begin{cases} z \equiv 3 \pmod{7} \\ z \equiv 4 \pmod{11} \end{cases}$$

is $z \equiv 59 \pmod{77}$. Hence, in $\mathbf{Z}_N$ we have

$$x_6 \equiv 23^2 + 59^2 \pmod{77}.$$

A check shows that $77 \mid (23^2 + 59^2 - 6)$. ◆

The reason why we might want Theorem 8.7 is that in considering $N = x^2 + y^2 + z^2 + w^2$, it is sufficient to assume N is squarefree. For if $N = q^2N'$ and N' is squarefree, then $N' = x^2 + y^2 + z^2 + w^2$ implies $N = (qx)^2 + (qy)^2 + (qz)^2 + (qw)^2$.

To motivate the next preliminary result (a technical lemma), observe that although the polynomial $x^2 + y^2 + z^2 + w^2$ is not decomposable over $\mathbf{Z}$, it is decomposable over $\mathbf{Z}[i]$, the domain of **Gaussian integers** (see remarks after Theorem 7.10):

$$N = (x + iy)(x - iy) + (z + iw)(z - iw).$$

Lemma 8.8.1 Let the matrix $\mathbf{A}$ be defined by

$$\mathbf{A} = \begin{pmatrix} N & c + di \\ c - di & m \end{pmatrix},$$

where $c, d \in \mathbf{Z}$ and $m, N \in \mathbf{N}$. If $\det \mathbf{A} = 1$, then $\mathbf{A} = \mathbf{BB}^*$ for some matrix $\mathbf{B}$ over $\mathbf{Z}[i]$, where $\mathbf{B}^*$ denotes the conjugate transpose of $\mathbf{B}$.

We shall return shortly to the proof of Lemma 8.8.1, but the following remarks may be helpful. Recall from linear algebra that matrix $\mathbf{M}^*$ is obtained from $\mathbf{M}$ by interchanging corresponding rows with columns and then taking the complex conjugate of all entries, as in this example:

$$\mathbf{M} = \begin{pmatrix} i & 1 + 2i \\ 4 & 3 - i \end{pmatrix} \qquad \mathbf{M}^* = \begin{pmatrix} -i & 4 \\ 1 - 2i & 3 + i \end{pmatrix}.$$

In the proof of Lemma 8.8.1, we will also need to know from linear algebra that if $\mathbf{M}, \mathbf{R}, \mathbf{S}$ are $n \times n$ matrices and $\mathbf{M} = \mathbf{RS}$, then $\det \mathbf{M} = (\det \mathbf{R})(\det \mathbf{S})$. Further, we have the relation $\det \mathbf{M}^* = (\det \mathbf{M})^*$ (Problem 8.21).

The Four-Square Problem is an immediate consequence of Lemma 8.8.1.

Theorem 8.8 (Four-Square Problem). Any integer $N > 0$ can be expressed as the sum of four squares.

Proof Assume N is squarefree. Then, from Theorem 8.7, $N - 1 \in \mathbf{Z}_N$ is a sum of two squares, that is,

$$N - 1 \equiv c^2 + d^2 \pmod{N}$$

with $0 \le c, d < N$. Equivalently, there is an integer $m > 0$ such that $-1 = c^2 + d^2 - mN$. Define the matrix $\mathbf{A}$ by

$$\mathbf{A} = \begin{pmatrix} N & c + di \\ c - di & m \end{pmatrix},$$

so that $\det \mathbf{A} = Nm - (c^2 + d^2) = 1$ from the preceding line. Hence, from Lemma 8.8.1 we can write

$$\mathbf{A} = \begin{pmatrix} N & c + di \\ c - di & m \end{pmatrix} = \mathbf{BB}^* = \begin{pmatrix} x + yi & z + wi \\ * & * \end{pmatrix}\begin{pmatrix} x - yi & * \\ z - wi & * \end{pmatrix}.$$

Verification of the 1, 1-element of $\mathbf{A}$ therefore gives

$$\begin{aligned} N &= (x + yi)(x - yi) + (z + wi)(z - wi) \\ &= x^2 + y^2 + z^2 + w^2. \quad \blacklozenge \end{aligned}$$

Note that Theorem 8.8 does not say that it is *necessary* to use four squares in order to represent a positive integer N, but merely that four squares are always *sufficient*. Some integers N can be represented by sums of three or two squares, or even by just one square.

Example 8.8

$$17 = 1^2 + 4^2 = 2^2 + 2^2 + 3^2$$

$$51 = 7^2 + 1^2 + 1^2 = 5^2 + 5^2 + 1^2 = 5^2 + 4^2 + 3^2 + 1^2$$

$$81 = 9^2 = 8^2 + 4^2 + 1^2 = 8^2 + 3^2 + 2^2 + 2^2. \quad \blacklozenge$$

We still owe a proof of Lemma 8.8.1. If $c = d = 0$, then $\det \mathbf{A} = Nm - (c^2 + d^2) = 1$ implies $N = m = 1$ (if $N, m > 0$). Hence, $\mathbf{B}$ can be taken to be the identity matrix $\mathbf{I}$. In what follows, assume that $c^2 + d^2 > 0$.

The proof is by mathematical induction on the value of $c^2 + d^2$, the initial case of $c^2 + d^2 = 0$ having just been treated. Given integers N_0, $m_0 > 0$ and integers c_0, d_0 such that $N_0 m_0 - c_0^2 - d_0^2 = 1$, assume that the lemma holds for all integers c, d satisfying $0 < c^2 + d^2 < c_0^2 + d_0^2$ and all integers N, $m > 0$ satisfying $Nm - c^2 - d^2 = 1$. Since N_0, m_0 are not both 1 (as this was handled previously), there are two cases to consider.

Case 1: $0 < N_0 \leq m_0$

Define the two matrices

$$\mathbf{A}_0 = \begin{pmatrix} N_0 & c_0 + d_0 i \\ c_0 - d_0 i & m_0 \end{pmatrix}, \qquad \mathbf{M} = \begin{pmatrix} 1 & 0 \\ x - yi & 1 \end{pmatrix},$$

where the integers x, y are to be specified. Let $\mathbf{A} = \mathbf{M}\mathbf{A}_0\mathbf{M}^*$; then $\det \mathbf{M} = \det \mathbf{A}_0 = 1$ implies that $\det \mathbf{A} = 1$. Explicitly, we find upon multiplying out the three matrices that $\mathbf{A}$ has the form

$$\mathbf{A} = \begin{pmatrix} N_0 & c + di \\ c - di & r \end{pmatrix}$$

where $c = c_0 + N_0 x$, $d = d_0 + N_0 y$, $r \in \mathbf{N}$. That $r > 0$ follows from $N_0 r = 1 + c^2 + d^2$.

We need to choose x, y so that $c^2 + d^2 < c_0^2 + d_0^2$. We must have either $N_0 < 2|c_0|$ or $N_0 < 2|d_0|$. This is automatic if $N_0 = 1$ because c_0, d_0 are not both zero. If $N_0 > 1$, then suppose to the contrary that $N_0 \geq 2 \max \{|c_0|, |d_0|\}$. Since $N_0 \leq m_0$, we have

$$c_0^2 + d_0^2 + 1 = N_0 m_0 \geq N_0^2 \geq 4 \, [\max \{|c_0|, |d_0|\}]^2$$

or, equivalently,

$$[\min \{|c_0|, |d_0|\}]^2 + 1 \geq 3[\max \{|c_0|, |d_0|\}]^2.$$

This is an impossibility, so either $N_0 < 2|c_0|$ or $N_0 < 2|d_0|$ must hold.

Integers x, y can now be conveniently chosen according to the four possible subcases:

$$\begin{cases} N_0 < 2c_0, & x = -1, y = 0, & c^2 + d^2 = (c_0 - N_0)^2 + d_0^2 < c_0^2 + d_0^2 \\ -N_0 > 2c_0, & x = 1, \quad y = 0, & c^2 + d^2 = (c_0 - N_0)^2 + d_0^2 < c_0^2 + d_0^2 \\ N_0 < 2d_0, & x = 0, \quad y = -1, & c^2 + d^2 = c_0^2 + (d_0 - N_0)^2 < c_0^2 + d_0^2 \\ -N_0 > 2d_0, & x = 0, \quad y = 1, & c^2 + d^2 = c_0^2 + (d_0 + N_0)^2 < c_0^2 + d_0^2. \end{cases}$$

Hence, $\mathbf{A}$ meets all of the conditions of the induction hypothesis, so there exists a matrix $\mathbf{B}$ such that $\mathbf{A} = \mathbf{BB}^*$, and we have

$$\mathbf{BB}^* = \mathbf{MA}_0\mathbf{M}^*.$$

Solving this matrix equation for $\mathbf{A}_0$, we obtain

$$\mathbf{A}_0 = (\mathbf{M}^{-1}\mathbf{B})(\mathbf{B}^*\mathbf{M}^{*-1}) = \mathbf{CC}^*,$$

where $\mathbf{C} = \mathbf{M}^{-1}\mathbf{B}$.

Case 2: $0 \le m_0 < N_0$

The argument is practically the same as that for Case 1, except that we now define

$$\mathbf{M} = \begin{pmatrix} 1 & x + yi \\ 0 & 1 \end{pmatrix},$$

and this leads to $c = c_0 + m_0 x$, $d = d_0 + m_0 y$. We leave you to work out the details (Problem 8.22).

By the Principle of Finite Induction, the lemma holds for all values of $c_0^2 + d_0^2$.

Example 8.9 Let $N_0 = 15$; a solution to $N_0 - 1 \equiv c^2 + d^2 \pmod{N_0}$ is $c_0 = 2$, $d_0 = 5$. If $N_0 m_0 = 1 + c_0^2 + d_0^2$, then $m_0 = 2$. We find that

$$\mathbf{A}_0 = \begin{pmatrix} 15 & 2 + 5i \\ 2 - 5i & 2 \end{pmatrix} = \begin{pmatrix} 1 + 3i & 1 + 2i \\ 1 & 1 \end{pmatrix}\begin{pmatrix} 1 - 3i & 1 \\ 1 - 2i & 1 \end{pmatrix} = \mathbf{CC}^*,$$

consistent with $15 = 1^2 + 3^2 + 1^2 + 2^2$. ◆

Lemma 8.8.1 is theoretically valuable, but does not offer much in the way of a practical approach to the factoring of a matrix $\mathbf{A}$. Hence, it is not an algorithm for expressing an integer $N > 0$ as a sum of four squares. The matrix factorization in Example 8.9 was worked out after $N_0 = 15$ had been written as the sum of four squares.

Problems

8.18. Illustrate Lemma 8.7.1 using $p = 13$ and $x = 11$. In $\mathbf{Z}_{13}$, in how many distinct ways (ordered pairs) can x be written as a sum of two squares? Illustrate Theorem 8.7 by showing that $102 \in \mathbf{Z}_{210}$ can be written as a sum of two squares.

8.19. The **unit circle problem** asks: In the field $\mathbf{Z}_p$ how many incongruent solutions (ordered pairs) are there to $x^2 + y^2 \equiv 1 \pmod{p}$?

(a) Select three $(4k + 1)$-primes p, and for each determine the number of solutions.
(b) Do the same for three $(4k + 3)$-primes. Formulate conjectures.

8.20. Show that the following is an identity:[2]

$$\left(\sum_{i=1}^{4} x_i^2\right)\left(\sum_{i=1}^{4} y_i^2\right) = \left(\sum_{i=1}^{4} x_i y_i\right)^2 + (x_1y_2 - x_2y_1 + x_3y_4 - x_4y_3)^2 + (x_1y_3 - x_3y_1 + x_4y_2 - x_2y_4)^2 + (x_1y_4 - x_4y_1 + x_2y_3 - x_3y_2)^2.$$

How does this relate to Theorem 8.8? Illustrate the use of this identity with the integer $N = 105$.

8.21. Prove that for 2×2 matrices $\mathbf{M}$ one has $\det(\mathbf{M}^*) = (\det \mathbf{M})^*$.

8.22. Work out the details of Case 2 of Lemma 8.8.1.

8.23. It was proved by **Jacobi** in 1828 that the number of ways a positive integer N can be written as a sum of four squares equals 8 times the sum of its divisors that are not divisible by 4. The four-square representations of N can involve squares of negative integers as well as of 0 and of positive integers. Also, permutations are to be regarded as important. For example, three distinct representations of 22 are

$$\begin{aligned} 22 &= 1^2 + 1^2 + 2^2 + 4^2 \\ &= 1^2 + 1^2 + 4^2 + 2^2 \\ &= 1^2 + 1^2 + (-4)^2 + 2^2. \end{aligned}$$

Verify Jacobi's result for $N = 11$ by writing out all of the possible representations.

8.24. As stated in the text, four squares are not always necessary in order to represent a positive integer N as a sum of squares. However, three squares are not

[2] Of a relation like this, the legendary **J.E. Littlewood** (1885–1977) used to say that the first time you see it, you refer to it as a trick. The second time you see it, it's a device. By the third time, it's simply an identity.

always sufficient. Prove that if $N \equiv 7 \pmod 8$, then N cannot be written as a sum of three squares.

8.25. Jacobi's result in Problem 8.23 is proved in (Andrews, Ekhad, and Zeilberger, 1993). Read this paper and make a report on its contents.

8.4 The Equation $x^4 + y^4 = z^4$

The easiest case of Fermat's Last Theorem (see Section 8.1) to prove is that of $n = 4$. We shall prove this case here because (a) it is a simple example of a representation problem of algebraic degree greater than 2, (b) it builds directly on material in Section 8.1, and (c) the standard proof is an elaboration of Fermat's own sketch of it, and is a wonderful example of the technique known as the **Method of Infinite Descent**,[3] which he invented.

Roughly speaking, the method may be thought of as mathematical induction in reverse. To disprove the existence of a solution to a Diophantine problem, one assumes that there is a solution. Then one shows that from this alleged solution one can construct a smaller one, and so on repetitively, until in a finite number of steps one reaches an alleged small solution that can be rejected by direct checking. In view of this, the initial assumption of even a single solution is false.

We begin, as did **Fermat**, by noting that if

$$x^4 + y^4 = z^2$$

has no solution in positive integers, then neither will $x^4 + y^4 = z^4$. Also, it is sufficient to seek a primitive solution of $x^4 + y^4 = z^2$, that is, a solution in which $(x, y, z) = 1$.

Theorem 8.9 The equation $x^4 + y^4 = z^2$ has no solution in positive integers.

Proof Assume that x_0, y_0, z_0 is a positive, primitive solution. Then from Theorem 8.1 z_0 is odd and we may take x_0 to be even and y_0 to be odd. Theorem 8.3 gives

$$\begin{cases} x_0^2 = 2st \\ y_0^2 = t^2 - s^2 \\ z_0 = t^2 + s^2, \end{cases}$$

[3] The name is Fermat's own. He first gave a bare sketch of the method in a memoir, "Rélation des nouvelles découvertes en la science des nombres," written in 1659 and sent to his friend **Pierre de Carcavi** (d. 1684) for transmission to **Christiaan Huygens** (1629–1695). Throughout his mathematical career **Fermat** was stingy in releasing details of his work. There is no doubt that the Method of Infinite Descent had been employed by him for many years prior to 1659, but no date for its initial use can be fixed. It is generally agreed by historians that it was with the Method of Infinite Descent that **Fermat** claimed to have a proof of his "Last Theorem." For these, and other remarks, see (Mahoney, 1994).

with $(s, t) = 1$, $0 < s < t$, and s, t of opposite parity. As y_0 is odd and $y_0^2 \equiv 1 \pmod 4$ necessarily, then we must have $t^2 \equiv 1 \pmod 4$ and $s^2 \equiv 0 \pmod 4$, that is, s is even and t is odd. Hence, set $s = 2a$, $(a, t) = 1$, so that $x_0^2 = 4at$. In view of $(a, t) = 1$, it follows that a, t are squares: $a = c^2$, $t = z_1^2$, $(z_1, c) = 1$. Consequently, $y_0^2 = (z_1^2)^2 - (2c^2)^2$ or, equivalently,

$$(2c^2)^2 + y_0^2 = (z_1^2)^2.$$

Theorem 8.3 can now be applied again to give

$$\begin{cases} 2c^2 = 2hk \\ y_0 = k^2 - h^2 \\ z_1^2 = k^2 + h^2, \end{cases}$$

with $(h, k) = 1$, $0 < h < k$, and h, k of opposite parity. Since now $c^2 = hk$ and $(h, k) = 1$, then h, k are both squares; set $h = x_1^2$ and $k = y_1^2$. We obtain the relation

$$x_1^4 + y_1^4 = z_1^2,$$

so that we have found another solution to $x^4 + y^4 = z^2$. It is a positive, primitive solution because $(h, k) = 1 \rightarrow (x_1, y_1) = 1 \rightarrow (x_1, y_1, z_1) = 1$. The solution x_1, y_1, z_1 is also a smaller solution than x_0, y_0, z_0 because

$$\begin{cases} x_0^2 > s > a = c^2 \geq h = x_1^2 \\ y_0^2 = (k^2 - h^2)^2 \geq [k^2 - (k-1)^2]^2 = (2k-1)^2 > k = y_1^2 \ (k > 1) \\ z_0 > t^2 = z_1^4 > z_1. \end{cases}$$

After a finite number of steps similar to the preceding steps, one must eventually reach $z_n^2 \leq 1000$. But simple trial shows that no square in $[1, 1000]$ can be expressed as a sum of two nonzero perfect fourth powers. It follows that the alleged primitive solution x_0, y_0, z_0 does not exist. ◆

Example 8.10 The nonzero perfect fourth powers not exceeding 1000 are 1, 16, 81, 256, 625. No two of these sum to give a perfect square. ◆

Problems

8.26. Why does it follow that the nonexistence of a solution in positive integers to $x^4 + y^4 = z^2$ implies a similar nonexistence of a solution to $x^4 + y^4 = z^4$? Why is it sufficient to seek only *primitive* solutions to $x^4 + y^4 = z^2$?

8.27. Let $n \in \mathbf{N}$; prove that $x^{4n} + y^{4n} = z^{4n}$ has no solution in positive integers.

8.28. Show that in order to prove Fermat's Last Theorem, it is sufficient to show that there are no solutions to $x^p + y^p = z^p$, $xyz \neq 0$, for every odd prime p.

8.29. Diophantine equations that are insoluble in $\mathbf{Z}$ may become soluble in finite integral domains. Show that $x^4 + y^4 = z^4$ is soluble (as a congruence) in $\mathbf{Z}_{19}$, but is insoluble in $\mathbf{Z}_{17}$.

8.30. The proof of the case $n = 3$ in Fermat's Last Theorem is a jump up in complexity from the $n = 4$ case. It uses facts about the domain $\mathbf{Z}[\omega]$ (see Section 6.7) or about number fields generally (see Chapter 9). Read these pages for background and then work through the details of a published proof of the $n = 3$ case (Gioia, 1970; Grosswald, 1984; Rose, 1994). Write up the results of your reading as an extra-credit assignment.

8.5 Other Representation Problems*

The preceding sections have considered the representations of integers by particular forms, such as $N = x_1^2 + x_2^2$ and $N = x_1^2 + x_2^2 + x_3^2 + x_4^2$. Clearly, there is no end to the variety of representational forms that one could entertain.

In a few cases, one can make definite statements with a minimum of effort. For example, it is easy to show that 2 is the only prime that can be represented as the sum of two positive cubes (Problem 8.31). Easier yet is the result that any integer N can be represented by one of the eight forms $N = \pm x_1^2 \pm x_2^2 \pm x_3^2$, for some integers x_1, x_2, x_3 (Problem 8.32). On the other hand, it is not known if an infinite number of primes can be represented by the forms $x^2 + 1$ or $x! \pm 1$. Nor is it known if there are an infinite number of primes among the Fibonacci numbers. It will probably take very deep methods of analysis, algebra, or other tools to solve these problems.

The Four-Square Problem is a particular case of a more general representation problem known as **Waring's Problem** (Ellison, 1971; Small, 1977; Nathanson, 1996). We may state Waring's Problem in the following way.

WARING'S PROBLEM

Given a positive integer n, is there a smallest positive integer r such that the Diophantine equation

$$N = x_1^n + x_2^n + \cdots + x_r^n$$

has a nonnegative solution $(x_1, x_2, \ldots, x_r)$ for all $N \geq 1$, and, if so, what is the value of r for each n?

* Sections marked with an asterisk can be skipped without impairing the continuity of the text.

It is customary to let $g(n)$ denote the function value at n that is the minimum positive integer r such that

$$N = x_1^n + x_2^n + \cdots + x_r^n$$

is solvable for all positive integers N. Thus, from Theorem 8.8 we have $g(2) = 4$. **Waring**[4] himself made a conjecture about $g(3)$ and $g(4)$. **Hilbert** proved in 1909 that $g(n)$ is finite for all n, but his proof gave no method for calculating $g(n)$.

Waring's Problem has not been completely solved, although a great deal of progress has been made on it. One easy result, however, is the following.

Theorem 8.10 $g(3) \geq 9$.

Proof Let $s = \lfloor (3/2)^n \rfloor$, where n is a fixed positive integer. Consider next the integer

$$m = 2^n s - 1.$$

Since $s \leq 3^n 2^{-n}$ for all $n \in \mathbf{N}$, then $m < 3^n - 1$. Thus, m cannot be represented as a sum of nth powers of integers 3 or larger. In fact, m can be written as

$$m = (s - 1)2^n + (2^n - 1)1^n,$$

and if this is written out term-by-term, we see that m is represented by $s - 1 + 2^n - 1$ nth powers and no fewer. Thus, $2^n + s - 2$ is a lower bound for $g(n)$. In particular, when $n = 3$, then $s = \lfloor 27/8 \rfloor = 3$ and $g(3) \geq 2^3 + 3 - 2 = 9$. ◆

On the basis of limited numerical evidence, **Waring** believed that $g(3)$ is actually 9. Table 8.2 gives some illustrative data. The data are, of course, only suggestive.

The data do seem to indicate that many integers require only five cubes or fewer, and that those integers that need eight or nine or possibly more cubes are in the minority. In fact, it was proved in 1939 that 23, 239 are the only integers that require nine cubes (Dickson, 1939b). Later, in 1943, **Yu. V. Linnik** showed that only finitely many integers require eight cubes.

It has become customary to let $G(n)$ denote the least value of r such that for all integers beyond some integer $N_0(n)$, the Diophantine equation

$$N = x_1^n + x_2^n + x_3^n + \cdots + x_r^n$$

is solvable. Clearly, one has $G(n) \leq g(n)$. Surprisingly, $G(n)$ is known for only a few n, whereas $g(n)$ is much better understood. For example, it is easy to show

[4] **Edward Waring** (1734–1798) was Lucasian professor of mathematics (the same chair as that occupied by **Newton** a century earlier) at Cambridge University. He made his conjectures in his book *Meditationes Algebraicae* (1770).

Table 8.2 Representations of Integers N as Sums of Positive Cubes

N	Representation	No. of Cubes
5	$1^3 + 1^3 + 1^3 + 1^3 + 1^3$	5
10	$2^3 + 1^3 + 1^3$	3
15	$2^3 + 1^3 + 1^3 + 1^3 + 1^3 + 1^3 + 1^3 + 1^3$	8
23	$2^3 + 2^3 + 1^3 + 1^3 + 1^3 + 1^3 + 1^3 + 1^3 + 1^3$	9
30	$3^3 + 1^3 + 1^3 + 1^3$	4
51	$3^3 + 2^3 + 2^3 + 2^3$	4
75	$4^3 + 2^3 + 1^3 + 1^3 + 1^3$	5
100	$4^3 + 3^3 + 2^3 + 1^3$	4
200	$4^3 + 4^3 + 4^3 + 2^3$	4
239	$4^3 + 4^3 + 3^3 + 3^3 + 3^3 + 3^3 + 1^3 + 1^3 + 1^3$	9
432	$6^3 + 6^3$	2
500	$5^3 + 5^3 + 5^3 + 5^3$	4
541	$7^3 + 5^3 + 4^3 + 2^3 + 1^3$	5

that $G(2) = 4$, but the determination of the value of $G(3)$ is as yet an unsolved problem in additive number theory.

Much current effort is being expended on finding progressively smaller upper bounds for various $G(n)$. In Table 8.3 we list some contemporary results. Some of these are likely to be revised downward in the near future.

Let us return to representations by cubes, and suppose that we want to place an upper bound on $g(3)$. Following Hardy and Wright (1979), we do this by first placing an upper bound on $G(3)$. The proof is entirely arithmetic in nature but is

Table 8.3 Recent Work on $G(n)$

n	$G(n)$
3	$G(3) \leq 7$[a]
4	$G(4) = 16$[b]
5	$G(5) \leq 17$[c]
6	$G(6) \leq 25$[c]
7	$G(7) \leq 33$[c]
8	$G(8) \leq 43$[c]
9	$G(9) \leq 51$[c]
10	$G(10) \leq 59$[c]

[a] G.L. Watson, "A Proof of the Seven Cube Theorem," *J. London Math. Soc.*, **26**, 153–156 (1951).
[b] H. Davenport, "On Waring's Problem for Fourth Powers," *Ann. Math.*, **40**, 731–747 (1939).
[c] R.C. Vaughan and T.D. Wooley, "Further Improvements in Waring's Problem," *Acta Math.*, **174**, 147–240 (1995).

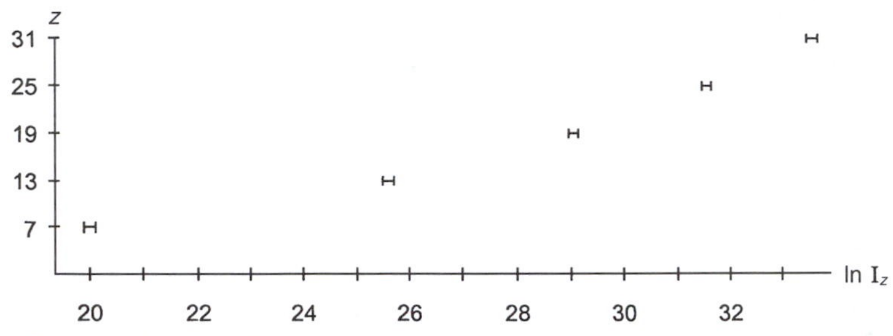

Figure 8.2 Early $\mathbf{I}_z$ intervals do not overlap, but do move closer together.

sophisticated. The idea is to look at intervals along the positive real line, and to show that all integers in any such interval are expressible as sums of a certain number of cubes.

Let $z = 6k + 1$, $k \in \mathbf{N}$. Define the interval $\mathbf{I}_z$ by

$$\mathbf{I}_z = [11z^9 + (z^3 + 1)^3 + 125z^3,\ 14z^9].$$

The reason for this choice will emerge shortly. However, we confirm that for $z \geq 3$, $14z^9 > 11z^9 + (z^3 + 1)^3 + 125z^3$, so that the above are indeed the left and right endpoints of intervals. For example, we have

$$\mathbf{I}_7 \approx [12(7)^9,\ 14(7)^9]$$

$$\mathbf{I}_{13} \approx [12(13)^9,\ 14(13)^9]$$

$$\mathbf{I}_{19} \approx [12(19)^9,\ 14(19)^9].$$

It is numerically obvious that for small z, the intervals are disjoint (Fig. 8.2). However, as z increases, the intervals move closer together. Eventually, the intervals begin to overlap. The following lemma is required.

Lemma 8.11.1 For all $z \geq 373$, adjacent intervals $\mathbf{I}_z$ overlap.

Proof Let $f(z) = 11z^9 + (z^3 + 1)^3 + 125z^3$ and $h(z) = 14z^9$, and let $x = z + 6$, $x \geq 13$. We want to show that $h(x - 6) > f(x)$ or that

$$14(x - 6)^9 > 12x^9 + 3x^6 + 128x^3 + 1.$$

This holds iff

$$\left(1 - \frac{6}{x}\right)^9 > \frac{6}{7} + \frac{3}{14x^3} + \frac{64}{7x^6} + \frac{1}{14x^9}.$$

We recall **Bernoulli's Inequality** from Problem 1.18: $(1 + u)^n > 1 + nu$ if $u > -1$, $u \neq 0$, $n > 1$. Hence, the previous inequality holds iff

$$1 - \frac{54}{x} > \frac{6}{7} + \frac{3}{14x^3} + \frac{64}{7x^6} + \frac{1}{14x^9}$$

or, equivalently,

$$x - 378 > \frac{3}{2x^2} + \frac{64}{x^5} + \frac{1}{2x^8}.$$

This is certainly true if $x \geq 379$, so $z \geq 373$. ◆

Example 8.11 Let $z = 373$; clearly $f(z + 6) > f(z)$. On the other hand,

$$f(z+6) < 12\left(\frac{379}{373}\right)^9 (373)^9 + 4\left(\frac{379}{373}\right)^6 (373)^6$$
$$= 13.85(373)^9 + (8.5 \times 10^{-8})(373)^9 < 14(373)^9.$$

Hence, $\mathbf{I}_{z+6}$ overlaps with $\mathbf{I}_z$. ◆

The value of f(373) is approximately 1.68×10^{24}. The lemma tells us that every integer $N \geq N_0 = 1.68 \times 10^{24}$ is therefore in some interval $\mathbf{I}_z$. Rather than looking at the entire real line from N_0 onward, we now focus on a single arbitrary interval.

Theorem 8.11 $G(3) \leq 13$.

Proof Let $N \geq N_0$ be arbitrary but fixed; by Lemma 8.11.1, N lies in some $\mathbf{I}_z$, $z = 6k + 1$, $k = 62, 63, 64, \ldots$. Define integers r, s, M as follows:

$$\begin{cases} 6r \equiv N \pmod{z^3}, & 1 \leq r \leq z^3 \\ s \equiv N - 4 \pmod{6}, & 0 \leq s \leq 5 \\ M = (r+1)^3 + (r-1)^3 + 2(z^3 - r)^3 + (sz)^3. \end{cases}$$

We note that r is unique because $(6, z^3) = d = 1$, so the congruence $6r \equiv N \pmod{z^3}$ has a single solution. The number M, as written, is a sum of five cubes. Further, we have

$$\begin{aligned} M &< (z^3 + 1)^3 + (z^3)^3 + 2(z^3)^3 + (5z)^3 \\ &= 3z^9 + (z^3 + 1)^3 + 125z^3 \\ &= f(z) - 8z^9 \\ &\leq N - 8z^9 \qquad \text{because } f(z) \leq N \leq h(z). \end{aligned}$$

Hence, $N - M - 8z^9$ is positive.

From the definition of M we have modulo z^3

$$\begin{aligned} M &\equiv (r+1)^3 + (r-1)^3 + 2(z^3 - r)^3 \pmod{z^3} \\ &\equiv (r+1)^3 + (r-1)^3 - 2r^3 \pmod{z^3} \\ &\equiv 6r \pmod{z^3} \\ &\equiv N \pmod{z^3} \\ &\equiv N - 8z^9 \pmod{z^3}. \end{aligned}$$

Thus, z^3 divides $N - M - 8z^9$.

The congruence $x^3 \equiv x \pmod 6$ holds for any integer x (Problem 3.14), so modulo 6 we find

$$\begin{aligned} M &\equiv (r+1) + (r-1) + 2(z^3 - r) + sz \pmod 6 \\ &\equiv 2z^3 + sz \pmod 6 \\ &\equiv (2+s)z \pmod 6 \\ &\equiv 2 + s \pmod 6 \qquad \text{because } z = 6k+1 \\ &\equiv N - 2 \pmod 6 \qquad \text{from the definition of } s \\ &\equiv N - 8 \pmod 6 \\ &\equiv N - 8z^9 \pmod 6 \quad \text{again because } z = 6k+1. \end{aligned}$$

Thus, 6 divides $N - M - 8z^9$.

Since $(6, z^3) = 1$, then $6z^3$ divides $N - M - 8z^9$, that is, there is a positive integer m such that

$$N = M + 8z^9 + 6mz^3.$$

By Theorem 8.8 we can express m as

$$m = x_1^2 + x_2^2 + x_3^2 + x_4^2.$$

The terms $8z^9 + 6mz^3$ can now be factored as

$$(z^3 + x_1)^3 + (z^3 - x_1)^3 + (z^3 + x_2)^3 + (z^3 - x_2)^3 \\ + (z^3 + x_3)^3 + (z^3 - x_3)^3 + (z^3 + x_4)^3 + (z^3 - x_4)^3,$$

a total of eight cubes. Since M is a sum of five cubes, then any integer $N \geq N_0$ can be expressed as a sum of at most $8 + 5 = 13$ cubes. ◆

The inequalities $G(3) \leq 13$ and $G(3) \leq g(3)$ do not allow us to conclude automatically that $g(3) \leq 13$. Nevertheless, an upper bound for $g(3)$ is now within easy reach.

Theorem 8.12 $g(3) \leq 13$.

Proof It is known from numerical tables that except for $N = 23$ and 239, all integers below 40,000 require only 8 cubes or fewer. From Theorem 8.11 we know that all integers $N \geq N_0 = 1.68 \times 10^{24}$ require no more than 13 cubes. We shall now examine the interval $240 \leq N < N_0$.

A preliminary result that is needed first is that if x is any positive integer and $y = \lfloor x^{1/3} \rfloor$, then $0 \leq (x^{1/3} - y) < 1$ and this leads to $x - y^3 < 3x^{2/3}$ (Problem 8.43).

Now let N be arbitrary, $240 \leq N < N_0$, and set $n = N - 240$, $n < 1.68 \times 10^{24}$. The following system of equations defines integers N_1, N_2, N_3, N_4, N_5:

$$\begin{cases} n = N - 240 \\ N_1 = n - \lfloor n^{1/3} \rfloor^3 \\ N_2 = N_1 - \lfloor N_1^{1/3} \rfloor^3 \\ N_3 = N_2 - \lfloor N_2^{1/3} \rfloor^3 \\ N_4 = N_3 - \lfloor N_3^{1/3} \rfloor^3 \\ N_5 = N_4 - \lfloor N_4^{1/3} \rfloor^3. \end{cases}$$

Addition of the six equations produces

$$N = 240 + N_5 + \lfloor n^{1/3} \rfloor^3 + \lfloor N_1^{1/3} \rfloor^3 + \lfloor N_2^{1/3} \rfloor^3 + \lfloor N_3^{1/3} \rfloor^3 + \lfloor N_4^{1/3} \rfloor^3.$$

From the preliminary result given earlier, we have

$$\begin{aligned} N_5 &< 3N_4^{2/3} < 3(3N_3^{2/3})^{2/3} < 3(3(3N_2^{2/3})^{2/3})^{2/3} \\ &< 3(3(3(3N_1^{2/3})^{2/3})^{2/3})^{2/3} < 3(3(3(3(3n^{2/3})^{2/3})^{2/3})^{2/3})^{2/3}. \end{aligned}$$

But $n < 1.68 \times 10^{24}$, so substitution of this in the last inequality leads to $N_5 < 27{,}100$. Hence, $240 + N_5 = 27{,}340 < 40{,}000$, so $240 + N_5$ is a sum of at most 8 cubes. The preceding equation for N contains five additional cubic terms, and therefore N is a sum of at most 13 cubes. This result, combined with Theorem 8.11, implies $g(3) \leq 13$. ◆

In case you are wondering, the value of $g(3)$ is actually 9. This was established via more involved methods by **A. Wieferich** in 1909. The best guess for $G(3)$ so far is that $G(3) = 4$ or 5. Incidentally, $g(4) = 19$ was only established as recently as 1992; progress is slow.

Problems

8.31. Let p be a prime.

(a) Show that $p = x_1^3 + x_2^3$ has a solution in positive integers iff $p = 2$.
(b) Show that $p = x^2 + 5x + 2$ has no integral solution for any odd p.

8.32. Let $N \in \mathbf{Z}$; prove that there exist integers x_1, x_2, x_3 such that for some choice of signs one has $N = \pm x_1^2 \pm x_2^2 \pm x_3^2$.

8.33. Prove that a prime p can be represented as $x_1^3 - x_2^3$, $x_1, x_2 \in \mathbf{N}$, iff $p = 3x_2(x_2 + 1) + 1$.

8.34. The number 50 is the smallest integer N that can be written as the sum of two squares in two different ways:

$$50 = 5^2 + 5^2 = 1^2 + 7^2.$$

The integer N that is the smallest integer representable as the sum of two cubes in two different ways,

$$N = x_1^3 + x_2^3 = x_3^3 + x_4^3,$$

is less than 2000. It is sometimes referred to informally as the **Hardy taxicab number** (Kanigel, 1991). Use the computer to discover N.

8.35. It is known that the integer $N = 635{,}318{,}657$ is representable as the sum of two fourth powers in two different ways. Use the computer to discover them.

8.36. It is desired to represent an odd prime p by the form

$$p = 2x^2 - y^2.$$

Make a list of several primes and determine which ones can be so represented.

(a) Then formulate a conjecture, based on your data.
(b) Test the conjecture by considering some larger primes; use the computer, if necessary, in order to find solutions x, y.
(c) If possible, prove your conjecture.

8.37. A theorem says that some integers N less than 77 and all integers exceeding 77 can be represented as sums of distinct positive integers whose reciprocals sum to 1 (Graham, 1963):

$$\begin{cases} N = a_1 + a_2 + \cdots + a_k \\ 1 = \dfrac{1}{a_1} + \dfrac{1}{a_2} + \cdots + \dfrac{1}{a_k} \\ 1 < a_1 < a_2 < \cdots < a_k. \end{cases}$$

(a) Show that if $k = 3$, there is only one integer $N < 77$ that can be so represented.
(b) Find a representable integer $N < 77$ for the case $k = 4$.

8.38. (a) Consider the integer $N = 50$. It can be expressed as

$$N = 3^3 + 2^3 + 2^3 + 1^3 + 1^3 + 1^3 + 1^3 + 1^3 + 1^3 + 1^3.$$

Does this contradict $g(3) = 9$?

(b) Find a lower bound for $g(4)$, and then find an integer N that requires that many fourth-powers for its representation.

8.39. Prove that $G(2) = 4$.

8.40. The text has said that $G(3)$ is probably 4 or 5. In fact, prove that $G(3) > 3$ must hold. [Hint: Consult Table F in Appendix II.]

8.41. Prove **Bernoulli's Inequality**[5] (see Problem 1.18) if you did not do so previously.

8.42. It has been conjectured that only seven integers require as many as 19 biquadrates; curiously, these seven are all less than 1000. Find them by computer.

8.43. In Theorem 8.12 we needed the result that $0 \leq x^{1/3} - \lfloor x^{1/3} \rfloor < 1$ for any $x \in \mathbf{N}$. Show this. Then show by simple factoring that $x - \lfloor x^{1/3} \rfloor^3 < 3x^{2/3}$.

8.44. Let x_1, x_2, x_3, x_4 be positive integers.

(a) Verify the following identity (more Littlewood lingo):

$$\begin{aligned} 6(x_1^2 + x_2^2 + x_3^2 + x_4^2)^2 = {} & [(x_1 + x_2)^4 + (x_1 - x_2)^4] \\ & + [(x_1 + x_3)^4 + (x_1 - x_3)^4] \\ & + [(x_1 + x_4)^4 + (x_1 - x_4)^4] \\ & + [(x_2 + x_3)^4 + (x_2 - x_3)^4] \\ & + [(x_2 + x_4)^4 + (x_2 - x_4)^4] \\ & + [(x_3 + x_4)^4 + (x_3 - x_4)^4]. \end{aligned}$$

(b) Now let $x = 6N + r$, $0 \leq r \leq 5$, where x, N, r are nonnegative integers. Show that $g(4) \leq 53$.

(c) Show that if $x \geq 81$, then $r \equiv 0, 1, 2, 81, 16$, or $17 \pmod 6$. Hence, for $x \geq 81$ the result of part (b) can be improved to $g(4) \leq 50$. This upper bound is not very good, since it is already known that $g(4) = 19$.

8.45. Although every positive integer N can be expressed as the sum of four squares, not every positive integer can be expressed as the sum of four *positive*

[5] After the Swiss mathematician **Jakob Bernoulli** (1654–1705).

squares. For example, $N = 6$ cannot be so expressed. Nor can every positive integer be expressed as the sum of five positive squares. Prove that any integer $N \geq 170$ can be expressed as the sum of five positive squares. [Hint: Let $N = 169 + n, n \geq 1$, and look at 169 carefully.]

8.46. Representation problems are sometimes simpler in finite fields.

(a) Find a finite field $\mathbf{Z}_p, p > 5$, in which 4 cannot be represented as the sum of two cubes.
(b) Let $x > 1$ be an integer. Show that there is a finite field $\mathbf{Z}_p$, $p > x$, in which x can be represented as the sum of two cubes.

8.47. Show that the following representation problem involving cubes has only one solution (a, b, c, d) in positive integers

$$a^3 + b^3 + c^3 = d^3, \qquad b = a + 1$$
$$c = b + 1$$
$$d = c + 1,$$

and find this solution.

RESEARCH PROBLEMS

1. It was proved by **J.-R. Chen** in 1965 that g(5) = 37. Typically, most integers do not require so many fifth powers for their representation. Investigate on the computer and see if you can uncover the smallest integer that does *require* 37 fifth powers.

2. The Diophantine equation $x^2 - dy^2 = n$ has been much studied (especially for $n = 1$, where it is known as **Pell's equation**). Now consider the analogous Diophantine equation $x^3 - dy^3 = n$, $d, n \in \mathbf{N}$. A nontrivial solution is a solution (x, y) in integers such that $xy \neq 0$. Investigate various choices of d, n to see which ones yield nontrivial solutions.

3. Let a be a given nonzero integer. Are there any solutions (x, n) to the Diophantine equation $x^2 + (2a - 1) = a^n$? Investigate this for different choices of a. See if you can prove anything here.

4. Investigate which integers in the domain $\mathbf{Z}[i]$ of Gaussian integers are expressible as sums of at most four squares.

References

Andrews, G.E., Ekhad, S.B., and Zeilberger, D., "A Short Proof of Jacobi's Formula for the Number of Representations of an Integer as a Sum of Four Squares," *Amer. Math. Monthly*, **100**, 274–276 (1993). Jacobi's formula is simpler than one might have expected; the proof here uses some material from the theory of partitions.

Apostol, T.M., *Mathematical Analysis*, 2nd ed., Addison–Wesley, Reading, MA, 1974. See pp. 206–209 for useful background material on the theory of infinite products.

Dence, J.B., and Dence, T.P., "An Application of the Calculus to Number Theory," *Math. Comp. Educ.*, **26**, 125–128 (1992). An interesting use of some standard theorems from undergraduate calculus to estimate the value of the multiplicative constant in the asymptotic formula for the number of integers expressible as sums of two squares.

Dickson, L.E., *Modern Elementary Theory of Numbers*, University of Chicago Press, Chicago, 1939a, pp. 88–96. The proof of the three-square problem involves material from the theory of positive ternary quadratic forms.

Dickson, L.E., "All Integers except 23 and 239 Are Sums of Eight Cubes," *Bull. Amer. Math. Soc.*, **45**, 588–591 (1939). This was an important result which, however, was soon superseded by Linnik's result that G(3) $\leq$ 7.

Ellison, W.J., "Waring's Problem," *Amer. Math. Monthly*, **78**, 10–36 (1971). Glance at this survey article in order to see the great amount of hard analysis that is needed in order to make any headway on such an easily stated theorem.

Fraser, O., "On Representing a Square as the Sum of Three Squares," *Amer. Math. Monthly*, **76**, 922–923 (1969). If n is a positive integer, then $n^2 = x^2 + y^2 + z^2$ has a solution (x, y, z) in positive integers iff n is not of the form 2^k or $2^k \cdot 5$.

Gioia, A.A., *The Theory of Numbers: An Introduction*, Markham Publishing Co., Chicago, 1970, pp. 123–128. See these pages for the proof of the $n = 3$ case of Fermat's Last Theorem. This excellent, small book was ahead of its time and can be highly recommended.

Graham, R.L., "A Theorem on Partitions," *J. Austral. Math. Soc.*, **3**, 435–441 (1963). Interesting paper on partitions of integers into summands whose reciprocals sum to 1; the source of our Problem 8.37

Grosswald, E., *Topics from the Theory of Numbers*, 2nd ed., Birkhäuser, Boston, 1984, pp. 293–298. An essentially parallel treatment to that in Gioia of the $n = 3$ case of Fermat's Last Theorem.

Grosswald, E., *Representations of Integers as Sums of Squares*, Springer, New York, 1985. See the first 65 pages for classical results on the representations of integers as sums of two, three, or four squares.

Hardy, G.H., and Wright, E.M., *An Introduction to the Theory of Numbers*, 5th ed., Oxford University Press, Oxford, England, 1979. See Chapter XXI (pp. 317–339) for a good discussion of representations by cubes and higher powers. Our proof of g(3) $\leq$ 13 is taken from these pages. Interested readers will find the notes at the end of Hardy and Wright's chapter useful.

Kanigel, R., *The Man Who Knew Infinity—A Life of the Genius Ramanujan*, Charles Scribner's Sons, New York, 1991, pp. 311–312. The story of the Hardy taxicab number is well known, but this wonderful book is worth reading from cover to cover.

Mahoney, M.S., *The Mathematical Career of Pierre de Fermat, 1601–1665*, 2nd ed., Princeton University Press, Princeton, NJ, 1994, pp. 36–37, 332–359. The first citation is to the *Ad logisticen speciosam notae priores* of **Viète**, and the second covers Fermat's "Rélation" to **Carcavi** in which he discloses a bit of his Method of Infinite Descent. Mahoney's monograph is filled with interesting material.

Mordell, L.J., *Diophantine Equations*, Academic Press, London, 1969, pp. 175–178. **Louis Joel Mordell** (1888–1972) was an eminent English number theorist whose speciality was Diophantine analysis. See the forementioned pages for a somewhat different approach (from that in Dickson or Grosswald) to the three-square problem. Mordell, the mathematician, is described informally in L.J. Mordell, *Reflections of a Mathematician*, Canadian Mathematical Congress, Montreal, 1959.

Nagell, T., *Introduction to Number Theory*, John Wiley & Sons, New York, 1951, pp. 122–123. Our proof of Thue's Remainder Theorem is taken from these pages.

Nathanson, M.B., *Additive Number Theory: The Classical Bases*, Springer, New York, 1996. Skim read this for material on Waring's Problem and Goldbach's Conjecture; advanced.

Ribenboim, P., *13 Lectures on Fermat's Last Theorem*, Springer, New York, 1979. Although the proof of Fermat's Last Theorem is now a done deal, this book is still worth a look. In it we read, for example, that by 1978 **S.S. Wagstaff** had shown (via much hard work on mainframe computers of the day) that Fermat's Last Theorem was true for all prime exponents less than 125,000. This work could only bolster the opinion of most individuals that Fermat's Last Theorem is true for all $n > 2$.

Rose, H.E., *A Course in Number Theory*, 2nd ed., Oxford University Press, Oxford, England, 1994, pp. 281–283. See these pages for a terse proof of the $n = 3$ case of Fermat's Last Theorem.

Shiu, P., "Counting Sums of Two Squares: The Meissel–Lehmer Method," *Math. Comp.*, **47**, 351–360 (1986). A somewhat detailed but efficient algorithm for calculating the number of integers not exceeding N that are sums of two squares.

Small, C., "Waring's Problem," *Math. Mag.*, **50**, 12–16 (1977). A much more informal summary than the technical review by Ellison of what was known about Waring's Problem at the time of Small's article.

Small, C., "A Simple Proof of the Four-Squares Theorem," *Amer. Math. Monthly*, **89**, 59–61 (1982). A neat algebraic proof that we used for our Theorem 8.8.

Chapter 9
An Introduction to Number Fields

9.1 Numbers Algebraic over a Field

In Section 3.4 some properties of fields were surveyed briefly. For many purposes in number theory these fields are too general. Interesting results and problems exist in the more specialized class of fields known as number fields. We adopt the following definition (Pollard and Diamond, 1975).

> **DEF.** A **number field F** is a commutative ring that contains **Z**, is contained in $\mathcal{C}$, and whose nonzero elements form a multiplicative group.

It follows that since **F** contains the multiplicative inverses of all the nonzero integers, then **F** must contain **Q**. The field **Q** is the simplest number field; the fields $\mathcal{R}$ and $\mathcal{C}$ are also number fields.

The concept of a number field can be used to partition the numbers in $\mathcal{C}$ into two classes: numbers that are algebraic and numbers that are transcendental, with respect to some number field **F**.

> **DEF.** A number $\theta \in \mathcal{C}$ is said to be **algebraic over F** if it is a zero of some nontrivial polynomial over **F**. Any number that is not algebraic over **F** is termed **transcendental over F**.

Example 9.1 Let $\mathbf{F} = \mathbf{Q}$, $\theta = \frac{\sqrt{3}}{2}$. Then θ satisfies $x^2 - \frac{3}{4} = 0$, so θ is algebraic over **Q** (θ also satisfies $4x^2 - 3 = 0$). Note that in this case $\theta \notin \mathbf{Q}$. On

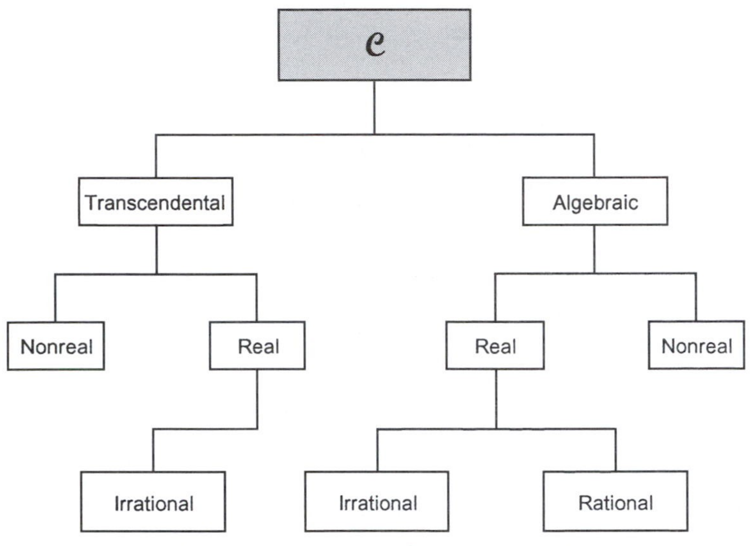

Figure 9.1 A division of numbers in $\mathscr{C}$

the other hand, it can be proved that $\theta = \mathrm{e}$ is transcendental over **Q**. But, e satisfies $x - \mathrm{e} = 0$, so e is algebraic over $\mathscr{R}$. ◆

Being algebraic or transcendental is therefore a function of the choice of number field. When the prepositional phrase "over **F**" is deleted, then the terms *algebraic number* and *transcendental number* automatically mean over **Q**. The numbers 0, 1, $\frac{1}{2}$, $\sqrt[3]{2} + i$, and each of the zeros of $x^5 - x + 1$ are algebraic numbers. The numbers e, π, ln 2 are transcendental numbers. Proving that a number is transcendental is almost always a difficult affair. There are no general procedures. Note that all real transcendental numbers are necessarily irrational (not in **Q**), whereas real algebraic numbers can be either rational or irrational (Fig. 9.1).

To return more generally to numbers algebraic over **F**, we observe that θ can certainly satisfy several polynomials if θ is algebraic. If θ satisfies $\mathrm{p}(x)$, we can clearly take $\mathrm{p}(x)$ to be monic,[1] since otherwise we could simply divide through by the (nonzero) leading coefficient in $\mathrm{p}(x)$. Further, of all possible monic polynomials that have θ as a zero, let us stipulate $\mathrm{p}(x)$ to be one of lowest degree. Such a $\mathrm{p}(x)$ is called a **minimal polynomial** for θ over **F**.

Theorem 9.1 If θ is algebraic over **F**, then θ has a unique minimal polynomial.

[1] A **monic** polynomial is one whose leading coefficient is 1.

Proof Suppose, to the contrary, that θ satisfies two monic polynomials of minimal degree n

$$p_1(x) = x^n + \sum_{i=1}^{n} a_i x^{n-i}$$

$$p_2(x) = x^n + \sum_{i=1}^{n} b_i x^{n-i},$$

where not every a_i equals the corresponding b_i. Let k be the largest power of x for which $a_{n-k} \neq b_{n-k}$. Define

$$p_3(x) = \frac{1}{a_{n-k} - b_{n-k}}[p_1(x) - p_2(x)].$$

Then this is monic of degree $k < n$. But $p_1(\theta) = p_2(\theta) = 0$, so $p_3(\theta) = 0$. This contradicts $p_1(x)$, $p_2(x)$ being minimal, so we must have $a_i = b_i$ for $i = 1, 2, \ldots, n$. ◆

Example 9.2 Show that $x^3 - 2$ is the minimal polynomial for $\sqrt[3]{2}$. Since $\sqrt[3]{2}$ is irrational (by an argument as in Example 2.4), the minimal polynomial for $\sqrt[3]{2}$ cannot be linear. Suppose that $\sqrt[3]{2}$ satisfies the quadratic $x^2 + ax + b$. Then $\sqrt[3]{4} + a\sqrt[3]{2} + b = 0$ and also $2 + a\sqrt[3]{4} + b\sqrt[3]{2} = 0$. Elimination of $\sqrt[3]{4}$ gives $\sqrt[3]{2} = (2 - ab)/(a^2 - b)$. For $b \neq a^2$ the right-hand side is rational, but the left-hand side is irrational, a contradiction. Hence, $\sqrt[3]{2}$ cannot satisfy any quadratic. Since it clearly satisfies $x^3 - 2$, this is unique by Theorem 9.1 and is the minimal polynomial. ◆

There is another way to do Example 9.2, and that involves the following interesting corollary of Theorem 9.1. The statement of this corollary, and its method of proof, are reminiscent of Theorem 5.1 (or see Lemma 4.8.3).

Corollary 9.1.1 If $p(x)$ is the minimal polynomial for θ and if θ satisfies the polynomial $q(x)$ over $\mathbf{F}$, then $p(x) \mid q(x)$.

Proof Let degree $p(x) = n$; according to the preceding theorem, $p(x)$ is unique. By Theorem 4.1 there are then unique polynomials $a(x), b(x) \in \mathbf{F}[x]$ such that

$$q(x) = a(x)p(x) + b(x)$$

and degree $b(x) < n$ or $b(x) = 0$. At $x = \theta$, we have $q(\theta) = p(\theta) = 0$, so $b(\theta) = 0$. But $b(x)$ cannot be a nonzero polynomial of degree $< n$ because $p(x)$ is the *minimal* polynomial for θ. Hence, the only possibility is that $b(x) = 0$, so $q(x) = a(x)p(x)$, that is, $p(x) \mid q(x)$. ◆

Example 9.3 Rework Example 9.2 using Corollary 9.1.1.

Suppose that the minimal polynomial for $\theta = \sqrt[3]{2}$ is the quadratic $x^2 + ax + b$. By Theorem 4.1 we obtain, uniquely,

$$x^3 - 2 = (x^2 + ax + b)(x - a) + \mathrm{r}(x),$$

where $\mathrm{r}(x) = [(a^2 - b)x + (ab - 2)]$. Corollary 9.1.1 then requires that $\mathrm{r}(x) = 0$ identically. This leads to $a^2 - b = 0$ and $ab - 2 = 0$. Elimination of b from these gives $a^3 = 2$, or $a = \sqrt[3]{2}$ and $b = \sqrt[3]{4}$. But $a, b \notin \mathbf{Q}$ (again, as in Example 2.4), so $x^3 - 2$ has no quadratic divisor over $\mathbf{Q}$, and $x^3 - 2$ must be the minimal polynomial for $\sqrt[3]{2}$. In Problem 9.6 we will lead you to an even quicker way of doing Example 9.2. ◆

In order to appreciate the next corollary of Theorem 9.1, we need the concept of an irreducible polynomial. This should be familiar to you, in a more restricted setting, from high-school algebra.

DEF. Let $\mathrm{f}(x), \mathrm{g}(x) \in \mathbf{F}[x]$ have degree ≥ 1. Then $\mathrm{f}(x)$ is said to be **divisible** by $\mathrm{g}(x)$, written $\mathrm{g}(x) \mid \mathrm{f}(x)$, if there is an $\mathrm{h}(x) \in \mathbf{F}[x]$ of degree ≥ 1 such that $\mathrm{g}(x)\mathrm{h}(x) = \mathrm{f}(x)$. If $\mathrm{f}(x) \in \mathbf{F}[x]$ is of degree n over $\mathbf{F}$ and can be represented as $g(x)\mathrm{h}(x) = \mathrm{f}(x)$, where $\mathrm{g}(x)$, $\mathrm{h}(x) \in \mathbf{F}[x]$ and both are of positive degree less than n, then $\mathrm{f}(x)$ is termed **reducible over F**. Otherwise, $\mathrm{f}(x)$ is **irreducible over F**.

Example 9.4 In $\mathbf{Q}[x]$, $\mathrm{g}(x) = x + \frac{1}{2}$ divides $\mathrm{f}(x) = x^3 + \frac{1}{2}x^2 + 4x + 2$ since $\mathrm{g}(x)\mathrm{h}(x) = \mathrm{f}(x)$, where $\mathrm{h}(x) = x^2 + 4$. Also, in $\mathbf{Q}[x]$, $\mathrm{f}(x) = x^2 - \frac{1}{4}$ is reducible over $\mathbf{Q}$, but $x^2 + 1$ and $x^2 - \frac{1}{3}$ are irreducible over $\mathbf{Q}$ because $x^2 + 1 = (x + i)(x - i)$ and $x^2 - \frac{1}{3} = (x + \frac{\sqrt{3}}{3})(x - \frac{\sqrt{3}}{3})$. ◆

Corollary 9.1.2 An irreducible polynomial of degree n over $\mathbf{F}$ has n distinct zeros in $\mathscr{C}$.

Proof Let $\mathrm{f}(x) = \Sigma_{i=0}^{n} a_i x^{n-i}$; it has n zeros in $\mathscr{C}$ according to the Fundamental Theorem of Algebra. It is sufficient to show that there are no double roots of $\mathrm{f}(x) = 0$.

Suppose r is a double root of $\mathrm{f}(x) = 0$. We can then rewrite $\mathrm{f}(x)$ as

$$\mathrm{f}(x) = \mathrm{a}_0(x - r)^2\left[x^{n-2} + \sum_{i=3}^{n} b_i x^{n-i}\right], \ (a_0 \neq 0)$$

where each $b_i \in \mathbf{F}$. Since $\mathrm{f}(x)$ is irreducible over $\mathbf{F}$ by hypothesis, then $a_0^{-1}\,\mathrm{f}(x)$ must be the minimal polynomial for r according to Corollary 9.1.1. Upon differentiation of $\mathrm{f}(x)$ we obtain

$$\begin{aligned}\mathrm{f}'(x) &= 2a_0(x - r)\left[x^{n-2} + \sum_{i=3}^{n} b_i x^{n-i}\right] \\ &\quad + a_0(x - r)^2\left[(n - 2)x^{n-3} + \sum_{i=3}^{n-1} (n - i)b_i x^{n-1-i}\right] \\ &= na_0x^{n-1} + \cdots + a_{n-1}.\end{aligned}$$

But now $(na_0)^{-1}\mathrm{f}'(x)$ is monic and of degree $n - 1$, and yet $(na_0)^{-1}\mathrm{f}'(r) = 0$, a contradiction since $a_0^{-1}\mathrm{f}(x)$ was minimal. It follows that r cannot be a double root and so $\mathrm{f}(x)$ has no multiple roots. ◆

DEF. If θ is algebraic over $\mathbf{F}$ and $\mathrm{p}(x)$, its minimal polynomial, is of degree n, then θ is said to be of **degree n over F**. The zeros $\theta_1 = \theta, \theta_2, \ldots, \theta_n$ are called the **conjugates of θ over F** (or, simply, the **conjugates** when $\mathbf{F} = \mathbf{Q}$).

Example 9.5 In Example 9.2 the conjugates of $\sqrt[3]{2}$ are

$$\sqrt[3]{2},\ \sqrt[3]{2}\frac{(-1 + i\sqrt{3})}{2},\ \sqrt[3]{2}\frac{(-1 - i\sqrt{3})}{2}.$$

As required by Corollary 9.1.2, they are distinct. ◆

Obviously, if $\theta_1 = \theta$ is algebraic over $\mathbf{F}$, then so are $\theta_2, \theta_3, \ldots, \theta_n$, if these are the other conjugates of θ over $\mathbf{F}$. Conjugates figure importantly in a proof to be given shortly of a general result on field extensions.

Problems

9.1. Define $\mathbf{S} = \{a + b\sqrt{3} : a, b \in \mathbf{Q}\}$. Is $\mathbf{S}$ a number field? Define $\mathbf{S}' = \{a + b\sqrt{-3} : a, b \in \mathbf{Q}\}$. Is $\mathbf{S}'$ a number field?

9.2. Is the set $\mathbf{S} = \{a + b\sqrt[3]{2} : a, b \in \mathbf{Q}\}$ a number field? Discuss. Is the set $\mathbf{S}' = \{a + b\sqrt{2} + c\sqrt[3]{2} : a, b, c \in \mathbf{Q}\}$ an example of a number field?

9.3. Show that each of the following is algebraic over **Q**:

(a) $(1 + i)/3$. (b) $1/\sqrt[3]{6}$. (c) $\sqrt{2} + \sqrt[3]{2}$.

9.4. Accept that e is transcendental. Are the numbers $\sqrt{2}$ e, e + 1, e^{-1} also transcendental? Is the product of two transcendental numbers necessarily transcendental? Do the transcendental numbers constitute a subfield of $\mathscr{C}$?

9.5. Find the minimal polynomials over **Q** for each of the following:

(a) $(1 + i\sqrt{3})/2$. (b) $\sqrt{2} + \sqrt[4]{2}$. (c) $\sqrt{(2i)}$.

9.6. **Eisenstein's Irreducibility Criterion** says: Let $f(x) = \Sigma_{i=0}^{n} a_i x^{n-i}$ be in **Z**[x]. If a prime p exists such that $p \nmid a_0$, $p^2 \nmid a_n$, $p \mid a_i$ for $1 \leq i \leq n$, then f(x) is irreducible over **Q**.

(a) Show how Eisenstein's Irreducibility Criterion obviates the work in Example 9.3.
(b) Show that $x^3 + 5x^2 - 10$ is irreducible over **Q**, and that $x^4 + 4$ is reducible over **Q**.
(c) If p(x) is the minimal polynomial of θ over **Q**, show that p(x) is irreducible over **Q**.

9.7. Two polynomials f(x), g(x) $\in$ **F**[x] are **relatively prime** (or **coprime**) if their only common divisors are some of the nonzero elements in **F**. Prove that if f(x), g(x) $\in$ **F**[x] are coprime, then there are polynomials a(x), b(x) $\in$ **F**[x] such that a(x)f(x) + b(x)g(x) = 1. [Hint: Look at the set **S** = {A(x)f(x) + B(x)g(x): A(x), B(x) $\in$ **F**[x]}.]

9.8. Find a monic polynomial p(x) of degree 3 over **Q** such that p(0) = 1 and $p(\sqrt{2}) = 0$.

9.9. Find a monic polynomial p(x) of degree 4 over **Q** such that $p(\sqrt{2} + \sqrt{5}) = 0$.

9.10. In each case show that g(x) | f(x) in **Q**[x]:

(a) $g(x) = x + \frac{1}{3}$. $f(x) = 243x^5 + 1$.
(b) $g(x) = x^2 + x + 1$. $f(x) = x^6 - 1$.
(c) $g(x) = -\frac{3}{4}x^3 + x - \frac{1}{2}$. $f(x) = \frac{1}{4}x^7 - x^6 - \frac{67}{36}x^3 + \frac{20}{9}x^2 - \frac{1}{2}$.

9.11. Conjecture: A polynomial p(x) $\in$ **Q**[x] that is reducible over **Q** has a zero in **Q**. Prove or give a counterexample.

9.12. Let **F** be a number field and f(x) $\in$ **F**[x] be of degree 2 or 3. Show that f(x) is reducible over **F** iff it has a zero in **F**.

9.13. Prove that if $\theta \neq 0$ and is algebraic over **F**, then all of the conjugates of θ are nonzero. Also show that θ^* (the complex conjugate of θ) is a conjugate of θ over **F**.

9.14. Suppose that the algebraic number θ has as its minimal polynomial $x^n + a_1x^{n-1} + a_2x^{n-2} + \cdots + a_n$, where each $a_i \in \mathbf{F}$ and $a_1 \neq 0$. Show that there is another algebraic number of degree n over **F** whose minimal polynomial has no term of algebraic degree $n - 1$.

9.15. Recall the following definition from earlier studies in mathematics:

> **DEF.** An infinite set **S** is **countable (denumerable)** if its members can be placed in a one-to-one correspondence with **N**.

Now prove that the algebraic numbers in any number field **F** are countable. To do this it is sufficient to show that the totality of all algebraic numbers (in $\mathscr{C}$) is countable. Let $a_0x^n + a_1x^{n-1} + a_2x^{n-2} + \cdots + a_n$ be a nonconstant polynomial over **Z**. Every algebraic number is a zero of some polynomial like this. Define the **rank** R of the polynomial by $R = n + \sum_{i=0}^{n} |a_i|$. The minimum value of R is obviously 2. Use R to help you to devise a scheme that lists the algebraic numbers in some order.

9.2 Extensions of Number Fields

Extensions of fields are, in an obvious sense, the opposite of subfields.

> **DEF.** Let **F** be a number field. Then any number field **K** that contains **F** is called an **extension** of **F**. All number fields are extensions of **Q**.

Here is one way to construct an extension field. Let $\theta \notin \mathbf{F}$ be algebraic over **F**.[2] Then define $\mathbf{K} = \mathbf{F}(\theta)$ to be the *smallest* field that contains both **F** and θ (Fig. 9.2). In general, there will be several fields, $\mathbf{K}_1$, $\mathbf{K}_2$, and so on, larger than **K** that contain both **F** and θ (don't forget $\mathscr{C}$ itself!), but $\mathbf{F}(\theta)$ denotes the smallest of them.

[2] The statement $\theta \notin \mathbf{F}$ is not really a restriction. For if $\theta \in \mathbf{F}$, then adjunction of θ to **F** produces just **F**; this is uninteresting.

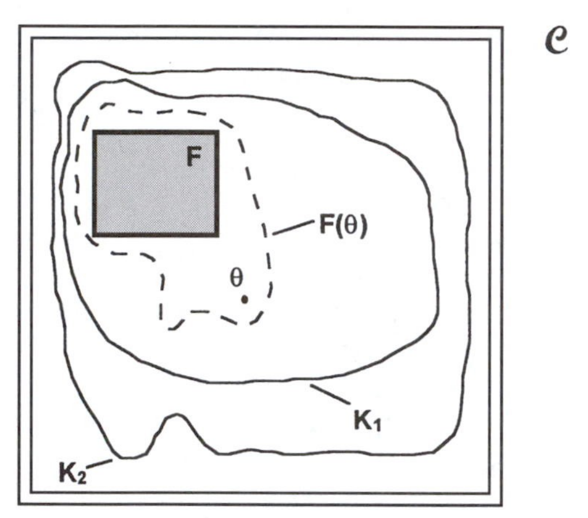

Figure 9.2 The field $\mathbf{F}(\theta)$ is the smallest field that contains $\mathbf{F}$ and θ

One says that $\mathbf{K} = \mathbf{F}(\theta)$ is the field formed by **adjunction** of θ to the field $\mathbf{F}$. The extension field here is called a **simple algebraic extension of F** because only a *single* algebraic number θ has been adjoined to $\mathbf{F}$.

The most obvious question at this point is what the elements of $\mathbf{F}(\theta)$ look like. The following illustration steers us in the right direction.

Example 9.6 Consider $\mathbf{K} = \mathbf{Q}(\sqrt[3]{2})$. Since $\mathbf{K}$ is a field, it is closed under arithmetic operations. Hence, not only does it contain $\theta = \sqrt[3]{2}$, but also $\sqrt[3]{4} = \theta^2$, $\sqrt[3]{4} + 2\sqrt[3]{2} = \theta^2 + 2\theta$, $5\sqrt[3]{4} - 3\sqrt[3]{2} + \frac{1}{2} = 5\theta^2 - 3\theta + \frac{1}{2}$, and so on. We see that these are all polynomials in θ over $\mathbf{Q}$. ◆

Theorem 9.2 Let $\mathbf{F}(\theta)$ be a simple algebraic extension of the number field $\mathbf{F}$. Then any element $\alpha \in \mathbf{F}(\theta)$ can be represented as a polynomial in θ over $\mathbf{F}$.

Proof Define the set $\mathbf{S}$ by

$$\mathbf{S} = \left\{\frac{f(\theta)}{g(\theta)} : f(x),\ g(x) \in \mathbf{F}[x],\ g(\theta) \neq 0\right\}.$$

It is easily verified that $\mathbf{S}$ is a field. Since $\mathbf{S}$ clearly contains $\mathbf{F}$, θ by appropriate choices of $f(x)$ and $g(x)$, then $\mathbf{F}(\theta) \subset \mathbf{S}$. However, any subfield of $\mathbf{F}(\theta)$ that contains both $\mathbf{F}$ and θ must also contain all elements of the form

$$a_0\,\theta^n + a_1\,\theta^{n-1} + \cdots + a_n,$$

where each $a_i \in \mathbf{F}$ and $n \in \mathbf{N}$, because of closure upon addition and multiplication. So $\mathbf{F}(\theta)$ must contain all of these elements as well as all quotients of such elements since $\mathbf{F}(\theta)$ is a field. Thus, we have $\mathbf{S} \subset \mathbf{F}(\theta)$, and this together with the previous set inclusion implies that $\mathbf{F}(\theta) = \mathbf{S}$.

Now let $\alpha \in \mathbf{F}(\theta)$; we can represent α as a quotient,

$$\alpha = \frac{f(\theta)}{g(\theta)}, \quad (g(\theta) \neq 0).$$

Let $p(x)$ be the minimal polynomial over $\mathbf{F}$ for θ. It is irreducible and, further, $p(x) \nmid g(x)$, for otherwise $g(\theta) = 0$ would hold, a contradiction. Hence, $p(x)$ and $g(x)$ are coprime, so by Problem 9.7 there are unique polynomials $a(x)$, $b(x) \in \mathbf{F}[x]$ such that

$$a(x)p(x) + b(x)g(x) = 1.$$

When $x = \theta$ this reduces to $b(\theta) = 1/[g(\theta)]$, and therefore

$$\alpha = f(\theta)b(\theta) = k(\theta).$$

Obviously, $k(\theta)$ is a polynomial in θ over $\mathbf{F}$ since $f(\theta)$, $b(\theta)$ are polynomials in θ over $\mathbf{F}$. ◆

Corollary 9.2.1 If θ is of degree n over $\mathbf{F}$, then $\alpha \in \mathbf{F}(\theta)$ is representable as a polynomial in θ over $\mathbf{F}$ of degree at most $n - 1$. Further, this polynomial is unique.

Proof Let $\alpha = f(\theta)/g(\theta) = k(\theta)$; from Theorem 4.1 we have

$$k(x) = q(x)p(x) + r(x),$$

where $p(x)$ is the minimal polynomial over $\mathbf{F}$ for θ, and either the degree $r(x) <$ degree $p(x) = n$ or $r(x) = 0$. When $x = \theta$, the equation reduces to $k(\theta) = r(\theta)$, so degree $k(x) \leq n - 1$.

Suppose $\alpha = k(\theta) = k_1(\theta)$, where $k(x)$, $k_1(x)$ are each of degree $n - 1$ at most. Then $(k - k_1)(\theta) = 0$ and so θ is a zero of a polynomial over $\mathbf{F}$ of degree less than n. This is a contradiction since $p(x)$ is minimal. Hence, $k(\theta)$ is unique. ◆

Example 9.7 Let $\mathbf{F}(\theta) = \mathbf{Q}(\sqrt[3]{2})$. Here, $\theta = \sqrt[3]{2}$ is of degree 3 over $\mathbf{Q}$ since the minimal polynomial for θ is $x^3 - 2$ (Example 9.2). Therefore, by Corollary 9.2.1 the elements of $\mathbf{Q}(\sqrt[3]{2})$ are numbers of the form $A + B\sqrt[3]{2} + C\sqrt[3]{4}$, $A, B, C \in \mathbf{Q}$. ◆

It is easy to show by simple arguments of irrationality that $\sqrt{3}$ is not an element of the extension field $\mathbf{Q}(\sqrt{2})$ (Problem 9.19). Hence, it is worthwhile to consider the adjunction of both $\sqrt{2}$ and $\sqrt{3}$ to $\mathbf{Q}$. But this adjunction can be imagined in two ways:

(1) Adjoin $\sqrt{2}$, $\sqrt{3}$ simultaneously, that is, construct the smallest field that contains $\mathbf{Q}$, $\sqrt{2}$, $\sqrt{3}$; call this field $\mathbf{Q}(\sqrt{2}, \sqrt{3})$.
(2) Adjoin $\sqrt{2}$ first to $\mathbf{Q}$ to give $\mathbf{Q}(\sqrt{2})$; then adjoin $\sqrt{3}$, that is , construct the smallest field that contains $\mathbf{Q}(\sqrt{2})$ and $\sqrt{3}$; call this field $\mathbf{Q}(\sqrt{2})(\sqrt{3})$.

Are the two fields just described the same? We make this definition:

DEF. Let $\alpha_1, \alpha_2, \ldots, \alpha_n$ be algebraic over the number field $\mathbf{F}$. Then the number field $\mathbf{K} = \mathbf{F}(\alpha_1, \alpha_2, \ldots, \alpha_n)$ obtained by adjoining $\alpha_1, \alpha_2, \ldots, \alpha_n$ simultaneously to $\mathbf{F}$ is called a **multiple algebraic extension** of $\mathbf{F}$ if $n > 1$.

Theorem 9.3 Let $\alpha, \beta \notin \mathbf{F}$ be algebraic over $\mathbf{F}$. Then $\mathbf{F}(\alpha, \beta) = \mathbf{F}(\alpha)(\beta)$.

Proof As usual, we show that two sets are equal by showing that each is a subset of the other.

$\mathbf{F}(\alpha)(\beta)$ is a subfield of $\mathscr{C}$ containing $\mathbf{F}(\alpha)$ and β, so $\mathbf{F}(\alpha)(\beta)$ contains $\mathbf{F}$, α, β. Hence, $\mathbf{F}(\alpha, \beta) \subset \mathbf{F}(\alpha)(\beta)$.

On the other hand, $\{\alpha\} \subset \{\alpha\} \cup \{\beta\}$ and $\{\beta\} \subset \{\alpha\} \cup \{\beta\}$, so $\mathbf{F}(\alpha) \subset \mathbf{F}(\alpha, \beta)$ and $\{\beta\} \subset \mathbf{F}(\alpha, \beta)$. Therefore, $\mathbf{F}(\alpha)(\beta) \subset \mathbf{F}(\alpha, \beta)$. The two set inclusions imply $\mathbf{F}(\alpha, \beta) = \mathbf{F}(\alpha)(\beta)$. ◆

Corollary 9.3.1 Let $\alpha_1, \alpha_2, \ldots, \alpha_n$ be algebraic over $\mathbf{F}$. Then $\mathbf{F}(\alpha_1, \alpha_2, \ldots, \alpha_n) = \mathbf{F}(\alpha_1)(\alpha_2) \ldots (\alpha_n)$.

Proof By mathematical induction. ◆

The corollary says that adjunction to $\mathbf{F}$ of n numbers algebraic over $\mathbf{F}$ is an unambiguous process.

Our last result in this section makes a striking connection between simple and multiple algebraic extensions. For example, suppose we wish to adjoin $\sqrt{2}$ and $\sqrt{3}$ to $\mathbf{Q}$. The multiple algebraic extension so obtained must contain all of the rationals, $\sqrt{2}$ and $\sqrt{3}$, and all additive or multiplicative combinations of these such as $1 - \sqrt{2}$ and $\sqrt{2} \cdot \sqrt{3} = \sqrt{6}$.

But now let $\theta = \sqrt{2} + \sqrt{3}$. This is algebraic over $\mathbf{Q}$ since θ satisfies $x^4 - 10x^2 + 1 = 0$. We find easily $\sqrt{3} = (\theta + \theta^{-1})/2$, $\sqrt{2} = (\theta - \theta^{-1})/2$, and $\sqrt{6} = (\theta^2 - 5)/2$. Thus, each element in the multiple extension $\mathbf{Q}(\sqrt{2}, \sqrt{3})$ can be expressed by additions and multiplications of the rationals and of the algebraic number θ, that is, $\mathbf{Q}(\sqrt{2}, \sqrt{3}) \subset \mathbf{Q}(\theta)$. Also, every integral power of θ can be expressed in terms of $\sqrt{2}$, $\sqrt{3}$, $\sqrt{6}$, and rationals, all of which are elements of $\mathbf{Q}(\sqrt{2}, \sqrt{3})$. So $\mathbf{Q}(\theta) \subset \mathbf{Q}(\sqrt{2}, \sqrt{3})$, and the two extensions are identical. The following theorem generalizes this.

Theorem 9.4 (The Primitive Element Theorem). If $\mathbf{F}$ is a number field and α, β are algebraic over $\mathbf{F}$, then there exists a number $\gamma \in \mathbf{F}(\alpha, \beta)$ such that $\mathbf{F}(\alpha, \beta) = \mathbf{F}(\gamma)$.

Proof Let the minimal polynomials for α, β be $f(x)$, $g(x)$, respectively, and suppose these to have degrees m, n. The conjugates of α, β over $\mathbf{F}$ are α_1 $(= \alpha)$, $\alpha_2, \ldots, \alpha_m$ and β_1 $(= \beta)$, $\beta_2, \ldots, \beta_n$, respectively. By Corollary 9.1.2 all of the β_j's are distinct. Therefore, the equation

$$\alpha_i + x\beta_j = \alpha_1 + x\beta_1 = \alpha + x\beta$$

has a unique solution $x \in \mathscr{C}$ for each $i \neq 1$ and each $j \neq 1$, namely,

$$x = \frac{\alpha_i - \alpha}{\beta - \beta_j}.$$

As $\mathbf{F}$ is infinitely large, we can certainly find a number $c \in \mathbf{F}$ such that $\alpha_i + c\beta_j \neq \alpha + c\beta$ for all $i, j \neq 1$. Define the number γ by

$$\gamma = \alpha + c\beta.$$

Since $c \in \mathbf{F}$, then $\gamma \in \mathbf{F}(\alpha, \beta)$ and so $\mathbf{F}(\gamma) \subset \mathbf{F}(\alpha, \beta)$. We now want to show set inclusion in the opposite direction.

We have $f(\gamma - c\beta) = f(\alpha) = 0$ because $f(x)$ is the minimal polynomial for α. Also, $g(\beta) = 0$ for an analogous reason. Thus, β is a common zero in $\mathscr{C}$ of $f(\gamma - cx)$ and $g(x)$. This is their only common zero , for if another zero of $g(x)$ were also a zero of $f(\gamma - cx)$, say $f(\gamma - c\beta_k) = 0$, then $\gamma - c\beta_k$ would have to equal one of the α_i's. This contradicts the way that c was selected.

Let $p(x)$ be the minimal polynomial over $\mathbf{F}(\gamma)$ for β. The polynomial $f(\gamma - cx)$ has coefficients in $\mathbf{F}(\gamma)$, and the polynomial $g(x)$, whose coefficients are in $\mathbf{F}$, then has coefficients automatically in $\mathbf{F}(\gamma)$. Corollary 9.1.1 now applies and we obtain $p(x) \mid f(\gamma - cx)$ and $p(x) \mid g(x)$. The polynomial $p(x)$ cannot be of the second degree or higher because if it were, then $g(x)$, $f(\gamma - cx)$ would have two or more common zeros in $\mathscr{C}$. So, $p(x)$ is linear, $p(x) = ax + b$, where $a, b \in \mathbf{F}(\gamma)$. But now $p(\beta) = 0$, so we find $\beta = -b/a$, which is a number in $\mathbf{F}(\gamma)$.

Finally, $\alpha = \gamma - c\beta$ is clearly also in $\mathbf{F}(\gamma)$. Then $\alpha, \beta \in \mathbf{F}(\gamma)$ implies that $\mathbf{F}(\alpha, \beta) \subset \mathbf{F}(\gamma)$. The two set inclusions yield $\mathbf{F}(\alpha, \beta) = \mathbf{F}(\gamma)$, and we are done. ◆

Example 9.8 Let $\mathbf{F} = \mathbf{Q}$, $\alpha = \sqrt{2}$, $\beta = \sqrt{3}$, $f(x) = x^2 - 2$, $g(x) = x^2 - 3$. The conjugates of α are $\sqrt{2}$, $-\sqrt{2}$ and the conjugates of β are $\sqrt{3}$, $-\sqrt{3}$. A number $c \in \mathbf{Q}$ such that

$$\alpha_i + c\beta_j \neq \sqrt{2} + c\sqrt{3}$$

for $i, j \neq 1$ is $c = 1$, for example. Then $\gamma = \alpha + c\beta = \sqrt{2} + \sqrt{3}$, and we recover the example discussed prior to the statement of the theorem. Clearly, other values for c would work also. ◆

The number γ in Theorem 9.4 is a **primitive element** of $\mathbf{F}(\gamma)$; it generates the extension of $\mathbf{F}$ to $\mathbf{F}(\gamma)$. In general, as suggested by Example 9.8, a given extension field $\mathbf{F}(\gamma)$ has several primitive elements.

Problems

9.16. Consider the $\mathbf{K}$ of Example 9.6. Show that the multiplicative inverse of $1 + \sqrt[3]{2}$ belongs to $\mathbf{K}$.

9.17. Let α be a nonreal root of $x^5 + x + 1 = 0$. Consider the field $\mathbf{Q}(\alpha)$. In it, express the multiplicative inverse of $\alpha^2 - \alpha - 2$ as a polynomial in α.

9.18. Let λ be a nonreal fourth root of 2 and let $\mathbf{F} = \mathbf{Q}$. Theorem 9.2 implies that $\alpha = (1 - \lambda)/(\lambda^3 - \lambda + 2)$ belongs to $\mathbf{Q}(\lambda)$. Corollary 9.2.1 implies that α can be represented as a cubic in λ over $\mathbf{Q}$. Find this cubic.

9.19. In general, many numbers in $\mathscr{C}$ will not be present in a given algebraic extension $\mathbf{F}(\theta)$. Use a simple argument to show that $\sqrt{3} \notin \mathbf{Q}(\sqrt{2})$.

9.20. Let $\mathbf{S}$ be a collection of number fields.

(a) If $\mathbf{F}, \mathbf{F}' \in \mathbf{S}$, show that $\mathbf{F} \cap \mathbf{F}' \in \mathbf{S}$.
(b) Let $\mathbf{F}_0$ be a fixed number field, let θ be a number algebraic over $\mathbf{F}_0$, and let $\mathbf{S}$ be the set of all number fields $\mathbf{F}$ that contain $\mathbf{F}_0, \theta$. What is $\bigcap_{\mathbf{F} \in \mathbf{S}} \mathbf{F}$?

9.21. Prove Corollary 9.3.1.

9.22. (a) Show that $i \in \mathbf{Q}(i + \sqrt{3})$.
(b) Show that $\cos(\pi/8)$ is algebraic over $\mathbf{Q}$. Now consider the extension field $\mathbf{Q}(\cos(\pi/8))$. Does $\tan(\pi/8)$ belong to it?

9.3 Transcendental Elements and Degree of an Extension

A simple algebraic extension is a bare-bones field. The definition suggests this, but there are simply lots of different kinds of numbers that are not in $\mathbf{F}(\theta)$. We recall from Section 9.1 that a number transcendental over $\mathbf{F}$ is one that is not the zero of any nontrivial polynomial over $\mathbf{F}$.

Theorem 9.5 If θ is algebraic over the number field $\mathbf{F}$, then $\mathbf{F}(\theta)$ contains no numbers that are transcendental over $\mathbf{F}$.

Proof Suppose $\alpha \in \mathbf{F}(\theta)$ and θ is of degree n. By Corollary 9.2.1 the numbers $1, \alpha, \alpha^2, \ldots, \alpha^n$ can be expressed as polynomials in θ over $\mathbf{F}$ of degree $n - 1$ at most,

$$\alpha^k = \sum_{j=0}^{n-1} c_{jk}\theta^j, \qquad k = 0, 1, \ldots, n.$$

We suppose the c_{jk}'s are known.

Now consider this system of n linear homogeneous equations in the $n + 1$ unknowns $a_0, a_1, \ldots, a_n$:

$$\begin{cases} c_{00}a_0 + c_{01}a_1 + c_{02}a_2 + \cdots + c_{0n}a_n = 0 \\ c_{10}a_0 + c_{11}a_1 + c_{12}a_2 + \cdots + c_{1n}a_n = 0 \\ \qquad \vdots \\ c_{n-1,0}a_0 + c_{n-1,1}a_1 + c_{n-1,2}a_2 + \cdots + c_{n-1,n}a_n = 0. \end{cases}$$

Because the number of unknowns exceeds the number of equations, the system has a nontrivial solution $(a_0, a_1, \ldots, a_n)$ in $\mathbf{F}$. Now form the sum

$$\begin{aligned} a_0\alpha^0 + a_1\alpha^1 + \cdots + a_n\alpha^n &= \sum_{k=0}^{n}\left(a_k \sum_{j=0}^{n-1} c_{jk}\theta^j\right) = \sum_{j=0}^{n-1}\sum_{k=0}^{n} a_k c_{jk}\theta^j \\ &= \sum_{j=0}^{n-1} \theta^j \left(\sum_{k=0}^{n} a_k c_{jk}\right) \\ &= 0 \end{aligned}$$

from the preceding. Thus, α is a zero of the polynomial

$$a_n x^n + a_{n-1}x^{n-1} + \cdots + a_1 x + a_0,$$

so α is algebraic over $\mathbf{F}$. As α was arbitrary, then no numbers in $\mathbf{F}(\theta)$ are transcendental and, further, all the numbers in $\mathbf{F}(\theta)$ are of degree not exceeding the degree of θ. ◆

Example 9.9 Let $\mathbf{Q}(\sqrt{2}, \sqrt{3}) = \mathbf{Q}(\sqrt{2} + \sqrt{3})$. Substitution shows that $\theta = \sqrt{2} + \sqrt{3}$ is a zero of $x^4 - 10x^2 + 1$. This polynomial is found to be irreducible over $\mathbf{Q}$ (Problem 9.23). Hence, by Theorem 9.5 every number in $\mathbf{Q}(\sqrt{2}, \sqrt{3})$ is of degree 4 or less over $\mathbf{Q}$. ◆

Theorem 9.5 says that from the single algebraic number θ there is generated a corpus of new numbers algebraic over $\mathbf{F}$, namely, the set complement $\mathbf{F}(\theta) \sim \mathbf{F}$. The elements of $\mathbf{F}$ itself are, of course, automatically algebraic over $\mathbf{F}$. If a different algebraic number θ' had been chosen, then $\mathbf{F}(\theta') \sim \mathbf{F}$ could have been different from $\mathbf{F}(\theta) \sim \mathbf{F}$. Is there some structure to the set of all conceivable numbers algebraic over $\mathbf{F}$? Amazingly, there is.

Corollary 9.5.1 Let $\mathbf{F}$ be a number field and $\mathbf{A}$ be the set of all numbers algebraic over $\mathbf{F}$. Then $\mathbf{A}$ is a field.

Proof Let $\alpha, \beta \in \mathbf{A}$ be arbitrary. Then α, β also belong to $\mathbf{F}(\alpha, \beta)$. Since $\mathbf{F}(\alpha, \beta)$ is a field, then $\alpha + \beta$, $\alpha - \beta$, $\alpha\beta$, α/β (when $\beta \neq 0$) belong to $\mathbf{F}(\alpha, \beta)$ too. But $\mathbf{F}(\alpha, \beta) = \mathbf{F}(\theta)$ for some $\theta \in \mathbf{F}(\alpha, \beta)$ that is algebraic over $\mathbf{F}$, according to Theorem 9.4. By Theorem 9.5 everything in $\mathbf{F}(\theta)$ is algebraic over $\mathbf{F}$, so $\alpha + \beta$, $\alpha - \beta$, $\alpha\beta$, α/β (when $\beta \neq 0$) are algebraic over $\mathbf{F}$. Hence, they belong to $\mathbf{A}$, and we see that addition and multiplication are binary operations on $\mathbf{A}$. If α satisfies $p(x) = 0$, then $-\alpha$ satisfies $p(-x) = 0$ and α^{-1} satisfies $x^m p(x^{-1}) = 0$ when $\alpha \neq 0$, where m is the degree of α over $\mathbf{F}$. Therefore, $-\alpha$, $\alpha^{-1} \in \mathbf{A}$, and $\mathbf{A}$ thus contains all of its additive and multiplicative identities. Commutativity, associativity, and the distributive laws are inherited from $\mathscr{C}$. It follows that $\mathbf{A}$ is a field. ◆

We have mentioned briefly transcendental elements. What happens if a number α transcendental over $\mathbf{F}$ is adjoined to $\mathbf{F}$? The field obtained is distinct because, again, lots of different kinds of numbers will be missing. For example, $\sqrt{2}$ is not an element of $\mathbf{Q}(\pi)$ (Problem 9.24). The relationship of $\mathbf{Q}(\pi)$ to $\mathbf{Q}$ is clearly different in some way from that of, say, $\mathbf{Q}(\sqrt{2})$ to $\mathbf{Q}$. This leads to the concepts of an extension field as a vector space and of the degree of an extension.

If $\mathbf{F}$ is a number field and $\mathbf{K}$ is an extension of $\mathbf{F}$, one can view $\mathbf{K}$ as a **vector space** over $\mathbf{F}$. What this means is that for every $x, y \in \mathbf{K}$ and every $a, b \in \mathbf{F}$, these hold:

1. $\mathbf{K}$ is an Abelian group under addition.
2. $ax \in \mathbf{K}$.

3. $a(x + y) = ax + ay$, $(a + b)x = ax + bx$.
4. $a(bx) = (ab)x$.
5. $1 \cdot x = x$, where 1 is the unity of **F**.

It then makes sense to talk about a basis for **K**.

DEF. A **basis for K over F** is a *finite* set of elements $\{x_1, x_2, \ldots, x_n\}$ in **K** such that

(a) $a_1x_1 + a_2x_2 + \cdots + a_nx_n = 0$ implies that all a_i are 0.
(b) Each $x \in \mathbf{K}$ can be represented uniquely as $x = \sum_{i=1}^{n} a_ix_i$, $a_i \in \mathbf{F}$.

The size of the basis set for a vector space is called its **dimension**.

Example 9.10 Let $\mathbf{F} = \mathbf{Q}$ and $\mathbf{K} = \mathbf{Q}(\sqrt{6}, \sqrt{15})$; then **K** contains $\sqrt{6}$, $\sqrt{15}$ as elements. It also contains $\sqrt{10} = \sqrt{6}\sqrt{15}/3$. No other distinct surds can be produced by multiplications. Therefore, a basis for $\mathbf{Q}(\sqrt{6}, \sqrt{15})$ is $\{1, \sqrt{6}, \sqrt{10}, \sqrt{15}\}$. ◆

Since a vector space has a dimension, the following definition is reasonable.

DEF. The **degree** of **K** over **F** is the dimension of **K** as a vector space over **F**. We denote the degree of **K** over **F** by $[\mathbf{K}: \mathbf{F}]$. When $\mathbf{F} \subset \mathbf{K}$ and $[\mathbf{K}: \mathbf{F}]$ is a positive integer, then **K** is called a **finite extension** of **F**.

In Example 9.10 we have $[\mathbf{Q}(\sqrt{6}, \sqrt{15}): \mathbf{Q}] = 4$.

Our first result with the concept of degree of an extension is really nothing more than linear algebra.

Theorem 9.6 If **L** is a finite extension of **K** and if **K** is a finite extension of **F**, then **L** is a finite extension of **F** and $[\mathbf{L}: \mathbf{F}] = [\mathbf{L}: \mathbf{K}][\mathbf{K}: \mathbf{F}]$

Proof Let $[\mathbf{L}: \mathbf{K}] = m$ and $[\mathbf{K}: \mathbf{F}] = n$; let $\mathbf{B}_1 = \{u_1, u_2, \ldots, u_m\}$ be a basis for **L** over **K** and let $\mathbf{B}_2 = \{v_1, v_2, \ldots, v_n\}$ be a basis for **K** over **F**. For any $\alpha \in \mathbf{L}$ we have, uniquely

$$\alpha = \sum_{i=1}^{m} b_iu_i,$$

where each $b_i \in \mathbf{K}$. But as $\mathbf{B}_2$ is a basis for $\mathbf{K}$ over $\mathbf{F}$, then each b_i can be written uniquely as

$$b_i = \sum_{j=1}^{n} c_{ij} v_j,$$

where each $c_{ij} \in \mathbf{F}$. Thus, $\alpha = \Sigma_{j=1}^{n} \Sigma_{i=1}^{m} c_{ij} u_i v_j$ uniquely, so the mn products $\{u_i v_j\}$ span $\mathbf{L}$ as a vector space over $\mathbf{F}$.

Furthermore, the $u_i v_j$'s are linearly independent, for if $u_i v_j = \Sigma_{k,l} d_{kl} u_l v_k$ held for some $u_i v_j$, then the set of c_{ij}'s in the expansion of α would be different (and, thus, not unique). The two properties of the set $\{u_i v_j\}$ show that it is a basis for $\mathbf{L}$ over $\mathbf{F}$ and so $[\mathbf{L}: \mathbf{F}] = mn = [\mathbf{L}: \mathbf{K}][\mathbf{K}: \mathbf{F}]$. ◆

Example 9.11 Suppose $[\mathbf{L}: \mathbf{F}] = 4$. Then there is no intermediate field $\mathbf{K}$ such that $\mathbf{L} \supset \mathbf{K} \supset \mathbf{F}$ and $[\mathbf{L}: \mathbf{K}] = 3$. For if so, then $[\mathbf{K}: \mathbf{F}] = 4/3$ and this contradicts the requirement that vector space dimensionality be a positive integer. ◆

We can now establish rather easily a connection between a number $\alpha \in \mathbf{K}$ being transcendental over $\mathbf{F}$ and the degree of $\mathbf{K}$ over $\mathbf{F}$.

Theorem 9.7 Suppose $\alpha \in \mathbf{K}$ is not in $\mathbf{F}$ but is transcendental over $\mathbf{F}$. Then $[\mathbf{K}: \mathbf{F}]$ is not defined.[3]

Proof Let $\mathbf{K}_1 = \mathbf{F}(\alpha)$ so that $\mathbf{K} \supset \mathbf{K}_1 \supset \mathbf{F}$. Suppose, to the contrary, that $[\mathbf{K}: \mathbf{F}]$ is some positive integer. Then by Theorem 9.6, $[\mathbf{K}_1: \mathbf{F}]$ is also a positive integer; let $[\mathbf{K}_1: \mathbf{F}] = n$. The $n + 1$ numbers $\{1, \alpha, \alpha^2, \ldots, \alpha^n\}$ are all in $\mathbf{K}_1$. They must be linearly dependent over $\mathbf{F}$ since their number exceeds the size of the basis. Thus, in the equation

$$c_0 \alpha^n + c_1 \alpha^{n-1} + \cdots + c_0 = 0,$$

where each $c_i \in \mathbf{F}$, not all of the c_i's need be 0. Therefore, α is a zero of a polynomial over $\mathbf{F}$ of some finite degree. This is a contradiction to α being transcendental over $\mathbf{F}$. We conclude that $[\mathbf{K}: \mathbf{F}]$ is not defined. ◆

Example 9.12 The obvious example: $[\mathfrak{R}: \mathbf{Q}]$ is not defined. ◆

We also have an easy connection between the degree of a field extension $\mathbf{F}(\theta)$ and the degree of the minimal polynomial for θ. Take a look again

[3] Some authors write $[\mathbf{K}: \mathbf{F}] = \infty$.

at Corollary 9.2.1. It says that if θ is algebraic over $\mathbf{F}$ of degree n, then the set $\{1, \theta, \theta^2, \ldots, \theta^{n-1}\}$ is a basis for $\mathbf{F}(\theta)$. Hence, Theorem 9.8 follows.

Theorem 9.8 If θ is algebraic over $\mathbf{F}$, then $[\mathbf{F}(\theta): \mathbf{F}]$ is the degree of the minimal polynomial for θ over $\mathbf{F}$.

Example 9.13 In Example 9.9 we have $[\mathbf{Q}(\sqrt{2} + \sqrt{3}): \mathbf{Q}] = 4$ because the degree of the minimal polynomial for $\sqrt{2} + \sqrt{3}$ over $\mathbf{Q}$ is 4. ◆

Problems

9.23. Verify in Example 9.9 that $x^4 - 10x^2 + 1$ is irreducible over $\mathbf{Q}$.

9.24. Let the number $\alpha \in \mathbf{F}$. The description of $\mathbf{F}(\alpha)$ as the smallest field that contains α and $\mathbf{F}$ is a global description. A more constructive description is to say that $\mathbf{F}(\alpha)$ is the set of quotients of all polynomials in α over $\mathbf{F}$. Then how does this allow you to conclude that $\sqrt{2} \notin \mathbf{Q}(\pi)$? Note that Theorem 9.2 cannot be invoked because $\mathbf{Q}(\pi)$ is not an algebraic extension of $\mathbf{Q}$.

9.25. What is $[\mathscr{C}: \mathscr{R}]$? What is $[\mathscr{C}: \mathbf{Q}]$?

9.26. Show that if $\mathbf{K} \supset \mathbf{F}$, then $[\mathbf{K}: \mathbf{F}] = 1$ iff $\mathbf{K} = \mathbf{F}$.

9.27. Find bases over $\mathbf{Q}$ for these number fields:

(a) $\mathbf{Q}(\sqrt{2} + \sqrt{3})$.
(b) $\mathbf{Q}(\omega^2)$, where $\omega = (-1 + i\sqrt{3})/2$.
(c) $\mathbf{Q}(\sqrt{2}, \sqrt[3]{2})$.

9.28. Let $\mathbf{K}$ be a finite extension of $\mathbf{F}$ and let $\alpha \in \mathbf{K}$. Then prove that the degree of α over $\mathbf{F}$ divides the degree of $\mathbf{K}$ over $\mathbf{F}$.

9.29. Show that there are no irreducible polynomials in $\mathscr{R}[x]$ of degree 3 or higher.

9.30. Suppose $[\mathbf{K}: \mathbf{F}] = p$, a prime, and let $\alpha \notin \mathbf{F}$ be an element of $\mathbf{K}$. Show that the degree of α over $\mathbf{F}$ is p.

9.31. Let $\mathbf{K}$ be an extension of $\mathbf{F}$ and let $\mathbf{A}_K$ denote those elements of $\mathbf{K}$ that are algebraic over $\mathbf{F}$. Prove that $\mathbf{A}_K$ is a subfield of $\mathbf{K}$. Note that this theorem is not quite the same as Corollary 9.5.1.

9.32. Prove that if $\mathbf{K}$ is an extension of $\mathbf{F}$, and if $a, b \in \mathbf{K}$ are algebraic over $\mathbf{F}$ of degrees m, n, respectively, then $a + b$ is algebraic over $\mathbf{F}$ of degree mn at most.

9.4 Algebraic Integers

In the remainder of this chapter we deal only with finite extensions. Amazingly, such extensions are the same as simple algebraic extensions! For suppose $[\mathbf{K}: \mathbf{F}] = n$ and $\{\alpha_1, \alpha_2, \ldots, \alpha_n\}$ is a basis for **K** over **F**. Each α_i is then algebraic over **F** by the same argument used in Theorem 9.7. Thus, we have, $\mathbf{K} = \mathbf{F}(\alpha_1, \alpha_2, \ldots, \alpha_n)$ and from Theorem 9.4, extended by induction, we obtain $\mathbf{K} = \mathbf{F}(\gamma)$ for some $\gamma \in \mathbf{K}$.

Conversely, if **K** is a simple algebraic extension of **F**, then $\mathbf{K} = \mathbf{F}(\theta)$, where θ is of some finite degree n over **F**. By Theorem 9.8 we then have $[\mathbf{K}: \mathbf{F}] = n$.

Specializing now to $\mathbf{F} = \mathbf{Q}$, we introduce the following standard term.

DEF. A finite extension of **Q** is called an **algebraic number field**.

Algebraic number fields necessarily contain no transcendental elements (Theorem 9.5). Thus, $\mathfrak{R}$ is a number field but not an algebraic number field.

What shall we mean by *integers* in an algebraic number field $\mathbf{Q}(\theta)$? As expected, the conventional interpretation is given in terms of polynomials, despite the fact that on the surface this seems roundabout.

DEF. A number α in an algebraic number field $\mathbf{Q}(\theta)$ is an **algebraic integer** if every coefficient in the minimal polynomial for α belongs to **Z**.

What we will do is show that this definition implies properties of α that are "reasonable" for integers. Clearly, we are anticipating here that the set of integers in $\mathbf{Q}(\theta)$ contains the set of integers in **Q**.

(1) First, if α is an integer in $\mathbf{Q}(\theta)$, then so are all of its conjugates. This is reasonable but not compelling. It is clearly true for **Q**, but only in a vacuous sense, since in **Q** an integer has just one conjugate.

(2) If $\alpha \in \mathbf{Q}(\theta)$ is an integer there and $\alpha \in \mathbf{Q}$, then α is actually an integer in **Q**. Because α belongs to **Q**, its minimal polynomial is of first degree, that is, α is a zero of $p(x) = x - \alpha$. But since α is also an integer in $\mathbf{Q}(\theta)$, then each coefficient in $p(x)$ is an element of **Z**, so $\alpha \in \mathbf{Z}$.

(3) Next, we want to show that the integers in $\mathbf{Q}(\theta)$ form a ring. This is eminently desirable since the integers in **Q** constitute a ring (**Z**). To show this, we introduce the concept of an **R**-module (Chahal, 1988). Theorem 9.9, which follows shortly, is presented only for the purpose of providing a link in the proof of the statement that the algebraic integers in any algebraic number field form a ring.

DEF. Let $\langle \mathbf{R}, +, \cdot \rangle$ be a ring. A set $\mathbf{M}$, together with a binary operation of addition ($\oplus$) on $\mathbf{M}$ and an operation of premultiplication ($*$) by elements of $\mathbf{R}$, is a (left) **R-module** if

1. $\mathbf{M}$ is an Abelian group.
2. $a * x \in \mathbf{M}$ for all $a \in \mathbf{R}$ and all $x \in \mathbf{M}$.
3. $a * (x \oplus y) = (a * x) \oplus (a * y)$ and $(a + b) * x = (a * x) \oplus (b * x)$ for all $a, b \in \mathbf{R}$ and all $x, y \in \mathbf{M}$.
4. $a * (b * x) = (a \cdot b) * x$.

If, in addition, $\mathbf{R}$ is a ring with identity 1 and $1 * x = x$ for all $x \in \mathbf{M}$, then $\mathbf{M}$ is called a **unitary R-module**.

Note that $\mathbf{M}$ is not a subset of $\mathbf{R}$, usually. Elements of $\mathbf{M}$ and $\mathbf{R}$ may be completely different objects. A unitary $\mathbf{R}$-module becomes a vector space over $\mathbf{R}$ in the special case where $\mathbf{R}$ is actually a field.

Example 9.14 Let $\mathbf{R}$ be any ring; by definition $\langle \mathbf{R}, + \rangle$ is an Abelian group, so the preceding requirement (1) is satisfied. Choose $\mathbf{M} = \mathbf{R}$; then requirements (2) through (4) are also met. Hence, $\mathbf{R}$ is an $\mathbf{R}$-module over itself. ◆

Example 9.15 Let $\mathbf{R} = \mathbf{Z}$ and let $\mathbf{M}$ be the set of all even integers. Clearly, $\langle \mathbf{M}, + \rangle$ is an Abelian group. Also, for any $a \in \mathbf{Z}$ and any $x \in \mathbf{M}$, we have $ax \in \mathbf{M}$. Requirements (3) and (4) also hold, so $\mathbf{M}$ is a $\mathbf{Z}$-module, actually a unitary $\mathbf{Z}$-module, since $1 \in \mathbf{Z}$. ◆

Example 9.16 Let $\mathbf{R} = \mathbf{Z}$ and let $\mathbf{M}$ be any Abelian group $\langle \mathbf{G}, \oplus \rangle$. For any $a \in \mathbf{Z}$ and any $x \in \mathbf{G}$, define $a * x$ to be x^a. If $a = 0$, x^a is the identity 1 of $\mathbf{G}$. If $a > 0$, then x^a means $\underbrace{x \oplus x \oplus \cdots \oplus x}_{a \text{ terms}}$. Finally, if $a < 0$, then x^a is the group inverse of $x^{|a|}$. Requirement (2) is automatically upheld. Next,

$$\begin{aligned} a * (x \oplus y) &= (x \oplus y)^a \\ &= \underbrace{(x \oplus y) \oplus (x \oplus y) \oplus \cdots \oplus (x \oplus y)}_{a \text{ pairs}} \\ &= (\underbrace{x \oplus x \oplus \cdots \oplus x}_{a \text{ terms}}) \oplus (\underbrace{y \oplus y \oplus \cdots \oplus y}_{a \text{ terms}}) \\ &= x^a \oplus y^a \\ &= (a * x) \oplus (a * y), \end{aligned}$$

so the first part of requirement (3) is met. The remaining requirements are easily shown (verify!), so **G** is a unitary **Z**-module. ◆

Bases for vector spaces should have their analogs for **R**-modules, so the following definition seems immediate.

DEF. A set of numbers $\{x_1, x_2, \ldots, x_n\}$ in a unitary **Z**-module **M** is a **basis for M over Z**, and **M** is said to be a **finitely generated unitary Z-module**, if every number $x \in \mathbf{M}$ can be expressed uniquely as

$$x = a_1x_1 + a_2x_2 + \cdots + a_nx_n,$$

where each $a_i \in \mathbf{Z}$. Alternate symbolism is that

$$\mathbf{M} = \mathbf{Z}x_1 + \mathbf{Z}x_2 + \cdots + \mathbf{Z}x_n.$$

Theorem 9.9 The following are equivalent:

(1) The number α is an algebraic integer.
(2) $\mathbf{Z}[\alpha]$, the set of all polynomials over **Z** in α, is a finitely generated unitary **Z**-module.
(3) There is a finitely generated unitary **Z**-module **M** different from $\{0\}$ such that $\alpha\mathbf{M} \subset \mathbf{M}$.

Proof (1) → (2). We assume $\alpha \neq 0$. As α is an algebraic integer, say of degree n, then there are integers $a_0, a_1, \ldots, a_{n-1}$ such that

$$\alpha^n = a_0 + a_1\alpha + \cdots + a_{n-1}\alpha^{n-1}.$$

Now consider the finitely generated unitary **Z**-module $\mathbf{M} = \mathbf{Z} + \mathbf{Z}\alpha + \cdots + \mathbf{Z}\alpha^{n-1}$. This is a set of polynomials over **Z** in α of degree $n - 1$ or less, so $\mathbf{M} \subset \mathbf{Z}[\alpha]$.

Now $\alpha\mathbf{M} = \mathbf{Z}\alpha + \mathbf{Z}\alpha^2 + \cdots + \mathbf{Z}\alpha^n = \mathbf{Z}a_0 + \mathbf{Z}^{(1)}\alpha + \mathbf{Z}^{(2)}\alpha^2 + \cdots + \mathbf{Z}^{(n-1)}\alpha^{n-1}$ from the previous equation, where $\mathbf{Z}^{(i)} \subset \mathbf{Z}$ is the set of integers of the form $p + qa_i$. Thus, $\alpha\mathbf{M} \subset \mathbf{M}$, so $\alpha^n \in \mathbf{M}$ in particular. Continuing, $\alpha^2\mathbf{M} = (\mathbf{Z}a_0)\alpha + \mathbf{Z}^{(1)}\alpha^2 + \cdots + \mathbf{Z}^{(n-1)}\alpha^n$, and again from the first equation we deduce $\alpha^2\mathbf{M} \subset \mathbf{M}$, so $\alpha^{n+1} \in \mathbf{M}$. By induction, $\alpha^k \in \mathbf{M}$ for all $k \in \mathbf{N}$; therefore, $\mathbf{Z}[\alpha] \subset \mathbf{M}$. The two set inclusions together imply $\mathbf{Z}[\alpha] = \mathbf{M}$, which says that $\mathbf{Z}[\alpha]$ is a finitely generated unitary **Z**-module.

(2) → (3). Choose $\mathbf{M} = \mathbf{Z}[\alpha]$; the previous part has then given us $\alpha\mathbf{Z}[\alpha] \subset \mathbf{Z}[\alpha]$.

(3) → (1). Suppose the finitely generated unitary **Z**-module $\mathbf{M} \neq \{0\}$,

$$\mathbf{M} = \mathbf{Z}x_1 + \mathbf{Z}x_2 + \cdots + \mathbf{Z}x_n,$$

is such that $\alpha\mathbf{M} \subset \mathbf{M}$. For each $i = 1, 2, \ldots, n$, the number αx_i is then in $\mathbf{M}$ (because $x_i \in \mathbf{M}$ and $\alpha\mathbf{M} \subset \mathbf{M}$), so αx_i can be expressed as a linear combination over $\mathbf{Z}$ of the basis elements $x_1, x_2, \ldots, x_n$:

$$\alpha x_i = \sum_{j=1}^{n} a_{ij}x_j \qquad (a_{ij} \in \mathbf{Z},\ i = 1, 2, \ldots, n).$$

In matrix form this is $\alpha\mathbf{Iv} = \mathbf{Av}$, where $\mathbf{I}$ is the $n \times n$ identity matrix and the vector $\mathbf{v}$ is

$$\mathbf{v} = \begin{pmatrix} x_1 \\ x_2 \\ \vdots \\ x_n \end{pmatrix}.$$

Since $\mathbf{M} \neq \{0\}$, then $\mathbf{v} \neq 0$ and from linear algebra the system of equations has a nontrivial solution $\mathbf{v}$ iff $\det(\alpha\mathbf{I} - \mathbf{A}) = 0$. Expansion of the determinant gives a polynomial in α of degree n

$$\alpha^n + c_1\alpha^{n-1} + c_2\alpha^{n-2} + \cdots + c_n,$$

where each $c_i \in \mathbf{Z}$ because the elements of $\mathbf{I}$, $\mathbf{A}$ are integers. Hence, the polynomial is 0 and this shows that α is an algebraic integer. ◆

Example 9.17 Let $\alpha = 1 + i$. Then $\mathbf{Z}[\alpha]$ is a unitary $\mathbf{Z}$-module $\mathbf{M}$. This follows because $\mathbf{M}$ is, in fact, the set of **Gaussian integers**, $\{A + Bi\colon A, B \in \mathbf{Z}\}$, and each Gaussian integer can be written as a polynomial in α: $A + Bi = (A - B) + B\alpha$. Clearly, $\mathbf{M}$ is finitely generated since the set $\{1, \alpha\}$ is a basis for $\mathbf{M}$ over $\mathbf{Z}$. So by Theorem 9.9 α is an algebraic integer. Alternatively, we know this because the minimal polynomial for α is $x^2 - 2x + 2$ (verify!). ◆

Example 9.18 $\sqrt{3}$ is an algebraic integer because its minimal polynomial is $x^2 - 3$. By Theorem 9.9 $\mathbf{Z}[\sqrt{3}] = \{a + b\sqrt{3}\colon a, b \in \mathbf{Z}\}$ is a finitely generated $\mathbf{Z}$-module $\mathbf{M}$. Further, $\sqrt{3}\,\mathbf{M} = \sqrt{3}\mathbf{Z}[\sqrt{3}] = \{3b + a\sqrt{3}\colon a, b \in \mathbf{Z}\}$ is a subset of $\mathbf{Z}[\sqrt{3}]$ because the rational term in each element of $\sqrt{3}\,\mathbf{M}$ is always a multiple of 3. ◆

Theorem 9.10 Let $\mathbf{K}$ be an algebraic number field and $\mathbf{O}_K$ be the set of algebraic integers in $\mathbf{K}$. Then $\mathbf{O}_K$ is a ring.

Proof If α is an algebraic integer in $\mathbf{K}$, then so is $-\alpha$. For example, if α satisfies $x^n + a_1x^{n-1} + a_2x^{n-2} + \cdots + a_n = 0$ and n is odd, then $-\alpha$ satisfies $x^n - a_1x^{n-1} + a_2x^{n-2} - \cdots - a_n = 0$. Thus, $\mathbf{O}_K$ contains all of its additive inverses. Commutativity under addition in $\mathbf{O}_K$ is inherited from $\mathbf{K}$, as are the associative law for multiplication and the two distributive laws. It remains therefore to show that addition and multiplication are binary operations on $\mathbf{O}_K$.

Suppose $\alpha, \beta \in \mathbf{O}_K$ and not both are zero. By Theorem 9.9, $\mathbf{Z}[\alpha]$ and $\mathbf{Z}[\beta]$ are finitely generated unitary $\mathbf{Z}$-modules. The set $\mathbf{M} = \mathbf{Z}[\alpha, \beta]$ is also a unitary $\mathbf{Z}$-module. It is finitely generated since every integral power of α, β can be written as linear combinations over $\mathbf{Z}$ of basis elements $\{x_n\}$ and $\{y_n\}$, respectively, and thus every monomial $\alpha^i\beta^j$ can be written as a linear combination over $\mathbf{Z}$ of basis elements $\{x_n y_m\}$.

Now consider $\gamma = \alpha + \beta$. Multiplication of any polynomial in α, β (that is, any element of $\mathbf{Z}[\alpha, \beta]$) by γ merely produces another polynomial in α, β, and therefore $\gamma\mathbf{M} \subset \mathbf{M}$. By Theorem 9.9 again, this implies that γ is an algebraic integer. The argument is parallel if $\gamma = \alpha\beta$. Therefore, $\alpha + \beta, \alpha\beta \in \mathbf{O}_K$, and so $\mathbf{O}_K$ is a ring. ◆

We now complete our list of "reasonable" properties for algebraic integers in $\mathbf{Q}(\theta)$ that we began prior to Example 9.14.

(4) Given a nonzero rational number p/q, there exists a nonzero integer n such that $n(p/q)$ is an integer (namely, choose $n = q$). It is desirable to have something analogous for algebraic numbers and integers.

Theorem 9.11 If α is a nonzero algebraic number in an algebraic number field $\mathbf{Q}(\theta)$, there is a nonzero integer n such that $n\alpha$ is an algebraic integer in $\mathbf{Q}(\theta)$.

Proof Let the minimal polynomial for α be

$$x^k + a_1x^{k-1} + \cdots + a_k,$$

where each $a_i \in \mathbf{Q}$. Let D be the least common multiple of the denominators of $a_1, a_2, \ldots, a_k$, and define $c_i = Da_i$ for $i = 1, 2, \ldots, k$. Then we have the equation

$$D\alpha^k + c_1\alpha^{k-1} + \cdots + c_k = 0,$$

where each $c_i \in \mathbf{Z}$. This is equivalent to

$$(D\alpha)^k + c_1(D\alpha)^{k-1} + \cdots + c_kD^{k-1} = 0.$$

Set $D\alpha = \beta$; then $\beta^k + c_1\beta^{k-1} + c_2D\beta^{k-2} + \cdots + c_kD^{k-1} = 0$, and thus by choosing $n = D$ we see that $D\alpha = \beta$ is an algebraic integer in $\mathbf{Q}(\theta)$. ◆

Example 9.19 Consider $\mathbf{Q}(\sqrt{2} + \sqrt{3})$.

The number $\alpha = \frac{1}{2}\sqrt{2} + \sqrt{3}$ is an element of $\mathbf{Q}(\sqrt{2} + \sqrt{3})$ (confirm !), but is not an algebraic integer. Its minimal polynomial is $x^4 - 7x^2 + \frac{25}{4}$ (confirm !), that is, $16\alpha^4 - 112\alpha^2 + 100 = 0$ holds. Now choose $n = 2$; then we have $(2\alpha)^4 - 28(2\alpha)^2 + 100 = 0$. Since $x^4 - 28x^2 + 100$ is the minimal polynomial for 2α (confirm !), then 2α is an algebraic integer in $\mathbf{Q}(\sqrt{2} + \sqrt{3})$. ◆

Problems

9.33. Determine which of the following algebraic numbers in the indicated algebraic number fields are algebraic integers:

(a) $\sqrt[3]{2} + \sqrt[3]{4}$ in $\mathbf{Q}(\sqrt[3]{2})$.
(b) $(1 + \sqrt[3]{2})/2$ in $\mathbf{Q}(\sqrt[3]{2})$.
(c) $(6 + 5\sqrt{52})/4$ in $\mathbf{Q}(\sqrt{52})$).
(d) $(5 + 7\sqrt{13})/2$ in $\mathbf{Q}(\sqrt{13})$.
(e) $(1 + 2\sqrt{3})/4$ in $\mathbf{Q}(\sqrt{3})$.

9.34. Let $\alpha \in \mathbf{Q}(\theta)$ be an algebraic number that satisfies the polynomial $x^3 + \frac{5}{3}x - \frac{1}{6}$. Find an integer n such that $n\alpha$ is an algebraic integer in $\mathbf{Q}(\theta)$.

9.35. Let $\alpha, \beta \in \mathbf{Q}(\theta)$ be algebraic integers of degree m, n, respectively. Prove that $\alpha\beta$ is also an algebraic integer and that its degree is, at most, mn.

9.36. Let $\mathbf{K} = \mathbf{Q}(\theta)$ be an algebraic number field of degree n. A set of algebraic integers $\{\alpha_1, \alpha_2, \ldots, \alpha_m\}$ in $\mathbf{K}$ is said to be an **integral basis** of $\mathbf{K}$ if every algebraic integer $\beta \in \mathbf{K}$ can be written uniquely as

$$\beta = c_1\alpha_1 + c_2\alpha_2 + \cdots + c_m\alpha_m,$$

where each $c_i \in \mathbf{Z}$. Assume that a given $\mathbf{K}$ has an integral basis. Prove that this is also a basis for $\mathbf{K}$ over $\mathbf{Q}$. It is known that every algebraic number field has at least one integral basis (Pollard and Diamond, 1975).

9.37. Show that there are an infinite number of algebraic integers in $\mathbf{Q}(\sqrt[3]{2})$ whose decimal values lie strictly between 0 and 1.

9.5 Integers in a Quadratic Field

Best understood among the algebraic number fields $\mathbf{Q}(\theta)$ are the quadratic fields, where $[\mathbf{Q}(\theta): \mathbf{Q}] = 2$. Such fields have $\{1, \theta\}$ as a basis over $\mathbf{Q}$ and θ itself is a zero of a quadratic polynomial that is irreducible over $\mathbf{Q}$.

By Theorem 9.11 we might as well assume that θ is an algebraic integer. For if $n\theta$ is an algebraic integer ($n \in \mathbf{Z}$) but θ is not, the fields $\mathbf{Q}(\theta)$ and $\mathbf{Q}(n\theta)$ are still the same thing. So let θ satisfy $x^2 + ax + b = 0$, where $a, b \in \mathbf{Z}$. Then we have

$$\theta = \frac{-a \pm \sqrt{a^2 - 4b}}{2}.$$

The integer $a^2 - 4b$ can be written uniquely as c^2D, where $c, D \in \mathbf{Z}$ and D is squarefree. Then $\theta = -\frac{1}{2}a \pm \frac{1}{2}c\sqrt{D}$ and we obtain Theorem 9.12.

Theorem 9.12 Every quadratic field is of the form $\mathbf{Q}(\sqrt{\mathrm{D}})$, where D is a squarefree integer.

Note that there is nothing to prevent D from being negative; $\mathbf{Q}(\sqrt{-5})$ is a perfectly legitimate quadratic field. Also, from the preceding remarks, the pair $\{1, \sqrt{D}\}$ is a basis for $\mathbf{Q}(\sqrt{\mathrm{D}})$, so every element in $\mathbf{Q}(\sqrt{\mathrm{D}})$ can be written in the form $(p/q) + (p'/q')\sqrt{D}$, where $p, p', q, q' \in \mathbf{Z}$ and $q, q' \neq 0$. By writing the fractions over a common denominator d , which may be taken to be positive, we see that every element in $\mathbf{Q}(\sqrt{\mathrm{D}})$ has the form $(A + B\sqrt{\mathrm{D}})/d$, with $A, B, d, D \in \mathbf{Z}$, $\mathrm{d} \neq 0$.

We now ask, which numbers of this form are the algebraic integers in $\mathbf{Q}(\sqrt{\mathrm{D}})$? Not surprisingly, the answer depends on the value of D.

Theorem 9.13 The algebraic integers in $\mathbf{Q}(\sqrt{\mathrm{D}})$ are the numbers α given by

$$\alpha = \begin{cases} \dfrac{A + B\sqrt{D}}{2}, & A \equiv B \pmod 2, \quad D \equiv 1 \pmod 4) \\ A + B\sqrt{D}, & \text{all } A, B \in \mathbf{Z}, \quad D \not\equiv 1 \pmod 4). \end{cases}$$

Proof An algebraic number $\alpha \in \mathbf{Q}(\sqrt{\mathrm{D}})$ that is not an integer in $\mathbf{Q}$ is an algebraic integer in $\mathbf{Q}(\sqrt{\mathrm{D}})$ iff it satisfies a minimal polynomial of the form

$$x^2 + ax + b = 0,$$

where $a, b \in \mathbf{Z}$. From the previous remarks, we have $\alpha = (A + B\sqrt{D})/d$, so substitution into the equation yields

$$\begin{cases} A^2 + B^2D + adA + bd^2 = 0 \\ B(2A + ad) = 0. \end{cases}$$

As we are assuming that $B \neq 0$, then $ad = -2A$ and the first equation becomes

$$B^2D - A^2 + bd^2 = 0.$$

Let $n = (A, d)$, so $B^2D = A^2 - bd^2$ implies $n^2 \mid B^2D$. But D is squarefree, and thus $n^2 \mid B^2$ and $n \mid B$. Given α, we can always choose A, B, d to be relatively prime; therefore, if $n \mid A$ and $n \mid B$, we must have $n = (A, d) = 1$. But $ad = -2A$, so $d \mid (-2)$ and thus $d = 1$ or 2.

Case 1. $d = 1$.

Then since $A, B, \sqrt{D}$ are algebraic integers in $\mathbf{Q}(\sqrt{D})$, and so is $A + B\sqrt{D}$, according to Theorem 9.10.

Case 2. $d = 2$.

The number $\alpha = (A + B\sqrt{D})/2$ satisfies the equation

$$x^2 - Ax + \frac{A^2 - B^2 D}{4} = 0,$$

so α is an algebraic integer iff $(A^2 - B^2 D)/4 \in \mathbf{Z}$, that is, iff $A^2 \equiv B^2 D \pmod 4$. Since $A \equiv 1 \pmod 2$, then $B^2 D \equiv A^2 \equiv 1 \pmod 4$. This demands that $B \equiv 1 \pmod 2$, and thus $B^2 \equiv 1 \pmod 4$. The only possibility for D is that $D \equiv 1 \pmod 4$, and we are done. ◆

Example 9.20 In $\mathbf{Q}(i)$ the only integers are numbers of the form $A + Bi$, $A, B \in \mathbf{Z}$; these are the so-called **Gaussian integers**. In $\mathbf{Q}(\sqrt{5})$ the algebraic integers are the numbers

$$\alpha = \frac{A + B\sqrt{5}}{2}, \qquad A \equiv B \pmod 2,$$

such as $1 - \sqrt{5}$ and $(3 + 5\sqrt{5})/2$. ◆

Let us recall now that in Section 6.7 we looked briefly at the system of numbers $\mathbf{Z}[\omega]$, where $\omega = (-1 + \sqrt{-3})/2$. These are numbers of the form

$$\alpha = a + b\left[\frac{-1}{2} + \frac{1}{2}\sqrt{-3}\right]$$
$$= \frac{(2a - b) + b\sqrt{-3}}{2},$$

with $a, b \in \mathbf{Z}$. Now $\mathbf{Q}(\omega)$ is the same as $\mathbf{Q}(\sqrt{-3})$, so by Theorem 9.13 the quadratic integers in $\mathbf{Q}(\omega)$ are the numbers α given by

$$\alpha = \frac{A + B\sqrt{-3}}{2}, \qquad A \equiv B \pmod 2.$$

But $2a - b \equiv b \pmod 2$, so $\mathbf{Z}[\omega]$ is the ring (actually, integral domain) of integers in $\mathbf{Q}(\sqrt{-3})$.

Problems

9.38. Show that there are an infinite number of algebraic integers in $\mathbf{Q}(\sqrt{5})$ whose decimal values lie strictly between 0 and 1.

9.39. Consider the algebraic number field $\mathbf{Q}(\sqrt{-5})$. Show that 29 can be nontrivially factored (that is, without ± 1 as a factor) as the product of two algebraic integers in $\mathbf{Q}(\sqrt{-5})$. What do you conclude from this?

9.40. Consider the algebraic number field $\mathbf{Q}(\sqrt{-13})$. Show that 77 can be non-trivially factored into pairs of algebraic integers in three different ways. What does this suggest, and why is your interpretation of the result not yet completely justified?

9.41. Let $\epsilon > 0$ be arbitrarily small. Show that there is an algebraic integer $\alpha \in \mathbf{Q}(\sqrt{13})$ such that $0 < \alpha < \epsilon$.

9.6 Norms

Norms are defined and used in many places in mathematics, usually in different ways. For example, on a vector space equipped with an inner product, a norm is defined for a vector. This gives a measure of the "length" of the vector. No such interpretation is given to norms of algebraic numbers, yet such norms are still useful for sorting the algebraic numbers. The definition of norm of an algebraic number builds upon the existence of a basis (Section 9.3) for an algebraic number field $\mathbf{Q}(\theta)$.

DEF. Let $\{u_1, u_2, \ldots, u_n\}$ be a basis for $\mathbf{Q}(\theta)$ and let α be an algebraic number in $\mathbf{Q}(\theta)$. For each $i = 1, 2, \ldots, n$ we have

$$\alpha u_i = \sum_{u=1}^{n} \rho_{ij} u_j$$

where each $\rho_{ij} \in \mathbf{Q}$. In matrix form this is $\alpha \mathbf{u} = \boldsymbol{\rho}\mathbf{u}$. We define the **norm** of α, $N\alpha$, by

$$N\alpha = \det \boldsymbol{\rho}.$$

Example 9.21 Let $\mathbf{Q}(\theta) = \mathbf{Q}(\sqrt{5})$ and $\alpha = (1 + \sqrt{5})/2$. An ordered basis for $\mathbf{Q}(\theta)$ is $\{1, \sqrt{5}\}$. We then obtain

$$\alpha u_1 = \left[\frac{1 + \sqrt{5}}{2}\right](1) = \frac{u_1}{2} + \frac{u_2}{2}$$

$$\alpha u_2 = \left[\frac{1 + \sqrt{5}}{2}\right](\sqrt{5}) = \frac{5u_1}{2} + \frac{u_2}{2}$$

$$\boldsymbol{\rho} = \begin{bmatrix} 1/2 & 1/2 \\ 5/2 & 1/2 \end{bmatrix}.$$

Hence the norm of α is

$$N\alpha = \begin{vmatrix} 1/2 & 1/2 \\ 5/1 & 1/2 \end{vmatrix} = -1. \quad \blacklozenge$$

Example 9.22 A basis for $\mathbf{Q}(\omega)$, where $\omega = (-1 + \sqrt{-3})/2$, is $\{1, \sqrt{-3}\}$. Let $\alpha = a + b\omega = [(2a - b) + b\sqrt{-3}]/2$. Then we obtain

$$\begin{cases} \alpha u_1 = \left[\dfrac{(2a - b) + b\sqrt{-3}}{2}\right](1) = \dfrac{(2a - b)u_1}{2} + \dfrac{bu_2}{2} \\ \alpha u_2 = \left[\dfrac{(2a - b) + b\sqrt{-3}}{2}\right](\sqrt{-3}) = \dfrac{-3bu_1}{2} + \dfrac{(2a - b)u_2}{2}, \end{cases}$$

$$\boldsymbol{\rho} = \begin{bmatrix} \dfrac{2a - b}{2} & \dfrac{b}{2} \\ \dfrac{-3b}{2} & \dfrac{2a - b}{2} \end{bmatrix}.$$

Hence the norm of α is

$$N\alpha = \begin{vmatrix} \dfrac{2a - b}{2} & \dfrac{b}{2} \\ \dfrac{-3b}{2} & \dfrac{2a - b}{2} \end{vmatrix} = a^2 - ab + b^2.$$

This agrees with what was given in Section 6.7. $\blacklozenge$

Note in Example 9.21 that the norm is negative. This is not an undesirable feature, as it would be in the case of vectors. It is possible, however, that the norm is dependent upon one's choice of basis for $\mathbf{Q}(\theta)$. This is definitely undesirable. Should one basis be chosen over all others? What criterion should be used? Fortunately, we are saved by Theorem 9.14.

Theorem 9.14 In a given algebraic number field $\mathbf{Q}(\theta)$, the norm of an algebraic number α is independent of the choice of basis for $\mathbf{Q}(\theta)$.

Proof Let $[\mathbf{Q}(\theta): \mathbf{Q}] = n$, and $\{u_1, u_2, \ldots, u_n\}$ and $\{v_1, v_2, \ldots, v_n\}$ be two bases for $\mathbf{Q}(\theta)$ over $\mathbf{Q}$. Each v_j can be expressed as a linear combination of the u_i's,

$$v_j = \sum_{i=1}^{n} c_{ji} u_i \qquad (j = 1, 2, \ldots, n)$$

or $\mathbf{v} = \mathbf{cu}$, that is,

$$\begin{pmatrix} v_1 \\ v_2 \\ \vdots \\ v_n \end{pmatrix} = \begin{pmatrix} c_{11} c_{12} & \cdots & c_{1n} \\ c_{21} c_{22} & \cdots & c_{2n} \\ \vdots & & \vdots \\ c_{n1} c_{n2} & \cdots & c_{nn} \end{pmatrix} \begin{pmatrix} u_1 \\ u_2 \\ \vdots \\ u_n \end{pmatrix}.$$

Similarly, each u_j can be expressed in terms of the v_i's,

$$u_j = \sum_{i=1}^{n} C_{ji} v_i$$

or $\mathbf{u} = \mathbf{Cv}$. Hence, $\mathbf{u} = \mathbf{C}(\mathbf{cu}) = (\mathbf{Cc})\mathbf{u}$ and $\mathbf{v} = \mathbf{c}(\mathbf{Cu}) = (\mathbf{cC})\mathbf{v}$, so $\mathbf{Cc} = \mathbf{cC} = \mathbf{I}$ and $\mathbf{c}$, $\mathbf{C}$ are inverses (hence, invertible).

Now suppose that in the u-basis one has $\alpha\mathbf{u} = \boldsymbol{\rho}\mathbf{u}$. Then replace $\mathbf{u}$ by $\mathbf{Cv}$ to give $\alpha\mathbf{Cv} = \boldsymbol{\rho}\mathbf{Cv}$ and premultiply both sides by $\mathbf{C}^{-1}$,

$$\mathbf{C}^{-1}\alpha\mathbf{Cv} = \mathbf{C}^{-1}\boldsymbol{\rho}\mathbf{Cv}.$$

On the left-hand side $\mathbf{C}^{-1}$ commutes with α because the latter is just a scalar; the result is $\alpha\mathbf{v} = (\mathbf{C}^{-1}\boldsymbol{\rho}\mathbf{C})\mathbf{v}$. Thus, in the v-basis we see that

$$N\alpha = \det(\mathbf{C}^{-1}\boldsymbol{\rho}\mathbf{C}).$$

We use the standard result from determinant theory that the determinant of the product of two or more $n \times n$ matrices is the product of their separate determinants.

$$\begin{aligned} \det(\mathbf{C}^{-1}\boldsymbol{\rho}\mathbf{C}) &= (\det \mathbf{C}^{-1})(\det \boldsymbol{\rho})(\det \mathbf{C}) \\ &= (\det \boldsymbol{\rho})[(\det \mathbf{C}^{-1})(\det \mathbf{C})] \\ &= (\det \boldsymbol{\rho})(\det \mathbf{I}) \\ &= (\det \boldsymbol{\rho}). \end{aligned}$$

Hence, $N\alpha = \det \boldsymbol{\rho}$ in the v-basis and the theorem follows. ◆

An immediate, and probably most useful, property of the norm is that it is multiplicative.

Theorem 9.15 If $\alpha, \beta \in \mathbf{Q}(\theta)$, then $N(\alpha\beta) = (N\alpha)(N\beta)$.

Proof Suppose in some basis $\{u_1, u_2, \ldots, u_n\}$ we have $\alpha\mathbf{u} = \boldsymbol{\rho}\mathbf{u}$ and $\beta\mathbf{u} = \boldsymbol{\eta}\mathbf{u}$. Multiply the latter by α and then replace $\alpha\mathbf{u}$ on the right by $\boldsymbol{\rho}\mathbf{u}$,

$$(\alpha\beta)\mathbf{u} = \alpha\boldsymbol{\eta}\mathbf{u} = \boldsymbol{\eta}(\alpha\mathbf{u}) = \boldsymbol{\eta}(\boldsymbol{\rho}\mathbf{u}) = (\boldsymbol{\eta}\boldsymbol{\rho})\mathbf{u}.$$

Hence, we see $N(\alpha\beta) = \det(\boldsymbol{\eta}\boldsymbol{\rho}) = (\det \boldsymbol{\eta})(\det \boldsymbol{\rho}) = (\det \boldsymbol{\rho})(\det \boldsymbol{\eta}) = (N\alpha)(N\beta)$. ◆

Example 9.23 In $\mathbf{Q}(\sqrt{5})$ let $u_1 = 1$, $u_2 = \sqrt{5}$, $\alpha = 2 - \frac{1}{2}\sqrt{5}$, $\beta = -1 + 3\sqrt{5}$. Then we obtain

$$\begin{cases} \alpha u_1 = 2 - \frac{1}{2}\sqrt{5} \\ \quad\;\; = 2u_1 - \frac{1}{2}u_2 \\ \alpha u_2 = \frac{-5}{2} + 2\sqrt{5} \\ \quad\;\; = \frac{-5}{2}u_1 + 2u_2 \end{cases} \qquad \begin{cases} \beta u_1 = -1 + 3\sqrt{5} \\ \quad\;\; = -1u_1 + 3u_2 \\ \beta u_2 = 15 - \sqrt{5} \\ \quad\;\; = 15u_1 - 1u_2 \end{cases}$$

$$N\alpha = \begin{vmatrix} 2 & (-1/2) \\ -5/2 & 2 \end{vmatrix} = \frac{11}{4}, \qquad N\beta = \begin{vmatrix} -1 & 3 \\ 15 & -1 \end{vmatrix} = -44.$$

On the other hand, $\alpha\beta = (2 - \frac{1}{2}\sqrt{5})(-1 + 3\sqrt{5}) = \frac{-19}{2} + \frac{13}{2}\sqrt{5}$, so

$$(\alpha\beta)u_1 = \frac{-19}{2}u_1 + \frac{13}{2}u_2$$

$$(\alpha\beta)u_2 = \frac{65}{2} - \frac{19}{2}\sqrt{5}$$

$$= \frac{65}{2}u_1 - \frac{19}{2}u_2$$

$$N(\alpha\beta) = \begin{vmatrix} -19/2 & 13/2 \\ 65/2 & -19/2 \end{vmatrix} = \frac{361}{4} - \frac{845}{4} = -121 = \frac{11}{4}(-44). \quad ◆$$

Corollary 9.15.1 $N(\alpha) = 0$ iff $\alpha = 0$.

Proof ($\leftarrow$) Let $\alpha = 0$ and let $\{u_1, u_2, \ldots, u_n\}$ be a basis for $\mathbf{Q}(\theta)$. Since the u_i's are linearly independent, we have for each j that $\alpha u_j = 0 = \sum_{i=1}^{n} \rho_{ji}u_i$ implies all $\rho_{ji} = 0$. Hence, $\boldsymbol{\rho}$ is the zero matrix and $\det \boldsymbol{\rho} = 0$.

($\rightarrow$) Let $N\alpha = 0$ and assume $\alpha \neq 0$. Then α^{-1} exists, and from Theorem 9.15

$$(N\alpha)(N\alpha^{-1}) = 0 = N(1).$$

But if we set $\beta = 1$ and write $\beta\mathbf{u} = \boldsymbol{\gamma}\mathbf{u}$, then clearly

$$\boldsymbol{\gamma}\begin{pmatrix} 1 & 0 & \ldots & 0 \\ 0 & 1 & \ldots & 0 \\ \vdots & & & \vdots \\ 0 & 0 & \ldots & 1 \end{pmatrix} = \mathbf{I},$$

so $\det \boldsymbol{\gamma} = \det \mathbf{I} = N(1) = 1$, a contradiction. Hence, $\alpha = 0$. ◆

So far, we know how to evaluate $N\alpha$ only by choosing *some* basis for $\mathbf{Q}(\theta)$ and then working out the products of α by the basis elements. We are now going to develop an alternative method to calculate $N\alpha$ that does not require work with a specific base.

Let $\alpha \in \mathbf{Q}(\theta)$ and suppose α satisfies the minimal polynomial $p(x) = x^m + b_1x^{m-1} + b_2x^{m-2} + \cdots + b_m$. Define $\mathbf{K} = \mathbf{Q}(\alpha)$; by Corollary 9.2.1 an ordered basis for $\mathbf{K}$ is $\boldsymbol{\alpha} = \{1, \alpha, \alpha^2, \ldots, \alpha^{m-1}\}$. Then we obtain

$$\alpha \cdot \alpha^t = \begin{cases} \alpha^{t+1}, & 0 \le t \le m-2 \\ -b_m - b_{m-1}\alpha - \cdots - b_1\alpha^{m-1}, & t = m-1. \end{cases}$$

If $\alpha\boldsymbol{\alpha} = \boldsymbol{\rho\alpha}$, then the $m \times m$ matrix $\boldsymbol{\rho}$ is seen to be

$$\boldsymbol{\rho} = \begin{bmatrix} 0 & 1 & 0 & \ldots & & \\ 0 & 0 & 1 & 0 & \ldots & \\ 0 & 0 & 0 & 1 & 0 & \ldots \\ \vdots & & & & & \\ -b_m & -b_{m-1} & -b_{m-2} & \ldots & & -b_1 \end{bmatrix}.$$

Expand the determinant of $\boldsymbol{\rho}$ in minors down the first column. The result is

$$\det \boldsymbol{\rho} = (-1)^{m-1}(-b_m)\begin{vmatrix} 1 & 0 & 0 & \ldots & 0 \\ 0 & 1 & 0 & \ldots & 0 \\ 0 & 0 & 1 & \ldots & 0 \\ \vdots & & & & \\ 0 & 0 & 0 & \ldots & 1 \end{vmatrix} = (-1)^m b_m.$$

This is the norm of α if m equals the degree of $\mathbf{Q}(\theta)$ over $\mathbf{Q}$. But if $[\mathbf{Q}(\theta)\colon \mathbf{Q}] = n > m$, then $\mathbf{Q}(\theta) \supset \mathbf{K} \supset \mathbf{Q}$. By Theorem 9.6, n/m must be an integer since this is $[\mathbf{Q}(\theta)\colon \mathbf{K}]$. Let $\mathbf{Q}(\theta)$ have a basis over $\mathbf{K}$ of $\{\beta_1, \beta_2, \ldots, \beta_{n/m}\}$. Then $\mathbf{Q}(\theta)$ has an ordered basis over $\mathbf{Q}$ of

$$\mathbf{A} = \{\beta_1, \beta_1\alpha, \beta_1\alpha^2, \ldots, \beta_1\alpha^{m-1}; \beta_2, \beta_2\alpha, \beta_2\alpha^2, \ldots, \beta_2\alpha^{m-1};$$
$$\ldots; \beta_{n/m}, \beta_{n/m}\alpha, \beta_{n/m}\alpha^2, \ldots, \beta_{n/m}\alpha^{m-1}\}.$$

For each β_i, $1 \le i \le n/m$, we now find

$$\alpha \cdot (\beta_i\alpha^t) = \begin{cases} \beta_i\alpha^{t+1}, & 0 \le t \le m-2 \\ \beta_i(-b_m - b_{m-1}\alpha - \cdots - b_1\alpha^{m-1}), & t = m-1. \end{cases}$$

This time let $\alpha\mathbf{A} = \boldsymbol{\rho}\mathbf{A}$. The matrix for $\boldsymbol{\rho}$ has a block structure, each block corresponding to a different β_i. That is, the $n \times n$ matrix $\boldsymbol{\rho}$ splits into n/m matrices (each of dimensions $m \times m$) along the right diagonal:

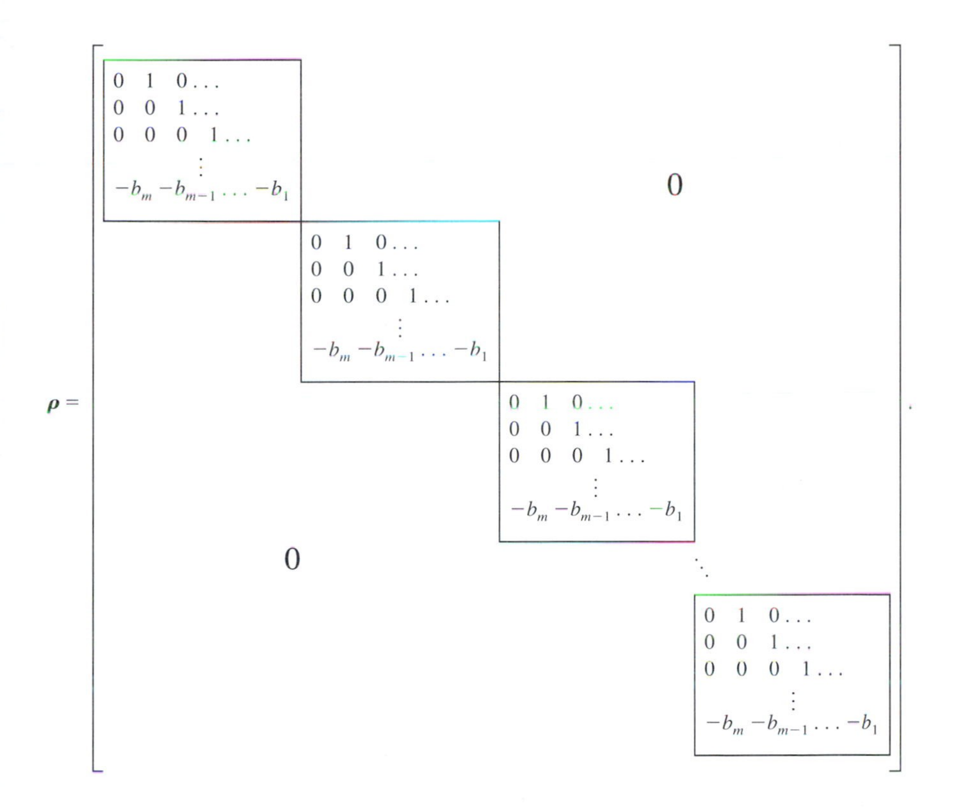

The determinant of $\boldsymbol{\rho}$ is the product of the determinants of the separate blocks, so $\det \boldsymbol{\rho} = [(-1)^m b_m]^{n/m}$. In summary, we have Theorem 9.16 and its corollary.

Theorem 9.16 Let $\alpha \in \mathbf{Q}(\theta)$, and let m be the degree of α over $\mathbf{Q}$ and n be the degree of $\mathbf{Q}(\theta)$ over $\mathbf{Q}$. If B is the constant term in the minimal polynomial for α, then $N\alpha = (-1)^n B^{n/m}$.

Corollary 9.16.1 If $\alpha \in \mathbf{Q}(\theta)$ is an algebraic integer, then $N\alpha$ is an integer.

Example 9.24 According to Example 9.9, $n = [\mathbf{Q}(\sqrt{2} + \sqrt{3}) \colon \mathbf{Q}] = 4$. Consider $\alpha = \sqrt{6}$. Its minimal polynomial is $x^2 - 6$, so $m = 2$. Hence, $N\alpha = (-1)^4(-6)^{4/2} = 36$. The same value is obtained if the procedure of Example 9.21 is followed. ◆

Problems

9.42. The algebraic integer α occurs in the two distinct algebraic number fields $\mathbf{Q}(\theta)$, $\mathbf{Q}(\theta')$. Must its norm be the same in the two fields?

9.43. Using only the definition of the norm, determine $N\alpha$ for each of the following choices of α from a suitable choice of basis for each field:

(a) $\alpha = (9 - \sqrt{5})/2$; $\mathbf{Q}(\sqrt{5})$.
(b) $\alpha = \sqrt{2} + \sqrt{6}$; $\mathbf{Q}(\sqrt{2} + \sqrt{3})$.
(c) $\alpha = (1 + \sqrt{-3})/2$; $\mathbf{Q}(\sqrt{-3})$.
(d) $\alpha = -\sqrt[3]{2} + \sqrt[3]{4}$; $\mathbf{Q}(\sqrt[3]{2})$.
(e) $\alpha = \sqrt[3]{2} - \sqrt[3]{4}$; $\mathbf{Q}(\sqrt[3]{2})$.
(f) $\alpha = 1 + \sqrt[5]{3} + \sqrt[5]{9}$; $\mathbf{Q}(\sqrt[5]{3})$.

9.44. Consider the algebraic number field $\mathbf{Q}(\sqrt[3]{2})$.

(a) Show that $\boldsymbol{\beta}_1 = \{1, \sqrt[3]{2}, \sqrt[3]{4}\}$, $\boldsymbol{\beta}_2 = \{2, \sqrt[3]{2} + \sqrt[3]{4}, -2\sqrt[3]{2} + \sqrt[3]{4}\}$ are bases for $\mathbf{Q}(\sqrt[3]{2})$.
(b) Find $\mathbf{c}$, $\mathbf{C}$, where $\mathbf{c}$ is the matrix that converts the $\boldsymbol{\beta}_1$ basis to the $\boldsymbol{\beta}_2$ basis, and $\mathbf{C} = \mathbf{c}^{-1}$.
(c) Express $\alpha = 2 - \frac{\sqrt[3]{2}}{2} + \frac{3\sqrt[3]{4}}{4}$ separately in the $\boldsymbol{\beta}_1$, $\boldsymbol{\beta}_2$ bases and determine $N\alpha$ in each case.
(d) Show explicitly that $N\alpha$ in the $\boldsymbol{\beta}_2$ basis is $\det(\mathbf{C}^{-1}\boldsymbol{\rho}\mathbf{C})$, where $\alpha\mathbf{u} = \boldsymbol{\rho}\mathbf{u}$ in the $\boldsymbol{\beta}_1$ basis.

9.45. Under what conditions can $N(\alpha) = N(-\alpha)$ for nonzero α? Under what conditions can $N(-\alpha) = -N(\alpha)$ for nonzero α?

9.46. Verify Theorem 9.15 for the following choices of α, β, $\mathbf{Q}(\theta)$:

(a) $\alpha = (9 - \sqrt{5})/2$, $\beta = \sqrt{5}$; $\mathbf{Q}(\sqrt{5})$.
(b) $\alpha = \sqrt{2} + \sqrt{6}$, $\beta = 2\sqrt{3}$; $\mathbf{Q}(\sqrt{2} + \sqrt{3})$.
(c) $\alpha = 1 + \sqrt[3]{2}$, $\beta = -1 - \sqrt[3]{4}$; $\mathbf{Q}(\sqrt[3]{2})$.
(d) $\alpha = (2 + i)/3$, $\beta = (-1 + 3i)/2$; $\mathbf{Q}(i)$.

9.47. Show that if $\alpha, \beta \in \mathbf{Q}(\sqrt{\mathrm{D}})$ are nonzero algebraic integers such that $\alpha \mid \beta$, then $N\alpha \mid N\beta$. The symbolism $\alpha \mid \beta$ means that for algebraic integers α, β there is an algebraic integer $\gamma \in \mathbf{Q}(\sqrt{\mathrm{D}})$ such that $\alpha\gamma = \beta$.

9.48. Show that the converse of Corollary 9.16.1 is false, that is, find an algebraic number $\alpha \in \mathbf{Q}(\theta)$ such that $N\alpha$ is an integer but α is not an algebraic integer. Then construct an example to show that the converse of the theorem in Problem 9.47 is also false.

9.7 Units and Primes in Algebraic Number Fields

Now that we know the algebraic integers in algebraic number fields, we want to look at two subsets of these integers: the units and the primes. Norms will be useful here.

DEF. Let α, β be two algebraic integers in $\mathbf{Q}(\theta)$. Then α **divides** β, written $\alpha \mid \beta$, if $\frac{\beta}{\alpha}$ is an algebraic integer in $\mathbf{Q}(\theta)$. By a **unit** ϵ we mean any algebraic integer in $\mathbf{Q}(\theta)$ such that $\epsilon \mid 1$. The algebraic integer π is a **prime** if (1) it is not 0 or a unit, and (2) when $\pi = \beta\gamma$ and β, γ are algebraic integers, then either β or γ is a unit (but not both).

These definitions reinforce similar statements that were given in Section 6.7. There we also introduced the concept of associates, which we repeat here.

DEF. Let α be any nonzero element of $\mathbf{Q}(\theta)$. Then $\boldsymbol{\beta}$ is an **associate** of $\boldsymbol{\alpha}$ if $\alpha = \beta\epsilon$, for some unit $\epsilon \in \mathbf{Q}(\theta)$.

The definition is unsymmetrical; we have not said "α, β are associates." However, it is easy to establish that the relation between α and β is a symmetrical one. In fact, we have Theorem 9.17.

Theorem 9.17 Being an associate in some algebraic number field $\mathbf{Q}(\theta)$ is an equivalence relation.

Proof Examine the three parts of the definition of an equivalence relation. ◆

Objects that obey an equivalence relation invariably possess structure as a set. In fact, it follows from Problem 3.59 that

Theorem 9.18 The units of an algebraic number field $\mathbf{Q}(\theta)$ constitute a multiplicative group.

This theorem explains why in the preceding definition of prime, β and γ cannot both be units. The product of two units is another unit, and a prime is not allowed to be a unit. Also, the definition of prime is consistent with that given in Section 6.7 since if $\pi = \beta\epsilon$, and π is a prime and ϵ is a unit, then β is an associate of π, so π has as divisors a unit and an associate. Further, if $\beta' \neq \pi$ is another associate of β, then $\beta = \beta'\epsilon'$ for some unit ϵ'. Hence, $\pi = (\beta'\epsilon')\epsilon = \beta'(\epsilon'\epsilon) = \beta'\epsilon''$ for some unit ϵ'' (Theorem 9.18), so π is also divisible by the unit ϵ'' and the associate β'. In general, the prime π is divisible by any of its associates, by any unit, and by no other algebraic integers in $\mathbf{Q}(\theta)$. This last line is a generalization of the statement "an integer p that is a prime in $\mathbf{Z}$ is divisible by $\pm p$, by ± 1 (the units), and by no other integers in $\mathbf{Z}$."

Example 9.25 Consider $\mathbf{Q}(\sqrt{-3})$. Then $\epsilon = (1 - \sqrt{-3})/2$, an algebraic integer (Theorem 9.13), is a unit because $\epsilon^{-1} = (1 + \sqrt{-3})/2$, which is also an algebraic integer. Clearly, ϵ^{-1} is a unit too. ◆

Example 9.26 In $\mathbf{Q}(\sqrt{-3})$, the number $\pi = (3 + \sqrt{-3})/2$ has only the following factorizations into algebraic integers (Problem 9.54):

$$\begin{aligned}\pi &= 1\left(\frac{3+\sqrt{-3}}{2}\right) = -1\left(\frac{-3-\sqrt{-3}}{2}\right) = \left(\frac{3-\sqrt{-3}}{2}\right)\left(\frac{1+\sqrt{-3}}{2}\right)\\ &= \left(\frac{-1-\sqrt{-3}}{2}\right)\left(\frac{-3+\sqrt{-3}}{2}\right) = \left(\frac{1-\sqrt{-3}}{2}\right)(\sqrt{-3})\\ &= \left(\frac{-1+\sqrt{-3}}{2}\right)(-\sqrt{-3}).\end{aligned}$$

In each case one, and only one, of the factors is a unit (which ones are they?), so π is a prime. ◆

It would certainly be useful to have a sure-fire criterion for identifying units. Here is where the norm can be of service. In Theorem 6.20 we stated that $\alpha \in \mathbf{Z}[\omega]$ is a unit iff $N\alpha = +1$. That was a special case associated with the fact that in $\mathbf{Q}(\omega)$ norms are never negative (verify!). The general case is given by Theorem 9.19.

Theorem 9.19 The algebraic integer $\alpha \in \mathbf{Q}(\theta)$ is a unit iff $N\alpha = \pm 1$.

Proof ($\rightarrow$) Suppose α is a unit in $\mathbf{Q}(\theta)$. Then $\alpha \mid 1$, so α^{-1} is an algebraic integer and $(N\alpha)(N\alpha^{-1}) = (N\alpha\alpha^{-1}) = N(1) = 1$. As N$\alpha$ is an integer (Corollary 9.16.1), we can write $N\alpha \mid 1$, whence $N\alpha = \pm 1$.

($\leftarrow$) Suppose $N\alpha = \pm 1$. All of the conjugates of α are algebraic integers since each satisfies the same minimal polynomial in $\mathbf{Z}[x]$. By Theorem 9.16 the constant term in this minimal polynomial is ± 1. If the conjugates are denoted $\alpha_1 = \alpha, \alpha_2, \ldots, \alpha_n$ then $\Pi_{i=1}^n \alpha_i = \pm 1$, so $\alpha \mid 1$ because the quotient is $\pm \Pi_{i=2}^n \alpha_i$, an algebraic integer. ◆

Corollary 9.19.1 If $\pi \in \mathbf{Q}(\theta)$ is an algebraic integer and $N\pi = p$, a prime, then π is a prime in $\mathbf{Q}(\theta)$.

Proof Let $\pi = \beta\gamma$ and suppose $N\pi = p$. Then either $N\beta$ or $N\gamma$ is ± 1, so by Theorem 9.19 either β or γ is a unit. By definition, π is then a prime. ◆

The converse of Corollary 9.19.1 is false (Problem 9.55). On the other hand, if π is a prime in $\mathbf{Q}(\theta)$ and π is an integer in $\mathbf{Z}$, then π is a prime in $\mathbf{Z}$.

Example 9.27 In a quadratic field $\mathbf{Q}(\sqrt{D})$, an integer $\alpha = x + y\sqrt{D}$ has norm $N\alpha = (x + y\sqrt{D})(x - y\sqrt{D})$ (verify!). Consider $\pi = 1 - 2\sqrt{-3} \in \mathbf{Q}(\sqrt{-3})$. Then $N\pi = (1 - 2\sqrt{-3})(1 + 2\sqrt{-3}) = 13$, so by Corollary 9.19.1 π is a prime in $\mathbf{Q}(\sqrt{-3})$. ◆

Actually, the remark immediately following Corollary 9.19.1 is quite interesting. It suggests that we can use the norm as an investigative tool. For example, look again at Problem 9.40. Two of the factorizations of 77 that you should have found are

$$77 = 7 \cdot 11 = (5 - 2\sqrt{-13})(5 + 2\sqrt{-13}).$$

Are these factorizations into primes in $\mathbf{Q}(\sqrt{-13})$? Suppose $5 - 2\sqrt{-13}$ were composite; then for some algebraic integers α, β, neither of which are units, one must have $5 - 2\sqrt{-13} = \alpha\beta$. It follows from Theorem 9.15 that

$$N\alpha \; N\beta = N(5 - 2\sqrt{-13}) = 77,$$

and so $N\alpha = 7$ and $N\beta = 11$, for example. Let $\alpha = A + B\sqrt{-13}$; then $A^2 + 13B^2 = 7$, and this clearly has no solution in $\mathbf{Z}$. Hence, α and β do not exist and $5 - 2\sqrt{-13}$ is prime in $\mathbf{Q}(\sqrt{-13})$, despite the fact that $N(5 - 2\sqrt{-13})$ is not a prime in $\mathbf{Z}$.

Further, we find that $7, 11, 5 + 2\sqrt{-13}$ are also primes in $\mathbf{Q}(\sqrt{-13})$. Hence, the two preceding factorizations of 77 are factorizations into primes! We have discovered, incidentally, that an analog of the Fundamental Theorem of Arithmetic (Theorem 2.1) does not hold in $\mathbf{Q}(\sqrt{-13})$. Nevertheless, we have now come to the brink of a whole body of questions regarding factorization of and divisibility among integers in algebraic number fields, a vast and incompletely known part of the theory of algebraic numbers. The following theorem, however, is an immediate analog of the prototype statement in **Z** (Theorem 1.5).

Theorem 9.20 Every algebraic integer $\alpha \in \mathbf{Q}(\theta)$, not zero or a unit or a prime, can be factored into a product of primes.

Proof Write $\alpha = \beta\gamma$, where β, γ are algebraic integers that are not units. If both are primes, then we are done. If β, for example, is not prime, then $1 < |N\beta| < |N\alpha|$ from Corollary 9.19.1. Now write $\beta = \delta\eta$, where δ, η are again algebraic integers that are not units. Here, $1 < |N\delta| < |N\beta|$, $1 < |N\eta| < |N\beta|$. The process continues until all factors of α are such that any further factorization of any of them yields a unit. The process is only finitely long since each successive factorization produces algebraic integers of successively smaller absolute norms, and these are bounded below by 1 (Theorem 9.19). ◆

Example 9.28 Consider $\alpha = 7 - 2\sqrt{6}$ in $\mathbf{Q}(\sqrt{6})$. Since $N\alpha = 25$, any factors of α must have norm $+5$ or -5, which is a prime. Then the factor will be a prime in $\mathbf{Q}(\sqrt{6})$. Algebraic integers $A + B\sqrt{6}$ of norm ± 5 in $\mathbf{Q}(\sqrt{6})$ satisfy $A^2 - 6B^2 = \pm 5$. One solution of this is $A = 1$, $B = -1$, and we find that $(1 - \sqrt{6})(1 - \sqrt{6}) = 7 - 2\sqrt{6}$. ◆

By Theorem 9.20 the algebraic integer $2 \in \mathbf{Q}(\theta)$ is either a prime π in $\mathbf{Q}(\theta)$ or has a prime factor π_1 in $\mathbf{Q}(\theta)$, so the set of primes in $\mathbf{Q}(\theta)$ is nonempty. If any finite set $\{\pi_1, \pi_2, \ldots, \pi_n\}$ represents the totality of primes in $\mathbf{Q}(\theta)$, then

$$N = 1 + \pi_1\pi_2 \ldots \pi_n$$

must either be prime or have a prime factor. If N is prime, it is clearly different from any of $\pi_1, \pi_2, \ldots, \pi_n$, and so the original set of primes was not complete. On the other hand, if N has a prime factor, it must be some π_j from $\{\pi_1, \pi_2, \ldots, \pi_n\}$, and so $\pi_j \mid 1$ also. But then this says that π_j is a unit, not a prime. In either case we arrive at a contradiction, and therefore we have the analog of Theorem 1.6.

Theorem 9.21 In any algebraic number field $\mathbf{Q}(\theta)$ there are an infinite number of prime algebraic integers.

For any algebraic number field $\mathbf{Q}(\theta)$ we are naturally interested in at least four aspects of the arithmetic in their rings of integers:

1. What are the units and how many are there?
2. Characterize the primes present.

3. Is there essential uniqueness of factorization of the composites into primes?
4. Do certain classic theorems in **N** such as the Euclidean algorithm or Fermat's Theorem carry over?

The investigation of these questions is naturally a lot easier when θ is of degree 2 over **Q** than when it is of higher degree. Consider item 1 in the preceding list. We already know that there are only four units in $\mathbf{Q}(i)$, and we can show that $\mathbf{Q}(\sqrt{-3})$ has six units (Problem 9.51), and that $\mathbf{Q}(\sqrt{2})$, $\mathbf{Q}(\sqrt{3})$ each have infinitely many units (Problem 9.58). But determination of the number of units in a cubic field is a very difficult problem. There are no general theorems to point the way. Nor are there any general theorems to aid in the characterization of the primes in an arbitrary algebraic number field.

It is not hard to prove that $\mathbf{Q}(i)$ possesses uniqueness of factorization (Ruchte and Ryden, 1973). It is just as easy to prove that $\mathbf{Q}(\sqrt{-3})$, $\mathbf{Q}(\sqrt{2})$, $\mathbf{Q}(\sqrt{3})$ also possess uniqueness of factorization. But, again, there are no general theorems to tell if an arbitrary $\mathbf{Q}(\theta)$ possesses the property or not. It is even an open question if there are an infinite number of algebraic number fields with uniqueness of factorization. It is known, as a result of work culminating in 1966 by **A. Baker** and **H.M. Stark**, that there are only nine imaginary quadratic fields $\mathbf{Q}(\sqrt{\mathrm{D}})$ that have the unique factorization property: $D = -1, -2, -3, -7, -11, -19, -43, -67$, and -163 (Problem 9.60). You can read a bit more at an elementary level about quadratic fields and uniqueness of factorization in (Stark, 1987).

Recognition of the failure of uniqueness of factorization in some algebraic number fields and the desire to somehow restore uniqueness led **E.E. Kummer** (1810–1893), and later **R. Dedekind** (1831–1916) to introduce into algebra the concept of **ideals**. You can read about them in many places (McCoy, 1948; Cohn, 1980; Grosswald, 1984).

We have barely scratched the surface on number fields. Time and space do not permit us to continue farther along the interesting path that we have been pursuing. You are urged to consult literature at the next level of sophistication beyond that of the present book (Pollard and Diamond, 1975; Hardy and Wright, 1979; Stewart and Tall, 1979; Ireland and Rosen, 1990; Ono, 1990).

Problems

9.49. Prove Theorem 9.17. If $A + B\sqrt{-5}$ is an algebraic integer in $\mathbf{Q}(\sqrt{-5})$, how many associates does it have?

9.50. Consider the field $\mathbf{Q}(\sqrt{-6})$. Using only the definition of unit, find all the units in $\mathbf{Q}(\sqrt{-6})$.

9.51. Show that there are only six units in the field $\mathbf{Q}(\omega)$, where $\omega = (-1 + \sqrt{-3})/2$.

9.52. Let $\alpha = a\sqrt[3]{2} + \sqrt[3]{4}$ be an element of $\mathbf{Q}(\sqrt[3]{2})$.

(a) Show that the magnitude of the norm of α is $|N\alpha| = |4 + 2a^3|$.
(b) Show that there are no units in $\mathbf{Q}(\sqrt[3]{2})$ that have the form α.
(c) Show that no two algebraic integers of the form α can have the same value of $|N\alpha|$.

9.53. Prove Theorem 9.18. Is the group of units in the field $\mathbf{Q}(\omega)$, where $\omega = (-1 + \sqrt{-3})/2$, a cyclic group?

9.54. In $\mathbf{Q}(\sqrt{-3})$ we wish to factor $\pi = (3 + \sqrt{-3})/2$ into algebraic integers: $\pi = \alpha\beta$. Show that there are only six pairs (α, β) of such factors.

9.55. Find a prime in $\mathbf{Q}(i)$ whose norm is not a prime in $\mathbf{Z}$.

9.56. Decide whether any of the following algebraic integers in the indicated fields are primes:

(a) $\alpha = 5$; $\mathbf{Q}(i)$.
(b) $\alpha = 2 + \sqrt{-6}$; $\mathbf{Q}(\sqrt{-6})$.
(c) $\alpha = 6 - 5\sqrt{2}$; $\mathbf{Q}(\sqrt{2})$.

Provide a prime factorization for any α that is not prime.

9.57. Here we look at a given algebraic integer in two different algebraic number fields.

(a) Show that $\mathbf{Q}(i)$ and $\mathbf{Q}(i + \sqrt{2})$ are distinct.
(b) Show that $\alpha = 1 + i$ is a prime in $\mathbf{Q}(i)$.
(c) Show that in $\mathbf{Q}(i + \sqrt{2})$ one has $N\alpha = 4$.
(d) Show that the algebraic number $\beta = [2 + \sqrt{2}\,(1 + i)]/2$ is an algebraic integer in $\mathbf{Q}(i + \sqrt{2})$ that is not a unit.
(e) Show that $\beta \mid \alpha$, and hence α is not a prime in $\mathbf{Q}(i + \sqrt{2})$.

9.58. Let k be a positive integer and define

$$(1 + \sqrt{2})^k = A_k + B_k\sqrt{2}.$$

(a) Show that if k is even, then (A_k, B_k) is a solution to $x^2 - 2y^2 = +1$, and if k is odd, then (A_k, B_k) is a solution to $x^2 - 2y^2 = -1$.
(b) Show that $\mathbf{Q}(\sqrt{2})$ has an infinite number of units.
(c) Prove that $\mathbf{Q}(\sqrt{3})$ has an infinite number of units.

9.59. Let $\alpha = A + B\sqrt[3]{2} + C\sqrt[3]{4} \in \mathbf{Q}(\sqrt[3]{2})$.

(a) Show that $N\alpha = A^3 + 2B^3 + 4C^3 - 6ABC$.
(b) Show that there are an infinite number of units in $\mathbf{Q}(\sqrt[3]{2})$.

9.60. Show that the Fundamental Theorem of Arithmetic does not hold in $\mathbf{Q}(\sqrt{-5})$.

9.61. Consult the paper (Ruchte and Ryden, 1973) dealing with the Fundamental Theorem of Arithmetic in $\mathbf{Q}(i)$. Work through the details of the proof in it and write a report on the result. This is a good project for extra credit.

9.62. Let p be an odd prime. Adjunction of $\zeta = e^{2\pi i/p}$, a primitive pth root of unity, to $\mathbf{Q}$ gives what is called a **cyclotomic field**, $\mathbf{Q}(\zeta)$. Cyclotomic fields have particular relevance for the treatment of many Diophantine equations.

(a) Show that $N\zeta = 1$. [Hint: Theorem 9.16 is useful.]
(b) Let $p = 3$; show that $1 + \zeta$ is a unit in $\mathbf{Q}(\zeta)$. Hence, can we conclude that there are infinitely many units in $\mathbf{Q}(\zeta)$?
(c) Again, let $p = 3$; show that $1 - \zeta$ is a prime in $\mathbf{Q}(\zeta)$.

RESEARCH PROBLEMS

1. It is known that in a real quadratic field $\mathbf{Q}(\sqrt{d})$ there is a unique unit $\epsilon_0 > 1$ such that every unit in $\mathbf{Q}(\sqrt{d})$ is given by $\pm\epsilon_0^n$, $n \in \mathbf{Z}$. The unique unit ϵ_0 is called the **fundamental unit** for the field $\mathbf{Q}(\sqrt{d})$. Investigate the cases $d = 5, 7, 13$ and see if you can discover the fundamental unit for each of these fields.

2. A quadratic field is called a **Euclidean field** if, given integers $\alpha, \beta \in \mathbf{Q}(\sqrt{d})$ with $\beta \neq 0$, there exist other integers τ and δ in $\mathbf{Q}(\sqrt{d})$ such that

$$\alpha = \beta\tau + \delta, \qquad |N\delta| < |N\tau|.$$

A Euclidean field is thus a field in which an analog of the Euclidean algorithm holds. Investigate the cases $d = -7, 10$ to see if either is a Euclidean field. In either case is there uniqueness of factorization? What is the connection?

3. Investigate the cubic field $\mathbf{Q}(\sqrt[3]{3})$ and see if it contains infinitely many units.

References

Chahal, J.S., *Topics in Number Theory*, Plenum Press, New York, 1988, pp. 72–74. Our treatment of Theorems 9.9 and 9.10 in Section 9.4 is taken from these pages.

Cohn, H., *Advanced Number Theory*, Dover Publications, New York, 1980, pp. 93–130. Good treatment of ideal theory in quadratic fields.

Grosswald, E., *Topics from the Theory of Numbers*, 2nd ed., Birkhäuser, Boston, 1984, pp. 219–231. A more concise, but more readable, introduction to ideal theory than that in Cohn. See also the author's Chapter 10 on the arithmetic of number fields.

Hardy, G.H., and Wright, E.M., *An Introduction to the Theory of Numbers*, 5th ed., Oxford University Press, Oxford, England, 1979, pp. 178–189, 204–232. See these three chapters for discussion mainly about quadratic fields.

Ireland, K., and Rosen, M., *A Classical Introduction to Modern Number Theory*, 2nd ed., Springer, New York, 1990, pp. 172–187. This brief chapter summarizes some of the general theory of algebraic numbers, but would need to be supplemented by much additional material in order to be appreciated.

McCoy, N.H., *Rings and Ideals*, Carus Monograph No. 8, Mathematical Association of America, Washington, DC, 1948. This well-written small monograph is essential reading for those wishing to pursue the use of ideals in number theory.

Ono, T., *An Introduction to Algebraic Number Theory*, Plenum Press, New York, 1990. Comparable to Pollard and Diamond (next citation), but contains more material.

Pollard, H., and Diamond, H.G., *The Theory of Algebraic Numbers*, 2nd ed., Carus Monograph No. 9, Mathematical Association of America, Washington, DC, 1975. Highly recommended monograph for further study. Pages 95–145 give a nice coverage of ideals in number theory.

Ruchte, M.F., and Ryden, R.W., "A Proof of Uniqueness of Factorization in the Gaussian Integers," *Amer. Math. Monthly*, **80**, 58–59 (1973). A very accessible proof of the title theorem. With essentially no change at all, the same proof applies to the uniqueness of factorization in $\mathbf{Q}(\sqrt{-3})$.

Stark, H.M., *An Introduction to Number Theory*, MIT Press, Cambridge, MA, 1987, pp. 257–316. Very readable chapter on quadratic fields by an eminent mathematician.

Stewart, I., and Tall, D., *Algebraic Number Theory*, Chapman and Hall, London, 1979. Relatively elementary; recommended.

Chapter 10
Partitions

10.1 Introduction to Partitions

Let us consider the division of an integer N into integral parts or **summands**. If the order of listing of these summands were important, then the integer 5, for example, would have 16 possible **compositions**.[1] Table 10.1 lists them according to increasing number of summands.

On the other hand, if the order of listing were not important, then the number of divisions, called the number of **partitions** $p(N)$, would drop to just 7, that is, $p(5) = 7$. When $N \geq 3$, $p(N)$ is always less than the number of ordered compositions.

Theorem 10.1 The number of compositions of the positive integer N is 2^{N-1}.

Proof Let the number of compositions of N be denoted $D(N)$; the case $D(1)$ is trivial and requires no further consideration. For $N \neq 1$, the compositions of N can be sorted into two types:

Type A: The first summand is 1

Type B: The first summand is not 1.

The number of compositions of type A is $D(N - 1)$. In each composition of type B reduce the first summand by 1. The resulting set of compositions is now a complete set of compositions of the number $N - 1$. Hence, the type A and type B compositions are equinumerous, and $D(N) = 2[D(N - 1)]$. But $D(2) = 2$, so $D(3) = 2 \cdot 2 = 4$, $D(4) = 2 \cdot 4 = 8$, and by induction $D(N) = 2^{N-1}$. ◆

[1] The term was coined by (Maj.) **Percy A. MacMahon** (1854–1929), a Maltese-born British mathematician whose specialty was combinatorial analysis.

Table 10.1 Compositions of the Integer 5

No. of Summands	Compositions	No. of Compositions
1	5	1
2	4 + 1, 1 + 4, 3 + 2, 2 + 3	4
3	3 + 1 + 1, 1 + 3 + 1, 1 + 1 + 3, 2 + 2 + 1, 2 + 1 + 2, 1 + 2 + 2	6
4	2 + 1 + 1 + 1, 1 + 2 + 1 + 1, 1 + 1 + 2 + 1, 1 + 1 + 1 + 2	4
5	1 + 1 + 1 + 1 + 1	1

Example 10.1 When $N = 5$, $D(N) = 2^{5-1} = 16$. The 16 compositions are listed in Table 10.1. ◆

Theorem 10.1 tells us that 2^{N-1} is an upper bound to $p(N)$. If we individually count the number of partitions of N, we arrive at the following first few values: $p(1) = 1$, $p(2) = 2$, $p(3) = 3$, $p(4) = 5$, $p(5) = 7$, $p(6) = 11$, $p(7) = 15$, and $p(8) = 22$. Some values for larger N are given in Table H of Appendix II in the rear of the book; a table to $N = 500$ is given in (Abramowitz and Stegun, 1965). There is no obvious pattern. Fig. 10.1 shows a partial graph of $p(N)$ versus N, from which it is clear that $p(N)$ rises rapidly with N. The precise form of the

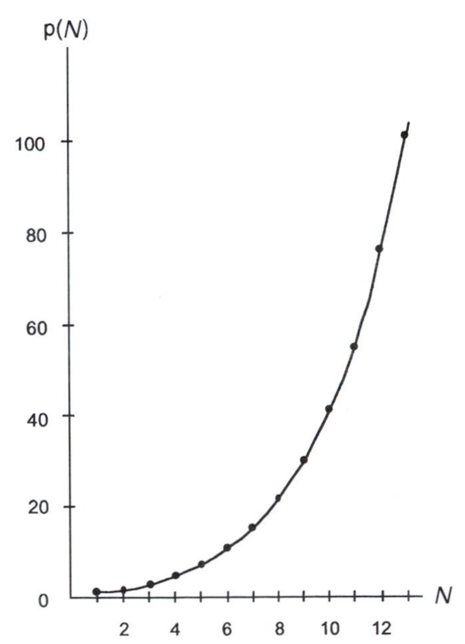

Figure 10.1 The number of partitions of N as a function of N

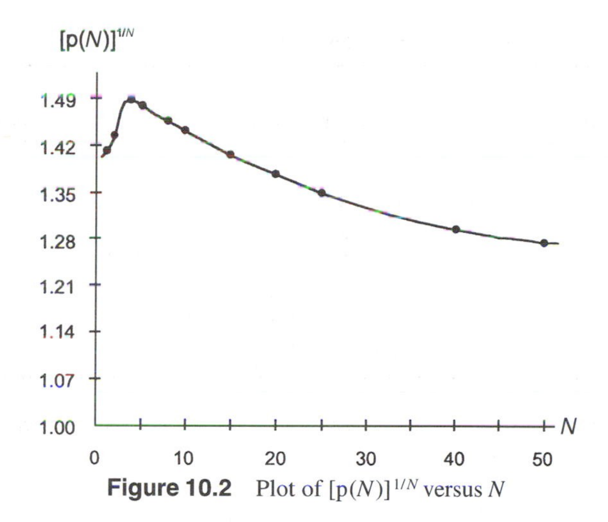

Figure 10.2 Plot of $[p(N)]^{1/N}$ versus N

function p(N) is complicated, however, and was not completely worked out until 1937 (Alder, 1969).

A limiting behavior of p(N) is suggested by Theorem 10.1. Since $p(N) < D(N) = 2^{N-1} < 2^N$, then $[p(N)]^{1/N} < 2$. It is not immediately clear to what value (if any) $[p(N)]^{1/N}$ approaches as $N \to \infty$, but Fig. 10.2 tends to support the conjecture that a limiting value is approached. The proof of this (Andrews, 1971a, 1994) provides us with an excellent opportunity to employ more ideas from analysis. The final result is preceded by two lemmas.

In what follows, $p_k(N)$ denotes the number of partitions of N into at most k summands. We make the definition that $p_k(0) = 1$ for any $k > 0$.

Lemma 10.2.1 For each $k > 0$, $p_k(N) \leq (N + 1)^k$ holds.

Proof Fix k, and let s_i denote the ith summand. In $N = s_1 + s_2 + \cdots + s_k$, each s_j would have any one of the $N + 1$ values in the interval $[0, N]$. Since there are k summands at most, there are $(N + 1)^k$ assignments at most. These include all of the k-partitions of all the integers from 0 to N, as well as some of the k-partitions of some numbers in the interval $[N + 1, kN]$. Hence, we have

$$p_k(N) \leq \sum_{i=0}^{N} p_k(i) \leq (N + 1)^k. \quad \blacklozenge$$

Example 10.2 Let $N = 6$, $k = 3$. In $6 = s_1 + s_2 + s_3$, each s_i can have any one of the seven values $0, 1, \ldots, 6$. The number of permutations of these numbers taken three at a time, and yielding sums ranging from 0 to 18, is $7 \times 7 \times 7 = (6 + 1)^3 = 343$. Clearly, these permutations include all of the partitions of 0 to 6 into at most three summands, as well as some of the 2- and

3-partitions of numbers 7, 8, . . . , 18. Included in the 343 partitions are the $p_3(6)$ partitions of 6 into at most three summands, so $p_3(6) < 343$. By actual count, $p_3(6) = 7$. ◆

Lemma 10.2.2 For each $k > 0$, $p(N) \leq p(N-1) + p_k(N) + p(N-k)$ holds.

Proof The partitions of N are of three types:

Type A: A partition contains at least one 1 as a summand.

Type B: A partition contains no 1's and has at most k summands.

Type C: A partition contains no 1's and has more than k summands.

From each partition of type A remove one 1 as a summand. The result is the set of partitions of the integer $N - 1$, $p(N-1)$ in number. The number of partitions of type B is clearly no greater than $p_k(N)$. For each partition of type C subtract one unit from each of the k smallest summands. The result is to give a set of partitions of $N - k$ that is certainly no greater than $p(N-k)$ in number. Hence, we have

$$\begin{aligned} p(N) &\leq |\text{type A}| + |\text{type B}| + |\text{type C}| \\ &\leq p(N-1) + p_k(N) + p(N-k). \end{aligned}$$ ◆

Although our principal interest in Lemma 10.2.2 is its use in the proof of the next theorem, we may note that the lemma can give us a sharper upper bound to $p(N)$ than is attainable from Theorem 10.1. Here is an illustration.

Example 10.3 Let $N = 7$ and $k = 2$, and suppose we know in advance that $p(5) = 7$ and $p(6) = 11$. We compute easily $p_2(7) = 4$. Hence, Lemma 10.2.2 gives us $p(7) \leq 11 + 4 + 7 = 22$, whereas Theorem 10.1 merely gives $p(7) \leq 2^6 = 64$. In fact, $p(7) = 15$. ◆

Theorem 10.2 $\text{Lim}_{N\to\infty} [p(N)]^{1/N} = 1$.

Proof If we can find a constant $K > 1$ corresponding to any $\epsilon > 0$ such that $1 \leq p(N) < K(1+\epsilon)^N$, for all N sufficiently large, then upon taking the positive Nth roots we obtain

$$1 \leq [p(N)]^{1/N} < K^{1/N}(1+\epsilon).$$

Now let $N \to \infty$; the result is $1 \leq \lim_{N\to\infty} [p(N)]^{1/N} < (1+\epsilon)$. Then since ϵ can be chosen arbitrarily small, $[p(N)]^{1/N}$ can be made arbitrarily close to 1.

To proceed, let $\epsilon > 0$ be given. Pick k sufficiently large so that $(1+\epsilon)^{k-1} > 2/\epsilon$. This is certainly possible since $(1+\epsilon)^{k-1}$ is an increasing function of k; in fact, $\lim_{k\to\infty} (1+\epsilon)^{k-1} = \infty$. Having chosen k, we then have from Lemma 10.2.1 that $p_k(N) \leq (N+1)^k$. Now consider

$$\lim_{N\to\infty} \frac{(N+1)^k}{(1+\epsilon)^{N-1}} = \lim_{N\to\infty} \frac{k(N+1)^{k-1}}{(1+\epsilon)^{N-1}\ln(1+\epsilon)},$$

by an application of l'Hôpital's Rule. The expression on the right-hand side is still of the form ∞/∞ because the exponent $(k-1) > 0$. Hence, $k-1$ more applications of l'Hôpital's Rule give

$$\lim_{N\to\infty} \frac{(N+1)^k}{(1+\epsilon)^{N-1}} = \lim_{N\to\infty} \frac{k!}{(1+\epsilon)^{N-1}[\ln(1+\epsilon)]^k} = 0.$$

This says that given an $\epsilon > 0$, we can find an $N_0 \in \mathbf{N}$ so large that for all $N \geq N_0$ one has

$$\frac{(N+1)^k}{(1+\epsilon)^{N-1}} < \frac{\epsilon}{2}$$

or, equivalently, $(N+1)^k < (\epsilon/2)(1+\epsilon)^{N-1}$ for $N \geq N_0$. In view of Lemma 10.2.1, we obtain $\mathrm{p}_k(N) < (\epsilon/2)(1+\epsilon)^{N-1}$. Now define K by

$$\mathrm{K} = 1 + \max_{0 \leq N \leq N_0} \left\{\frac{\mathrm{p}(N)}{(1+\epsilon)^N}\right\}.$$

Rearrangement yields the strict inequality

$$\mathrm{p}(N) < \mathrm{K}(1+\epsilon)^N$$

for all $N \leq N_0$.

What we want to do now is to extend the validity of this inequality beyond N_0; this sounds like the Principle of Finite Induction is needed. The preceding inequality holds for all $N \leq N_0$; assume it also holds for all $N < N^*$, where $N^* > N_0$. Then from Lemma 10.2.2 we have

$$\begin{aligned}
\mathrm{p}(N^*) &= \mathrm{p}(N^*-1) + \mathrm{p}_k(N^*) + \mathrm{p}(N^*-k) \\
&< \mathrm{K}(1+\epsilon)^{N^*-1} + \mathrm{p}_k(N^*) + \mathrm{K}(1+\epsilon)^{N^*-k} && \text{by the induction hypothesis} \\
&< \mathrm{K}(1+\epsilon)^{N^*-1} + \frac{\epsilon}{2}(1+\epsilon)^{N^*-1} + \mathrm{K}(1+\epsilon)^{N^*-k} && \text{from the preceding paragraph} \\
&= \mathrm{K}(1+\epsilon)^{N^*-1}\left[1 + \frac{\epsilon}{2\mathrm{K}} + (1+\epsilon)^{1-k}\right] \\
&< \mathrm{K}(1+\epsilon)^{N^*-1}\left[1 + \frac{\epsilon}{2} + (1+\epsilon)^{1-k}\right] && \text{as } K > 1 \\
&< \mathrm{K}(1+\epsilon)^{N^*-1}\left[1 + \frac{\epsilon}{2} + \frac{\epsilon}{2}\right] && \text{by the way } k \text{ was picked} \\
&= \mathrm{K}(1+\epsilon)^{N^*}.
\end{aligned}$$

So, since $p(N) < K(1 + \epsilon)^N$ for $N \le N_0$ and $p_k(N) < (\epsilon/2)(1 + \epsilon)^{N-1}$ for all $N \ge N_0$, then the preceding steps show that $p(N) < K(1 + \epsilon)^N$ holds for $N = N^* = N_0 + 1$ also. This in turn leads to $p(N) < K(1 + \epsilon)^N$ holding for $N = N_0 + 2$, and so on. By the Principle of Finite Induction, the inequality holds for all $N \ge N_0$. ◆

The importance of Theorem 10.2 is this. Because of the complexity of the actual formula for $p(N)$, that formula is not used to calculate $p(N)$ for modest N (say, $N = 50$). Nor is $p(N)$ usually determined by actually writing out all the partitions ($p(50) = 204{,}226$, for example!). Instead, relations from the theory of generating functions are used, and Theorem 10.2 is needed to justify them. The use of generating functions offers a general approach to partitions that far transcends the use of very elementary arguments. We take this up in Sections 10.2 and 10.3.

An elementary argument that does work for a few simple problems involves the use of **Ferrers graphs** (after the English mathematician **Norman M. Ferrers**, 1829–1903). A Ferrers graph for a partition is an arrangement of dots, each summand being shown as a horizontal display of dots. Thus, the partition $4 + 3 + 1 + 1$ for 9 is pictured as

```
• • • •
• • •
•
•
```

The **conjugate partition** of a given partition is the partition noted by reading the dots in the given partition in columns rather than in rows. So the conjugate partition of $4 + 3 + 1 + 1$ is $4 + 2 + 2 + 1$. Some partitions such as

```
• • • •
• • •
• •
•
```

have been called **self-conjugate** by **MacMahon** (1920).

Theorem 10.3 Let $p_k(N)$ denote the number of partitions of N into at most k summands, and let $\pi_k(N)$ stand for the number of partitions of N in which no summand is larger than k. Then $p_k(N) = \pi_k(N)$.

Proof Consider any one of the $p_k(N)$ partitions. It has no more than k rows. Then in the conjugate partition no column has more than k dots. Hence, the given partition from the set of $p_k(N)$ partitions has a counterpart from the set of $\pi_k(N)$ partitions.

On the other hand, if we start with one of the $\pi_k(N)$ partitions, then this has its counterpart among the $p_k(N)$ partitions. We thus have a one-to-one correspondence between the two sets, and since the sets are finite they are equal in size. ◆

Problems

10.1. The following conjecture in additive number theory is made: Every integer $N \geq 6$ of the form $4k + 2$ is the sum of two (not necessarily distinct) primes of the form $4k + 3$. Investigate this conjecture as far as you can.

10.2. Another partition conjecture is this: Every integer greater than 6 is the sum of two relatively prime positive integers greater than 1. Prove this conjecture.

10.3. A known theorem says that every even integer greater than 38 is the sum of two positive, odd, composite integers (Just and Schaumberger, 1973; Vaidya, 1975). Prove this. [Hint: Consider separately integers of the forms $10k$, $10k + 2$, $10k + 4$, $10k + 6$, and $10k + 8$.]

10.4. What connection do you see between the number of compositions of an integer N and Pascal's triangle?

10.5. (a) Explain how you know that $p(N)$ is a strictly increasing function, that is, $p(N + 1) > p(N)$ for any $N \in \mathbf{N}$.

(b) In view of Theorem 10.1, why is $p(N) < 2^{N-2}$, $N \geq 5$, plausible?

10.6. Without explicitly writing out all the partitions, show that $p(9) \leq 42$.

10.7. In fact, $p(9) = 30$. Write them all out and indicate the values of $p_k(9)$ for $k = 1, 2, \ldots, 9$.

10.8. Deduce a formula for $p_2(N)$ in terms of N. Consider as separate cases even N and odd N.

10.9. **Hardy**[2] and his brilliant Indian protégé **Ramanujan** (Berndt, 1989; Kanigel, 1991) showed in 1919 that $p(N)$ is given approximately by

$$p(N) \approx \frac{1}{4N\sqrt{3}} \exp\left\{\pi\sqrt{\frac{2N}{3}}\right\}.$$

Estimate $p(72)$ and compare with the exact value (see Table H in Appendix II).

[2] The name **G.H. Hardy** has already appeared a number of times in earlier chapters (as co-author of the famous number theory book with Wright). The man and his work are beautifully described in C.P. Snow's introduction in (Hardy, 1969). This little book should be required reading for any serious mathematics student.

10.10. Illustrate the content of Theorem 10.2 by choosing $\epsilon = 0.5$. Then determine k, N_0, and K, and look at values of $p(N)$ versus $K(1 + \epsilon)^N$. The table of partition numbers given in Table H of Appendix II will be useful here.

10.11. Illustrate Theorem 10.3 by writing out the $\pi_4(9)$ partitions of 9 and comparing them with the $p_4(9)$ partitions of 9 from Problem 10.7.

10.12. What geometrical feature characterizes the Ferrers graph for a self-conjugate partition? Use this idea to sketch the Ferrers graphs for all of the self-conjugate partitions of 15.

10.13. In view of Problem 10.12, prove that the number of self-conjugate partitions of an integer N is equal to the number of partitions of N into distinct odd summands.

10.2 Generating Functions

We now want to consider the generating function approach to partitions (Mott, Kandel, and Baker, 1986). The basic idea is simple and dates back to **Euler**.

DEF. Let $\{s_n\}$ be a sequence of real numbers. The power series $s_0 + s_1x + s_2x^2 + \ldots + s_nx^n + \ldots$ is called the **generating function** for the sequence.

Although the sequence $\{s_n\}$ might be finite, the usual case is when it is infinite. The generating function is then an infinite series, which we can write formally as $\Sigma_{n=0}^{\infty} s_nx^n$ without any immediate regard for convergence. In elementary applications a generating function can be regarded as just an alternative notation for the sequence $\{s_n\}$. The powers of x merely serve as place markers, enabling us to distinguish clearly among the members of the sequence.

If $\Sigma_{n=0}^{\infty} a_nx^n$ and $\Sigma_{n=0}^{\infty} b_nx^n$ are two generating functions, then we define[3] addition (and subtraction) of them and multiplication by a constant k by

$$\sum_{n=0}^{\infty} a_nx^n \pm \sum_{n=0}^{\infty} b_nx^n = \sum_{n=0}^{\infty} (a_n \pm b_n)x^n$$

$$k \sum_{n=0}^{\infty} a_nx^n = \sum_{n=0}^{\infty} (ka_n)x^n.$$

The two generating functions are equal if $a_n = b_n$ for all n.

[3] These are actually theorems, not definitions, when properly stated. That is, they can be justified rigorously; see any standard calculus book for this (Leithold, 1990). We take this justification for granted here. Nearly all of our arguments involving generating functions hold rigorously if $|x| < 1$.

Multiplication of two generating functions might be defined in various ways. The following way leads to conventional results when the generating functions are polynomials. Again, this definition is really a theorem that can be justified under certain conditions.

DEF. Given two generating functions $\Sigma_{n=0}^{\infty} a_n x^n$ and $\Sigma_{n=0}^{\infty} b_n x^n$, their **Cauchy product** is the generating function $\Sigma_{n=0}^{\infty} c_n x^n$, where each c_n is given by

$$c_n = \sum_{k=0}^{n} a_k b_{n-k}.$$

The Cauchy product is well defined because each c_n is a *finite* sum of real numbers.

Example 10.4 The Fibonacci numbers, $\{F_n\}_{n=0}^{\infty}$, are defined by $F_0 = 0$, $F_1 = 1$, $F_n = F_{n-1} + F_{n-2}$ for $n \geq 2$. Let g(x) denote the generating function for the Fibonacci numbers:

$$g(x) = \sum_{n=0}^{\infty} F_n x^n.$$

Then multiplication by the place marker x gives

$$xg(x) = \sum_{n=0}^{\infty} F_n x^{n+1} = \sum_{n=1}^{\infty} F_{n-1} x^n$$

$$x^2 g(x) = \sum_{n=1}^{\infty} F_{n-1} x^{n+1} = \sum_{n=2}^{\infty} F_{n-2} x^n.$$

From the definition of F_n, we then obtain

$$g(x) - x = xg(x) + x^2 g(x),$$

so the generating function for $\{F_n\}_{n=0}^{\infty}$ is

$$g(x) = \frac{x}{1 - x - x^2}. \quad \blacklozenge$$

Example 10.5 The power series $\Sigma_{n=0}^{\infty} x^n$ is the generating function for the sequence $\{s_n\}_{n=0}^{\infty} = \{1\}_{n=0}^{\infty}$. The Cauchy product of this generating function with itself is

$$\left[\sum_{n=0}^{\infty} x^n\right]\left[\sum_{n=0}^{\infty} x^n\right] = \sum_{n=0}^{\infty} c_n x^n$$

$$= 1 + 2x + 3x^2 + 4x^3 + \ldots$$

Thus, the Cauchy product is the generating function for the sequence of the members of **N**. ◆

We can interpret c_n in Example 10.5 as the number of pairs of exponents that sum to n, that is, c_n is the number of nonnegative integral solutions (x, y) of the Diophantine equation

$$x + y = n,$$

with order regarded as important. That is, if $n = 7$, then (4, 3) and (3, 4) are treated as distinct solutions. Another interpretation is that c_n in Example 10.5 is the number of ways of distributing n indistinguishable objects between two labeled boxes (Fig. 10.3).

The situation can be generalized. Given the Diophantine equation

$$x_1 + x_2 + \ldots + x_r = n,$$

the number of nonnegative integral solutions to it, with order regarded as important, is the coefficient c_n in the generating function

$$g(x) = \left(\sum_{k=0}^{\infty} x^k \right)^r.$$

To calculate c_n from this generating function, it is necessary to make repeated use of the definition of the Cauchy product. By analogy to Fig. 10.3, the c_n so calculated is the number of unrestricted ways of distributing n indistinguishable balls among r labeled boxes.

Example 10.6 How many solutions in nonnegative integers are there to $x + y + z = 7$, if order is important?

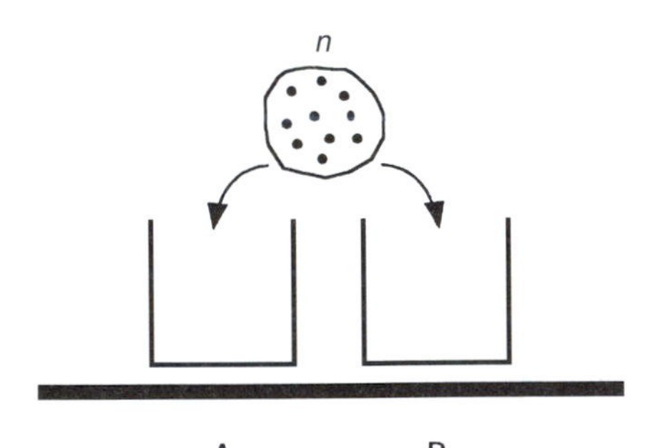

Figure 10.3 The coefficient of x^n in $(\sum_{n=0}^{\infty} x^k)^2$ is the number of unrestricted ways of distributing n balls between two labeled boxes.

Let $\Sigma_{n=0}^{\infty} a_n x^n$ denote the generating function $(\Sigma_{n=0}^{\infty} x^n)^2$, and $\Sigma_{n=0}^{\infty} b_n x^n$ be the generating function $\Sigma_{n=0}^{\infty} x^n$. We want

$$\sum_{n=0}^{\infty} c_n x^n = \left(\sum_{n=0}^{\infty} a_n x^n\right)\left(\sum_{n=0}^{\infty} b_n x^n\right).$$

From Example 10.5 we know that $a_k = (k + 1)$. Hence, we calculate c_7 from the definition of the Cauchy product as

$$\begin{aligned} c_7 &= a_0b_7 + a_1b_6 + a_2b_5 + a_3b_4 + a_4b_3 + a_5b_2 + a_6b_1 + a_7b_0 \\ &= 1(1) + 2(1) + 3(1) + 4(1) + 5(1) + 6(1) + 7(1) + 8(1) \\ &= 36. \quad \blacklozenge \end{aligned}$$

A still further generalization is possible. Suppose we ask for the number of ways of distributing 10 indistinguishable objects between two labeled boxes A and B, in which A can only contain 2, 4, 6, or 9 objects and B can only contain 1, 4, 6, or 8 objects. If there were no restrictions, the number of distributions would be 11, but with restrictions this number can only be reduced. The possibilities are shown in Fig. 10.4.

To set this up in terms of a generating function, let the exponents in the power series $A(x) = \Sigma_j x^j$ take on only the restrictive values for the number of objects in box A.

$$A(x) = x^2 + x^4 + x^6 + x^9.$$

Similarly, let the exponents in the power series $B(x) = \Sigma_{k=0} x^k$ take on only the restricted values for the number of objects in box B:

$$B(x) = x + x^4 + x^6 + x^8.$$

Then the generating function for the number of ways of distributing n indistinguishable objects between the two boxes, subject to the given restrictions stated earlier, is the Cauchy product

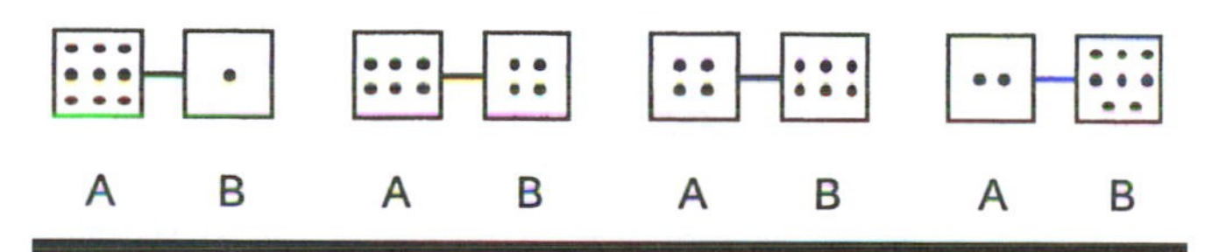

Figure 10.4 Distributing 10 indistinguishable objects between two labeled boxes with certain restrictions

$$g(x) = A(x) \cdot B(x)$$
$$= x^3 + x^5 + x^6 + x^7 + 2x^8 + 4x^{10} + 2x^{12} + x^{13} + x^{14} + x^{15} + x^{17}.$$

No. of ways of distributing 8 objects in two boxes subject to the restrictions is two.

No. of ways of distributing 10 objects in two boxes subject to the restrictions is four. (Fig. 10.4)

The distribution of objects in boxes, as in the preceding examples, is treated extensively in (MacMahon, 1920).

Example 10.7 Find the number of solutions of $x + y = 27$ in nonnegative integers if the first is a prime and the second is a multiple of 4.

The required generating function is given by

$$g(x) = \left(\sum_{i=1}^{\infty} x^{p_i}\right)\left(\sum_{n=1}^{\infty} x^{4n}\right), \qquad |x| < 1$$
$$= (x^2 + x^3 + x^5 + x^7 + x^{11} + x^{13} + \ldots) \cdot$$
$$(x^4 + x^8 + x^{12} + x^{16} + x^{20} + \ldots)$$
$$= x^6 + x^7 + x^9 + x^{10} + 2x^{11} + x^{13} + x^{14} + 3x^{15} + 2x^{17} + x^{18}$$
$$+ 3x^{19} + 3x^{21} + x^{22} + 4x^{23} + 3x^{25} + x^{26} + 5x^{27} + 3x^{29} + \ldots.$$
$$\uparrow$$

Hence, there are five solutions, explicitly, $3 + 24$, $7 + 20$, $11 + 16$, $19 + 8$, $23 + 4$. ◆

Note that the answer is automatically given for a situation in which order is no longer an issue because the two sets of summands—primes and multiples of 4—are disjoint.

One may have a need to divide one generating function by another. In real arithmetic, division of one number by another (nonzero) number can be viewed as a multiplication of the first by the multiplicative inverse of the second: $\frac{7}{6} = 7 \cdot 6^{-1}$. We apply this idea to generating functions by first considering the multiplicative inverse of a generating function. We say it one last time: The definition can be phrased more correctly as a theorem and justified rigorously under certain conditions.

DEF. If $A(x) = \Sigma_{n=0}^{\infty} a_n x^n$, then $B(x) = \Sigma_{n=0}^{\infty} b_n x^n$ is termed the **multiplicative inverse**, $[A(x)]^{-1}$, of $A(x)$ if in the Cauchy product $A(x) \cdot B(x) = \Sigma_{n=0}^{\infty} c_n x^n$ one has $c_0 = 1$ and $c_n = 0$ for all $n > 0$.

The Cauchy product of $A(x)$ and $B(x)$ is

$$A(x) \cdot B(x) = a_0 b_0 + (a_1 b_0 + a_0 b_1)x + (a_2 b_0 + a_1 b_1 + a_0 b_2)x^2 + (a_3 b_0 + a_2 b_1 + a_1 b_2 + a_0 b_3)x^3 + \ldots,$$

so $A(x) \cdot B(x) = 1$ leads to an infinite set of linear equations:

$$\begin{cases} a_0 b_0 = 1 \\ a_1 b_0 + a_0 b_1 = 0 \\ a_2 b_0 + a_1 b_1 + a_0 b_2 = 0 \\ a_3 b_0 + a_2 b_1 + a_1 b_2 + a_0 b_3 = 0 \\ \quad \vdots \end{cases}.$$

The first equation can be solved for b_0 iff $a_0 \neq 0$: $b_0 = 1/a_0$. All of the succeeding equations contain a term $a_0 b_n$, so that in each case solving for b_n necessitates a division by a_0.

Example 10.8 Consider the generating function $A(x) = \Sigma_{n=0}^{\infty} x^n$. We note that $a_0 = 1 \neq 0$, so the multiplicative inverse exists. Writing out the first few of the infinite family of equations, we find that $b_0 = 1$, $b_1 = -1$, and $b_n = 0$ for $n \geq 2$. Hence, the inverse of $A(x)$ is just $1 - x$. Was this a surprise? ◆

We now define the division of one generating function by another as follows:

DEF. Let $A(x) = \Sigma_{n=0}^{\infty} a_n x^n$ and $B(x) = \Sigma_{n=0}^{\infty} b_n x^n$, and suppose $b_0 \neq 0$. Then $A(x)$ is said to be **divisible** by $B(x)$ if $C(x) = \Sigma_{n=0}^{\infty} c_n x^n$ is such that

$$A(x) \cdot [B(x)]^{-1} = C(x).$$

Example 10.9 Let $A(x) = \Sigma_{n=0}^{\infty} x^{p_n}$ from Example 10.7 and let $B(x) = \Sigma_{n=0}^{\infty} x^n$. From Example 10.8 we have $[B(x)]^{-1} = 1 - x$. Hence, from the previous definition we obtain

$$\frac{A(x)}{B(x)} = \left(\sum_{n=1}^{\infty} x^{p_n} \right)(1 - x) = x^2 - x^4 + x^5 - x^6 + x^7 - x^8 + \ldots.$$ ◆

Problems

10.14. For what sequence of numbers is this the generating function $g_1(x) = (1 - x)^{-3}$, $0 < |x| < 1$,? For what sequence of numbers is $g_2(x) = \ln[(1 + x)/(1 - x)]$, $0 < |x| < 1$, the generating function?

10.15. Write out the first six terms in the Cauchy product of the two generating functions given in the previous problem.

10.16. How many solutions in *nonnegative* integers are there to the Diophantine equation $w + x + y + z = 10$ if order is important? How many solutions in *positive* integers only are there to this same equation if order is important?

10.17. Show that the number of solutions in positive integers to the Diophantine equation $x + y + z = n$ is $(n - 1)(n - 2)/2$, if order is important.

10.18. Give the generating function for the number of nonnegative integral solutions to the Diophantine equation

$$\sum_{i=1}^{n} x_i = N, \qquad N \in \mathbf{N},$$

if order is important and if $x_i = 0$ or 1 for each i. How many solutions are there when $N = 3$, $n = 6$?

10.19. In how many ways can $2N$ be expressed as the sum of two positive even numbers if order is important? If order is not important? Illustrate with $N = 7$.

10.20. A die is in the form of a regular dodecahedron (12 faces) rather than a cube, and the faces are accordingly numbered from 1 to 12. Two such dice are tossed. Give the generating function for N_S, the number of ways the sum S can be obtained, if the first die shows an odd number and the second shows an even number. Compute N_{13}.

10.21. Show that the number of solutions in nonnegative integers to the Diophantine equation $w + x + y + z = N$, $N \geq -1$, is $(N + 1)(N + 2)(N + 3)/6$, if order is important.

10.22. A sequence of integers $\{X_n\}_{n=0}^{\infty}$ is defined as follows: $X_0 = 0$, $X_1 = 1$, $X_{n+1} = X_n + n$ for $n \geq 1$. Find the generating function for $\{X_n\}_{n=0}^{\infty}$. Also find an explicit formula for X_n.

10.23. Among the more famous sequences of numbers in number theory is the sequence $\{B_n\}_{n=0}^{\infty}$ of **Bernoulli numbers** (after the Swiss mathematician **Jakob Bernoulli**, 1654–1705). These numbers are defined by the generating function

$$\sum_{n=0}^{\infty} \frac{B_n}{n!} x^n = \frac{x}{e^x - 1}.$$

Work out the first five nonzero Bernoulli numbers. More on the Bernoulli numbers appears in Sections 11.5 and 11.6.

10.24. Multiply the defining equation for the Bernoulli numbers in Problem 10.23 by $e^x - 1$. Replace e^x by its Maclaurin series (which converges absolutely for all x), and then take the Cauchy product. Equate like powers of x on both sides and prove, finally, that for $n \geq 2$

$$B_n = \sum_{k=0}^{n} \binom{n}{k} B_k,$$

where $\binom{n}{k} = n!/[k!(n-k)!]$. From the generating function one obtains directly $B_0 = 1$. Use the recurrence relation to compute B_8.

10.25. Clearly, all of the Bernoulli numbers are rational. Let us write them in reduced terms. Assemble a computer program to work out B_2, B_4, B_6, B_8, B_{10}, B_{12}, B_{14}, and B_{16} in fractional form. Formulate from this limited set a conjecture about the denominators of the nonzero Bernoulli numbers beyond B_1.

10.26. Show that $\Sigma_{n=2}^{\infty} B_n x^n/n$ is an even function of x, and hence prove that $B_{2k+1} = 0$ for $k \geq 1$.

10.27. Consider $[\mathrm{g}(x)]^2 = \Sigma_{n=0}^{\infty} G_n x^n$, where $\mathrm{g}(x)$ is the generating function for the Fibonacci numbers and $G_0 = G_1 = 0$. The G_n's do not obey $G_n - G_{n-1} - G_{n-2} = 0$ as do the F_n's. However, by what should the "0" be replaced in order to give a true relation? Prove your conjecture.

10.28. The **generalized Fibonacci numbers,** $\{\hat{F}_n\}_{n=0}^{\infty}$, are defined by

$$\hat{F}_n = \begin{cases} a & \text{if } n = 0 \\ b & \text{if } n = 1 \\ \hat{F}_{n-1} + \hat{F}_{n-2} & \text{if } n \geq 2 \end{cases}$$

where $a, b \in \mathbf{Z}$. Work out the generating function $\mathrm{g}(x)$ for the $\hat{F}_n$'s.

10.29. Let the sequence $\{\mathrm{f}_n(x)\}_{n=0}^{\infty}$ be defined by $\mathrm{f}_n(x) = \Sigma_{k=0}^{n} F_{2k+2}^2 x^{2k}$, where the F_{2k+2}'s are Fibonacci numbers. Show that $\lim_{n\to\infty} \mathrm{f}_n\left(\frac{1}{4}\right)$ exists, and find its value.

10.30. Some generating functions are given, not in the form of a power series, but in the form of a **Dirichlet series**:

$$\mathrm{g}(s) = \sum_{n=1}^{\infty} \frac{\mathrm{f}(n)}{n^s},$$

where $\mathrm{f}(n)$ is an arithmetic function. When $\mathrm{g}(s)$ is the famous **Riemann zeta function,** $\zeta(s) = \Sigma_{n=1}^{\infty} n^{-s}$, $s > 1$, then $\mathrm{f}(n)$ is the function $\sigma(n)$.

Use the generating function to explicitly calculate $\sigma(6)$.

10.31. Another generating function given in the form of a Dirichlet series is

$$g(s) = \frac{1}{\zeta(s)} = \sum_{n=1}^{\infty} \frac{\mu(n)}{n^s}, \qquad s > 1.$$

Use this Dirichlet series to calculate $\mu(6)$ and $\mu(8)$.

10.3 The Generating Function for Partitions

Now that we have some background on generating functions, we can return to our immediate goal, which is to obtain the generating function for $p(N)$, the number of unrestricted partitions of N into positive summands.

We recall from Examples 10.5 and 10.8 that $\Sigma_{N=0}^{\infty} x^N = 1/(1 - x)$ is the generating function for the sequence $\{s_N\}_{N=0}^{\infty} = \{1\}_{N=0}^{\infty}$. A reinterpretation is that each coefficient c_N $(= 1)$ in the expansion of $\Sigma_{N=0}^{\infty} x^N$ is the number of partitions of N in which no summand exceeds 1; let $\pi_1(N)$ denote this number.

$$\frac{1}{1 - x} = \sum_{N=0}^{\infty} \pi_1(N)x^N.$$

We make the special definition $\pi_1(0) = 1$; clearly, $\pi_1(N) = 1$ also for $N \geq 1$.

By extension, we let $\pi_2(N)$ denote the number of partitions of N into positive summands in which no summand exceeds 2. Table 10.2 gives some early results.

Let $g(x)$ be the generating function for the sequence $\{\pi_2(N)\}_{N=0}^{\infty}$. Drawing on Example 10.5, we observe

$$\begin{aligned} g(x) &= 1 + x + 2x^2 + 2x^3 + 3x^4 + 3x^5 + 4x^6 + 4x^7 + \ldots \\ &= (1 + 2x + 3x^2 + 4x^3 + 5x^4 + 6x^5 + 7x^6 + 8x^7 + \ldots) \\ &\quad - (0 + x + x^2 + 2x^3 + 2x^4 + 3x^5 + 3x^6 + 4x^7 + \ldots) \\ &= \frac{1}{(1 - x)^2} - xg(x). \end{aligned}$$

Table 10.2 Partition Numbers of N in Which No Summand Exceeds 2

N	$\pi_2(N)$
1	1
2	2
3	2
4	3
5	3
6	4
7	4
8	5

Solving for $g(x)$, we obtain conjecturally

$$g(x) = \frac{1}{(1-x)^2(1+x)} = \prod_{j=1}^{2} \frac{1}{1-x^j} = \sum_{N=0}^{\infty} \pi_2(N)x^N.$$

In view of this result and the previous one for $\{\pi_1(N)\}_{N=0}^{\infty}$, the following conjecture seems warranted. We let $\pi_k(N)$ stand for the number of partitions of N into positive summands in which no summand exceeds k; as usual, $\pi_k(0) = 1$ by convention.

Lemma 10.4.1 The generating function for $\pi_k(N)$ is $\Pi_{j=1}^{k} 1/(1-x^j)$.

Proof The generating function $\Sigma_{i=0}^{\infty} x^{ji}$ is $1/(1-x^j)$ (for $|x| < 1$). Making repeated use of the Cauchy product, we have (with proper justification)

$$\begin{aligned}\prod_{j=1}^{k} \frac{1}{1-x^j} &= \left[\sum_{i=0}^{\infty} x^i\right]\left[\sum_{i=0}^{\infty} x^{2i}\right]\cdots\left[\sum_{i=0}^{\infty} x^{ki}\right]\\ &= 1 + x^1 + (x^{2\cdot 1} + x^{1\cdot 2}) + (x^{3\cdot 1} + x^{1\cdot 1 + 1\cdot 2} + x^{1\cdot 3})\\ &\quad + (x^{4\cdot 1} + x^{2\cdot 1 + 1\cdot 2} + x^{2\cdot 2} + x^{1\cdot 1 + 1\cdot 3} + x^{1\cdot 4}) + \ldots.\end{aligned}$$

For illustrative purposes here, we have assumed that k is some integer greater than 3. The exponents of the terms in the last pair of parentheses shown all have a value of $N = 4$ and represent the five different ways of writing N as a linear combination of positive integers, none of which exceeds 4:

$$N = N_1 \cdot 1 + N_2 \cdot 2 + N_3 \cdot 3 + N_4 \cdot 4.$$

Here, the coefficient N_j means the number of times that j is used in a given linear combination. Thus, the exponent $2 \cdot 1 + 1 \cdot 2$ in $x^{(2\cdot 1 + 1\cdot 2)}$ corresponds to the choices (for $N = 4$) $N_1 = 2, N_2 = 1, N_3 = 0, N_4 = 0$.

In general, corresponding to a given exponent N there are several terms, each of which corresponds to a way of writing N as a sum,

$$N = N_1 \cdot 1 + N_2 \cdot 2 + N_3 \cdot 3 + \cdots + N_k \cdot k.$$

But each of these ways is just a partition of N into summands, none of which exceeds k. It follows that

$$\prod_{j=1}^{k} \frac{1}{1-x^j} = \sum_{N=0}^{\infty} \pi_k(N)x^N. \quad \blacklozenge$$

Example 10.10 From the partial expansion of $\prod_{j=1}^{4} 1/(1-x^j)$, we find $\pi_4(7) = 11$. ◆

It now seems clear that if we desire the number of unrestricted partitions of N, then k should not be held to any fixed finite value. If $k \to \infty$, then $\pi_k(N) \to p(N)$ for any N. The use of the Root Test from calculus together with Theorem 10.2

now shows that the series $\Sigma_{N=0}^{\infty} \text{p}(N)x^N$ converges for $|x| < 1$. What we have to show is that this infinite series is identical to the following infinite product: $\prod_{j=1}^{\infty} 1/(1 - x^j) = \lim_{k\to\infty} \prod_{j=1}^{k} 1/(1 - x^j)$ (Andrews, 1971b).

Theorem 10.4 (Euler). If $|x| < 1$, then $\Pi_{j=1}^{\infty} 1/(1 - x^j) = \Sigma_{N=0}^{\infty} \text{p}(N)x^N$, and thus the infinite product is the generating function for the number of partitions p(N).

Proof Fix x arbitrarily so that $|x| < 1$. Then from Lemma 10.4.1 we have immediately

$$\left| \sum_{N=0}^{\infty} \text{p}(N)x^N - \prod_{j=1}^{k} \frac{1}{1 - x^j} \right| = \left| \sum_{N=0}^{\infty} \text{p}(N)x^N - \sum_{N=0}^{\infty} \pi_k(N)x^N \right|.$$

Whenever $N \le k$, the two numbers p(N) and $\pi_k(N)$ are automatically the same (why?). Accordingly, these terms in the two series on the right-hand side cancel out. Whenever $N > k$, then $\text{p}(N) > \pi_k(N)$.

$$\left| \sum_{N=0}^{\infty} \text{p}(N)x^N - \prod_{j=1}^{k} \frac{1}{1 - x^j} \right| = \left| \sum_{N=k+1}^{\infty} \{\text{p}(N) - \pi_k(N)\}x^N \right|$$
$$\le \sum_{N=k+1}^{\infty} \{\text{p}(N) - \pi_k(N)\}|x|^N.$$

The inequality follows by mathematical induction from the simplest case of a sum of two real numbers (Problem 10.35):

TRIANGLE INEQUALITY

If $x, y \in \mathfrak{R}$, then $|x + y| \le |x| + |y|$.

Next, $\text{p}(N) > \pi_k(N)$ implies that $\text{p}(N) - \pi_k(N)$ is positive and is less than p(N) itself.

$$\left| \sum_{N=0}^{\infty} \text{p}(N)x^N - \prod_{j=1}^{k} \frac{1}{1 - x^j} \right| \le \sum_{N=k+1}^{\infty} \text{p}(N)|x|^N.$$

Let $\epsilon > 0$ be given. Since the Root Test has told us that the limit of the series on the right as $k \to \infty$ is 0, then we can find a $k_0 \in \mathbf{N}$ such that $\Sigma_{N=k_0+1}^{\infty} \text{p}(N)\ |x|^N < \epsilon$. It follows that

$$\left| \sum_{N=0}^{\infty} \text{p}(N)x^N - \prod_{j=1}^{k_0} \frac{1}{1 - x^j} \right| < \epsilon,$$

and this is the same thing as $\prod_{j=1}^{\infty} 1/(1 - x^j) = \sum_{N=0}^{\infty} \text{p}(N)x^N$. ◆

Example 10.11 Calculate p(6).

Let

$$A(x) = \frac{1}{1-x}\frac{1}{1-x^2} = \sum_{k=0}^{\infty} x^k \sum_{k=0}^{\infty} x^{2k}$$
$$= 1 + x + 2x^2 + 2x^3 + 3x^4 + 3x^5 + 4x^6 + \ldots$$

$$\text{and } B(x) = \frac{1}{1-x^3}\frac{1}{1-x^4} = \sum_{k=0}^{\infty} x^{3k} \sum_{k=0}^{\infty} x^{4k}$$
$$= 1 + x^3 + x^4 + x^6 + x^7 + x^8 + \ldots.$$

The Cauchy product of A(x), B(x) is

$$C(x) = A(x) \cdot B(x) = 1 + (a_1b_0 + a_0b_1)x + (a_2b_0 + a_1b_1 + a_0b_2)x^2 + \ldots$$
$$= 1 + x + 2x^2 + 3x^3 + 5x^4 + 6x^5 + 9x^6 + \ldots.$$

Next let

$$D(x) = \frac{1}{1-x^5}\frac{1}{1-x^6} = \sum_{k=0}^{\infty} x^{5k} \sum_{k=0}^{\infty} x^{6k} = 1 + x^5 + x^6 + \ldots,$$

and the Cauchy product of C(x), D(x) is

$$E(x) = C(x) \cdot D(x) = 1 + (c_1d_0 + c_0d_1)x + (c_2d_0 + c_1d_1 + c_0d_2)x^2 + \ldots$$
$$= 1 + x + 2x^2 + 3x^3 + 5x^4 + 7x^5 + \underset{\uparrow}{11x^6} + \ldots.$$

Since $1/(1 - x^7)$ begins $1 + x^7 + \ldots$ and we only want p(6), we are done and p(6) = 11. ◆

Problems

10.32. Compute the value of $\pi_4(10)$.

10.33. (a) Without using generating functions, find an explicit expression for $\pi_2(N)$. Then compare your answer with the calculated values obtained from the first several terms of the generating function for $\pi_2(N)$.

(b) Use the formula in part (a) to deduce $\lim_{N\to\infty} \pi_2(N)/p(N)$.

10.34. What is the generating function g(x) for the numbers of partitions of integers N into at most k positive summands? Use this to compute $p_3(11)$.

10.35. Prove the Triangle Inequality for pairs of real numbers. Then extend it by induction to any number of real numbers.

10.36. Why was it necessary to use the Triangle Inequality in the proof of Theorem 10.4? Under what circumstance could its use be avoided?

10.37. Extend Example 10.11 and compute p(10).

10.38. A power series is defined by the generating function

$$g_1(x) = \frac{1}{\prod_{n=1}^{r}(1 - x^{2n})} = \sum_{N=0}^{\infty} c_N x^N.$$

What is the interpretation of the coefficients c_n in this power series? Illustrate with $N = 10, r = 4$.

A power series is defined by the generating function

$$g_2(x) = \frac{1}{\prod_{n=1}^{\infty}(1 - x^{n^2})} = \sum_{N=0}^{\infty} b_N x^N.$$

What is the interpretation of the coefficients b_N in this power series? Illustrate with $N = 9$.

10.39. In the generating function $g(x) = \sum_{N=0}^{\infty} p(N)x^N$, let $x = \frac{1}{2}$. Define $p(0) = 1$, as usual.

(a) Show that $g(\frac{1}{2}) > 3$.

(b) Prove that if $j \geq 2$, then $\frac{1}{1 - (1/2)^j} = 1 + \frac{1}{2^j - 1} < 1 + \frac{1}{(j - 1)^2}$.

(c) In view of part (b), show that

$$g\left(\frac{1}{2}\right) < \frac{1024}{315} \exp\left\{\frac{\pi^2}{6} - \frac{49}{36}\right\} \approx 4.32.$$

Thus, $g(\frac{1}{2})$ is bounded.

10.4 Some Partition Identities*

Let $\mathbf{S}_1, \mathbf{S}_2$ be two sets of integers and let $p(\mathbf{S}, N)$ stand generically for the number of partitions of an integer N when its summands are drawn exclusively from the set **S**. A **partition identity** is then an equality of the form

$$p(\mathbf{S}_1, N) = p(\mathbf{S}_2, N)$$

* Sections marked with an asterisk can be skipped without impairing the continuity of the text.

Table 10.3 Number of Partitions of N into Distinct Summands

N	$p(\mathbf{S}_1, N)$	N	$p(\mathbf{S}_1, N)$
1	1	5	3
2	1	6	4
3	2	7	5
4	2	8	6

for all N. We have already seen one example of a partition identity, namely, Theorem 10.3. We chanced upon that theorem by merely playing with dots; in general, we shall not be so lucky with other identities. However, knowledge of Theorem 10.4 and its lemma and of their methods of proof will enable us to be more systematic (Andrews, 1994).

Suppose $\mathbf{S}_1$ is all of $\mathbf{N}$, and for a given N we are not allowed to use any member of $\mathbf{S}_1$ more than once as a summand. For example, if $N = 6$, we could write $6 = 4 + 2$, but $6 = 3 + 3$ is not allowed. Thus, $p(\mathbf{S}_1, N)$ means the number of partitions of N into distinct summands. Values of $p(\mathbf{S}_1, N)$ for the first few N are easily worked out by counting and are given in Table 10.3.

We now seek a second set $\mathbf{S}_2$ that will give us a partition identity. Clearly, $1 \in \mathbf{S}_2$ since $N = 1$ has only one partition from any set $\mathbf{S}$ that contains 1. However, $2 \notin \mathbf{S}_2$ because if both 1 and 2 belonged to $\mathbf{S}_2$, then we could form two partitions of $N = 2$ with them, and yet $p(\mathbf{S}_1, 2)$ is only 1. Continuing, $3 \in \mathbf{S}_2$ (so $\mathbf{S}_2$ now contains 1, 3), and with 1 and 3 we can construct two partitions of $N = 3$, namely 3 and $1 + 1 + 1$. Next, $4 \notin \mathbf{S}_2$, for if it did then $\mathbf{S}_2$ would contain 1, 3, 4 and we could construct three partitions of $N = 4$, namely 4, $3 + 1$, $1 + 1 + 1 + 1$; but Table 10.3 indicates $p(\mathbf{S}_1, 4) = 2$ only.

You can see the strategy by now. We leave it to you to reason through the conclusions that $5, 7 \in \mathbf{S}_2$ and $6, 8 \notin \mathbf{S}_2$. The pattern appears to be that $\mathbf{S}_2 = \mathbf{O}$, the set of odd integers, and we conjecture the following theorem.

Theorem 10.5 (Euler). The number of partitions of N into distinct summands is equal to the number of partitions of N into odd summands.

Proof Theorem 10.4 gives the number of unrestricted partitions of N because the exponents j span all of $\mathbf{N}$. If the j's are restricted to odd integral values, then we obtain the generating function for the number of partitions of N into odd summands:

$$\sum_{N=0}^{\infty} p(\mathbf{O}, N)x^N = \prod_{j=1}^{\infty} \frac{1}{1 - x^{2j-1}}.$$

On the other hand, we saw in Section 10.2 that if the exponents a_j in $A(x) = \Sigma_j x^a$ are taken from a set $\mathbf{A}$ and the exponents b_j in $B(x) = \Sigma_j x^b$ are taken from a set **B**, then A(x)B(x) is the generating function for the number of partitions of N into two summands: one from **A** and the other from **B**. If, now, we desire no limit to the number of summands, then an infinite product of functions like A(x), B(x) is required. Further, if we require each summand to be distinct, then the exponents in A(x), B(x), C(x), . . . must be disjoint sets. This can be arranged by setting A(x) = $1 + x$, B(x) = $1 + x^2$, C(x) = $1 + x^3$, and so on.

It follows that

$$\sum_{N=0}^{\infty} \mathrm{p}(\mathbf{S_1}, N)x^N = \prod_{j=1}^{\infty} (1 + x^j).$$

By elementary factoring of $1 - x^{2j}$ we can rewrite the equation as

$$\sum_{N=0}^{\infty} \mathrm{p}(\mathbf{S_1}, N)x^N = \prod_{j=1}^{\infty} \left\{\frac{1 - x^{2j}}{1 - x^j}\right\}$$

$$= \prod_{j=1}^{\infty} (1 - x^{2j}) \prod_{j=1}^{\infty} \frac{1}{1 - x^j}.$$

The last line must be justified because we have altered the order of multiplication of infinitely many factors (Problem 10.54).

Now write the last infinite product on the right in terms of odd j's and even j's, again with justification required:

$$\prod_{j=1}^{\infty} \frac{1}{1 - x^j} = \prod_{j=1}^{\infty} \frac{1}{(1 - x^{2j})(1 - x^{2j-1})}$$

$$= \frac{1}{\prod_{j=1}^{\infty} (1 - x^{2j})} \prod_{j=1}^{\infty} \frac{1}{1 - x^{2j-1}}.$$

Combination of this with the equation before yields

$$\sum_{N=0}^{\infty} \mathrm{p}(\mathbf{S_1}, N)x^N = \prod_{j=1}^{\infty} \frac{1}{1 - x^{2j-1}}$$

$$= \sum_{N=0}^{\infty} \mathrm{p}(\mathbf{O}, N)x^N. \quad \blacklozenge$$

Is it not astounding that two such different conditions as "odd" and "distinct" should always yield the same number of partitions of an integer N? Probably **Euler** thought so too.

Example 10.12 The partitions of 6 into distinct summands are 6, 5 + 1, 4 + 2, 3 + 2 + 1. The partitions of 6 into odd summands are 5 + 1, 3 + 3, 3 + 1 + 1 + 1, 1 + 1 + 1 + 1 + 1 + 1. The two sets of partitions are the same size. ◆

Table 10.4 Number of Partitions of N into Summands with None Being Used More than Twice

N	$p(\mathbf{S}_1, N)$	N	$p(\mathbf{S}_1, N)$
1	1	6	7
2	2	7	9
3	2	8	13
4	4	9	16
5	5	10	22

In the spirit of the partition identity that we have just worked out fully, let us look at a more complicated example. This time let $\mathbf{S}_1 = \mathbf{N}$, and in any partition of an integer N we shall not use the same summand more than *twice*. Thus, $6 = 4 + 1 + 1$ is permitted, but $6 = 3 + 1 + 1 + 1$ is not. Values of $p(\mathbf{S}_1, N)$ for the first few N are given in Table 10.4.

We again seek a second set $\mathbf{S}_2$ that will give us a partition identity. It is immediate that $1, 2 \in \mathbf{S}_2$. If $3 \in \mathbf{S}_2$, then three partitions of 3 would be possible, whereas we desire only two; hence $3 \notin \mathbf{S}_2$. If $4 \in \mathbf{S}_2$, then $4 = 2 + 2 = 2 + 1 + 1 = 1 + 1 + 1 + 1$, which is the correct number of partitions of 4. If $5 \in \mathbf{S}_2$, then $5 = 4 + 1 = 2 + 2 + 1 = 2 + 1 + 1 + 1 = 1 + 1 + 1 + 1 + 1$, which is also the correct number of partitions of 5. Continuing, if $\mathbf{S}_2 = \{1, 2, 4, 5, 6\}$, then eight partitions of 6 are possible ($6 = 5 + 1 = 4 + 2 = 4 + 1 + 1 = 2 + 2 + 2 = 2 + 2 + 1 + 1 = 2 + 1 + 1 + 1 + 1 = 1 + 1 + 1 + 1 + 1 + 1$), whereas we desire only seven; hence $6 \notin \mathbf{S}_2$.

The reasoning is pursued for $N = 7, 8, 9, 10$, and one finds that $\mathbf{S}_2 = \{1, 2, 4, 5, 7, 8, 10, \ldots\}$. The pattern appears to be that 3 and its multiples are missing. If this is so, then we need the generating function for the number of partitions of N in which no summand is a multiple of 3. Since all integers are of the form $3j$, $3j - 1$, or $3j - 2$, the desired partition function is given by an obvious modification of Theorem 10.4:

$$\sum_{N=0}^{\infty} p(\mathbf{S}_1, N)x^N = \prod_{j=1}^{\infty} \frac{1}{1 - x^{3j-1}} \frac{1}{1 - x^{3j-2}}.$$

The first few terms in the expansion of this infinite product yield coefficients that agree with the entries in Table 10.4.

We still need a generating function for $p(\mathbf{S}_1, N)$. Some kind of modification of the generating function for $p(\mathbf{S}_1, N)$ in the preceding example seems reasonable. We might replace $1 + x^j$ everywhere there by $1 + x^j + x^j = 1 + 2x^j$, but this very quickly leads to numbers that are much too large. Replacing $1 + x^j$ everywhere by $(1 + x^j)^2$ seems more reasonable and gives numbers that are better but still too large. In this case, factors of the form $1 + 2x^j + x^{2j}$ may not be right because x^j has been weighted too heavily by the multiplier 2. We are led to conjecture Theorem 10.6 (Andrews, 1976).

Theorem 10.6 The number of partitions of N in which no summand appears more than twice is given by the generating function

$$\sum_{N=0}^{\infty} \mathrm{p}(\mathbf{S_1}, N)x^N = \prod_{j=1}^{\infty} (1 + x^j + x^{2j}).$$

Proof Let the elements of $\mathbf{S_1} = \mathbf{N}$ be listed as $n_1, n_2, n_3, \ldots$ so that

$$\prod_{j=1}^{\infty} (1 + x^j + x^{2j}) = (1 + x^{n_1} + x^{2n_1})(1 + x^{n_2} + x^{2n_2})(1 + x^{n_3} + x^{2n_3}) \ldots .$$

Each term in the expansion of the right-hand side would be obtained by choosing one factor from each of the infinite number of trinomials. For example, one might choose as factors x^{2n_1}, x^{n_2}, x^{2n_3}, 1, x^{n_5}, and 1's for all remaining factors. These give as one term in the expansion of the right-hand side

$$x^{2n_1+n_2+2n_3+n_5},$$

and so one partition of 15 in which no summand appears more than twice is

$$15 = (1 + 1) + 2 + (3 + 3) + 5.$$

Let N be selected; then no summand of any partition of N can exceed N. The expansion of the infinite product then need not extend beyond the first N trinomials. The multiplication of a finite number of trinomials and the rearrangement of terms during this multiplication are valid arithmetic operations. The general term so obtained has the form

$$\sum_{0 \le c_1 \le 2} \sum_{0 \le c_2 \le 2} \cdots \sum_{0 \le c_N \le 2} x^{(c_1 n_1 + c_2 n_2 + \cdots + c_N N)}.$$

Since each c_j lies in the interval [0, 2], no integer n_j is used more than twice in the exponent. Hence, for any integer $\mathbf{I} \le N$, the collection of all general terms like the preceding one with $c_1 n_1 + c_2 n_2 + \cdots + c_N N = \mathbf{I}$ represents all possible partitions of $\mathbf{I}$ with no summand employed more than twice. ◆

Example 10.13 Calculate $\mathrm{p}(\mathbf{S_1}, 5)$.

From Theorem 10.6 we find stepwise:

1. $(1 + x + x^2)(1 + x^2 + x^4) = 1 + x + 2x^2 + x^3 + 2x^4 + x^5 + x^6.$
2. $(1 + x + 2x^2 + x^3 + 2x^4 + x^5 + x^6)(1 + x^3 + x^6) =$
$1 + x + 2x^2 + 2x^3 + 3x^4 + 3x^5 + 3x^6 + \ldots .$
3. $(1 + x^4 + x^8)(1 + x^5 + x^{10}) = 1 + x^4 + x^5 + x^8 + x^9 + \ldots .$
4. $(1 + x + 2x^2 + 2x^3 + 3x^4 + 3x^5 + 3x^6 + \ldots)$
$\cdot (1 + x^4 + x^5 + x^8 + x^9 + \ldots)$
$= 1 + x + 2x^2 + 2x^3 + 4x^4 + 5x^5 + 6x^6 + \ldots .$
↑

Hence, $\mathrm{p}(\mathbf{S_1}, 5) = 5$. ◆

We are now ready to give a proof of the desired partition identity involving the sets $\mathbf{S}_1$, $\mathbf{S}_2$ that were defined previously.

Theorem 10.7 The number of partitions of N into summands that are congruent to either 1 modulo 3 or 2 modulo 3 is equal to the number of partitions of N in which no summand is used more than twice.

Proof We have from Theorem 10.6 and by simple factoring

$$\sum_{N=0}^{\infty} \mathrm{p}(\mathbf{S}_1, N)x^N = \prod_{j=1}^{\infty} (1 + x^j + x^{2j})$$
$$= \prod_{j=1}^{\infty} \frac{1 - x^{3j}}{1 - x^j}.$$

Whenever $j = 3k$ in the denominator, there is an identical factor in the numerator at $j = k$. Hence, on the assumption that in the infinite product factors may be permuted without any change in value, one can reorganize the infinite product so as to contain a sequence of factors of 1:

$$\left\{\frac{1 - x^3}{1 - x^3}, \frac{1 - x^6}{1 - x^6}, \frac{1 - x^9}{1 - x^9}, \ldots\right\}.$$

The result is that all factors in the numerator of the infinite product are cancelled by factors somewhere in the denominator and we are left with

$$\sum_{N=0}^{\infty} \mathrm{p}(\mathbf{S}_1, N)x^N = \prod_{j=1}^{\infty} \frac{1}{1 - x^{3j-1}} \frac{1}{1 - x^{3j-2}} = \sum_{N=0}^{\infty} \mathrm{p}(\mathbf{S}_2, N)x^N. \quad \blacklozenge$$

Theorem 10.7 is a special case of a somewhat more general theorem (Glaisher, 1883). You might be able to guess the form of this more general result. Other partition identities are discussed in Andrews (1987).

Problems

10.40. Let $\mathbf{S}_1 = \mathbf{N}$ but in any partition of N do not allow any even summand to be used more than once. Conjecture a partition identity.

10.41. Let $\mathbf{S}_1 = \mathbf{N}$ but in any partition of N do not allow any summand to be used *exactly once* (it can, however, be used two or more times or not at all). Conjecture a partition identity.

10.42. Let $\mathbf{S}_1 = \mathbf{N}$ but in any partition of N any pair of summands, when there is such a pair, must have a difference of at least 2. Conjecture a partition identity.

10.43. Let $\mathbf{S}_1 = \mathbf{N}$ but in any partition of N any pair of summands, when there is such a pair, must have a difference of at least 2 and consecutive odd summands cannot appear. Conjecture a partition identity.

10.44. Let $\mathbf{S}_1 = \mathbf{N}$ but in any partition of N any pair of summands must have a difference of at least 4. Is a partition identity possible?

10.45. Let $\mathbf{S}_1 = \mathbf{N}$ but in any partition of N there must be either one or two odd summands, no more and no fewer. Is a partition identity possible?

10.46. Let $\mathbf{S}_1 = \mathbf{O}$ (the odd elements of $\mathbf{N}$) but in any partition of N use of two or more *consecutive* odd summands is prohibited. Is a partition identity possible?

10.47. We tend to believe that as N increases one must have the inequality $\mathrm{p}(\mathbf{S}, N + 1) \geq \mathrm{p}(\mathbf{S}, N)$. Show that it is possible to define a set $\mathbf{S}$ together with a restriction on its use so that for some N one has $\mathrm{p}(\mathbf{S}, N + 1) < \mathrm{p}(\mathbf{S}, N)$.

10.48. A theorem says: Fix $m \in \mathbf{N}$; then the number of partitions of N in which only summands incongruent to 0 (mod 2^m) may be repeated equals the number of partitions of N in which no summand appears more than $2^{m+1} - 1$ times. Illustrate this theorem with several cases.

10.49. Illustrate Theorem 10.5 by using both generating functions to calculate the number of partitions of 8 according to the two restrictions.

10.50. On the assumption that the various manipulations are valid, show that the following equality holds:

$$\prod_{j=1}^{\infty} \frac{1}{1 - x^{4j-1}} \frac{1}{1 - x^{4j-2}} \frac{1}{1 - x^{4j-3}} = \prod_{j=1}^{\infty} (1 + x^j + x^{2j} + x^{3j}).$$

Interpret this equality in terms of partitions.

10.51. A famous partition identity known as the **First Rogers–Ramanujan Identity** says that the number of partitions of N with summands that differ by at least 2 is equal to the number of partitions of N with summands that are congruent to 1 modulo 5 or 4 modulo 5. Illustrate this theorem with $N = 12$. Conjecture the generating function for the number of partitions of N with summands that are congruent to 1 or 4 (mod 5), and then apply it to $N = 12$ as a check on your answer. The First Identity, as well as a Second Identity, were proved first by the English mathematician **L.J. Rogers** in 1894 and since then by several other individuals (Gupta, 1970; Andrews and Baxter, 1989).

10.52. In this problem and the next two we look more closely at infinite products. As implied in the text, an infinite product $\Pi_{n=1}^{\infty} a_n$ is said to **converge** if the limit of the sequence of partial products is some nonzero finite number A: $\lim_{N\to\infty} \Pi_{n=1}^{N} a_n = A$. Supply reasons in the following proof that if $1 > a_n \geq$

0 and the series $\Sigma_{n=1}^{\infty} a_n$ converges, then $\Pi_{n=1}^{\infty} (1 - a_n)$ converges (Andrews, 1971b).

STATEMENTS	*REASONS*
1. There exists an $N \in \mathbf{N}$ such that $\Sigma_{n=N}^{\infty} a_n < \frac{1}{2}$.	1. Why?
2. We have $(1 - a_N)(1 - a_{N+1}) \geq 1 - a_N - a_{N+1}$	2. Why?
3. More generally, for any $m \geq N$, $\prod_{n=N}^{m} (1 - a_n) \geq 1 - \sum_{n=N}^{m} a_n$.	3. Why?
4. Now define the partial products $p_m = \prod_{n=1}^{m} (1 - a_n)$. Then we have $\frac{p_m}{p_{N-1}} = (1 - a_N)(1 - a_{N+1}) \ldots (1 - a_m) \geq 1 - \sum_{n=N}^{m} a_n > \frac{1}{2}$.	4. Why?
5. For any $k \in \mathbf{N}$, p_k is positive, so p_m is bounded below by the positive number $\frac{1}{2}p_{N-1}$.	5. Why?
6. The p_m's form a decreasing sequence, that is, $p_m - p_{m+1} \geq 0$.	6. Why?
7. Hence, the sequence $\{p_m\}_{m=1}^{\infty}$ converges to some positive value, that is $\lim_{m \to \infty} \prod_{n=1}^{m} (1 - a_n) = A > 0$.	7. Why?

10.53. Supply reasons in the following proof that the converse of Problem 10.52 holds, that is, if $0 \leq a_n < 1$ and $\Pi_{n=1}^{\infty} (1 - a_n)$ converges, then $\Sigma_{n=1}^{\infty} a_n$ is also convergent.

STATEMENTS	*REASONS*
1. Let $c > 0$ be the limit for $\lim_{N \to \infty} \prod_{n=1}^{N} (1 - a_n)$.	1. Why?
2. Then for any N one has $c < (1 - a_1)(1 - a_2) \ldots (1 - a_N)$.	2. Why?
3. For any $x \in \mathfrak{R}$, $1 - x \leq e^{-x}$.	3. Why?

STATEMENTS	*REASONS*
4. Hence, we have $$c \leq \exp\{-a_1 - a_2 - \cdots - a_N\}$$ or equivalently, $$\ln c < -\sum_{n=1}^{N} a_n.$$	4. Why?
5. Define the partial sums $S_N = \Sigma_{n=1}^{N} a_n$. Then the partial sums form an increasing sequence that is bounded above.	5. Why?
6. So, $\lim_{N\to\infty} S_N$ exists, and thus $\Sigma_{n=1}^{\infty} a_n$ is convergent.	6. Why?

10.54. Supply reasons in the following proof that if $|a_N| < 1$ for real a_N, and $\Sigma_{n=1}^{\infty} |a_n|$ converges, then any rearrangement of the order of the factors in $\Pi_{n=1}^{\infty} (1 + a_n)$ has no effect on the value of the infinite product.

STATEMENTS	*REASONS*
1. For real a_n, $|a_n| < 1$, we have $$\ln(1 + a_n) = a_n - \frac{a_n^2}{2} + \frac{a_n^3}{3} - \frac{a_n^4}{4} + \cdots.$$	1. Why?
2. It follows that $$\begin{aligned} |\ln(1 + a_n)| &\leq |a_n| + \left|\frac{a_n^2}{2}\right| + \left|\frac{a_n^3}{3}\right| + \left|\frac{a_n^4}{4}\right| + \cdots \\ &\leq |a_n| + \left|\frac{a_n^2}{2}\right| \sum_{k=0}^{\infty} |a_n|^k. \end{aligned}$$	2. Why?
3. This yields $$|\ln(1 + a_n)| \leq |a_n| + \left|\frac{a_n^2}{2}\right| [1 - |a_n|]^{-1}.$$	3. Why?
4. There exists an $N \in \mathbf{N}$ such that for all $n \geq N$ one has $|a_n| < \frac{1}{2}$.	4. Why?
5. Hence, for $n \geq N$, line 3 becomes $$|\ln(1 + a_n)| \leq \left(\frac{3}{2}\right) |a_n|.$$	5. Why?
6. Consequently, $\Sigma_{n=1}^{\infty} |\ln(1 + a_n)|$ converges, that is, $\Sigma_{n=1}^{\infty} \ln(1 + a_n)$ converges absolutely.	6. Why?

STATEMENTS	*REASONS*
7. The terms in $\Sigma_{n=1}^{\infty} \ln(1 + a_n)$ can be rearranged in any way whatsoever with no effect on the sum.	7. Why?
8. Consequently, the factors in $\Pi_{n=1}^{\infty} (1 + a_n)$ can be rearranged in any way whatsoever with no effect on the product.	8. Why?

10.55. Let $\mathbf{S}_1$ be the set of positive integers congruent to 1 or 5 modulo 6. Let $\mathbf{S}_2 = \mathbf{N}$, and interpret $p(\mathbf{S}_2, N)$ to mean the number of partitions of N in which the difference of any pair of summands ≥ 3 *and* in any partition consecutive multiples of 3 do not appear. For example, $1 + 6 + 9$ is an unacceptable partition of 16, but $3 + 12$ is an acceptable partition of 15. Write down the generating function for $p(\mathbf{S}_1, N)$, and then verify that the values of $p(\mathbf{S}_1, N)$ it gives for $N = 3, 4, \ldots, 20$ equal the corresponding values of $p(\mathbf{S}_2, N)$. The partition identity for all N was proved in 1926 by the German mathematician **Issai Schur** (1875–1941). A proof can be found in (Andrews, 1994).

10.5 Euler's Pentagonal Number Theorem

By now it is apparent that even though we have a systematic procedure in Theorem 10.4 for determining the number of partitions of N, implementation of this procedure for all but the smallest of N is still quite laborious. An improved procedure is on the way.

You may have noticed in your work with Theorem 10.4 that multiplication of the first several factors in $\Pi_{j=1}^{\infty} (1 - x^j)$ tends to give polynomials, most or all of whose nonzero coefficients are $+1$'s and -1's. For example,

$$\prod_{j+1}^{6} (1 - x^j) = 1 - x - x^2 + x^5 + 2x^7 - x^9 - x^{10} - x^{11} - x^{12} + 2x^{14} + x^{16} - x^{19} - x^{20} + x^{21}.$$

Actually, several further multiplications would show that many of the powers of x have zero coefficients.

A term in the expansion of the infinite product is obtained by multiplying together either a 1 or a $(-x^j)$ from each binomial. If most coefficients turn out to be 0, this means that for the coefficient of x^N, those products of the form

$$\underbrace{(-x^{a_1})(-x^{a_2}) \ldots (-x^{a_r})}_{\text{even number of factors}} = (-1)^{2k} x^N, \qquad r = 2k,$$

with $a_1 + a_2 + \cdots + a_r = N$, and those products of the form

$$\underbrace{(-x^{b_1})(-x^{b_2}) \ldots (-x^{b_s})}_{\text{odd number of factors}} = (-1)^{2k+1}x^N, \qquad s = 2k + 1,$$

with $b_1 + b_2 + \cdots + b_s = N$, tend to occur an equal number of times, whereupon $(-1)^{2k}x^N + (-1)^{2k+1}x^N = 0$. The exceptions to this statement tend to give coefficients of either $+1$ or -1, as noted.

Thus, what we are looking at are partitions of an integer N into an even number of distinct summands and partitions of N into an odd number of distinct summands. Let $\mathrm{p_E}(N)$, $\mathrm{p_O}(N)$ denote the numbers of these partitions, respectively. The conjecture is that

$$\mathrm{p_E}(N) - \mathrm{p_O}(N) = \begin{cases} 0 & \text{most of the time} \\ (-1)^j & \text{the rest of the time,} \end{cases}$$

where the nature of the integer j remains to be determined.

Euler discovered this result, and proved it with skillful manipulation of infinite products. We shall present, however, a beautiful proof employing Ferrers graphs that appeared in 1881 and is due to the early American mathematician **Fabian Franklin** (1853–1939) (Wilson,1939). The idea is to seek a one-to-one correspondence that is "almost correct" between graphs with an even number of distinct summands and graphs with an odd number of distinct summands.

We shall write partitions in order of decreasing summands, and analogously draw graphs with the largest number of dots at the top. Let **P** be a partition of N into r distinct parts, so that $N = a_1 + a_2 + \cdots + a_r$, and $a_1 > a_2 > \cdots > a_r$. Let S(**P**) $= a_r$ be the smallest summand of **P** and let L(**P**) be the largest integer k such that all of $a_1 = a_2 + 1$, $a_2 = a_3 + 1, \ldots, a_{k-1} = a_k + 1$ hold; if $a_1 \neq a_2 + 1$, set L(**P**) = 1. In general, L(**P**) $\leq r$ since there are only r summands.

There are now two cases, depending on the relationship between S(**P**) and L(**P**).

Case 1. S(**P**) $\leq$ L(**P**)

This case is illustrated by the Ferrers graph of a partition for $N = 28$ shown in Fig. 10.5. Now add 1 to the first S(**P**) summands and delete the last summand; this results in a transfer of S(**P**) dots upward. At the same time, a partition of $r = 5$ (= odd) distinct summands has been converted into one of $r = 4$ (= even) distinct summands.

Case 2. S(**P**) $>$ L(**P**)

This case is illustrated by the Ferrers graph of $N = 28$ shown in Fig. 10.6. Now subtract 1 from each of the first L(**P**) summands and create a new summand of L(**P**) dots; this results in a transfer of L(**P**) dots downward. At the same time, a partition of $r = 6$ (= even) distinct summands has been converted into one of $r = 7$ (= odd) distinct summands.

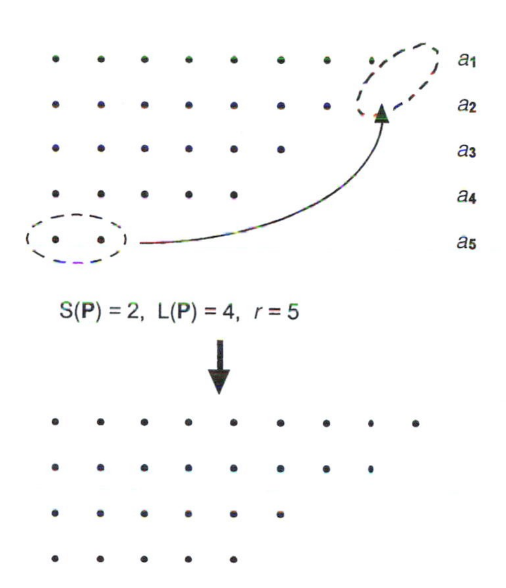

8+7+6+5+2 ⟶ 9+8+6+5

Figure 10.5 Illustrating Case 1 in Franklin's proof

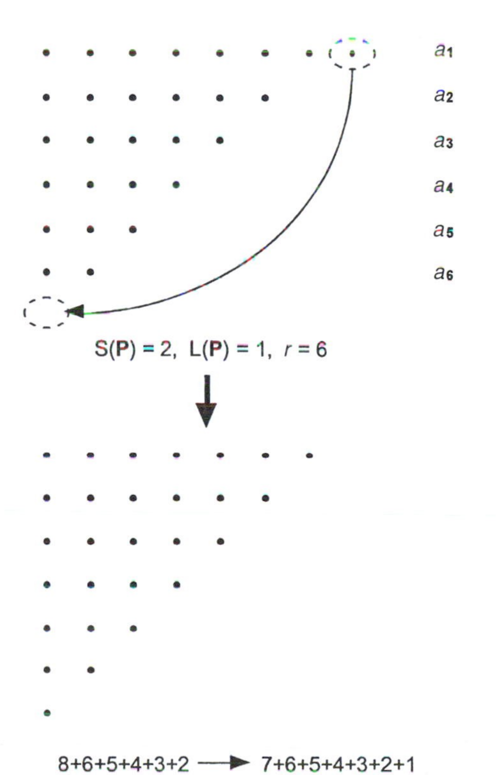

Figure 10.6 Illustrating Case 2 in Franklin's proof

The two inequalities S(**P**) $\leq$ L(**P**) and S(**P**) $>$ L(**P**) cover all possibilities as far as the relationship between S(**P**) and L(**P**) is concerned. Since all partitions into distinct summands must contain either an odd or an even number of summands, it would seem that we have shown that any odd partition can be converted into an even partition and vice versa.

Certainly this is true geometrically when L(**P**) $< r$, for if S(**P**) $\leq$ L(**P**), then there is room at the top to accommodate all dots that are moved upward upon deleting the entire bottom row. If S(**P**) $>$ L(**P**), then the bottom row is long enough so that when dots are moved downward one will not get two rows of the same number of dots.

But what about when L(**P**) $= r$? If S(**P**) $<$ L(**P**), the partition can be handled under Case 1. But if S(**P**) $=$ L(**P**), there is a problem, as shown by the example in Figure 10.7.

What has happened is that there is insufficient room at the top of the first graph to accommodate all of the dots moved upward, and as a consequence the old and the new Ferrers graphs have the same number of rows. There has been no change in parity of the partition (even $\rightarrow$ even). Breakdown here occurs for those N satisfying

$$N = \mathrm{S}(\mathbf{P}) + [\mathrm{S}(\mathbf{P}) + 1] + [\mathrm{S}(\mathbf{P}) + 2] + \cdots + [\mathrm{S}(\mathbf{P}) + \mathrm{S}(\mathbf{P}) - 1]$$

$$= \sum_{k=r}^{2r-1} k$$

$$= r\frac{(3r - 1)}{2}.$$

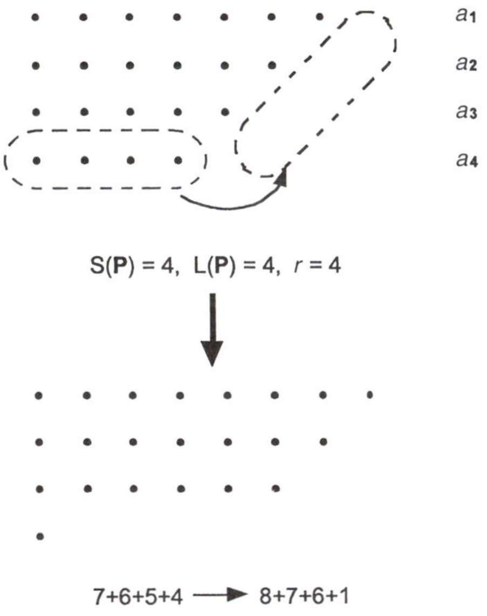

Figure 10.7 A case where one-to-one correspondence breaks down

When r is even, an even partition fails to convert to an odd partition, and when r is odd, an odd partition fails to convert to an even partition. Thus, we can write

$$\mathrm{p_E}(N) - \mathrm{p_O}(N) = (-1)^r$$

if $N = r(3r - 1)/2$.

The remaining possibility is when $\mathrm{L}(\mathbf{P}) = r$ and $\mathrm{S}(\mathbf{P}) > \mathrm{L}(\mathbf{P})$. If $\mathrm{S}(\mathbf{P}) - \mathrm{L}(\mathbf{P}) \geq 2$, we can always handle the partition under Case 2. But if $\mathrm{S}(\mathbf{P}) - \mathrm{L}(\mathbf{P}) = 1$, we again encounter a problem, as shown in Fig. 10.8.

In this instance, the bottom row of **P** was not long enough to avoid getting two rows of equal length when dots were moved downward. Breakdown occurs for those N satisfying

$$\begin{aligned} N &= \mathrm{S}(\mathbf{P}) + [\mathrm{S}(\mathbf{P}) + 1] + [\mathrm{S}(\mathbf{P}) + 2] + \cdots + [\mathrm{S}(\mathbf{P}) + \mathrm{L}(\mathbf{P}) - 1] \\ &= \sum_{k=r+1}^{2r} k \\ &= r\frac{(3r + 1)}{2}. \end{aligned}$$

Again, when r is even (odd), an even (odd) partition fails to convert to an odd (even) partition, and we can again write

$$\mathrm{p_E}(N) - \mathrm{p_O}(N) = (-1)^r$$

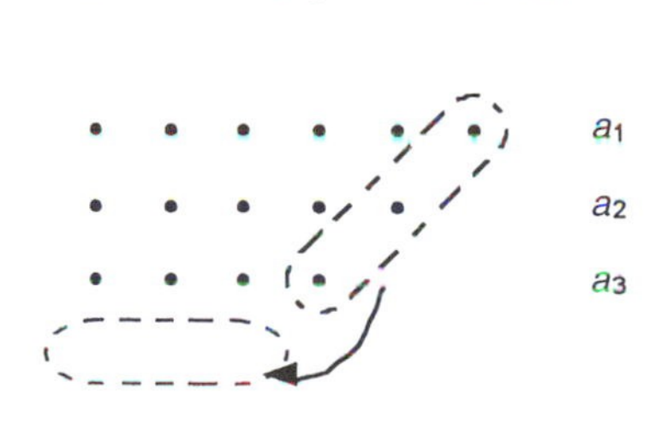

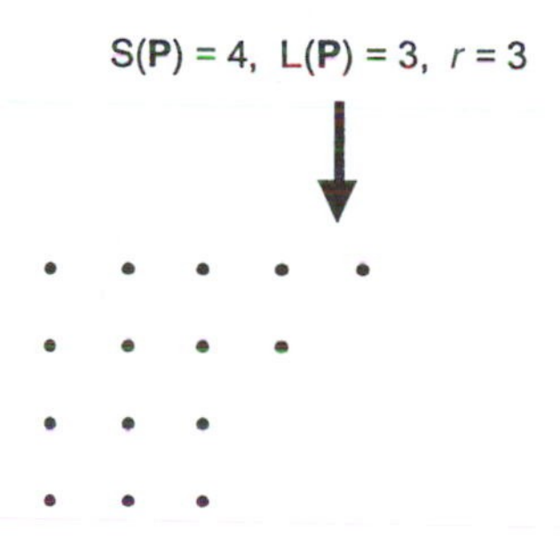

Figure 10.8 Another case of one-to-one correspondence breakdown

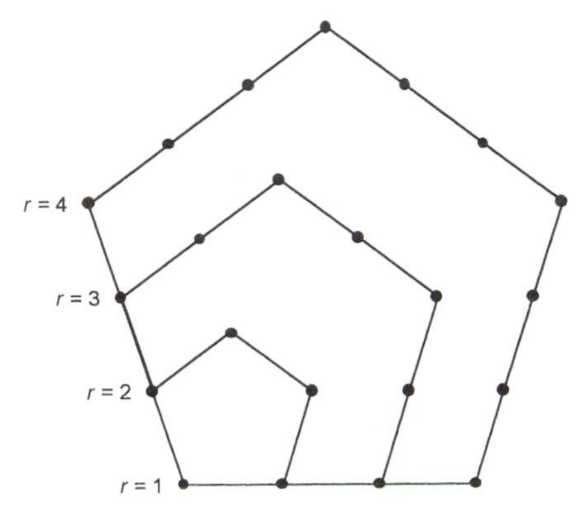

Figure 10.9 The pentagonal numbers

if $N = r(3r + 1)/2$. This troublesome case and the previous one can never occur simultaneously, that is, there is no $N > 0$ such that $r(3r - 1)/2 = r'(3r' + 1)/2$ for some $r, r' \in \mathbf{N}$. We can combine the two results in the statement

$$\mathrm{p_E}(N) - \mathrm{p_O}(N) = \begin{cases} (-1)^r & \text{if } N = r\dfrac{(3r \pm 1)}{2} \\ 0 & \text{otherwise.} \end{cases}$$

Euler observed the interesting fact that integers of the form $r(3r - 1)/2$ tally the total numbers of dots in concentric figures of pentagons (Fig. 10.9). These numbers (e.g., 1, 5, 12, 22, 35, . . .) are referred to as **pentagonal numbers**. The integers $r(3r + 1)/2$ do not tally dots in such diagrams; nevertheless, by association they are also sometimes called pentagonal numbers.[4]

Finally, let us return once again to the infinite product mentioned at the start of this section. Putting together all that we have learned about odd and even partitions, we have Theorem 10.8 (Shanks, 1951).

Theorem 10.8 (Euler's Pentagonal Number Theorem). If $|x| < 1$, then

$$\prod_{j=1}^{\infty} (1 - x^j) = 1 + \sum_{n=1}^{\infty} (-1)^n x^{n(3n+1)/2} + \sum_{n=1}^{\infty} (-1)^n x^{n(3n-1)/2}.$$

Example 10.14 Suppose that we desire the first 10 nonvanishing powers of x in the infinite product on the left-hand side of the equation in Theorem 10.8. Taking the first five terms in each infinite series on the right-hand side, we obtain

[4] Collectively, the integers $r(3r \pm 1)/2$ are also called **Euler's integers** in some quarters; they have interesting properties (Ehrhart, 1984).

$$\prod_{j=1}^{\infty} (1 - x^j) = 1 - x - x^2 + x^5 + x^7 - x^{12} - x^{15}$$

$$+ x^{22} + x^{26} - x^{35} - x^{40} + \dots . \quad \blacklozenge$$

Euler proved his Pentagonal Number Theorem in 1750 by an inductive method, but the result had appeared earlier in the *Ars conjectandi* (1713) of **Jakob Bernoulli**. This important early work on probability also introduced to the mathematical community the now widely used Bernoulli numbers (see Problem 10.23).

In the last section of this chapter we will show you why so much effort has been expended to obtain Euler's Pentagonal Number Theorem.

Problems

10.56. Obtain $\Pi_{j=1}^{10} (1 - x^j)$ by brute force and record how many powers of x are missing, how many have coefficients $+1$, and how many have coefficients -1.

10.57. Assume $|x| < 1$. Take the polynomial in Problem 10.56 and divide it into 1. What is the combinatorial significance of the coefficients of $x, x^2, x^3, \dots, x^{10}$ in the quotient?

10.58. Write out explicitly these two subgroups of partitions of 8: those partitions using an even number of distinct summands, and those partitions using an odd number of distinct summands. Do the same for $N = 12$. What difference between this case and the $N = 8$ case do you see?

10.59. For the partitions of 8 in Problem 10.58, show the one-to-one correspondence between the Ferrers graphs in the two groups.

10.60. Prove that $r(3r - 1)/2 = r'(3r' + 1)/2$ holds only when $r = r' = 0$.

10.61. Prove, in general, that for Fig. 10.9 the total number of dots enclosed by r concentric pentagons is $r(3r - 1)/2$.

10.62. More generally, let $\{P(2n + 1, k)\}_{k=1}^{\infty}$ be the **polygonal numbers** of order $2n + 1$; for example, $\{P(5, k)\}_{k=1}^{\infty}$ are the pentagonal numbers (those given by concentric pentagons).

(a) Show that any $(2n + 1)$-gonal number is given explicitly by $(n \geq 1)$

$$P(2n + 1, k) = \frac{k}{2}[(2n - 1)k + (3 - 2n)].$$

(b) Similarly, let $\{P(2n, k)\}_{k=1}^{\infty}$ be the polygonal numbers of even order $(2n)$. Show that any $(2n)$-gonal number is given explicitly by $(n \geq 2)$

$$P(2n, k) = k[(n - 1)k - (n - 2)].$$

(c) Show that $P(2n, k)P(2n, k + 2) + (n - 2)^2$ is a perfect square. Illustrate with the octagonal numbers (Dence, 1995).

10.63. Use the Pentagonal Number Theorem to show $\prod_{j=1}^{\infty} (2^j - 1)/2^j \approx 0.289$.

10.64. Write the first 15 pentagonal numbers $\{1, 5, 12, 22, \ldots\}$ in a row in ascending order. Now form the differences of successive members, and then form the differences of these differences. Prove the pattern that results.

10.65. A sequence $\{a_n\}$ of integers is said to have **parity of period *k*** if k is the smallest positive integer such that a_{n+k} is even when a_n is even and a_{n+k} is odd when a_n is odd. Prove that the sequence of increasing Euler integers (p. 394) has a parity of period 8.

10.66. Prove that if $24N + 1$ is a perfect square k^2, then N is an Euler integer.

10.67. Read the paper (Konecny, 1972); then apply that author's results to the calculation of the number of partitions of 20 into four distinct summands.

10.68. There is an unexpected connection between the number of unrestricted partitions of N, the number of partitions of N into distinct summands, and the pentagonal numbers. This connection is given in (Ehrhart, 1984). Look it up, and together with an extension of your work in Problem 10.67, illustrate the connection with $N = 20$.

10.69. Let $F(x) = \prod_{j=1}^{\infty} 1/(1 - x^j)$ and $\theta(x) = \prod_{j=1}^{\infty} (1 - x^j)$, so that $F(x) = 1/\theta(x)$. Now differentiate both sides, replace infinite products by series expansions, write the Cauchy product on the right-hand side, and arrive at

$$\sum_{N=1}^{\infty} Np(N)x^N = \sum_{N=1}^{\infty} x^N \sum_{m=0}^{N} p(m)\sigma(N - m),$$

where $\sigma(N - m)$ means the sum of all the divisors of $N - m$. Then obtain, finally,

$$Np(N) = \sum_{m=0}^{N-1} p(m)\sigma(N - m).$$

10.70. Program the result in Problem 10.69 and use it to calculate p(50).

10.71. Consult the reference (Shanks, 1951). This gives a strictly algebraic derivation of Euler's Pentagonal Number Theorem. The paper is short and should be accessible to you with a little effort. Digest it and write up a summary in your own words.

10.6 A Recursive Formula for Unrestricted Partitions

Although Problem 10.70 is a decided improvement over Theorem 10.4 as an algorithm for computing unrestricted partition numbers, it still involves more work than is necessary. But with the Pentagonal Number Theorem it is now an easy step to our goal of obtaining an efficient algorithm for calculating partitions. Combination of Theorems 10.4 and 10.8 gives

$$1 = \sum_{k=0}^{\infty} \mathrm{p}(k)x^k \left[1 + \sum_{n=1}^{\infty} (-1)^n x^{n(3n+1)/2} + \sum_{n=1}^{\infty} (-1)^n x^{n(3n-1)/2} \right]$$

$$= \sum_{k=0}^{\infty} \mathrm{p}(k)x^k \sum_{n=-\infty}^{\infty} (-1)^n x^{n(3n+1)/2}.$$

The exponents of x^k are nonnegative integers and the exponents of $x^{n(3n+1)/2}$ are also nonnegative integers. We write the Cauchy product of the two series as

$$1 = \sum_{N=0}^{\infty} \left\{ \sum_{k+\frac{n(3n+1)}{2}=N} (-1)^n \mathrm{p}(k) \right\} x^N.$$

Since the discrete variable k is not independent of N, n, we can eliminate it and write

$$1 = \sum_{N=0}^{\infty} \left\{ \sum_{\frac{n(3n+1)}{2}\leq N} (-1)^n \mathrm{p}\left[N - \frac{n(3n+1)}{2} \right] \right\} x^N.$$

The variable x is arbitrary, so the two sides must have identical coefficients for each power of x. If $N = 0$, then n can only be 0, so the expression in the braces (which must equal the 1 of the left-hand side) reduces to $\mathrm{p}(0)$. Our usual convention is that $\mathrm{p}(0) = 1$.

When $N = 1$, then only $n = 0, -1$ satisfy $(n/2)(3n + 1) \leq N$, and the expression in the braces (which now must equal 0) becomes

$$(-1)^0\mathrm{p}(1 - 0) + (-1)^{-1}\mathrm{p}(1 - 1) = 0,$$

or $\mathrm{p}(1) - \mathrm{p}(0) = 0$. In general, for $N > 0$ we have

$$\sum_{0\leq\frac{n(3n+1)}{2}\leq N} (-1)^n \mathrm{p}\left[N - \frac{n(3n+1)}{2} \right] = 0.$$

As $n(3n + 1)/2$ assumes the values $\{0, 1, 2, 5, 7, 12, 15, \ldots\}$ in succession, the n's are ticked off in sequence: $0, -1, 1, -2, 2, -3, 3$, and so on. The signs of the terms in the preceding finite series therefore occur in an orderly manner, and we have Theorem 10.9 as the final result.

Theorem 10.9 Let $\{n\}$ be the sequence $\{0, -1, 1, -2, 2, -3, 3, \ldots\}$. Then $\mathrm{p}(N) - [\mathrm{p}(N-1) + \mathrm{p}(N-2)] + [\mathrm{p}(N-5) + \mathrm{p}(N-7)] - [\mathrm{p}(N-12) + \mathrm{p}(N-15)] + \cdots + (-1)^n \mathrm{p}[N - n(3n+1)/2] = 0$, the series breaking off before the first n for which $n(3n + 1)/2 > N$.

Theorem 10.9 is our desired recurrence relation for the calculation of partition numbers. It could clearly be programmed on a computer and used to compute easily p(100), for example. Here is an example of a hand calculation to illustrate the algorithm.

Example 10.15 Given that $\mathrm{p}(0) = 1$, $\mathrm{p}(1) = 1$, $\mathrm{p}(2) = 2$, $\mathrm{p}(3) = 3$, $\mathrm{p}(4) = 5$, $\mathrm{p}(5) = 7$, $\mathrm{p}(6) = 11$, and $\mathrm{p}(7) = 15$, compute p(12).

From Theorem 10.9 we obtain, successively,

$$\begin{aligned} &N=8:\ n = 0, -1, 1, -2, 2 && \mathrm{p}(8) = \mathrm{p}(8-1) + \mathrm{p}(8-2) - [\mathrm{p}(8-5) + \mathrm{p}(8-7)] \\ &&& \quad = \mathrm{p}(7) + \mathrm{p}(6) - \mathrm{p}(3) - \mathrm{p}(1) \\ &&& \quad = 22 \end{aligned}$$

$$\begin{aligned} &N=9:\ n = 0, -1, 1, -2, 2 && \mathrm{p}(9) = \mathrm{p}(8) + \mathrm{p}(7) - \mathrm{p}(4) - \mathrm{p}(2) \\ &&& \quad = 30 \end{aligned}$$

$$\begin{aligned} &N=10:\ n = 0, -1, 1, -2, 2 && \mathrm{p}(10) = \mathrm{p}(9) + \mathrm{p}(8) - \mathrm{p}(5) - \mathrm{p}(3) \\ &&& \quad = 42 \end{aligned}$$

$$\begin{aligned} &N=11:\ n = 0, -1, 1, -2, 2 && \mathrm{p}(11) = \mathrm{p}(10) + \mathrm{p}(9) - \mathrm{p}(6) - \mathrm{p}(4) \\ &&& \quad = 56 \end{aligned}$$

$$\begin{aligned} &N=12:\ n = 0, -1, 1, -2, 2, -3 && \mathrm{p}(12) = \mathrm{p}(11) + \mathrm{p}(10) - \mathrm{p}(7) - \mathrm{p}(5) + \mathrm{p}(0) \\ &&& \quad = 77. \quad \blacklozenge \end{aligned}$$

Theorem 10.9 was used by **MacMahon** in 1918 to compute p(200) = 3,972,999,029,388. This result was used for comparison purposes by **Hardy** and **Ramanujan** in their work of 1919 (Problem 10.9). Their analytical theory produced a value for p(200) that was within 0.1 unit of the correct figure!

Problems

10.72. Write a computer program to calculate partition numbers using Theorem 10.9. Then use it to examine the following congruence property that was first proposed by **Ramanujan** (Drost, 1997):

$$p(5k + 4) \equiv 0 \pmod 5.$$

For test cases, look at $k = 5, 10, 15, 20, 25, 30$.

10.73. Find another partition congruence in the spirit of the one in Problem 10.72. Look at several test cases for it, including values of the argument that extend beyond Table H in Appendix II.

10.74. Let $p_k^*(N)$ denote the number of partitions of N into summands not divisible by k, where $k \geq 2$. Let m satisfy $0 \leq m \leq k - 1$. Then there is the theorem (Thanigasalam, 1974): If $p(kN + m) \equiv 0 \pmod k$ for $N = 0, 1, 2, \ldots$, then $p_k^*(kN + m) \equiv 0 \pmod k$ for $N = 0, 1, 2, \ldots$. Read the cited short paper and work through the details of the proof. Illustrate the theorem using the congruence in Problem 10.72.

RESEARCH PROBLEMS

1. Let $Q(N)$ denote the number of partitions of N into odd, distinct summands. For example, $9 = 5 + 3 + 1$, and thus $Q(9) = 2$. Investigate the possible congruence $p(N) \equiv Q(N) \pmod 2$ for many choices of N.

2. It is of interest to examine the divisibility properties of the values of $p(N)$. Investigate

(a) Possible choices of N for which $13 \mid p(N)$.
(b) Possible choices of N for which $p(N)$ is prime.
(c) Possible choices of N for which $p(N) \equiv 1 \pmod N$.
(d) Possible choices of N for which $p(N) \equiv 4 \pmod N$.

3. Let $\mathbf{S}_1$ be the set of summands, all of which are odd and greater than 1. Let $\mathbf{S}_2$ be the set of summands, all of which are distinct and are not powers of 2 ($1 = 2^0 \notin \mathbf{S}_2$). Investigate the partition numbers $p(\mathbf{S}_1, N)$, $p(\mathbf{S}_2, N)$. Also, compare these numbers with the number of partitions of N into unequal summands such that the two largest summands differ by exactly 1.

References

Abramowitz, M., and Stegun, I.A., *Handbook of Mathematical Functions*, Dover Publications, New York, 1965, pp. 836–839. This table (to $N = 500$) gives the number of unrestricted partitions and the number of partitions into distinct parts.

Alder, H.L., "Partition Identities—From Euler to the Present," *Amer. Math. Monthly*, **76**, 733–746 (1969). A survey article that will give you the flavor of research on hunting for partition identities.

Andrews, G.E., "A Combinatorial Proof of a Partition Function Limit," *Amer. Math. Monthly*, **78**, 276–278 (1971a). Our proofs of Theorem 10.2 and the two preceding lemmas are taken from this short paper.

Andrews, G.E., *Number Theory*, W.B. Saunders, Philadelphia, 1971b. See pp. 160–166 for material relating to Euler's partition formula, on which we based our Theorem 10.4 and our Problem 10.52.

Andrews, G.E., *Encyclopedia of Mathematics and Its Applications*, Vol. 2, "The Theory of Partitions," Addison–Wesley, Reading, MA, 1976. Chapters 1 and 2 are a nice introduction to infinite series-product generating functions for partitions. Our Theorem 10.6 comes from there.

Andrews, G.E., "Further Problems on Partitions," *Amer. Math. Monthly*, **94**, 437–439 (1987). One interesting result: $p(\mathbf{S}_1, N) = p(\mathbf{S}_2, N - 1)$, where $S_1 = \{x: x = \text{odd}, \pm 4, \pm 6, \pm 8, \pm 10 \pmod{32}\}$ and $\mathbf{S}_2 = \{x: x = \text{odd}, \pm 2, \pm 8, \pm 12, \pm 14 \pmod{32}\}$.

Andrews, G.E., *Number Theory*, Dover Publications, New York, 1994. The journal article that presented Andrews' proof of our Theorem 10.2 is reprinted in Appendix A of this available Dover republication of Andrews' original number theory text. See also Chapters 12–14 on partitions.

Andrews, G.E., and Baxter, R.J., "A Motivated Proof of the Rogers–Ramanujan Identities," *Amer. Math. Monthly*, **96**, 401–409 (1989). The proof here is not difficult to follow, but relies on a famous result of **Jacobi** called the **Triple Product Identity**, which we do not discuss in this book.

Berndt, B.C., "Srinivasa Ramanujan," *Amer. Scholar*, **58**, 234–244 (1989). This nontechnical article (with no references) describes the life and work of India's most famous scientist.

Dence, J.B., "Problem No. 67," *Missouri J. Math. Sci.*, **7**, 141–142 (1995). The source and solution of our Problem 10.62.

Drost, J.L., "A Shorter Proof of the Ramanujan Congruence Modulo 5," *Amer. Math. Monthly*, **104**, 963–964 (1997). A proof of the relation $p(5m + 4) \equiv 0 \pmod 5$.

Ehrhart, E., "Euler's Integers," *Fibonacci Quart.*, **22**, 218–228 (1984). Neat article on pentagonal numbers (also known as Euler's integers). However, be warned that there are some errors in this paper.

Glaisher, J.W.L., "A Theorem in Partitions," *Messenger Math.*, **12**, 158–170 (1883). The source of our Theorem 10.7 and of much more.

Gupta, H., "Partitions—A Survey," *J. Res. Nat. Bur. Stand.*, **74B**, 1–29 (1970). A survey article, with many interesting facts in it, of the same scope as that by Alder.

Hardy, G.H., *A Mathematician's Apology*, Cambridge University Press, Cambridge, England, 1969. Beautifully written by a master, even if you do not agree with all he has to say.

Just, E., and Schaumberger, N., "A Curious Property of the Integer 38," *Math. Mag.*, **46**, 221 (1973). The source of our Problem 10.3. See also the reference by A.M. Vaidya.

Kanigel, R., *The Man Who Knew Infinity*, Charles Scribner's Sons, New York, 1991. A fascinating biography of the elusive genius **Ramanujan**.

Konecny, V., "A Recursive Formula for the Number of Partitions of an Integer N Into m Unequal Integral Parts," *Math. Mag.*, **45**, 91–94 (1972). Interesting, easy-to-follow article, whose result is easily stated; the source of our Problem 10.67.

Leithold, L., *The Calculus of a Single Variable with Analytic Geometry*, 6th ed., Harper & Row, New York, 1990. Read pp. 742–781 for a review of several standard theorems on power series.

MacMahon, P.A., *An Introduction to Combinatory Analysis*, Cambridge University Press, Cambridge, England, 1920. This is a delightful, short (71-page) introduction to the author's major two-volume work on the subject that appeared in 1915–1916.

Mott, J.L., Kandel, A., and Baker, T.P., *Discrete Mathematics for Computer Scientists and Mathematicians*, Prentice–Hall, Englewood Cliffs, NJ, 1986. Read pp. 237–329 for a fine summary on recursive relations and generating functions.

Shanks, D., "A Short Proof of an Identity of Euler," *Proc. Amer. Math. Soc.*, **2**, 747–749 (1951). The source of our Problem 10.71.

Thanigasalam, K., "Congruence Properties of Certain Restricted Partitions," *Math. Mag.*, **47**, 154–156 (1974). If $k \geq 2$ is given and $\mathrm{p}_k^*(N)$ denotes the number of partitions of N into parts not divisible by k, then $\mathrm{p}_k^*(km + n) \equiv 0 \pmod{k}$, $m = 0, 1, 2, \ldots$, holds if $\mathrm{p}(km + n) \equiv 0 \pmod{k}$, $0 \leq n \leq k - 1$; see our Problem 10.74.

Vaidya, A.M., "On Representing Integers as Sums of Odd Composite Integers," *Math. Mag.*, **48**, 221–223 (1975). An extension of the result of Just and Schaumberger, namely, if p and q are distinct odd primes, the largest even integer not expressible as a sum of an odd composite multiple of p and an odd composite multiple of q is $2pq + p + q$. See our Problem 10.3.

Wilson, A.H., "A Forgotten Mathematician," *Scripta Math.*, **6**, 121–123 (1939). A short note on the American mathematician **Fabian Franklin**.

Chapter 11
Recurrence Relations

11.1 Introduction to Recurrence Relations

It frequently happens that the members of a sequence are connected by a relationship among various of them. This is important enough to be made the basis of a definition.

> **DEF.** Let $\{s_n\}_{n=0}^{\infty}$ be a sequence of mathematical objects (e.g., real numbers, functions, matrices). A **recurrence relation** is any definite relation that connects each s_n to one or more members before it in the sequence.

Here are some simple cases for sequences of *numbers*.

Example 11.1 Let $\{s_n\}_{n=0}^{\infty}$ be the sequence of factorials. Then the recurrence relation is

$$s_n = \begin{cases} 1 & \text{if } n = 0 \\ ns_{n-1} & \text{if } n > 0. \end{cases}$$

Or, let $\{s_{2n+1}\}_{n=0}^{\infty}$ be the sequence of coefficients in the Maclaurin series expansion of $\sin x$. Then since this expansion is

$$\sin x = \sum_{n=0}^{\infty} (-1)^n \frac{x^{2n+1}}{(2n+1)!},$$

the recurrence relation is

$$s_{2n+1} = \begin{cases} 1 & \text{if } n = 0 \\ \dfrac{-s_{2n-1}}{2n(2n+1)} & \text{if } n > 0. \end{cases} \quad ◆$$

Similarly, the members of a sequence of *functions* may be connected to each other by some definite relationship (Problem 11.9). For example, the following sequence is important in analysis and physics.

Example 11.2 The members of the sequence of orthogonal **Legendre polynomials** are connected by the recurrence relation

$$\mathrm{P}_n(x) = \begin{cases} 1 & \text{if } n = 0 \\ x & \text{if } n = 1 \\ \left(2 - \dfrac{1}{n}\right) x\mathrm{P}_{n-1}(x) - \left(1 - \dfrac{1}{n}\right) \mathrm{P}_{n-2}(x) & \text{if } n \geq 2. \end{cases}$$

The next few members of the sequence after $\mathrm{P}_1(x)$ are $\mathrm{P}_2(x) = (3x^2 - 1)/2$, $\mathrm{P}_3(x) = x(5x^2 - 3)/2$, and $\mathrm{P}_4(x) = (35x^4 - 30x^2 + 3)/8$. ◆

A recurrence relation can sometimes form the basis of the *definition* of a sequence. More commonly, however, a recurrence relation is used for *analytical* or *computational* purposes. For example, we saw from Theorem 10.9 how a recurrence relation could be used to calculate $\mathrm{p}(N)$, the number of unrestricted partitions of the positive integer N. The recurrence relation there is definitely of computational importance; it would not make a good definition of $\mathrm{p}(N)$ because it lacks transparency and simplicity.

Our focus in this chapter is on certain types of recurrence relations among real numbers. These kinds of recurrence relations sometimes arise naturally in simple combinatorial problems (Problems 11.1–11.5).

For example, suppose we wish to count the number of binary words that do not contain the bit pattern 11. We let s_n denote the number of binary words of length n that do not contain 11 (for $n = 1$, we do not consider the word 0). Table 11.1 gives the first few cases. It is convenient, for use later on, to define $s_0 = 0$. Does the sequence of s_n's begin to look familiar?

Fix n for the moment and let the representation

$$\begin{array}{ccccccccc} X = 1 & 0 & a & b & c & d & \ldots & \_ \\ 1 & 2 & 3 & 4 & 5 & 6 & & n \end{array}$$

Table 11.1 Binary Words of Length n That Do Not Contain the Bit Pattern 11

n	Words	s_n
1	1	1
2	10	1
3	100, 101	2
4	1000, 1001, 1010	3
5	10000, 10001, 10010, 10100, 10101	5

denote one of the s_n binary words without an 11. If $a = 0$, then we can associate the word X with the word

$$X' = \underset{2}{1} \quad \underset{3}{0} \quad \underset{4}{b} \quad \underset{5}{c} \quad \underset{6}{d} \quad \ldots \quad \underset{n}{\_},$$

which is one of the s_{n-1} binary words without an 11-bit pattern. On the other hand, if $a = 1$, then associate X with

$$X'' = \underset{3}{1} \quad \underset{4}{b} \quad \underset{5}{c} \quad \underset{6}{d} \quad \ldots \quad \underset{n}{\_},$$

which is one of the s_{n-2} binary words without an 11-bit pattern. The correspondences are invertible, so we obtain for the recurrence relation

$$s_n = s_{n-1} + s_{n-2}.$$

To solve this recurrence relation means to find a general formula for the nth term in the sequence $\{s_n\}_{n=1}^{\infty}$. Unfortunately, there is no general method that can be used for all recurrence relations. We show now two specialized techniques, one algebraic and one analytic, and apply them to the solution of the preceding recurrence relation.

Method 1

The recurrence relation $s_n = s_{n-1} + s_{n-2}$ is **homogeneous**, that is, it does not contain a constant term. Because of this we investigate the possibility that $s_n = x^n$ for all n, where x is some as yet unspecified constant. Substitution into the recurrence relation and cancellation of a factor x^{n-2} throughout gives

$$x^2 - x - 1 = 0.$$

This equation has roots $x = (1 \pm \sqrt{5})/2$, so the recurrence relation has two independent solutions, namely,

$$s_n^{(1)} = \left[\frac{1 + \sqrt{5}}{2}\right]^n, \qquad s_n^{(2)} = \left[\frac{1 - \sqrt{5}}{2}\right]^n.$$

Next, we show that any linear combination of these two solutions is also a solution of the recurrence relation. Let $s_n = c_1 s_n^{(1)} + c_2 s_n^{(2)}$ for all $n \geq 1$, where c_1, c_2 are arbitrary constants independent of n. Then

$$\begin{aligned} s_{n-1} + s_{n-2} &= [c_1 s_{n-1}^{(1)} + c_2 s_{n-1}^{(2)}] + [c_1 s_{n-2}^{(1)} + c_2 s_{n-2}^{(2)}] \\ &= c_1[s_{n-1}^{(1)} + s_{n-2}^{(1)}] + c_2[s_{n-1}^{(2)} + s_{n-2}^{(2)}] \\ &= c_1 s_n^{(1)} + c_2 s_n^{(2)} \\ &= s_n, \end{aligned}$$

so s_n is indeed a solution. We desire the s_n's to be integers, in accordance with Table 11.1. The entire sequence is determined by the recurrence relation, together with the values of s_1, s_2. Hence, we have

$$s_1 = 1 = c_1\left[\frac{1+\sqrt{5}}{2}\right] + c_2\left[\frac{1-\sqrt{5}}{2}\right]$$

$$s_2 = 1 = c_1\left[\frac{1+\sqrt{5}}{2}\right]^2 + c_2\left[\frac{1-\sqrt{5}}{2}\right]^2.$$

These two equations lead uniquely to $c_1 = \sqrt{5}/5$, $c_2 = -\sqrt{5}/5$. The final solution of the recurrence relation is therefore

$$s_n = \frac{[(1+\sqrt{5})/2]^n - [(1-\sqrt{5})/2]^n}{\sqrt{5}}.$$

This formula is remarkable because it is integral for all $n \in \mathbf{N}$. This is not at all obvious by casual inspection. See Problem 11.6 for further examples of Method 1. Notice, from the preceding, that $s_0 = 0$, as we defined earlier.

Example 11.3 We compute by direct expansion

$$\begin{aligned} s_6 &= \frac{[(1+\sqrt{5})/2]^6 - [(1-\sqrt{5})/2]^6}{\sqrt{5}} \\ &= \frac{1}{\sqrt{5}}\left(\frac{1 + 6\sqrt{5} + 75 + 100\sqrt{5} + 375 + 150\sqrt{5} + 125}{64} - \right. \\ &\qquad \left. \frac{1 - 6\sqrt{5} + 75 - 100\sqrt{5} + 375 - 150\sqrt{5} + 125}{64}\right) \\ &= \frac{1}{\sqrt{5}}\left(\frac{576 + 256\sqrt{5}}{64} - \frac{576 - 256\sqrt{5}}{64}\right) \\ &= \frac{1}{\sqrt{5}}(4\sqrt{5} + 4\sqrt{5}) \\ &= 8. \end{aligned}$$

This result agrees with s_6, as computed from the recurrence relation: $s_6 = s_5 + s_4 = 5 + 3 = 8$ (see Table 11.1). ◆

Method 2

We suppose that the members of the sequence $\{s_n\}_{n=1}^{\infty}$ are given by some generating function $g(x)$. It is convenient to write the generating function in the form

$$g(x) = \sum_{n=1}^{\infty} \frac{s_n}{n!} x^n$$

because we wish to differentiate this series and obtain a new series $g'(x)$ that is similar in form to $g(x)$. Applying the Ratio Test to the previous series, we have

$$\lim_{n\to\infty}\left[\left|\frac{s_n}{n!}x^n\right| \Big/ \left|\frac{s_{n-1}}{(n-1)!}x^{n-1}\right|\right] = |x| \lim_{n\to\infty}\left|\frac{s_n}{s_{n-1}}\frac{1}{n}\right|.$$

Since $s_n = s_{n-1} + s_{n-2}$ and $s_{n-1} > s_{n-2}$ for all $n > 3$, then

$$\frac{s_n}{s_{n-1}} = \frac{s_{n-1} + s_{n-2}}{s_{n-1}} < \frac{2s_{n-1}}{s_{n-1}} = 2,$$

and so

$$|x| \lim_{n\to\infty}\left|\frac{s_n}{s_{n-1}}\frac{1}{n}\right| \le 2|x| \lim_{n\to\infty}\frac{1}{n} = 0.$$

Hence, g(x) is a convergent power series for all real x and term-by-term differentiation is justified.

Differentiation gives

$$g'(x) = \sum_{n=1}^{\infty}\frac{s_n}{(n-1)!}x^{n-1} = \sum_{n=0}^{\infty}\frac{s_{n+1}}{n!}x^n,$$

and if we define $s_0 = 0$, as before, then

$$g(x) + g'(x) = \sum_{n=0}^{\infty}\frac{s_n + s_{n+1}}{n!}x^n = \sum_{n=0}^{\infty}\frac{s_{n+2}}{n!}x^n.$$

A second differentiation gives

$$g''(x) = \sum_{n=1}^{\infty}\frac{s_{n+1}}{(n-1)!}x^{n-1} = \sum_{n=0}^{\infty}\frac{s_{n+2}}{n!}x^n,$$

and thus we arrive at the differential equation

$$g(x) + g'(x) = g''(x).$$

The auxiliary equation of this linear differential equation with constant coefficients is $y^2 - y - 1 = 0$. Its roots are (as noted in Method 1) $y = (1 \pm \sqrt{5})/2$. Hence, the general solution of the differential equation can be written as (Rainville and Bedient, 1981)

$$\begin{aligned} g(x) &= c_1 \exp\left[\frac{(1+\sqrt{5})x}{2}\right] + c_2 \exp\left[\frac{(1-\sqrt{5})x}{2}\right] \\ &= c_1\left[1 + \frac{(1+\sqrt{5})}{2}x + \frac{(1+\sqrt{5})^2}{8}x^2 + \dots\right] \\ &\quad + c_2\left[1 + \frac{(1-\sqrt{5})}{2}x + \frac{(1-\sqrt{5})^2}{8}x^2 + \dots\right] \\ &= (c_1 + c_2) + \frac{x}{2}[c_1(1+\sqrt{5}) + c_2(1-\sqrt{5})] + \dots . \end{aligned}$$

The first two terms of the generating function are

$$\frac{s_0}{0!}x^0 + \frac{s_1}{1!}x^1 = 0 + x,$$

so we must have

$$\begin{cases} c_1 + c_2 = 0 \\ c_1\dfrac{(1 + \sqrt{5})}{2} + c_2\dfrac{(1 - \sqrt{5})}{2} = 1. \end{cases}$$

This system of equations yields $c_1 = 1/\sqrt{5}$, $c_2 = -1/\sqrt{5}$.

It follows that g(x) is given by

$$g(x) = \sum_{n=0}^{\infty} \frac{\{[(1 + \sqrt{5})/2]^n - [(1 - \sqrt{5})/2]^n\}}{\sqrt{5}\, n!} x^n$$

and comparison of this with $\Sigma_{n=1}^{\infty} s_n x^n/n!$ gives, finally,

$$s_n = \frac{[(1 + \sqrt{5})/2]^n - [(1 - \sqrt{5})/2]^n}{\sqrt{5}}.$$

See Problem 11.8 for further examples of Method 2.

Example 11.4 Consider the sequence defined by $s_0 = 3$, $s_n = 2s_{n-1} + 1$ for $n > 0$. Let $g(x) = \Sigma_{n=0}^{\infty} s_n x^n/n!$; then upon differentiation, we obtain

$$g'(x) = \sum_{n=1}^{\infty} \frac{s_n}{(n-1)!} x^{n-1} = \sum_{m=0}^{\infty} \frac{s_{m+1}}{m!} x^m,$$

and replacement of s_{m+1} by $2s_m + 1$ leads to the linear differential equation $g'(x) = 2g(x) + e^x$. Review how to solve this (Rainville and Bedient, 1981); the solution is $g(x) = 4e^{2x} - e^x$, so $s_n = 4 \cdot 2^n - 1$ (verify!). ◆

Problems

11.1. Let **S** be a set of n elements, and let s_n denote the number of subsets of **S** that are nonempty. Deduce a recurrence relation for s_n.

11.2. Let s_n denote the number of binary words of length n that do not contain the bit pattern 001. Deduce a recurrence relation for s_n.

11.3. Let s_n denote the number of diagonals in a convex polygon of n sides. Deduce a recurrence relation for s_n.

11.4. Let s_n denote the maximum number of regions into which the plane is divided by n straight lines. Deduce a recurrence relation for s_n.

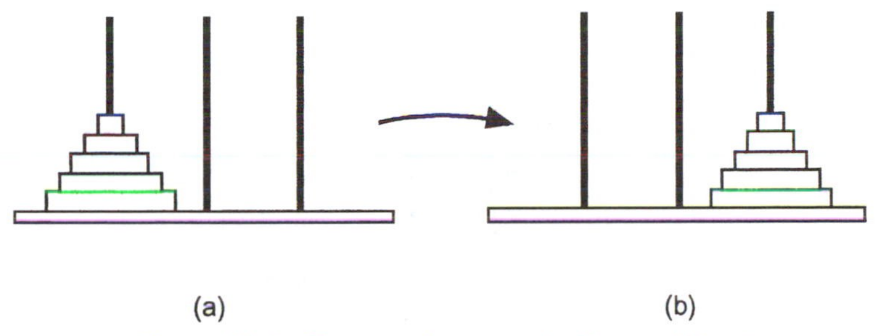

Figure 11.1 The game known as the Tower of Hanoi

11.5. The puzzle known as the **Tower of Hanoi** was created in 1883 by **Edouard Lucas**. The idea is to move the rings shown in Figure 11.1(a) to another peg, subject to the following rules:

1. Rings are to be moved one at a time.
2. During the manipulations no ring can be placed atop another of smaller diameter.
3. All rings must end up stacked as shown in Figure 11.1(b).

Let s_n denote the minimum number of moves needed to effect the transformation shown in the figure in the case where there are n rings. Deduce a recurrence relation for s_n.

11.6. Solve each of the following recurrence relations in a manner analogous to that used in Method 1:

(a) $s_n = 5s_{n-1} - 3s_{n-2}, \quad s_0 = 1,\ s_1 = 2.$
(b) $s_n = 19s_{n-2} + 30s_{n-3}, \quad s_0 = 1,\ s_1 = 2,\ s_2 = 3.$
(c) $s_n = s_{n-1} - s_{n-2} + s_{n-3}, \quad s_0 = 1,\ s_1 = 2,\ s_2 = 3.$

11.7. Show that the sequence in Problem 11.6(c) is periodic, that is, find the smallest positive integer k such that $s_{n+k} = s_n$ for all $n \geq 0$.

11.8. Solve each of the following recurrence relations by the use of generating functions:

(a) $s_n = 2s_{n-1} + 3, \quad s_0 = -1.$
(b) $s_n = s_{n-1} + 2s_{n-2} - 1, \quad s_0 = 0,\ s_1 = 2.$

11.9. Here we look at one example of a recurrence relation among *functions*. The important **Hermite differential equation**, prominent in physics, is really an infinite family of differential equations,

$$\frac{d^2H_n(x)}{dx^2} - 2x\frac{dH_n(x)}{dx} + 2nH_n(x) = 0$$
$$(n = 0, 1, 2, 3, \ldots).$$

The functions $\{\mathrm{H}_n(x)\}_{n=0}^{\infty}$ that satisfy these differential equations are the so-called **Hermite polynomials**. They also satisfy the recurrence relation

$$\mathrm{H}_{n+2}(x) = 2x\mathrm{H}_{n+1}(x) - 2(n+1)\mathrm{H}_n(x),$$

which can be derived directly from the differential equation. A conventional initialization of the sequence of polynomials is $\mathrm{H}_0(x) = 1$, $\mathrm{H}_1(x) = 2x$.

(a) The generating function for the Hermite polynomials must be a function of two variables, $\mathrm{g}(x, u)$, where x is the argument of the polynomials and u is the expansion variable in the infinite series representation of $\mathrm{g}(x, u)$. Show that

$$x\mathrm{g}(x, u) = u\mathrm{g}(x, u) + \frac{1}{2}\left(\frac{\partial \mathrm{g}}{\partial u}\right)_x.$$

(b) Solve the preceding differential equation for $\mathrm{g}(x, u)$, making use of the sequence initialization.
(c) From the generating function $\mathrm{g}(x, u)$ work out $\mathrm{H}_5(x)$, and show that this agrees with what you compute from the recurrence relation.

11.10. Let $\{\mathrm{H}_n(x)\}_{n=0}^{\infty}$ be the sequence of polynomials of Problem 11.9. Prove that if n is odd, then $\mathrm{H}_n(0) = 0$, and if n is even, then $\mathrm{H}_n(0) \neq 0$.

11.2 A General Theorem on Linear Recurrence Relations

Nearly all of our examples of recurrence relations have been **linear relations**, that is, relations in which every term is either a constant or a constant times the first power of a member of the sequence $\{s_n\}_{n=0}^{\infty}$. The general, linear, homogeneous recurrence relation of order r with constant coefficients is

$$s_n = c_1 s_{n-1} + c_2 s_{n-2} + \cdots + c_r s_{n-r},$$

where the c_i's are constants. The following theorem is a handy and satisfying result to know about such relations.

Theorem 11.1 The generating function for a sequence of numbers $\{s_n\}_{n=0}^{\infty}$ that are connected by a linear, homogeneous recurrence relation with constant coefficients is a rational algebraic expression.

Proof Let the sequence $\{s_n\}_{n=0}^{\infty}$ be connected by the rth-order, linear, homogeneous recurrence relation

$$s_n = \sum_{k=1}^{r} c_k s_{n-k}, \quad n \geq r$$

and let the generating function for $\{s_n\}_{n=0}^{\infty}$ be

$$g(x) = \sum_{n=0}^{\infty} s_n x^n.$$

We need to show that $g(x)$ is absolutely convergent on some interval $-R < x < R$ of nonzero width.

Define $C = \max\{|c_1|, |c_2|, \ldots, |c_r|\}$ and $S = |s_0| + |s_1| + \cdots + |s_{r-1}|$. We have immediately $|s_j| \le C(C+1)^{j-r}S$ for $j = r$. Assume it is also true for integers j satisfying $r < j \le k$. Then,

$$\begin{aligned}
|s_{k+1}| &\le \sum_{j=1}^{r} |c_j s_{k+1-j}| \\
&\le C \sum_{j=1}^{r} |s_{k+1-j}| \\
&\le C^2 S \sum_{j=1}^{r} (C+1)^{k+1-j-r} \qquad \text{by the induction hypothesis} \\
&= C^2 S \left\{ \frac{(C+1)^{k-r}[1-(C+1)^{-r}]}{1-(C+1)^{-1}} \right\} \\
&= CS(C+1)^{k+1-r}[1-(C+1)^{-r}] \\
&\le CS(C+1)^{k+1-r}.
\end{aligned}$$

Hence, by the Principle of Finite Induction $|s_n| \le C(C+1)^{n-r}S$ holds for all integers $n \ge r$.

We see then that for $R = (C+1)^{-1}$, and $0 \le |x| < R$, one has

$$g(x) \le \sum_{n=r}^{\infty} CS(C+1)^{n-r}|x|^n = CS(C+1)^{-r} \sum_{n=r}^{\infty} y^n, \quad |x| = yR,$$

where $0 < y < 1$. The geometric series converges, so by the Comparison Test $g(x)$ converges absolutely for $|x| < R$ (it may converge over a wider interval, but we don't need this). Hence, rearrangement of an infinite number of terms in $g(x)$ is permitted for $|x| < (C+1)^{-1}$.

Choose any such nonzero x; define the polynomial

$$P(x) = 1 - \sum_{k=1}^{r} c_k x^k.$$

Then the product $P(x)g(x)$ can be written, after rearrangement of an infinite number of terms, as

$$\begin{aligned}
\mathrm{P}(x)\mathrm{g}(x) &= \sum_{n=0}^{\infty} s_n x^n - \sum_{n=0}^{\infty}\sum_{k=1}^{r} s_n c_k x^{n+k} \\
&= \sum_{j=0}^{\infty} s_j x^j - \sum_{j=1}^{r-1}\sum_{k=1}^{j} s_{j-k} c_k x^j - \sum_{j=r}^{\infty}\sum_{k=1}^{r} s_{j-k} c_k x^j, \quad (n = j - k) \\
&= s_0 + \sum_{j=1}^{r-1}\left[s_j - \sum_{k=1}^{j} s_{j-k} c_k\right] x^j + \sum_{j=r}^{\infty}\left[s_j - \sum_{k=1}^{r} s_{j-k} c_k\right] x^j \\
&= s_0 + \sum_{j=1}^{r-1}\left[s_j - \sum_{k=1}^{j} s_{j-k} c_k\right] x^j
\end{aligned}$$

because of the recurrence relation. Hence, $\mathrm{P}(x)\mathrm{g}(x)$ is a polynomial $\mathrm{c}(x)$ of degree $r - 1$ at most, so $\mathrm{g}(x) = \mathrm{c}(x)/\mathrm{P}(x)$. ◆

Example 11.5 Let $s_n = s_{n-1} + s_{n-2}$, $s_0 = 0$, $s_1 = 1$, as in Table 11.1. Then $r = 2$ and $\mathrm{P}(x) = 1 - x - x^2$, and we have

$$\mathrm{c}(x) = 1 + x(s_1 - s_0 c_1) = x,$$

so $\mathrm{g}(x) = x/(1 - x - x^2)$. See Example 10.4. ◆

Example 11.6 Let $s_n = s_{n-1} - 2s_{n-2} + s_{n-3}$, $s_0 = 0$, $s_1 = 1$, $s_2 = 2$. Then $r = 3$ and $\mathrm{P}(x) = 1 - x + 2x^2 - x^3$, and

$$\begin{aligned}
\mathrm{c}(x) &= x(s_1 - s_0 c_1) + x^2(s_2 - s_1 c_1 - s_0 c_2) \\
&= x + x^2,
\end{aligned}$$

so $\mathrm{g}(x) = x(1 + x)/(1 - x + 2x^2 - x^3)$. Partial expansion of the generating function gives

$$\mathrm{g}(x) = x + 2x^2 - 3x^4 - x^5 + 5x^6 + 4x^7 + \ldots,$$

and the coefficients here agree with the first six nonzero values for the s_n's as computed directly from the recurrence relation. ◆

Problems

11.11. Work out the generating functions $\mathrm{g}(x)$ for the following sequences of numbers:

(a) The sequence in Problem 11.6(a).
(b) $\{s_n\}_{n=0}^{\infty}$, where $s_0 = 1$, $s_1 = 1$, $s_2 = 2$, $s_n = 3s_{n-1} - 3s_{n-2} + s_{n-3}$ for $n \geq 3$.

(c) $\{s_n\}_{n=0}^{\infty}$, where $s_0 = -2$, $s_1 = 1$, $s_2 = 3$, $s_3 = 5$, $s_4 = 7$, and $s_n = s_{n-2} + 2s_{n-3} + s_{n-5}$ for $n \geq 5$.

11.12. (a) Consider the periodic sequence $\{s_n\}_{n=0}^{\infty} = \{1, 2, 3, 1, 2, 3, \ldots\}$. What is the generating function, and what is the recurrence relation?

(b) Consider the structured sequence $\{s_n\}_{n=0}^{\infty} = \{1, 2, 3, 4, 2, 4, 6, 8, 4, 8, 12, 16, 8, 16, 24, 32, \ldots\}$. What is the generating function, and what is the recurrence relation?

11.13. Consider the linear recurrence relation $s_n = \Sigma_{k=1}^{r} c_k s_{n-k}$, $n \geq r$, for which the P(x) of Theorem 11.1 is

$$\mathrm{P}(x) = 1 - \sum_{k=1}^{r} c_k x^k.$$

Define a polynomial f(x) by

$$\mathrm{f}(x) = x^r - \sum_{k=1}^{r} c_k x^{r-k}.$$

(a) Show that $\mathrm{P}(x) = x^r\,\mathrm{f}(x^{-1})$.

(b) Suppose the zeros $\{\alpha_i\}$ of f(x) are distinct. Show that

$$\mathrm{P}(x) = \prod_{i=1}^{r} (1 - \alpha_i x).$$

(c) Show that the generating function g(x) for the sequence $\{s_n\}_{n=0}^{\infty}$ is expressible in the form

$$\mathrm{g}(x) = \sum_{i=1}^{r} \frac{\beta_i}{(1 - \alpha_i x)},$$

where the β_i's are constants.

(d) Finally, show that each s_n is expressible as

$$s_n = \sum_{i=1}^{r} \beta_i \alpha_i^n,$$

that is, the nth member of the sequence is a fixed linear combination of nth powers of the zeros of the polynomial f(x).

(e) Apply the result of part (d) to the sequence $\{s_n\}_{n=0}^{\infty}$, where $s_0 = 2$, $s_1 = 4$, $s_2 = 6$, $s_n = \frac{5}{2}s_{n-1} - 3s_{n-2} + s_{n-3}$ for $n \geq 3$, and write an explicit formula for s_n.

11.3 Lucas Sequences

A systematic study of linear homogeneous second-order recurrence relations was initiated by **Edouard Lucas**, who published two important papers on the subject in 1878. The work was continued and improved upon about 35 years later by **R.D. Carmichael** (Carmichael, 1913). By now, the following terminology has become standard.

DEF. Let the integers P, Q satisfy $P^2 - 4Q > 0$. Then the sequence $\{\mathrm{U}_n(P, Q)\}_{n=0}^{\infty}$ defined by

$$\mathrm{U}_n(P, Q) = \begin{cases} 0 & \text{if } n = 0 \\ 1 & \text{if } n = 1 \\ P\mathrm{U}_{n-1}(P, Q) - Q\mathrm{U}_{n-2}(P, Q) & \text{if } n \geq 2 \end{cases}$$

is known as the **Lucas sequence with parameters *P* and *Q***. The associated sequence $\{\mathrm{V}_n(P, Q)\}_{n=0}^{\infty}$ given by

$$\mathrm{V}_n(P, Q) = \begin{cases} 2 & \text{if } n = 0 \\ P & \text{if } n = 1 \\ P\mathrm{V}_{n-1}(P, Q) - Q\mathrm{V}_{n-2}(P, Q) & \text{if } n \geq 2 \end{cases}$$

is called the **companion Lucas sequence with parameters *P* and *Q***. When $P = 1$ and $Q = -1$, the sequences $\{\mathrm{U}_n(P, Q)\}_{n=0}^{\infty}$ and $\{\mathrm{V}_n(P, Q)\}_{n=0}^{\infty}$ are the **Fibonacci numbers** and the **Lucas numbers**, respectively. When $P = 2$ and $Q = -1$, the sequence $\{\mathrm{U}_n(P, Q)\}_{n=0}^{\infty}$ is the set of **Pell numbers**.

Partial listings of members of some Lucas and companion Lucas sequences are given in Table I(1) in Appendix II.

It follows from Problem 11.13 that for $n \geq 0$

$$\begin{cases} \mathrm{U}_n(P, Q) = c_1\alpha^n + c_2\beta^n \\ \mathrm{V}_n(P, Q) = d_1\alpha^n + d_2\beta^n, \end{cases}$$

where $(x - \alpha)(x - \beta) = x^2 - Px + Q$. Using the initializations given previously in the definition of the sequences, we obtain

$$\begin{cases} \mathrm{U}_n(P, Q) = \dfrac{1}{\sqrt{P^2 - 4Q}}[\alpha^n - \beta^n] \\ \mathrm{V}_n(P, Q) = \alpha^n + \beta^n, \end{cases}$$

where by convention $\alpha > \beta$. These formulas are called the **Binet formulas**.[1]

[1] After the French mathematician **Jacques P.M. Binet** (1786–1856), who rediscovered (in 1843) the explicit formula for the Fibonacci numbers that had been known to **Abraham DeMoivre** (1667–1754) as early as 1718.

Because of the simplicity of the recurrence relation for the Lucas and companion Lucas sequences, these sequences possess an enormous number of arithmetical relationships. A great many of the fascinating properties of the Fibonacci numbers, for example, are merely special cases of these relationships (Honsberger, 1985). The Binet formulas are very useful in deriving these relationships.

Let $D = P^2 - 4Q$, $\alpha - \beta = \sqrt{D}$, and $Q = \alpha\beta$; for brevity we write $U_n = \mathrm{U}_n(P, Q)$ and $V_n = \mathrm{V}_n(P, Q)$. It follows that

$$U_{2n} = \frac{\alpha^{2n} - \beta^{2n}}{\sqrt{D}} = \frac{\alpha^n - \beta^n}{\sqrt{D}}(\alpha^n + \beta^n) = U_n V_n$$

$$V_{2n} = \alpha^{2n} + \beta^{2n} = (\alpha^n + \beta^n)^2 - 2\alpha^n\beta^n = V_n^2 - 2Q^n.$$

Similarly, the following useful identities are easily established:

$$\begin{aligned} U_m V_n + U_n V_m &= \frac{\alpha^m - \beta^m}{\sqrt{D}} \cdot (\alpha^n + \beta^n) + \frac{\alpha^n - \beta^n}{\sqrt{D}} \cdot (\alpha^m + \beta^m) \\ &= \frac{(\alpha^{m+n} - \alpha^n\beta^m + \alpha^m\beta^n - \beta^{m+n})}{\sqrt{D}} \\ &\quad + \frac{(\alpha^{m+n} - \alpha^m\beta^n + \alpha^n\beta^m - \beta^{m+n})}{\sqrt{D}} \\ &= \frac{2(\alpha^{m+n} - \beta^{m+n})}{\sqrt{D}} \\ &= 2U_{m+n} \end{aligned}$$

$$\begin{aligned} V_m V_n + D U_m U_n &= (\alpha^m + \beta^m)(\alpha^n + \beta^n) + D\frac{(\alpha^m - \beta^m)(\alpha^n - \beta^n)}{\sqrt{D}\qquad\sqrt{D}} \\ &= (\alpha^{m+n} + \alpha^m\beta^n + \alpha^n\beta^m + \beta^{m+n}) \\ &\quad + (\alpha^{m+n} - \alpha^n\beta^m - \alpha^m\beta^n + \beta^{m+n}) \\ &= 2(\alpha^{m+n} + \beta^{m+n}) \\ &= 2V_{m+n}. \end{aligned}$$

The members of Lucas and companion Lucas sequences also possess a wide variety of number-theoretic properties, including divisibility properties (Ribenboim, 1996). The following theorem is illustrative; the crux of the proof is an identity that is much in the same spirit as the preceding ones and that is best established from the Binet formulas.

Theorem 11.2 If $n, k \in \mathbf{N}$, then $U_{kn} \equiv 0 \pmod{U_n}$.

Proof We have for $m \geq n$

$$U_{m+n} = \frac{\alpha^{m+n} - \beta^{m+n}}{\sqrt{D}} = \frac{\alpha^m - \beta^m}{\sqrt{D}} \cdot (\alpha^n + \beta^n) - \left[\frac{\alpha^m\beta^n - \alpha^n\beta^m}{\sqrt{D}}\right]$$

$$= U_m V_n - (\alpha\beta)^n \cdot \left[\frac{\alpha^{m-n} - \beta^{m-n}}{\sqrt{D}}\right]$$

$$= U_m V_n - Q^n U_{m-n}. \qquad (*)$$

Now the theorem is true for $k = 1$; when $k = 2$, take $m = n$ and (*) yields $U_{2n} = U_n V_n$, so $U_n \mid U_{2n}$. When $k = 3$, take $m = 2n$, and (*) now yields $U_{3n} = U_{2n}V_n - Q^n U_n$. But $U_n \mid U_{2n}$, so $U_n \mid U_{3n}$ follows automatically. The general theorem then follows by mathematical induction. ◆

For example, let $P = 4$ and $Q = 3$. We compute the first few members of the Lucas sequence and display them in Table 11.2. Upon checking, we find that $1093 \mid 2391484$ and $1093 \mid 5230176601$, and also that $121 \mid 29524$, $121 \mid 7174453$, and $121 \mid 1743392200$.

Our final example in this section shows how one can expand a U_n in powers of the "independent variable" Q, somewhat like a binomial series, and with V_n's playing the role of binomial coefficients.

Theorem 11.3 Let $U_n(P, Q) = U_n$ and $V_n(P, Q) = V_n$, and assume n is odd: $n = 2m + 1$. Then

$$U_n = Q^m + V_2 Q^{m-1} + V_4 Q^{m-2} + \cdots + V_{n-1}.$$

Table 11.2 The First Members of the Lucas Sequence $\{U_n(4, 3)\}_{n=1}^{\infty}$

n	U_n	n	U_n	n	U_n
1	1	8	3280	15	7174453
2	4	9	9841	16	21523360
3	13	10	29524	17	64570081
4	40	11	88573	18	193710244
5	121	12	265720	19	581130733
6	364	13	797161	20	1743392200
7	1093	14	2391484	21	5230176601

Proof From the Binet formula $V_{2k} = \alpha^{2k} + \beta^{2k}$ and from $Q = \alpha\beta$, we obtain

$$Q^m + \sum_{k=1}^{m} V_{2k}Q^{m-k} = \alpha^m\beta^m + \sum_{k=1}^{m} [\alpha^{m+k}\beta^{m-k} + \alpha^{m-k}\beta^{m+k}].$$

In each term of the indicated series on the right, the sum of the exponents is $2m$, but in no term is the α-exponent equal to the β-exponent (this would occur for $k = 0$). This possibility, however, is realized by the term $\alpha^m\beta^m$. Thus, since $(x^r - y^r)/(x - y) = \Sigma_{k=0}^{r-1} x^{r-1-k}y^k$ from algebra, we see that

$$Q^m + \sum_{k=1}^{m} V_{2k}Q^{m-k} = \sum_{k=0}^{2m} \alpha^{2m-k}\beta^k = \frac{\alpha^{2m+1} - \beta^{2m+1}}{\alpha - \beta} = U_n. \quad \blacklozenge$$

The theorem, as given, is restricted to odd subscripts n; however, there is a similar companion result for even n (Problem 11.33). It and Theorem 11.3 imply immediately Corollary 11.3.1.

Corollary 11.3.1 For $n \geq 1$, $U_n \equiv V_{n-1} \pmod{Q}$.

Example 11.7 The first few members of the companion Lucas sequence associated with the sequence in Table 11.2 are $V_1 = 4$, $V_2 = 10$, $V_3 = 28$, $V_4 = 82$, $V_5 = 244$. Then since $U_6 = 364$ and $Q = 3$, we have $364 \equiv 244 \pmod{3}$, as required by Corollary 11.3.1. ◆

Problems

11.14. Let $P = 3$, $Q = -1$; work out the first 10 members of the sequence $\{U_n(P, Q)\}_{n=0}^{\infty}$ and $\{V_n(P, Q)\}_{n=0}^{\infty}$. Also show that U_3, U_4, V_3, V_4 are correctly obtained by using, alternatively, the Binet formulas.

11.15. Show explicitly that the following are true for the indicated members of the sequences in Problem 11.14:

(a) $U_3V_6 + U_6V_3 = 2U_9$.
(b) $U_9 \equiv 0 \pmod{U_3}$.
(c) $U_7 = Q^3 + V_2Q^2 + V_4Q + V_6$.
(d) $U_8 \equiv V_7 \pmod{Q}$.

11.16. Derive the Binet formulas given in this section for $U_n(P, Q)$ and $V_n(P, Q)$.

11.17. This problem is adapted from (Vajda, 1989), but is probably older. A staircase has n steps. One can climb it by taking any odd number of steps at a time; the moves can be made in any allowable sequence. Let W_n denote the number of ways to climb the staircase. Show that W_n is given by a Binet formula.

11.18. Show that there is only one choice of P, Q such that $\mathrm{U}_n(P, Q) = \mathrm{V}_n(P, Q) - 2$ for all n.

11.19. Let $P = 1, Q = -2$; show that for all $n \geq 1$ one has $(U_n, U_{n+1}) = 1$. Generalizations of this are known whenever $(P, Q) = 1$ (McDaniel, 1991). When $P = 1, Q = -2$, is $(V_n, V_{n+1}) = 1$ also true?

11.20. Prove the following relations involving Fibonacci and Lucas numbers:

a) $L_n^2 = L_{2n} + 2(-1)^n$.
b) $F_{n-1} + F_{n+1} = L_n$.
c) $L_{n-1} + L_{n+1} = 5F_n$.
d) $L_n^2 - 5F_n^2 = 4(-1)^n$.

11.21. Show that $F_n = \lceil \alpha^n/\sqrt{5} \rceil$, where $\lceil x \rceil$ is the **ceiling function** of x and means the smallest integer equal to or exceeding x. Illustrate with F_{25}.

11.22. If $P^2 - 4Q > 0$ and $P > 0$, show that for $n > 0$ one has the pair of inequalities

$$\begin{cases} \mathrm{U}_{n+1}(P, Q) > \dfrac{P}{2}\mathrm{U}_n(P, Q) \\ \mathrm{V}_{n+1}(P, Q) > \dfrac{P}{2}\mathrm{V}_n(P, Q) \end{cases}$$

(André-Jeannin, 1995).

11.23. Let $\{P_n\}_{n=0}^{\infty}$ be the sequence of Pell numbers. Prove that $\Sigma_{k=1}^{n} P_k = \frac{1}{2}[P_{n+1} + P_n - 1]$ (Kostal, 1995).

11.24. Write Pascal's Triangle in the form as shown here. Discover a relationship between entries in the triangle and running sums of the Fibonacci numbers (e.g., 1, 2, 4, 7, 12, . . .).

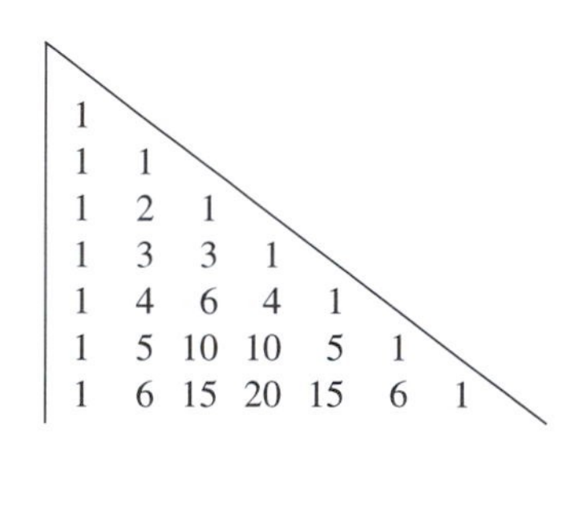

1						
1	1					
1	2	1				
1	3	3	1			
1	4	6	4	1		
1	5	10	10	5	1	
1	6	15	20	15	6	1

11.25. Identities among Fibonacci numbers that involve third or higher powers are not common. Prove the following ones, however:

(a) $F_{n+1}^3 + F_n^3 - F_{n-1}^3 = F_{3n}$.
(b) $F_{n+2}^3 - 3F_{n+1}^3 - 6F_n^3 + 3F_{n-1}^3 + F_{n-2}^3 = 0$.

11.26. (a) Investigate the conjecture that if n is an odd prime, then the Fibonacci number F_n is a prime.

(b) Investigate the nature of the quantity S,

$$S = \sum_{k=0}^{n} \binom{n}{k} F_k, \qquad \binom{n}{k} = \frac{n!}{k!(n-k)!},$$

for different choices of n.

(c) Write a computer program to work out the first 50 or so Fibonacci numbers. Pick a small prime $p > 5$ and determine which Fibonacci numbers are divisible by p. Formulate a conjecture from what you observe.

11.27. Derive the following formula, which was obtained by **Lucas** in one of his papers from 1878:

$$\sum_{n=1}^{\infty} \frac{Q^{2^{n-1}}}{U_{2^n}} = \beta.$$

In view of this, what is the sum $\Sigma_{n=0}^{\infty} 1/P_{2^n}$, where the P_{2^n}'s are the Pell numbers (Dence, 1995)?

11.28. Prove the following relations involving the Fibonacci numbers:

a) $\Sigma_{k=1}^{n} F_{4k} = F_{2n+1}^2 - 1$ and $\Sigma_{k=0}^{n} F_{4k+2} = F_{2n+2}^2$.
b) $\Sigma_{k=1}^{n} (-1)^{k+1} F_k = 1 - (-1)^n F_{n-1}$.
c) $\Sigma_{k=1}^{n} F_k^2 < 2F_n^2$ $(n \geq 3)$.

11.29. Reprove Theorem 11.2 without using mathematical induction.

11.30. Less well known than the Fibonacci numbers, but equally interesting, are the **Fibonacci polynomials**, $\{\mathrm{F}_n(x)\}_{n=1}^{\infty}$, defined by $\mathrm{F}_1(x) = 1$, $\mathrm{F}_2(x) = x$, $\mathrm{F}_n(x) = x\mathrm{F}_{n-1}(x) + \mathrm{F}_{n-2}(x)$ for $n \geq 3$.

(a) Find a Binet formula for the $\mathrm{F}_n(x)$'s.
(b) Prove that $[\mathrm{F}_n(x)]^2 - \mathrm{F}_{n-1}(x)\mathrm{F}_{n+1}(x) = (-1)^{n-1}$, $n \geq 2$.
(c) Find a formula for the sum $\Sigma_{k=1}^{n} [\mathrm{F}_k(x)]^2$.
(d) Find a generating function $\mathrm{g}(x, u)$ for the $\mathrm{F}_n(x)$'s, where u is the expansion variable in the infinite series.
(e) Prove that for $n > 1$ all the zeros of $\mathrm{F}_n(x)$ lie along the imaginary axis.

11.31. Let $Q = -1$ and define the matrix $\mathbf{M}$ by

$$\mathbf{M} = \begin{pmatrix} P & 1 \\ 1 & 0 \end{pmatrix}.$$

Show that for any positive integer n

$$\mathbf{M}^n = \begin{pmatrix} \mathrm{U}_{n+1}(P, Q) & \mathrm{U}_n(P, Q) \\ \mathrm{U}_n(P, Q) & \mathrm{U}_{n-1}(P, Q) \end{pmatrix}.$$

11.32. Let p be an odd prime, and suppose that $p \mid P$ but $p \nmid Q$. Show that $\mathrm{U}_k(P, Q) \equiv 0 \pmod{p}$ iff k is even. On the other hand, if $p \nmid P$ but $p \mid Q$, then show that $\mathrm{U}_k(P, Q) \not\equiv 0 \pmod{p}$ for all $k \geq 1$.

11.33. The companion theorem to Theorem 11.3 reads as follows: Let $n = 2m$, $\mathrm{U}_n(P, Q) = U_n$ and $\mathrm{V}_n(P, Q) = V_n$. Then

$$U_n = V_1 Q^{m-1} + V_3 Q^{m-2} + V_5 Q^{m-3} + \cdots + V_{n-1}.$$

Prove it, and thus conclude the proof of Corollary 11.3.1.

11.34. Let $n = 2m$; prove the following expansion:

$$P^n = \binom{n}{m} Q^m + \binom{n}{m-1} Q^{m-1} V_2 + \binom{n}{m-2} Q^{m-2} V_4 + \cdots + V_n,$$

and thus show (for even n, although it is true for all $n \geq 1$)

$$V_n \equiv P^n \pmod{Q}.$$

This is a companion result to Corollary 11.3.1.

11.35. Let $P, Q \in \mathbf{Z}$, where $P > 0$ is not a perfect square and $(P, Q) = 1$, $P - 4Q > 0$. Let α, β be the zeros of $x^2 - \sqrt{P}x + Q$, $\alpha > \beta$. Then the sequences $\{W_n\}_{n=1}^{\infty}$, $\{X_n\}_{n=1}^{\infty}$, defined by $W_n = (\alpha^n - \beta^n)/(\alpha - \beta)$ and $X_n = \alpha^n + \beta^n$, are termed the **Lehmer** and **companion Lehmer sequences**, respectively (Lehmer, 1930).

(a) Show that W_0, W_{2n+1}, X_{2n} are integers, but W_{2n}, X_{2n+1} are irrational.
(b) Let $P = 5$, $Q = -3$. Prove that $X_{4n+2} \equiv 0 \pmod{11}$.

11.4 Stirling's Numbers

A Lucas sequence or a companion Lucas sequence is a function of a single variable n. In this section we look at two sequences that are functions of two discrete variables (Graham, Knuth, and Patashnik, 1989).

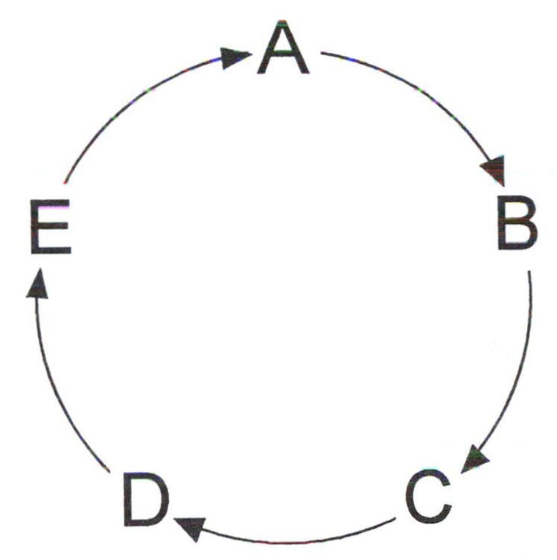

Figure 11.2 This 5-cycle has the equivalent notations [A,B,C,D,E], [B,C,D,E,A], [C,D,E,A,B], [D,E,A,B,C] and [E,A,B,C,D].

> **DEF.** A **k-cycle** ($k \in \mathbf{N}$) is a cyclic arrangement of k objects that we shall agree to read in a clockwise fashion and that has no unique starting point.

If we specify a k-cycle by a compact linear notation, then such a cycle has k equivalent notations (Fig. 11.2).

In how many ways can four distinguishable objects be partitioned into two cycles? There are two possible classes of partitions:

Class 1: a 1-cycle and a 3-cycle.

Class 2: two 2-cycles.

Fig. 11.3 diagrams the possibilities for the two classes.

Similarly, you can discover that there are six ways of dividing four objects into one cycle, six ways of dividing four objects into three cycles, and trivially one way of dividing four objects into four cycles.

> **DEF.** Let n, k be positive integers, $k \leq n$. The number of ways of dividing n distinct objects into k cycles is called the **Stirling's number**[2] **of the first kind**, $S(n, k)$.

[2] **James Stirling** (1692–1770) was an important Scottish mathematician in the generation after **Newton** (Gillispie, 1976). He worked in the calculus and analytic geometry. The Maclaurin series, for example, was actually discovered by **Stirling** some 25 years before **Maclaurin** published it. For more information on **Stirling**, see (Tweddle, 1988).

CLASS 1		CLASS 2	
[A]	[B,C,D]	[A,B]	[C,D]
[A]	[B,D,C]	[A,C]	[B,D]
[B]	[A,C,D]	[A,D]	[B,C]
[B]	[A,D,C]		
[C]	[A,B,D]		
[C]	[A,D,B]		
[D]	[A,B,C]		
[D]	[A,C,B]		

Figure 11.3 There are 11 ways of dividing four objects into two cycles.

Thus, from Fig. 11.3 we have $S(4, 2) = 11$. Simple counting of permutations gives $S(n, 1) = (n - 1)!$ and $S(n, n - 1) = n(n - 1)/2$. For other choices of k besides $1, n - 1$ the values of $S(n, k)$ are less obvious, so we seek a recurrence relationship among the Stirling's numbers of the first kind.

Theorem 11.4 $S(n, k) = (n - 1)S(n - 1, k) + S(n - 1, k - 1),\ n, k \in \mathbf{N}.$

Proof Consider an arbitrary object A. If A constitutes a 1-cycle, then A could be associated with any one of the $S(n - 1, k - 1)$ modes of subdivision of the remaining $n - 1$ objects. If A is not a 1-cycle, then A has been inserted into some one of the k cycles of the other $n - 1$ objects. There are m ways to insert a new object into an m-cycle, so summation over all of the k cycles gives $n - 1$ ways to insert a new object into any of the k cycles for each mode of subdivision of $n - 1$ objects. As there are $S(n - 1, k)$ modes of subdivision of $n - 1$ objects into k cycles, this gives $(n - 1)S(n - 1, k)$ modes of subdivision of n objects into k cycles in which A does not appear as a 1-cycle. ◆

Theorem 11.4, together with the two special cases mentioned prior to the theorem, allows us to construct rapidly Table 11.3 showing values of $S(n, k)$. A larger listing is found in Table I(2) of Appendix II, and a still larger listing is given in (Abramowitz and Stegun, 1965). Additionally, it is convenient to define $S(0, 0) = 1$ and $S(n, 0) = 0$ for $n > 0$.

Table 11.3 A Short Table of Stirling's Numbers of the First Kind, $S(n, k)$

$n \backslash k$	1	2	3	4	5	6
1	1					
2	1	1				
3	2	3	1			
4	6	11	6	1		
5	24	50	35	10	1	
6	120	274	225	85	15	1

The Stirling's numbers of the first kind show up in an important algebraic connection. Let us compute the **Pochhammer symbol** defined by $(x)_n = x(x-1) \cdot (x-2) \ldots (x-n+1)$,[3] for a few choices of the positive integer n:

$$(x)_2 = x(x-1) = x^2 - x$$

$$(x)_3 = (x)_2(x-2) = x^3 - 3x^2 + 2x$$

$$(x)_4 = (x)_3(x-3) = x^4 - 6x^3 + 11x^2 - 6x.$$

We notice in each case that the coefficients of the various powers of x are (in absolute terms) Stirling's numbers of the first kind.

Theorem 11.5 $(x)_n = \sum_{k=1}^{n} (-1)^{n+k} S(n, k) x^k, n \geq 1.$

Proof The theorem is true for $n = 1, 2$; we assume it to be true also for $n = N$ and all k satisfying $1 \leq k \leq N$. Let b_k be the coefficient of x^k in $(x)_N$ and let c_k be the coefficient of x^k in $(x)_{N+1}$. Then

$$(x)_{N+1} = (x)_N(x - N),$$

so we have for $1 \leq k \leq N$

$$\begin{aligned} c_k &= b_{k-1} - Nb_k \\ &= (-1)^{N+k-1} S(N, k-1) - N(-1)^{N+k} S(N, k) \\ &= (-1)^{N+k-1}[S(N, k-1) + NS(N, k)] \\ &= (-1)^{N+k-1} S(N+1, k) \\ &= (-1)^{(N+1)+k} S(N+1, k), \end{aligned}$$

[3] This function is sometimes also referred to as a **falling factorial power**, is notated $x^{\underline{n}}$, and is read "x to the n falling." The Pochhammer symbol given in Problem 4.62 is sometimes called a **rising factorial power**, written $x^{\overline{n}}$.

after using Theorem 11.4. Hence, Theorem 11.5 holds for all integers $n \geq 1$ by the Principle of Finite Induction. ◆

Example 11.8 Let C(n, k) denote a binomial coefficient. Then

$$C(n, k) = \frac{n!}{k!(n-k)!} = \frac{(n)_k}{k!},$$

and from Theorem 11.5 this becomes

$$C(n, k) = \frac{1}{k!} \sum_{j=1}^{k} (-1)^{k+j} S(k, j) n^j.$$

For $n = 9$, $k = 6$, we obtain from Theorem 11.3

$$\begin{aligned} C(9, 6) &= \frac{1}{6!}[-S(6, 1) \cdot 9 + S(6, 2) \cdot 9^2 - S(6, 3) \cdot 9^3 \\ &\quad + S(6, 4) \cdot 9^4 - S(6, 5) \cdot 9^5 + S(6, 6) \cdot 9^6] \\ &= \frac{1}{720}[-1080 + 22194 - 164025 + 557685 \\ &\quad - 885735 + 531441] \\ &= 84. \end{aligned}$$ ◆

Example 11.9 In Theorem 11.5 set $x = -1$. Then

$$\begin{aligned} (x)_n &= x(x-1)(x-2) \ldots (x-n+1) \\ &= -1(-2)(-3) \ldots (-n) \\ &= (-1)^n n!, \end{aligned}$$

$$\begin{aligned} \text{so } (-1)^n n! &= \sum_{k=1}^{n} (-1)^{n+k} S(n, k)(-1)^k \\ &= (-1)^n \sum_{k=1}^{n} (-1)^{2k} S(n, k) \end{aligned}$$

$$\text{and} \qquad n! = \sum_{k=1}^{n} S(n, k).$$

For example, $120 + 274 + 225 + 85 + 15 + 1 = 6!$. ◆

Stirling himself studied a second interesting sequence of numbers. Instead of partitioning n objects into cycles, let us just partition the n objects into nonempty

CLASS 1		CLASS 2	
{A}	{B,C,D}	{A,B}	{C,D}
{B}	{A,C,D}	{A,C}	{B,D}
{C}	{A,B,D}	{A,D}	{B,C}
{D}	{A,B,C}		

Figure 11.4 There are seven ways of dividing four objects into two nonempty sets

subsets. Fig. 11.4 shows the possible ways of dividing four objects into two nonempty subsets. Class-1 types are those modes of subdivision in which a singleton set is present; class-2 modes of subdivision contain no singleton sets.

Similarly, there are six ways of dividing four objects into three nonempty sets, and trivially one way of dividing four objects into either one nonempty set or four nonempty sets (verify these assertions).

DEF. Let n, k be positive integers, $k \leq n$. The number of ways of dividing n distinct objects into k nonempty sets is called the **Stirling's number of the second kind**, $s(n, k)$.[4]

Thus, from Fig. 11.4 we have $s(4, 2) = 7$. Clearly, $s(n, n) = s(n, 1) = 1$ for any n. Another easy case is $s(n, 2) = 2^{n-1} - 1$, $n \geq 2$ (Problem 11.40), but for other choices of k the trends are less obvious. A recurrence relation is again needed.

Theorem 11.6 $s(n, k) = ks(n - 1, k) + s(n - 1, k - 1), \qquad n, k \in \mathbf{N}.$

Proof Pattern the proof after that of Theorem 11.4. Consider as separate cases those where a fixed object A is a singleton set and where it is not. ◆

In Table 11.4 we show some early values of the Stirling's numbers of the second kind; a larger listing is in Table I(3) of Appendix II as well as in (Abramowitz and Stegun, 1965). Additionally, it is convenient to define $s(0, 0) = 1$ and $s(n, 0) = 0$ for $n > 0$.

The Stirling's numbers of the second kind also show up in an algebraic setting. Let us express the first few powers of x as linear combinations of Pochhammer

[4] Notation for both kinds of Stirling's numbers is highly author dependent.

Table 11.4 A Short Table of Stirling's Numbers of the Second Kind, s(*n*, *k*)

n \ k	1	2	3	4	5	6
1	1					
2	1	1				
3	1	3	1			
4	1	7	6	1		
5	1	15	25	10	1	
6	1	31	90	65	15	1

symbols (falling factorial power type). This is the inverse of what we did with the Stirling's numbers of the first kind:

$$\begin{cases} x^2 = (x^2 - x) + x = (x)_2 + (x)_1 \\ x^3 = (x^3 - 3x^2 + 2x) + 3(x^2 - x) + x = (x)_3 + 3(x)_2 + (x)_1 \\ x^4 = (x^4 - 6x^3 + 11x^2 - 6x) + 6(x^3 - 3x^2 + 2x) + 7(x^2 - x) + x \\ \quad = (x)_4 + 6(x)_3 + 7(x)_2 + (x)_1. \end{cases}$$

The following analog of Theorem 11.5 seems warranted.

Theorem 11.7 $x^n = \Sigma_{k=1}^{n} \mathrm{s}(n, k)(x)_k, n \geq 1.$

Proof We have $(x - k)(x)_k = (x - k)[x(x - 1) \ldots (x - k + 1)] = x(x - 1) \ldots (x - k) = (x)_{k+1}$, so we obtain the identity

$$x(x)_k = (x)_{k+1} + k(x)_k.$$

The theorem holds for $n = 1$; assume it holds for $n = m - 1$. Then

$$\begin{aligned} x^m = x(x^{m-1}) &= \sum_{k=1}^{m-1} \mathrm{s}(m-1, k)x(x)_k \\ &= \sum_{k=1}^{m-1} \mathrm{s}(m-1, k)(x)_{k+1} \\ &\quad + \sum_{k=1}^{m-1} \mathrm{s}(m-1, k)k(x)_{k+1} \\ &= \sum_{j=2}^{m} \mathrm{s}(m-1, j-1)(x)_j \\ &\quad + \left[\sum_{k=1}^{m-1} \mathrm{s}(m, k)(x)_k - \sum_{k=1}^{m-1} \mathrm{s}(m-1, k-1)(x)_k\right] \end{aligned}$$

from Theorem 11.6. Cancellation of terms from the first and third summations leads to

$$x^m = s(m-1, m-1)(x)_m + \sum_{k=1}^{m-1} s(m, k)(x)_k - s(m-1, 0)(x)_1$$

$$= \sum_{k=1}^{m} s(m, k)(x)_k$$

because $s(m-1, 0) = 0$. Hence, the theorem holds for all $n \in \mathbf{N}$ by the Principle of Finite Induction. ◆

Interesting properties of the Stirling's numbers of the second kind flow from either Theorem 11.6 or 11.7. A summation property of them is given in the next Example.

Example 11.10 Let x be a continuous variable and let D_x denote differentiation with respect to x. Then

$$D_x(x)_n = nx^{n-1} - \frac{n(n-1)^2}{2}x^{n-2} + \cdots + (-1)^{n-1}(n-1)!,$$

so $[D_x(x)_n]_{x=0} = (-1)^{n-1}(n-1)!$. From Theorem 11.7 we have

$$nx^{n-1} = \sum_{k=1}^{n} s(n, k)D_x(x)_k,$$

and setting $x = 0$ we obtain

$$0 = \sum_{k=1}^{n} (-1)^{k-1}(k-1)!s(n, k).$$

For example, if $n = 5$

$$1 - 15 + 2(25) - 6(10) + 24(1) = 0. \quad ◆$$

Problems

11.36. Compute by hand and compare S(8, 3), s(8, 3), C(8, 3). Do the same for S(8, 6), s(8, 6), C(8, 6).

11.37. Write a computer program to calculate values of $S(n, k)$. Use it to show that $\Sigma_{k=1}^{12} S(12, k) = 12!$, as Example 11.9 demands.

11.38. Let $C(n, k)$ be a binomial coefficient, $n, k \in \mathbf{N}$, $n \geq k$. Show that $S(n, k) \geq C(n, k)$ holds for nearly all (n, k) pairs. Under what conditions does strict inequality hold?

11.39. Let $p > 2$ be a prime. Show that $S(p - 1, 2) \equiv 1 \pmod{p}$ is equivalent to $S(p, 2) \equiv 0 \pmod{p}$.

11.40. Show that $s(n, 2) = 2^{n-1} - 1, n \geq 2$.

11.41. How many ordered subsets $\{a, b, c\}$ are there of the set $\{2, 3, 4, 5, \ldots\}$ such that $abc = 2 \cdot 3 \cdot 5 \cdot 7 \cdot 11 \cdot 13$ and a is odd?

11.42. A **surjection** is a mapping from a set $\mathbf{X}$ *onto* a set $\mathbf{Y}$. Prove that the number of surjections from a set $\mathbf{X}$ (with n elements) onto a set $\mathbf{Y}$ (with k elements) is $k!s(n, k)$.

11.43. Prove the following summation formula for the Stirling's numbers of the second kind:

$$s(n, k) = \frac{1}{k!} \sum_{j=0}^{k-1} (-1)^j \binom{k}{j} (k - j)^n.$$

Then program this and compute $s(10, k)$ for $k = 1, 2, \ldots, 10$.

11.44. Define the sequence of operators $\{\hat{\theta}_n\}_{n=1}^{\infty}$ by $\hat{\theta}_n = x^n d^n/dx^n$. Prove that

$$\hat{\theta}_1^n = \sum_{k=1}^{n} s(n, k)\hat{\theta}_k$$

holds for all $n \in \mathbf{N}$ (Goldstein, 1934).

11.45. The following is a kind of orthogonality relation between the Stirling's numbers of the first and second kinds:

$$\sum_{j=k}^{n} S(j, k)s(n, j)(-1)^{j+k} = 0.$$

Prove it, then illustrate it for some ordered pair (n, k).

11.46. Derive an expression for the sum

$$S = \sum_{k=1}^{n} \binom{k}{3}.$$

11.5 Bernoulli Numbers

Near the end of his professional career, the Swiss mathematician **Jakob Bernoulli** (1654–1705) became interested in the general formulas for the sums of powers of the positive integers. He deduced (but did not prove) the general pattern of these formulas, and simultaneously introduced the so-called Bernoulli numbers, $\{B_n\}_{n=0}^{\infty}$.[5]

We first mentioned these numbers in Problems 10.23 through 10.26. Instead of following Bernoulli's approach, we shall interestingly obtain the Bernoulli numbers by a connection with the Stirling's numbers of the second kind. This is still one more route to these fascinating numbers in addition to the pathways implicit in (Lehmer, 1988).

We recall from Theorem 11.7

$$\begin{aligned} x^n &= \sum_{k=0}^{n} \mathrm{s}(n,\ k)(x)_k \\ &= \sum_{k=0}^{n} \mathrm{s}(n,\ k)k!\binom{x}{k} \\ &= \sum_{k=0}^{n} \mathrm{s}(n,\ k)k!\left[\binom{x+1}{k+1} - \binom{x}{k+1}\right], \end{aligned}$$

where the **generalized binomial coefficient** $\binom{x}{k}$ is defined as

$$\binom{x}{k} = \begin{cases} \dfrac{x(x-1)(x-2)\ .\ .\ .\ (x-k+1)}{k!} & \text{if } k \in \mathbf{N} \\ 1 & \text{if } k = 0, \end{cases}$$

and the Pascal-like sum rule

$$\binom{x}{k} + \binom{x}{k+1} = \binom{x+1}{k+1}$$

holds (verify!). The terms in the summation expression for x^n form the basis for the definition of a sequence of polynomials. We define the **Bernoulli polynomials**, $\{B_n(x)\}_{n=1}^{\infty}$, by

$$B_n(x) = n \sum_{k=0}^{n-1} \mathrm{s}(n-1,\ k)k!\binom{x}{k+1} + \mathrm{C}_n,$$

where C_n is an additive constant to be determined.

[5] **Bernoulli** wrote up this work in a book, *Ars conjectandi* ("The Art of Conjecture"), which was published posthumously in 1713 (Struik, 1986). In it he states that from his formulas he was able to calculate *intra semiquadrantem horae* (inside of half of a quarter-hour) the sum of the 10th powers of the first 1000 positive integers as 91, 409, 924, 241, 424, 243, 424, 241, 924, 242, 500!

Example 11.11 We compute

$$B_3(x) = 3\left[s(2, 0)\ 0!\binom{x}{1} + s(2, 1)\ 1!\binom{x}{2} + s(2, 2)\ 2!\binom{x}{3}\right] + C_3$$

$$= 3\left[0 + 1 \cdot 1 \cdot \frac{x(x-1)}{2!} + 1 \cdot 2 \cdot \frac{x(x-1)(x-2)}{3!}\right] + C_3$$

$$= x^3 - \frac{3}{2}x^2 + \frac{1}{2}x + C_3. \quad \blacklozenge$$

Combination of the summation expression for x^n with the definition of the Bernoulli polynomials gives

$$x^n = \frac{1}{n+1}[B_{n+1}(x+1) - B_{n+1}(x)].$$

By differentiation of this we obtain

$$nx^{n-1} = \frac{1}{n+1}[B'_{n+1}(x+1) - B'_{n+1}(x)].$$

$$= B_n(x+1) - B_n(x),$$

so

$$\frac{1}{n+1}B'_{n+1}(x+1) - B_n(x+1) = \frac{1}{n+1}B'_{n+1}(x) - B_n(x).$$

As x is arbitrary, this shows that the polynomial $[1/(n+1)]\,B'_{n+1}(x) - B_n(x)$, $n = 1, 2, \ldots$, has period 1. But the only kind of polynomial that is periodic is a constant one, so

$$\frac{1}{n+1}B'_{n+1}(x) - B_n(x) = D_n.$$

Let us choose the constants C_n so that each $D_n = 0$ and $B_n(x) = B'_{n+1}(x)/(n+1)$. When $n = 1$, the equation $nx^{n-1} = B_n(x+1) - B_n(x)$ reduces to $1 = B_1(x+1) - B_1(x)$ and this implies that $B_1(x) = x + C_1$, so $B_0(x) = [1/(0+1)]\,B'_1(x) = 1$. In the definition of the $B_n(x)$'s we see that $B_n(1) = C_n$ for $n \geq 2$. Hence, when $B'_2(x) = 2B_1(x) = 2x + 2C_1$ is integrated, we obtain

$$B_2(x) = x^2 + 2C_1x + C_2,$$

so $B_2(1) = C_2 = 1 + 2C_1 + C_2$, and $C_1 = -\frac{1}{2}$.

Similarly, integration of $B'_3(x) = 3B_2(x) = 3x^2 - 3x + 3C_2$ gives $B_3(x) =$

Table 11.5 The First Five Bernoulli Polynomials

n	$B_n(x)$
0	1
1	$x - \frac{1}{2}$
2	$x^2 - x + \frac{1}{6}$
3	$x^3 - \frac{3}{2}x^2 + \frac{1}{2}x$
4	$x^4 - 2x^3 + x^2 - \frac{1}{30}$

$x^3 - \frac{3}{2}x^2 + 3C_2x + C_3$, so at $x = 1$ we have $1 - \frac{3}{2} + 3C_2 + C_3 = C_3$ and $C_2 = \frac{1}{6}$. The C_n's are called the **Bernoulli numbers**, and henceforth we designate them as $B_n = C_n$. Table 11.5 gives the first five Bernoulli polynomials, and Figure 11.5 presents their graphs.

The preceding discussion shows that the Bernoulli numbers are recursive in nature, and the steps shown there suggest Theorem 11.8.

Theorem 11.8

$$\sum_{k=0}^{n-1} B_k \binom{n}{k} = 0 \quad \text{for } n \geq 2.$$

Proof Consecutive integrations of $B'_{n+1}(x) = (n + 1)B_n(x)$ lead to

$$\frac{1}{n!}B_n(x) = \frac{x^n}{n!} + \frac{B_1}{1!}\frac{x^{n-1}}{(n-1)!} + \frac{B_2}{2!}\frac{x^{n-2}}{(n-2)!} + \cdots + \frac{B_{n-1}}{(n-1)!}\frac{x}{1!} + \frac{B_n}{n!}.$$

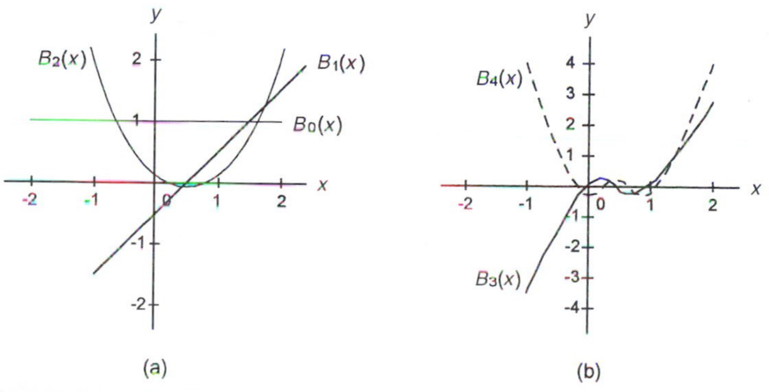

Figure 11.5 Graphs (a) of the Bernoulli polynomials $B_0(x)$, $B_1(x)$, $B_2(x)$ and (b) the Bernoulli polynomials $B_3(x)$, $B_4(x)$

This follows by induction since it is clearly true for $n = 1$, and replacement of $B_n(x)$ by $[1/(n + 1)] B'_{n+1}(x)$ gives after one more integration

$$\frac{1}{(n+1)!} B_{n+1}(x) = \frac{x^{n+1}}{(n+1)!} + \frac{B_1}{1!}\frac{x^n}{n!} + \frac{B_2}{2!}\frac{x^{n-1}}{(n-1)!} + \cdots + \frac{B_n}{n!}\frac{x}{1!} + \frac{B_{n+1}}{(n+1)!}.$$

In the first equation let $x = 1$; $B_n(1) = B_n$ then yields

$$0 = \frac{1}{n!} + \frac{B_1}{1!\,(n-1)!} + \frac{B_2}{2!\,(n-2)!} + \cdots + \frac{B_{n-1}}{(n-1)!\,1!},$$

and the theorem follows by multiplication of this by $n!$. ◆

Theorem 11.8, together with $B_0 = 1$, completely determines the sequence $\{B_n\}_{n=0}^{\infty}$. We may therefore take the theorem as the *definition* of the Bernoulli numbers. Contrast this definition with that given in Problem 10.23. Note that unlike the case of the Fibonacci or Stirling's numbers, the order of the recurrence relation for the Bernoulli numbers changes continually. This feature makes it difficult to find a simple closed form for the Bernoulli numbers. Table 11.6 gives some early values of these numbers. Larger listings appear in Table I(5) of Appendix II and in (Abramowitz and Stegun, 1965).

Example 11.12 From Theorem 11.8 and Table 11.6, we obtain

$$B_0\binom{13}{0} + B_1\binom{13}{1} + \sum_{n=1}^{6} B_{2n}\binom{13}{2n} = 0.$$

Hence, B_{12} is calculated to be

$$\begin{aligned} B_{12} &= \frac{1}{13}\left[-B_0 - 13B_1 - \sum_{n=1}^{5} B_{2n}\binom{13}{2n}\right] \\ &= \frac{1}{13}\left[-1 + \frac{13}{2} - \left(13 - \frac{143}{6} + \frac{286}{7} - \frac{429}{10} + \frac{65}{3}\right)\right] \\ &= \frac{1}{13}\left[\frac{-691}{210}\right] \\ &= \frac{-691}{2730}. \end{aligned}$$ ◆

Table 11.6 A Short Table of Nonzero Bernoulli Numbers

n	B_n	n	B_n	n	B_n
0	1	6	1/42	14	7/6
1	−1/2	8	−1/30	16	−3617/510
2	1/6	10	5/66	18	43867/798
4	−1/30	12	−691/2730	20	−174611/330

Bernoulli's original goal of discovering the general pattern for the formulas of the sums of the nth powers of the first several positive integers is now an easy consequence of the content of Theorem 11.8. The result is called **Bernoulli's Identity** (Williams, 1997). In what follows we let $T_n(k) = 1^n + 2^n + \ldots + (k-1)^n$.

Corollary 11.8.1 For $n \in \mathbf{N}$ one has $T_n(k) = \frac{1}{n+1}\sum_{j=0}^{n} B_j\binom{n+1}{j}k^{n+1-j}$.

Proof We have

$$\begin{aligned}\sum_{x=1}^{k-1} x^n &= \frac{1}{n+1}\sum_{x=1}^{k-1}[B_{n+1}(x+1) - B_{n+1}(x)]\\ &= \frac{1}{n+1}[B_{n+1}(k) - B_{n+1}(1)]\\ &= \frac{1}{n+1}[B_{n+1}(k) - B_{n+1}]\end{aligned}$$

because the series is a telescoping series. Making use of the second equation in the proof of Theorem 11.8, we obtain

$$\begin{aligned}T_n(k) &= \frac{1}{n+1}\left[\sum_{j=0}^{n+1} B_j\binom{n+1}{j}k^{n+1-j} - B_{n+1}\right]\\ &= \frac{1}{n+1}\sum_{j=0}^{n} B_j\binom{n+1}{j}k^{n+1-j}. \quad \blacklozenge\end{aligned}$$

Example 11.13 Find the sum $\sum_{i=1}^{50} i^4$.

In Corollary 11.8.1 let $k = 51, n = 4$. Then

$$\begin{aligned}T_4(51) &= \frac{1}{5}\left[B_0\binom{5}{0}51^5 + B_1\binom{5}{1}51^4 + B_2\binom{5}{2}51^3\right.\\ &\quad \left. + B_3\binom{5}{3}51^2 + B_4\binom{5}{4}51\right]\\ &= \frac{1}{5}(51)\left[1\cdot 51^4 - \frac{5}{2}\cdot 51^3 + \frac{10}{6}\cdot 51^2 - \frac{5}{30}\cdot 1\right]\\ &= \frac{1}{5}(51)\left(\frac{19313725}{3}\right)\\ &= 65{,}666{,}665. \quad \blacklozenge\end{aligned}$$

Example 11.14 Bernoulli's boast concerned the number $T_{10}(1001)$. Show that it ends in 500.

From Corollary 11.8.1 we have $11T_{10}(1001) = \sum_{k=0}^{5} B_{2k}\binom{11}{2k}(1001)^{11-2k} + B_1\binom{11}{1}(1001)^{10}$. Multiplying both sides by the least common multiple (LCM) of the denominators of the B_{2k}'s (LCM = 2310) and reducing both sides modulo 10,000, we obtain (verify!) $541T_{10}(1001) \equiv 500 \pmod{1000}$. The unique solution of this is $T_{10}(1001) \equiv 500 \pmod{1000}$. ◆

A recurrence relation is not a very useful mode of expression of the Bernoulli numbers for analytic purposes. Let us then seek a generating function f(x) for them:

$$\mathrm{f}(x) = \sum_{n=0}^{\infty} B_n \frac{x^n}{n!}$$

$$= 1 - \frac{1}{2}x + \frac{1}{12}x^2 - \frac{1}{720}x^4 + \frac{1}{30240}x^6 - \dots .$$

Let $\mathrm{F}(x) = 1 + Ax + Bx^2 + Cx^3 + Dx^4 + \dots$ be the multiplicative inverse of f(x), assumed valid in some restricted domain. Then

$$1 = \mathrm{f}(x)\mathrm{F}(x) = 1 + x\left(A - \frac{1}{2}\right) + x^2\left(\frac{1}{12} - \frac{1}{2}A + B\right) + x^3\left(\frac{1}{12}A - \frac{1}{2}B + C\right)$$

$$+ x^4\left(-\frac{1}{720} + \frac{1}{12}B - \frac{1}{2}C + D\right) + \dots$$

and equating coefficients of like powers of x, we obtain $A = \frac{1}{2}$, $B = \frac{1}{6}$, $C = \frac{1}{24}$, $D = \frac{1}{120}$. Hence,

$$\mathrm{F}(x) = 1 + \frac{1}{2}x + \frac{1}{6}x^2 + \frac{1}{24}x^3 + \frac{1}{120}x^4 + \dots$$

$$= \frac{\left(1 + x + \frac{1}{2}x^2 + \frac{1}{6}x^3 + \frac{1}{24}x^4 + \dots\right) - 1}{x},$$

so we conjecture Theorem 11.9.

Theorem 11.9

$$\frac{x}{\mathrm{e}^x - 1} = \sum_{n=0}^{\infty} B_n \frac{x^n}{n!}.$$

Proof Let a sequence of real numbers $\{b_n\}_{n=0}^{\infty}$ have the generating function $\Sigma_{n=0}^{\infty} b_n x^n/n!$, and define $b_0 = 1$. Then

$$
\begin{aligned}
(e^x - 1) &= \sum_{n=0}^{\infty} b_n \frac{x^n}{n!} = \sum_{k=1}^{\infty} \frac{x^k}{k!} \sum_{n=0}^{\infty} b_n \frac{x^n}{n!} \\
&= \sum_{m=1}^{\infty} x^m \sum_{k=1}^{m} \frac{b_{m-k}}{(m-k)!\,k!}, \qquad (m = n + k) \\
&= \sum_{m=1}^{\infty} \frac{1}{m!} x^m \left[\sum_{k=1}^{m} \binom{m}{m-k} b_{m-k} \right] \\
&= \sum_{m=1}^{\infty} \frac{1}{m!} x^m \left[\sum_{n=0}^{m-1} \binom{m}{n} b_n \right], \qquad (n = m - k) \\
&= \frac{x}{1!} \binom{1}{0} b_0 + \sum_{m=2}^{\infty} \frac{1}{m!} x^m \left[\sum_{n=0}^{m-1} \binom{m}{n} b_n \right] \\
&= x \qquad \text{by hypothesis.}
\end{aligned}
$$

If this is to hold for arbitrary x, then for all $m > 1$ we must have

$$
\sum_{n=0}^{m-1} \binom{m}{n} b_n = 0.
$$

By Theorem 11.8 this implies that the sequence $\{b_n\}_{n=0}^{\infty}$ is the sequence of the Bernoulli numbers. ◆

The entries in Table 11.6 suggest that for $n \geq 1$, $B_{2n+1} = 0$ and that the algebraic signs of B_{2n} alternate. The first assertion here follows quickly from Theorem 11.9.

Corollary 11.9.1 For $n \geq 1$, $B_{2n+1} = 0$.

Proof We have from Theorem 11.9

$$
\frac{x}{e^x - 1} = \sum_{n=2}^{\infty} B_n \frac{x^n}{n!} + B_0 + B_1 x
$$

or

$$
\frac{x}{e^x - 1} + \frac{x}{2} = 1 + \sum_{n=2}^{\infty} B_n \frac{x^n}{n!}.
$$

On the left-hand side, replace x by $-x$; we obtain

$$\frac{-x}{e^{-x}-1}-\frac{x}{2}=\frac{-xe^{x}}{1-e^{x}}-\frac{x}{2}=\frac{-x(e^{x}-1)}{1-e^{x}}+\frac{x}{e^{x}-1}-\frac{x}{2}$$

$$=\frac{x}{2}+\frac{x}{e^{x}-1}.$$

Hence, $[(x/2)+x/(e^{x}-1)]$ is an even function of x, so only even powers of x can survive in the infinite series. This implies that $B_n = 0$ for $n > 2$ odd. ◆

The second half of the preceding assertion, that the signs of B_{2n} alternate, requires more work. There are several proofs available; we prefer one by the late English mathematician **L.J. Mordell** because it uses nothing more sophisticated than Theorem 11.9 (Mordell, 1973).

Theorem 11.10 For $n \geq 1$, $(-1)^{n-1}B_{2n} > 0$.

Proof We have from Theorem 11.9

$$\frac{x}{e^{x}+1}=\frac{x}{e^{x}-1}-\frac{2x}{e^{2x}-1}=\sum_{n=0}^{\infty}(1-2^{n})B_n\frac{x^n}{n!}.$$

Multiplication of both sides by $x/(e^{x}-1)$ gives

$$\frac{x}{2}\frac{2x}{e^{2x}-1}=\sum_{k=0}^{\infty}B_k\frac{x^k}{k!}\sum_{n=0}^{\infty}(1-2^{n})B_n\frac{x^n}{n!}$$

or

$$\sum_{r=0}^{\infty}2^{r-1}B_r\frac{x^{r+1}}{r!}=\sum_{k=0}^{\infty}B_k\frac{x^k}{k!}\sum_{n=0}^{\infty}(1-2^{n})B_n\frac{x^n}{n!}.$$

From Corollary 11.9.1, the terms on the left that survive are those for $r = 1$ and for all even r. Hence, the coefficients of the even powers of x (except that of x^2) on the right must vanish, that is, the coefficient of x^{2m} on the right satisfies

$$0=\sum_{j=1}^{m}(1-2^{2j})\frac{B_{2j}}{(2j)!}\frac{B_{2m-2j}}{(2m-2j)!}$$

$$=(1-2^{2m})\frac{B_{2m}}{(2m)!}+\sum_{j=1}^{m-1}(1-2^{2j})\frac{B_{2j}}{(2j)!}\frac{B_{2m-2j}}{(2m-2j)!}.$$

We now reason by induction; $(-1)^{n-1}B_{2n} > 0$ is true for $n = 1$. Assume it is also true for all n satisfying $1 \leq n \leq m-1$. Then B_{2j} has the algebraic sign $(-1)^{j-1}$, B_{2m-2j} has the sign $(-1)^{m-j-1}$, and each term in the summation therefore has the sign $(-1)^{1+j-1+m-j-1} = (-1)^{m-1}$. It follows that $(1 - 2^{2m})B_{2m}/(2m)!$ has algebraic sign $(-1)^m$, so B_{2m} has sign $(-1)^{m-1}$. By the Principle of Finite Induction $(-1)^{n-1}B_{2n} > 0$ holds for all $n \in \mathbf{N}$. ◆

A principal use of the Bernoulli numbers is in the **Euler–Maclaurin summation formula**, which is useful in a practical sense for the replacement of an infinite series by an integral plus a sum of a few terms. We do not present a treatment of the summation formula in this book, as the theory is somewhat delicate, but we refer the interested reader to a brief discussion elsewhere (Bromwich, 1926). The summation formula has numerous applications.

There is a vast literature on the Bernoulli numbers; we haven't even scratched the surface. The reader who wishes to learn more about these interesting numbers will find plenty to choose from in (Dilcher, Skula and Slavutskii, 1991).

Problems

11.47. Compute B_{22}, B_{24} from Theorem 11.8.

11.48. Extend Table 11.5 by adding $B_5(x)$, $B_6(x)$, $B_7(x)$.

11.49. Deduce $T_5(k)$ by means of Corollary 11.8.1.

11.50. Is it true that in any Bernoulli polynomial $B_n(x)$ the algebraic signs of the terms alternate? How many nonzero terms are present in $B_{13}(x)$?

11.51. Is it true that $B_{13}(x)$ possesses an even number of nonzero real zeros?

11.52. **Fermat** in 1636, predating **Jakob Bernoulli**, gave several formulas for sums of powers of the integers (Boyer, 1943). The following is a recursive way of deriving these formulas (Kelly, 1984). Start with

$$(1 + r)^{n+1} = \sum_{i=0}^{n+1} \binom{n+1}{i} r^i.$$

In this, set $r = k - 1, k - 2, \ldots, 1$, successively.

(a) Write the resulting equations, then add them, taking due account of extensive cancellations.

(b) Let $T_n(k) = 1^n + 2^n + \cdots + (k - 1)^n$. Show that the result in part (a) becomes

$$k^{n+1} = 1 + \sum_{i=0}^{n} \binom{n+1}{i} T_i(k).$$

(c) Using (b), work out the formulas for $T_n(k)$ for $n = 2, 3, \ldots, 7$. What is the coefficient of k^n in $T_n(k)$? What is the coefficient of k^{n-1} in $T_n(k)$? Are these two answers what you should have expected?

11.53. Let $t \in \mathcal{R}$ be nonzero. Prove that

$$\sum_{j=0}^{k-1} \mathrm{e}^{jt} = \sum_{n=0}^{\infty} T_n(k) \frac{t^n}{n!}.$$

11.54. Derive the relations

$$\begin{cases} T_n(k) = \displaystyle\int_0^k B_n(x)\, dx \\ B_{n+1}(x) = (n+1) \displaystyle\int_0^x B_n(t)\, dt + B_{n+1}. \end{cases}$$

Using these relations together with information in Table 11.5, deduce $T_5(k)$. Compare your work with that in Problem 11.49.

11.55. Show that ln 3 is an approximate root of the equation $x^4 - 60x^2 + 720x - 720 = 0$, and indicate how this is suggested by Theorem 11.9.

11.56. Show that the correct values of B_0, B_1, B_2 are obtained when $x/(\mathrm{e}^x - 1)$ is expanded in a Taylor series about $x = 0$.

11.57. Compare the coefficient of x in the definition of the Bernoulli polynomials and in the first equation in the proof of Theorem 11.8, and thereby derive this expression for the Bernoulli numbers:

$$B_n = \sum_{k=0}^{n} s(n, k)(-1)^k \frac{k!}{k+1}, \qquad n \geq 0.$$

After referring to Table 11.4, calculate B_6.

11.58. As a follow-up to the previous problem, prove that

$$B_n = \sum_{k=1}^{n} \frac{1}{k+1} \sum_{m=1}^{k} (-1)^m \binom{k}{m} m^n, \qquad n \geq 1.$$

How does this show that the denominator of B_n is at most the least common multiple of the set $\{2, 3, 4, \ldots, n+1\}$? In this regard, look at the denominator of B_{12}.

11.59. A result in the literature (Leeming, 1989) says that

$$|B_{2n}| < 5\sqrt{\pi n} \left(\frac{n}{\pi \mathrm{e}} \right)^{2n}, \qquad n \geq 2.$$

According to one form of **Stirling's approximation** for factorials,

$$n! \sim \sqrt{2\pi n}\left(\frac{n}{\mathrm{e}}\right)^n.$$

Show that $\lim_{n\to\infty} |B_{2n}|/(2n)! = 0$. What does this tell you about the convergence of the series $\Sigma_{n=0}^{\infty} B_n x^n/n!$?

11.60. It is also known that $|B_{2n}| > 2(2n)!(2\pi)^{-2n}$, $n \geq 2$. What can you now say about the convergence of $\Sigma_{n=0}^{\infty} B_n x^n/n!$?

11.61. Show that

$$\coth u = \sum_{n=0}^{\infty} \frac{2^{2n} B_{2n} u^{2n-1}}{(2n)!}, \qquad u \neq 0,$$

in some suitable region of convergence. Use this to estimate coth 1 and csch 1.

11.62. Show that the Bernoulli polynomials are given by the generating function

$$\frac{t\mathrm{e}^{xt}}{\mathrm{e}^t - 1} = \sum_{n=0}^{\infty} B_n(x)\frac{t^n}{n!}.$$

The series is known to converge if $|t| < 2\pi$. Then show that

$$2^x = \frac{1}{\ln 2} + \sum_{n=1}^{\infty} B_n(x)\frac{(\ln 2)^{n-1}}{n!}.$$

How well do the two sides agree if $x = 1$ and we use only the first four nonzero terms of the series?

11.63. The approximation for $n!$ given in Problem 11.59 can be proved in several ways.

(a) One way is described in (Johnsonbaugh, 1981). Read the proof there and write it out in your own words.
(b) How does it follow that the inequality

$$\sqrt{2\pi n}\left(\frac{n}{\mathrm{e}}\right)^n < n!$$

holds for all positive integers n?

11.64. It is always of interest to study the zeros of the members of any sequence of polynomials.

(a) Prove that $B_{2n+1}(x)$ always has at least two nonnegative real zeros if $n \geq 1$.
(b) Find the nonnegative real zeros of $B_1(x)$, $B_3(x)$, $B_5(x)$, $B_7(x)$. Formulate a conjecture.
(c) It is known that for $B_{2n}(x)$, $n \geq 1$, there is just one real zero in the interval $(\frac{1}{6}, \frac{1}{4})$ and just one real zero in the interval $(\frac{3}{4}, \frac{5}{6})$. Confirm this for $B_4(x)$ and $B_6(x)$. Write a computer program to help you in obtaining the zeros for parts (b) and (c).

11.65. Prove that $B_n(1 - x) = (-1)^n B_n(x)$, $n \geq 0$. Consequently, show that $B_{2n+1}(\frac{1}{2}) = 0$.

11.66. Show that in the interval (0, 1), $B_{2n}(x)$ must have an odd number of local extrema. Illustrate this by graphing $B_8(x)$.

11.67. Pursuing Problem 11.66 further, we suppose that $B_{2n+1}(x)$ has a zero $\alpha \neq \frac{1}{2}$ in (0, 1). Show that this implies that $B_{2n-1}(x)$ also has a zero $\alpha' \neq \frac{1}{2}$ in (0, 1). Hence, conclude that $B_3(x)$ must have such a zero; as this is a contradiction, infer that the complete subset of zeros of $B_{2n+1}(x)$ in [0, 1] is $\{0, \frac{1}{2}, 1\}$.[6]

11.68. The Bernoulli numbers can be represented by determinants (Haussner, 1894):

$$B_{2n} = \frac{2n}{2^{2n}(2^{2n}-1)} \begin{vmatrix} 1 & 0 & 0 & \cdots & 0 & 1 \\ \binom{3}{1} & 1 & 0 & \cdots & 0 & 1 \\ \binom{5}{1} & \binom{5}{3} & 1 & \cdots & 0 & 1 \\ & & \vdots & & & \\ \binom{2n-3}{1} & \binom{2n-3}{3} & \binom{2n-3}{5} & \cdots & 1 & 1 \\ \binom{2n-1}{1} & \binom{2n-1}{3} & \binom{2n-1}{5} & \cdots & \binom{2n-1}{2n-3} & 1 \end{vmatrix}.$$

Verify this for B_8.

[6] **Fermat** would have appreciated this proof. It is just an application of his technique known as the **Method of Infinite Descent** (see Section 8.4).

11.6 Connection with the Riemann Zeta Function

We conclude this chapter by establishing a relationship between the Bernoulli numbers and the famous zeta function. The zeta function, $\zeta(s)$, was first investigated by **Euler** (Ayoub, 1974); this work ranks as among his most brilliant. A century later the German mathematician **Georg F.B. Riemann** (1826–1866) studied the zeta function further.

Let us recall from Problem 10.30 the definition

$$\zeta(s) = \sum_{n=1}^{\infty} \frac{1}{n^s}, \qquad \text{Re } s > 1.$$

In this section we shall restrict s to be real; the Integral Test requires $s > 1$ in order for the series to converge. Our goal is to prove

$$\zeta(2n) = \frac{(-1)^{n-1}(2\pi)^{2n}B_{2n}}{2(2n)!}, \qquad n \geq 1$$

a special case of which is the well known relation

$$\sum_{n=1}^{\infty} \frac{1}{n^2} = \frac{\pi^2}{6}.$$

Our treatment follows the first proof in (Berndt, 1975), for which we need the following lemma.

Lemma 11.11.1 Let $\mathrm{f}''(x)$ be continuous on $[0, a]$, $0 < a < \pi$, and suppose that $\mathrm{f}(0) = 0$. Then

$$\lim_{N\to\infty} \int_0^a \mathrm{f}(x)\frac{\sin(Nx)}{\sin x}\, dx = 0.$$

Proof Define $\mathrm{g}(x) = \mathrm{f}(x)/\sin x$, and integrate once by parts to give

$$\int_0^a \mathrm{g}(x)\sin(Nx)\, dx = \left.\frac{-\mathrm{g}(x)\cos(Nx)}{N}\right|_0^a + \frac{1}{N}\int_0^a \mathrm{g}'(x)\cos(Nx)\, dx.$$

By l'Hôpital's rule,

$$\lim_{x\to 0^+} \mathrm{g}(x) = \lim_{x\to 0^+} \frac{\mathrm{f}(x)}{\sin x} = \lim_{x\to 0^+} \frac{\mathrm{f}'(x)}{\cos x} = \mathrm{f}'(0),$$

which exists by hypothesis, so

$$\lim_{N\to\infty} \left.\frac{-\mathrm{g}(x)\cos(Nx)}{N}\right|_0^a = \lim_{N\to\infty}\left[\frac{-\mathrm{g}(a)\cos(Na) + \mathrm{f}'(0)}{N}\right] = 0.$$

Further, since $g'(x) = [\sin x\ f'(x) - \cos x\ f(x)]/\sin^2 x$, we have by another application of l'Hôpital's rule

$$\lim_{x \to 0^+} g'(x) = \lim_{x \to 0^+} \frac{\sin x\ f''(x) + \sin x\ f(x)}{2 \sin x \cos x} = \frac{f''(0)}{2},$$

which also exists by hypothesis. Thus, the integrand $g'(x) \cos(Nx)$ is bounded and continuous on $[0, a]$, so

$$\lim_{N \to \infty} \int_0^a g'(x) \cos(Nx)\, dx = 0,$$

and this completes the proof.[7] ◆

Now let $k \in \mathbf{N}$ and define the sequence of numbers $\{\mathbf{I}_n(k)\}_{n=1}^{\infty}$ by

$$\mathbf{I}_n(k) = \int_0^{1/2} \{B_{2n}(x) - B_{2n}\} \cos(2\pi kx)\, dx.$$

One integration by parts using the identity $B'_{2n}(x) = 2n\ B_{2n-1}(x)$ from Section 11.5 yields

$$\mathbf{I}_n(k) = \frac{-2n}{2\pi k} \int_0^{1/2} B_{2n-1}(x) \sin(2\pi kx)\, dx.$$

A second integration by parts, with $B'_{2n-1}(x) = (2n - 1)B_{2n-2}(x)$, gives

$$I_n(k) = \frac{-2n}{2\pi k} \left[-B_{2n-1}(x) \frac{\cos 2\pi kx}{2\pi k} \Bigg|_0^{1/2} + \frac{2n-1}{2\pi k} \int_0^{1/2} B_{2n-2}(x) \cos(2\pi kx)\, dx \right]$$

$$= \frac{-2n(2n-1)}{(2\pi k)^2} \int_0^{1/2} B_{2n-2}(x) \cos(2\pi kx)\, dx, \qquad n > 1.$$

Here, we have made use of the fact that $B_n(0) = 0$ for odd $n > 1$, and $B_n(\frac{1}{2}) = 0$ for odd n (Problem 11.65). When $n = 1$, the two integrations by parts give directly $\mathbf{I}_1(k) = 1/(2\pi k)^2$.

When $n > 1$, we perform $2n - 3$ more integrations by parts. The end result is (verify!)

$$\mathbf{I}_n(k) = \frac{(-1)^n (2n)!}{(2\pi k)^{2n-1}} \int_0^{1/2} B_1(x) \sin(2\pi kx)\, dx.$$

$$= \frac{(-1)^{n-1}(2n)!}{2(2\pi k)^{2n}}.$$

We use this to establish the main result of this section.

[7] Lemma 11.11.1 is a weakened version of a famous result in real analysis known as the **Riemann–Lebesgue lemma** (Spiegel, 1969). It is important in Fourier series.

Theorem 11.11 $\zeta(2n) = [(-1)^{n-1}(2\pi)^{2n}B_{2n}]/[2(2n)!],\ n \geq 1.$

Proof We have from the previous development

$$\frac{(-1)^{n-1}(2n)!}{2(2\pi k)^{2n}} = \int_0^{1/2} \{B_{2n}(x) - B_{2n}\} \cos(2\pi kx)\, dx.$$

Sum both sides from $k = 1$ to $k = N$:

$$\frac{(-1)^{n-1}(2n)!}{2(2\pi)^{2n}} \sum_{k=1}^{N} \frac{1}{k^{2n}} = \int_0^{1/2} \{B_{2n}(x) - B_{2n}\} \sum_{k=1}^{N} \cos(2\pi kx)\, dx.$$

To perform the summation on the right, we recall **Euler's Formula** from complex arithmetic: $e^{ix} = \cos x + i \sin x$. Hence, the series that follows is a finite geometric series with constant ratio $e^{2\pi ix}$. The sum is thus

$$\begin{aligned}\sum_{k=1}^{N} e^{2\pi kix} &= \sum_{k=0}^{N} e^{2\pi kix} - 1\\ &= \frac{1 - e^{2\pi(N+1)ix}}{1 - e^{2\pi ix}} - 1\\ &= \frac{1 - \cos(2\pi(N+1)x) - i\sin(2\pi(N+1)x)}{1 - \cos(2\pi x) - i\sin(2\pi x)} - 1.\end{aligned}$$

We rationalize the fraction by multiplying numerator and denominator by $1 - \cos(2\pi x) + i\sin(2\pi x)$. The result is, after some trigonometric manipulations (verify!),

$$\sum_{k=1}^{N} e^{2\pi kix} = \left[\frac{\sin(2N+1)\pi x}{2\sin \pi x} - \frac{1}{2}\right] + i\left[\frac{-\cos(2N+1)\pi x}{2\sin \pi x} + \frac{\cot \pi x}{2}\right].$$

The real part on the right-hand side is, from Euler's Formula, the sum of the cosine terms that appear in the left-hand side,

$$\sum_{k=1}^{N} \cos 2\pi kx = \frac{\sin(2N+1)\pi x}{2\sin \pi x} - \frac{1}{2},$$

so

$$\frac{(-1)^{n-1}(2n)!}{2(2\pi)^{2n}} \sum_{k=1}^{N} \frac{1}{k^{2n}} = \int_0^{1/2} \{B_{2n}(x) - B_{2n}\}\left[\frac{\sin(2N+1)\pi x}{2\sin \pi x} - \frac{1}{2}\right] dx. \qquad (*)$$

We now let $N \to \infty$. The left-hand side of (*) converges because $2n > 1$; we obtain

$$\frac{(-1)^{n-1}(2n)!}{2(2\pi)^{2n}}\zeta(2n).$$

On the right-hand side we apply Lemma 11.11.1, in which we let $f(\pi x) = \{B_{2n}(x) - B_{2n}\}/2$. This gives for the right-hand side of (*)

$$-\frac{1}{2}\int_0^{1/2} \{B_{2n}(x) - B_{2n}\}\, dx.$$

From Section 11.5 we use $B'_{2n+1}(x)/(2n + 1) = B_{2n}(x)$, giving finally

$$\begin{aligned}\frac{(-1)^{n-1}(2n)!}{2(2\pi)^{2n}}\zeta(2n) &= \frac{-1}{2(2n + 1)}\int_0^{1/2} d\,[B_{2n+1}(x)] + \frac{B_{2n}}{2}\int_0^{1/2} dx \\ &= 0 + \frac{B_{2n}}{4}\end{aligned}$$

from Problem 11.65. The theorem is equivalent to this last line. ◆

Example 11.15 Theorem 11.11 gives us immediately

$$\sum_{k=1}^{\infty}\frac{1}{k^4} = \frac{\pi^4}{90}, \qquad \sum_{k=1}^{\infty}\frac{1}{k^6} = \frac{\pi^6}{945}, \qquad \sum_{k=1}^{\infty}\frac{1}{k^8} = \frac{\pi^8}{9450}.$$

A major unsolved problem in number theory is the fact that no closed forms are known for the sum of the reciprocals of any of the odd powers of the positive integers.[8] ◆

Example 11.16 Write Theorem 11.11 as

$$\begin{aligned}|B_{2n}| &= \frac{2(2n)!}{(2\pi)^{2n}}\zeta(2n) \\ &= \frac{2}{(2\pi)^{2n}}\left(\frac{2n}{\mathrm{e}}\right)^{2n}\sqrt{4\pi n}\left[\frac{(2n)!}{\sqrt{4\pi n}(2n/\mathrm{e})^{2n}}\right]\zeta(2n)\end{aligned}$$

or

$$\frac{|B_{2n}|}{4\sqrt{\pi n}(n/\pi\mathrm{e})^{2n}} = \frac{(2n)!}{\sqrt{4\pi n}(2n/\mathrm{e})^{2n}}\zeta(2n).$$

[8] Even worse, except for $\zeta(3)$, it is not even known if $\zeta(2n + 1)$ is rational or irrational (Van der Poorten, 1979)!

On the right-hand side, we have from Problem 11.59

$$\lim_{n\to\infty} \frac{(2n)!}{\sqrt{4\pi n}(2n/\mathrm{e})^{2n}} = 1$$

and $\lim_{n\to\infty} \zeta(2n) = 1$. Hence, $|B_{2n}| \sim 4\sqrt{\pi n}(n/\pi \mathrm{e})^{2n}$. Letting $f(n) = 4\sqrt{\pi n}(n/\pi \mathrm{e})^{2n}$, we obtain the following selected values:

n	$\lvert B_{2n}\rvert$	$f(n)$	$\lvert B_{2n}\rvert/f(n)$
6	.2531	.2513	1.0072
10	529.12	526.92	1.0042
14	2.7298×10^7	2.7217×10^7	1.0030
15	6.0158×10^8	5.9991×10^8	1.0028

See also (Leeming, 1977). ◆

Problems

11.69. Confirm that $\zeta(s)$ is defined for positive s iff $s > 1$.

11.70. In Lemma 11.11.1 let $f(x) = x$ and $a = \pi/2$. Evaluate the integrals

$$\int_0^a f(x)\frac{\sin(Nx)}{\sin x}\,dx \quad \text{for} \quad N = 1, 2, 3, 4, 6.$$

11.71. In Theorem 11.11 verify that

$$\sum_{k=1}^{N} \cos(2\pi kx) = \frac{\sin(2N+1)\pi x}{2\sin \pi x} - \frac{1}{2}.$$

11.72. Let $\zeta(2n)$ be represented by $A\pi^{2n}/B$, $A, B \in \mathbf{N}$. The examples of $n = 1, 2, 3, 4$ have $A = 1$. Can it ever happen that $A > 1$?

11.73. Regarding the asymptotic expression in Example 11.16, show that it is actually a lower bound for $|B_{2n}|$, that is,

$$|B_{2n}| > 4\sqrt{\pi n}\left(\frac{n}{\pi \mathrm{e}}\right)^{2n}, \text{ for } n \geq 1.$$

11.74. Show that $\zeta(2n) < 2n/(2n-1)$ for $n \geq 1$, and hence

$$|B_{2n}| < \frac{8n^2(2n-2)!}{(2\pi)^{2n}}.$$

11.75. We have $\sin x = (\mathrm{e}^{ix} - \mathrm{e}^{-ix})/2i$ and $\cos x = (\mathrm{e}^{ix} + \mathrm{e}^{-ix})/2$.

(a) Show that $x \cot x = ix + 2ix/(e^{2ix} - 1)$, where $0 < |x| < \pi$.
(b) Then that $x \cot x = \Sigma_{n=0}^{\infty} (-1)^n 2^{2n} B_{2n}/(2n)!\, x^{2n}$, $0 < |x| < \pi$.
(c) From this show that

$$\cot x = \frac{1}{x} - 2 \sum_{n=1}^{\infty} \frac{\zeta(2n)}{\pi^{2n}} x^{2n-1}, \quad 0 < |x| < \pi.$$

(d) Finally, show that $\Sigma_{n=1}^{\infty} \zeta(2n)/2^{2n-1} = 1$.
(e) How closely do the two sides agree if only the first five terms of the series are summed?

11.76. This problem continues the spirit of the previous one.

(a) Prove that $\tan x = \cot x - 2 \cot 2x$.
(b) Hence, show that

$$\tan x = \sum_{n=1}^{\infty} (-1)^{n-1} \frac{B_{2n}}{(2n)!} 2^{2n}(2^{2n} - 1)x^{2n-1}, \qquad \left(|x| < \frac{\pi}{2}\right).$$

(c) Show that the coefficients of x, x^3, x^5 in this series agree with what one finds in the Maclaurin series expansion of $\tan x$.
(d) Finally, show that $\frac{1}{2}\pi = \Sigma_{n=1}^{\infty} \zeta(2n)(2^{2n} - 1)/4^{2n-1}$.
(e) How closely do the two sides agree if only the first five terms of the series are summed?

11.77. Show that $|B_{2n}/B_{2n-2}| \sim n(2n - 1)/2\pi^2$. This result was given by **Ramanujan** in 1911, but was obtained differently. Illustrate the result numerically.

11.78. The **Euler polynomials**, $\{E_n(x)\}_{n=0}^{\infty}$, are the unique polynomials with the property that for any $x \in \mathfrak{R}$,

$$x^n = \frac{E_n(x + 1) + E_n(x)}{2}.$$

(a) Deduce the first five Euler polynomials.
(b) The **Euler numbers**, $\{E_n\}_{n=0}^{\infty}$, are defined as $E_n = 2^n E_n(\frac{1}{2})$. Obtain the first three nonzero Euler numbers.
(c) By a technique similar to that used in this section, the Euler numbers can be given an infinite series representation (Dence, 1997):

$$E_{2n} = 2(-1)^n(2n)! \left(\frac{2}{\pi}\right)^{2n+1} \sum_{k=0}^{\infty} \frac{(-1)^k}{(2k + 1)^{2n+1}}.$$

Evaluate the sum $1 + \frac{1}{125} - \frac{1}{343} - \frac{1}{1331} + \frac{1}{2197} + \frac{1}{4913} - \cdots$.

RESEARCH PROBLEMS

1. Sequences of integers that have no obvious pattern often acquire such a pattern when the sequences are mapped into finite rings. Investigate what happens when various Lucas sequences, $\{U_n(P, Q)\}_{n=0}^{\infty}$, and various companion Lucas sequences, $\{V_n(P, Q)\}_{n=0}^{\infty}$, are mapped into various rings $\mathbf{Z}_n$.

2. All of the Bernoulli numbers beyond B_0 are fractions. Plausibly, one might be able to write $B_n = c_n - \Sigma_k (1/k)$, $c_n \in \mathbf{Z}$, for each Bernoulli number B_n, where the summation is over some subset of positive integers $k \geq 2$. Investigate if this representational form really does appear to be possible for the Bernoulli numbers.

3. A prime p is termed **regular** if p does not divide the numerators of B_2, B_4, B_6, . . . , B_{p-3}. Otherwise, a prime is designated as **irregular**. In his attempt to prove Fermat's Last Theorem, **E.E. Kummer** needed to know the regular primes. Investigate the primes on computer and uncover the first three irregular primes.

4. Refer to Problem 11.78. Investigate the mapping of the Euler numbers, $\{E_n\}_{n=0}^{\infty}$, into the ring $\mathbf{Z}_9$. Try to prove what you observe.

References

Abramowitz, M., and Stegun, I.A. (eds.), *Handbook of Mathematical Functions*, Dover Publications, New York, 1965, pp. 810, 833–835. The first citation is to values of B_{2n} up to $n = 30$; the second is to values of S(n, k), s(n, k) up to $n = 25$.

André-Jeannin, R., "Problem B-752" (solved by A.N.' t Woord), *Fibonacci Quart.*, **33**, 182 (1995). The source of our Problem 11.22.

Ayoub, R., "Euler and the Zeta Function," *Amer. Math. Monthly*, **81**, 1067–1086 (1974). This is a dynamite article on one of the most brilliant parts of Euler's career.

Berndt, B.C., "Elementary Evaluation of $\zeta(2n)$," *Math. Mag.*, **48**, 148–154 (1975). Our proof of Theorem 11.11 is taken from the first (and neater) of the two proofs given in this interesting article.

Boyer, C.B., "Pascal's Formula for the Sums of Powers of the Integers," *Scripta Mathematica*, **9**, 237–244 (1943). The work by **Pascal** on these formulas postdated that of **Fermat** by some 18 years.

Bromwich, T.J. I'A., *Introduction to the Theory of Infinite Series*, 2nd ed., Macmillan, London, 1926, pp. 324–331. See these pages for a brief discussion of the Euler–Maclaurin summation formula.

Carmichael, R.D., "On the Numerical Factors of the Arithmetic Forms $\alpha^n \pm \beta^n$," *Ann. Math.*, **15**, 30–70 (1913). Contains very interesting theorems; skim read this.

Dence, J.B., "Problem 68" (solved by R. Prielipp and G.E. Bergum), *Missouri J. Math. Sci.*, **7**, 44–47 (1995). The source of our Problem 11.27.

Dence, J.B., "A Development of Euler Numbers," *Missouri J. Math. Sci.*, **9**, 148–155 (1997). Some interesting details on these first cousins of the Bernoulli numbers.

Dilcher, K., Skula, L., and Slavutskii, I.S. (eds.), *Queen's Papers in Pure and Applied Mathematics*, No. 87, "Bernoulli Numbers: Bibliography (1713–1990)," Queen's University, Kingston, Ontario, 1991. Contains about 2000 references, arranged in alphabetical order by author.

Gillispie, C.C. (ed.), *Dictionary of Scientific Biography*, Vol. XIII, Charles Scribner's Sons, New York, 1976, pp. 67–70. A brief biographical summary about **James Stirling**, whose total mathematical output was important but disappointingly small.

Goldstein, H.J., "An Application of Stirling's Numbers," *Amer. Math. Monthly*, **41**, 565–570 (1934). The result in Problem 11.44 was taken from this more extensive paper.

Graham, R.L., Knuth, D.E., and Patashnik, O., *Concrete Mathematics: A Foundation for Computer Science*, Addison–Wesley, Reading, MA, 1989, pp. 243–258. This is a superb book on mathematics; see the indicated pages for material on Stirling's numbers, and also on Eulerian numbers (which we do not discuss in the present book, but which are given in Table I (4) in Appendix II).

Haussner, R., "Independente Darstellung der Bernoulli'schen und Euler'schen Zahlen durch Determinanten," *Z. für Math. u. Phys.*, **39**, 183–188 (1894). Our Problem 11.68 is taken from this paper.

Honsberger, R., *Mathematical Gems III*, Mathematical Association of America, Washington, DC, 1985, pp. 102–138. A rich and delightful short chapter on Fibonacci and Lucas numbers; worth careful study.

Johnsonbaugh, R., "The Trapezoidal Rule, Stirling's Formula, and Euler's Constant," *Amer. Math. Monthly*, **88**, 696–698 (1981). The author uses the trapezoidal rule to show that $\lim_{n\to\infty} [n^n\sqrt{2\pi n}/(n!\mathrm{e}^n)] = 1$. Another nice treatment of this problem that uses only properties of integrals is in Marsaglia, G., and Marsaglia, J.C.W., "A New Derivation of Stirling's Approximation to $n!$," *Amer. Math. Monthly*, **97**, 826–829 (1990).

Kelly, C., "An Algorithm for Sums of Integer Powers," *Math. Mag.*, **57**, 296–297 (1984). A nice application of the binomial theorem (for positive integral powers).

Kostal, J.J., "Problem B-754" (solved by G.A. Bookhout), *Fibonacci Quart.*, **33**, 184 (1995). The source of our Problem 11.23.

Leeming, D.J., "An Asymptotic Estimate for the Bernoulli and Euler Numbers," *Can. Math. Bull.*, **20**, 109–111 (1977). The author derives the simple result $|B_{2n}| \sim n/\pi\mathrm{e}$ as $n \to \infty$, which is consistent with our result in Example 11.16 since $\lim_{n\to\infty} (4\sqrt{\pi n})^{1/2n} = 1$.

Leeming, D.J., "The Real Zeros of the Bernoulli Polynomials," *J. Approx. Theory*, **58**, 124–150 (1989). The inequality in our Problem 11.59 appears in Lemma 3.1 of Leeming's paper; this was a new result.

Lehmer, D.H., "An Extended Theory of Lucas' Functions," *Ann. Math.*, **31**, 419–448 (1930). With a little effort you should be able to follow much of what is in this paper.

Lehmer, D.H., "A New Approach to Bernoulli Polynomials," *Amer. Math. Monthly*, **95**,

905–911 (1988). Lehmer produces the Bernoulli polynomials by using Raabe's multiplication theorem (1851):

$$\frac{1}{m}\sum_{k=0}^{m-1} B_n\left(x+\frac{k}{m}\right) = m^{-n}B_n(mx).$$

McDaniel, W.L., "The G.C.D. in Lucas Sequences and Lehmer Number Sequences," *Fibonacci Quart.*, **29**, 24–29 (1991). Let m, n be positive integers and $d = (m, n)$. Then (U_m, V_n), (V_m, V_n) equal 1, 2 or V_d, depending on the powers of 2 that divide m and n.

Mordell, L.J., "The Sign of the Bernoulli Numbers," *Amer. Math. Monthly*, **80**, 547–548 (1973). The source of our proof of Theorem 11.10.

Rainville, E.D., and Bedient, P.E., *Elementary Differential Equations*, 6th ed., Macmillan, New York, 1981, pp. 36–40, 100–115. A review of some standard techniques for solving linear differential equations with constant coefficients.

Ribenboim, P., *The New Book of Prime Number Records*, Springer–Verlag, New York, 1996, pp. 53–74. Extensive summary of material on Lucas sequences, which beckons for your participation.

Spiegel, M.R., *Real Variables*, Schaum's Outline Series, McGraw–Hill, New York, 1969, p. 137. See this citation for a proof of a more general version of the Riemann–Lebesgue theorem (or lemma) than that in our Lemma 11.11.1. Compare with a still more general version in Whittaker, E.T., and Watson, G.N., *A Course of Modern Analysis*, 4th ed., Cambridge University Press, Cambridge, England, 1963, pp. 172–174.

Struik, D.J. (ed.), *A Source Book in Mathematics: 1200–1800*, Princeton University Press, Princeton, NJ, 1986, pp. 316–324. An extract (in English) of some of Bernoulli's work on series.

Tweddle, I., *James Stirling*, Scottish Academic Press, Edinburgh, Scotland, 1988. There are very few books available on the life and work of this eminent Scottish mathematician.

Vajda, S., *Fibonacci & Lucas Numbers, and the Golden Section*, Ellis Horwood Ltd., Chichester, England, 1989, p. 10. The source of our Problem 11.17. This book is packed with neat stuff; an appendix gives 106 formulas involving F_n's and L_n's.

Van der Poorten, A., "A Proof That Euler Missed—Apéry's Proof of the Irrationality of $\zeta(3)$," *Math. Intelligencer*, **1**, 196–203 (1979). The door is wide open for you to get your name in the mathematics history books. A proof of the irrationality of $\zeta(3)$ is given in Beukers, F., "A note of Irrationality of $\zeta(2)$ and $\zeta(3)$," *Bull. London Math. Soc.*, **II**, 268-272 (1979).

Williams, K.S., "Bernoulli's Identity without Calculus," *Math. Mag.*, **70**, 47–50 (1997). The author derives $\sum_{r=1}^{n-1} r^k = \frac{1}{k+1}\sum_{j=0}^{k}\binom{k+1}{j}B_j n^{k+1-j}$, $n, k \in \mathbf{N}$, from $B_m = \frac{-1}{m+1}\sum_{j=0}^{m-1}\binom{m+1}{j}B_j$. A related paper in which the Bernoulli numbers themselves are generated on a tree is Woon, S.C., "A Tree for Generating Bernoulli Numbers," *Math. Mag.*, **70**, 51–56 (1997). This paper and the one by Williams would make a suitable outside reading assignment.

TWO AFTER-DINNER DESSERTS

You've worked very hard reading this book, perhaps almost as hard as we did writing it. To bring your experience to a close after the main fare, we offer for your pleasure the following two tasty treats.

1. (In the manner of **Lewis Carroll**) Brothers Robert, Paul, Danny, and their dog Spikey are sitting at a table. On the table is a pile of pistachios.

1. Robert gives one pistachio to Spike Dog and takes a third of the remaining pistachios for himself. He gives the rest to Paul.
2. Paul gives one pistachio to Spike Dog and takes a third of the remaining pistachios for himself. He gives the rest to Dan.
3. Danny gives one pistachio to Spike Dog, keeps two-thirds of what is left, and gives the remainder back to Paul.

The boys can tell at a glance that the original pile contained more than 3 dozen nuts. What is the minimum size of the pile?

2. You've fraternized with the following eminent mathematicians. See if you can identify them from the following scanty biographical clues.

(a) Entered the University of Budapest at age 17 and published his first mathematical paper (on the distribution of primes) while a freshman.
(b) He calculated and did mathematics as effortlessly as men breathe, and even in total blindness during the last 17 years of his life he continued to produce.
(c) Equations named after him are required to have integral solutions, although he usually didn't find all of them.
(d) Began his career in number theory by discovering at age 3 (or thereabouts) an addition error in his father's payroll.
(e) La! la! la! A parliamentary councillor by day, a mathematician by night.
(f) American pioneer in computational number theory, and whose father and wife were also professional mathematicians.
(g) The more silent partner of a dynamic English duo that published jointly about 100 papers on mathematics (mainly analysis and number theory).

Appendix I
Notation*

General Mathematical Symbols

iff	If and only if (1.1)
$\rightarrow$	Implies (1.1)
$\leftrightarrow$	Implies and is implied by (1.1)
$\in$, $\notin$	Membership (or not) in a set (1.1)
$\mid$	Divides (1.3)
$\sim$	Negation of a proposition (1.1); asymptotic to (1.8); equivalent to (3.1)
$\lfloor\ \rfloor$	Greatest integer function (1.8)
$\lceil\ \rceil$	Ceiling function (11.3)
$(\cdot)$	Group binary operation (2.3)
$\equiv$	Is congruent to (3.1); is identically congruent to (4.1)
$*$	Convolution of two arithmetic functions (7.7)

Notation for Sets

$\mathbf{A}$	The set of all arithmetic functions (7.8); the set of all numbers algebraic over a field **F** (9.3)
$\mathbf{A}_m$	The set of all multiplicative functions (7.8)

* Numbers in parentheses indicate section where first used.

$\mathbf{A}_p^{(k)}$	The set of the least positive kth-power residues modulo the prime p (6.1)
$\mathscr{C}$	The set of all complex numbers (1.1)
$\mathbf{D}$	Integral domain (3.3)
$\mathbf{F}$	Field (3.4)
$\mathbf{F}(\theta)$	The adjunction of the number θ to the field $\mathbf{F}$ (9.2)
$\mathbf{F}(\alpha, \beta)$	The adjunction of the numbers α, β to the field $\mathbf{F}$ (9.2)
$\mathbf{F}[x]$	The set of all polynomials in x with coefficients in the field $\mathbf{F}$
$\mathbf{G}$	The set of elements comprising a group (2.3); the group itself (2.4)
$\langle \mathbf{G}, \cdot \rangle$	Group (2.3)
$g\mathbf{H}$	Left coset of $\mathbf{H}$ that contains the element g (3.8)
$\mathbf{H}$	Subgroup of $\mathbf{G}$ (2.3)
$\mathscr{I}$	The set of all irrational numbers (1.1)
$\mathbf{K}$	Extension field (3.4, 9.2)
$k\mathbf{Z}$	The set of all integral multiples of the nonzero integer k (2.4)
$\mathbf{M}$	$\mathbf{R}$-module (9.4)
$\mathbf{N}$	The set of all natural numbers (1.1)
$\mathbf{Q}$	The set of all rational numbers (1.1)
$\mathbf{Q}(\sqrt{D})$	Quadratic algebraic field (9.5)
$\mathscr{R}$	The set of all real numbers (1.1); binary relation (3.1)
$\mathbf{R}$	The set of elements comprising a ring (3.3); the ring itself (3.3)
$\langle \mathbf{R}, +, \cdot \rangle$	Ring (3.3)
$\mathbf{R}[x]$	The set of all polynomials in x with coefficients in the ring $\mathbf{R}$ (4.1)
$\mathbf{R}^{\times}$	The set of units in the ring $\langle \mathbf{R}, +, \cdot \rangle$ (3.4)
$\mathbf{S}_i$	Residue class containing the integer i (3.2)
$\mathbf{U}_A$	The set of arithmetic functions that have an inverse with respect to convolution (7.8)
$\mathbf{V}$	Klein 4-group (*Vierergruppe*) (2.3)
$\mathbf{Z}$	The set of all integers (1.1); the additive group of integers (2.4)

$\mathbf{Z}[x]$	The set of all polynomials in x with coefficients in $\mathbf{Z}$ (3.1)
$\mathbf{Z}_{n(p)}$	The finite ring of the residue classes modulo $n(p)$ (3.5)
$\mathbf{Z}_{\mathrm{p}}^{\times}$	A maximal set of integers incongruents to 0 modulo the prime p (e.g., $\{1, 2, \ldots, p-1\}$ (5.2)
$\mathbf{Z}[i]$	The set of all numbers of the form $a + bi$, where $a, b \in \mathbf{Z}$ and $i = \sqrt{-1}$, that is, the set of Gaussian integers (6.7, 9.5)
$\mathbf{Z}[\omega]$	The set of all numbers of the form $a + b\omega$, where $a, b \in \mathbf{Z}$ and $\omega = (-1 + \sqrt{-3})/2$ (6.7)

Numbers, Functions, and Sequences

$(\frac{A}{p})$	Legendre's symbol (6.2)
$(A \mid n)$.	Jacobi's symbol (6.3)
$(\frac{\alpha}{\pi})_3$	Cubic residue symbol (6.7)
$(\frac{\alpha}{\pi})_4$	Quartic residue symbol (6.7)
B	Brun's constant (1.8)
$\{\mathrm{B}_n\}$	The sequence of Bernoulli numbers (10.2, 11.5)
$\mathrm{B}(N)$	Number of integers not exceeding N that are expressible as the sum of two squares (8.1)
$\{\mathrm{B}_n(x)\}$	The sequence of Bernoulli polynomials (11.5)
$\mathrm{C}(x)$	Number of Carmichael numbers not exceeding x (3.9)
d, (x, y)	Greatest common divisor of x and y (2.4)
$\mathrm{D}(N)$	Number of ordered compositions of the positive integer N (10.1)
$[d_0; d_1, d_2, \ldots, d_n]$	Finite, simple continued fraction (3.11)
$\mathrm{DR}(N)$	Digital root of the positive integer N (3.1)
$\mathrm{DS}(N)$	Digital sum of the positive integer N (1.3)
ϵ_i	A unit of an algebraic number field (6.7)
$\epsilon(n)$	Identity function of the positive integer n (7.1)
$\{\mathrm{F}_n\}$	The sequence of Fermat numbers (1.8, 2.1); the sequence of Fibonacci numbers (1.1)
$\mathrm{F}(n)$	$\Sigma_{d\mid n}\mathrm{f}(d)$, where f is an arithmetic function (7.2)
$\{\mathrm{F}_n(x)\}$	The sequence of Fibonacci polynomials (11.3)
$\mathrm{f}^{-1}(n)$	The inverse with respect to convolution of the multiplicative function f, evaluated at the positive integer n (7.7)

$g(x)$	Generating function (10.2)
$g(n)$, $G(n)$	Waring's problem numbers (8.5)
HCI	Highly composite integer (2.1)
$\{H_n\}$	The sequence of hexagonal numbers (7.4)
$\{H_n(x)\}$	The sequence of Hermite polynomials (11.1)
$\mathbf{I}$	Group identity element (2.3)
$I(n)$	Abundance index of a positive integer n (7.4); arithmetical unity function of a positive integer n (7.7)
$\operatorname{ind}_g A$	Index of A to the base g (5.4)
$[\mathbf{K}: \mathbf{F}]$	Degree of the extension field $\mathbf{K}$ over the field $\mathbf{F}$ (9.3)
$\{L_n\}$	The sequence of the Lucas numbers (6.3, 11.3)
$\lambda(n)$	Liouville function of the positive integer n (7.1)
$\Lambda(n)$	von Mangoldt function of the positive integer n (7.6)
ℓi	Logarithmic integral (1.8)
M, $[x, y]$	Least common multiple of x and y (2.7)
$\mathbf{M}^*$	Conjugate transpose of a matrix $\mathbf{M}$ (8.2)
$\{M_n\}$	The sequence of Mersenne numbers (4.4)
$(m)_n$	Pochhammer's symbol (4.6)
$\{M_p\}$	The sequence of Mersenne primes (1.8)
$\mu(n)$	Möbius function of the positive integer n (7.1)
$N\alpha$	Norm of the algebraic number α (6.7)
$p(N)$	Number of unrestricted partitions of the positive integer N (10.1)
$p_k(N)$	Number of partitions of the positive integer N into at most k summands (10.1)
$p(\mathbf{S}, N)$	Number of partitions of the positive integer N when its summands are drawn from the set $\mathbf{S}$ (10.4)
$\operatorname{psp}(a)$	Pseudoprime to the base a (3.9)
$P_2(n)$	Number of twin primes $(p, p + 2)$ in the interval $3 \leq p + 2 \leq n + 2$ (1.8)
$P(n, k)$	The kth polygonal number of order n (10.5)
$(\frac{p}{q})_4$	Rational quartic residue symbol (6.7)
$P_p^{(k)}$	The product of the members of $\mathbf{A}_p^{(k)}$ (6.1)
P_k/Q_k	The kth-order convergent to a simple continued fraction (3.11)

$\phi(n)$	Euler phi-function of the positive integer n (3.7)
π	A prime in a complex algebraic number field (6.7)
π^{α}	Euler factor of any odd perfect number (7.4)
$\pi(n)$	Number of positive primes that do not exceed n (1.4); product-of-divisors function of the positive integer n (7.1)
$\pi_k(N)$	Number of partitions of the positive integer N into summands no larger than k (10.1)
$\{S(n, k)\}$	The sequence of Stirling's numbers of the first kind (11.4)
$\{s(n, k)\}$	The sequence of Stirling's numbers of the second kind (1.1, 11.4)
$\sigma(n)$	Sum-of-divisors function of the positive integer n (7.1)
$\sigma_u(n)$	Sum-of-unitary-divisors function of the positive integer n (7.4)
$\{T_n(x)\}$	The sequence of Chebyshev polynomials of the first kind (1.1)
$T_p^{(k)}$	The sum (total) of the members of $\mathbf{A}_p^{(k)}$ (6.1)
$\tau(n)$	Number-of-divisors function of the positive integer n (7.1)
$\Theta(n)$	$\Sigma_{d\mid n}\ \phi(n)$ (7.2)
$u(n)$	Unit function of the positive integer n (7.1)
$\{U_n(P, Q)\}$	The Lucas sequence with parameters P and Q (11.3)
$\{V_n(P, Q)\}$	The companion Lucas sequence with parameters P and Q (11.3)
$\{W_n\}$	The Lehmer sequence (11.3)
$\{X_n\}$	The associated Lehmer sequence (11.3)
$\zeta(s)$	Riemann zeta function (10.2)

Appendix II
Mathematical Tables

Table A. Constants

$\pi \approx 3.14159265358979323846$	
$\pi^{-1} \approx 0.31830988618379067153$	
$\pi^2 \approx 9.86960440108935861883$	
$\ln \pi \approx 1.14472988584940017414$	
$e \approx 2.71828182845904523536$	
$e^{-1} \approx 0.36787944117144232159$	
$e^2 \approx 7.38905609893065022723$	
$\ln 10 \approx 2.30258509299404568401$	
$\log_{10} e \approx 0.43429448190325182765$	
$\sqrt{2} \approx 1.41421356237309504880$	
$\sqrt[3]{2} \approx 1.25992104989487316476$	
$\sqrt{3} \approx 1.73205080756887729352$	
$\phi \approx 1.61803398874989484820$	(golden ratio)
$\ln 2 \approx 0.69314718055994530941$	
$\gamma \approx 0.57721566490153286061$	(Euler's Constant)
$B \approx 1.902160577783278$	(Brun's Constant)

Table B. Powers of Integers; Factorials

n	2^n	3^n	5^n
1	2	3	5
2	4	9	25
3	8	27	125
4	16	81	625
5	32	243	3,125
6	64	729	15,625
7	128	2,187	78,125
8	256	6,561	390,625
9	512	19,683	1,953,125
10	1,024	59,049	9,765,625
11	2,048	177,147	48,828,125
12	4,096	531,441	244,140,625
13	8,192	1,594,323	1,220,703,125
14	16,384	4,782,969	6,103,515,625
15	32,768	14,348,907	30,517,578,125
16	65,536	43,046,721	152,587,890,625
17	131,072	129,140,163	762,939,453,125
18	262,144	387,420,489	3,814,697,265,625
19	524,288	1,162,261,467	19,073,486,328,125
20	1,048,576	3,486,784,401	95,367,431,640,625
21	2,097,152	10,460,353,203	476,837,158,203,125
22	4,194,304	31,381,059,609	2,384,185,791,015,625
23	8,388,608	94,143,178,827	11,920,928,955,078,125
24	16,777,216	282,429,536,481	59,604,644,775,390,625
25	33,554,432	847,288,609,443	298,023,223,876,953,125

N	N^2	N^3	N^4	N^5
2	4	8	16	32
3	9	27	81	243
4	16	64	256	1,024
5	25	125	625	3,125
6	36	216	1,296	7,776
7	49	343	2,401	16,807
8	64	512	4,096	32,768
9	81	729	6,561	59,059
10	100	1,000	10,000	100,000
11	121	1,331	14,641	161,051
12	144	1,728	20,736	248,832
13	169	2,197	28,561	371,293
14	196	2,744	38,416	537,824
15	225	3,375	50,625	759,375
16	256	4,096	65,536	1,048,576
17	289	4,913	83,521	1,419,857
18	324	5,832	104,976	1,889,568
19	361	6,859	130,321	2,476,099
20	400	8,000	160,000	3,200,000
21	441	9,261	194,481	4,084,101
22	484	10,648	234,256	5,153,632
23	529	12,167	279,841	6,436,343
24	576	13,824	331,776	7,962,624
25	625	15,625	390,625	9,765,625
26	676	17,576	456,976	11,881,376
27	729	19,683	531,441	14,348,907
28	784	21,952	614,656	17,210,368
29	841	24,389	707,281	20,511,149
30	900	27,000	810,000	24,300,000
31	961	29,791	923,521	28,629,151
32	1,024	32,768	1,048,576	33,554,432
33	1,089	35,937	1,185,921	39,135,393
34	1,156	39,304	1,336,336	45,435,424
35	1,225	42,875	1,500,625	52,521,875

N	N^2	N^3	N^4	N^5
36	1,296	46,656	1,679,616	60,466,176
37	1,369	50,653	1,874,161	69,343,957
38	1,444	54,872	2,085,136	79,235,168
39	1,521	59,319	2,313,441	90,224,199
40	1,600	64,000	2,560,000	102,400,000
41	1,681	68,921	2,825,761	115,856,201
42	1,764	74,088	3,111,696	130,691,232
43	1,849	79,507	3,418,801	147,008,443
44	1,936	85,184	3,748,096	164,916,224
45	2,025	91,125	4,100,625	184,528,125
46	2,116	97,336	4,477,456	205,962,976
47	2,209	103,823	4,879,681	229,345,007
48	2,304	110,592	5,308,416	254,803,968
49	2,401	117,649	5,764,801	282,475,249
50	2,500	125,000	6,250,000	312,500,000

N	$N!$
1	1
2	2
3	6
4	24
5	120
6	720
7	5,040
8	40,320
9	362,880
10	3,628,800
11	39,916,800
12	479,001,600
13	6,227,020,800
14	87,178,291,200
15	1,307,674,368,000
16	20,922,789,888,000
17	355,687,428,096,000
18	6,402,373,705,728,000
19	121,645,100,408,832,000
20	2,432,902,008,176,640,000
21	51,090,942,171,709,440,000
22	1,124,000,727,777,607,680,000
23	25,852,016,738,884,976,640,000
24	620,448,401,733,239,439,360,000
25	15,511,210,043,330,985,984,000,000

Table C. The First 1000 Primes

2	31	73	127	179	233
3	37	79	131	181	239
5	41	83	137	191	241
7	43	89	139	193	251
11	47	97	149	197	257
13	53	101	151	199	263
17	59	103	157	211	269
19	61	107	163	223	271
23	67	109	167	227	277
29	71	113	173	229	281
283	353	419	467	547	607
293	359	421	479	557	613
307	367	431	487	563	617
311	373	433	491	569	619
313	379	439	499	571	631
317	383	443	503	577	641
331	389	449	509	587	643
337	397	457	521	593	647
347	401	461	523	599	653
349	409	463	541	601	659
661	739	811	877	947	1,019
673	743	821	881	953	1,021
677	751	823	883	967	1,031
683	757	827	887	971	1,033
691	761	829	907	977	1,039
701	769	839	911	983	1,049
709	773	853	919	991	1,051
719	787	857	929	997	1,061
727	797	859	937	1,009	1,063
733	809	863	941	1,013	1,069
1,087	1,153	1,229	1,297	1,381	1,453
1,091	1,163	1,231	1,301	1,399	1,459
1,093	1,171	1,237	1,303	1,409	1,471
1,097	1,181	1,249	1,307	1,423	1,481
1,103	1,187	1,259	1,319	1,427	1,483

1,109	1,193	1,277	1,321	1,429	1,487
1,117	1,201	1,279	1,327	1,433	1,489
1,123	1,213	1,283	1,361	1,439	1,493
1,129	1,217	1,289	1,367	1,447	1,499
1,151	1,223	1,291	1,373	1,451	1,511
1,523	1,597	1,663	1,741	1,823	1,901
1,531	1,601	1,667	1,747	1,831	1,907
1,543	1,607	1,669	1,753	1,847	1,913
1,549	1,609	1,693	1,759	1,861	1,931
1,553	1,613	1,697	1,777	1,867	1,933
1,559	1,619	1,699	1,783	1,871	1,949
1,567	1,621	1,709	1,787	1,873	1,951
1,571	1,627	1,721	1,789	1,877	1,973
1,579	1,637	1,723	1,801	1,879	1,979
1,583	1,657	1,733	1,811	1,889	1,987
1,993	2,063	2,131	2,221	2,293	2,371
1,997	2,069	2,137	2,237	2,297	2,377
1,999	2,081	2,141	2,239	2,309	2,381
2,003	2,083	2,143	2,243	2,311	2,383
2,011	2,087	2,153	2,251	2,333	2,389
2,017	2,089	2,161	2,267	2,339	2,393
2,027	2,099	2,179	2,269	2,341	2,399
2,029	2,111	2,203	2,273	2,347	2,411
2,039	2,113	2,207	2,281	2,351	2,417
2,053	2,129	2,213	2,287	2,357	2,423
2,437	2,539	2,621	2,689	2,749	2,833
2,441	2,543	2,633	2,693	2,753	2,837
2,447	2,549	2,647	2,699	2,767	2,843
2,459	2,551	2,657	2,707	2,777	2,851
2,467	2,557	2,659	2,711	2,789	2,857
2,473	2,579	2,663	2,713	2,791	2,861
2,477	2,591	2,671	2,719	2,797	2,879
2,503	2,593	2,677	2,729	2,801	2,887
2,521	2,609	2,683	2,731	2,803	2,897
2,531	2,617	2,687	2,741	2,819	2,903
2,909	3,001	3,083	3,187	3,259	3,343
2,917	3,011	3,089	3,191	3,271	3,347

2,927	3,019	3,109	3,203	3,299	3,359
2,939	3,023	3,119	3,209	3,301	3,361
2,953	3,037	3,121	3,217	3,307	3,371
2,957	3,041	3,137	3,221	3,313	3,373
2,963	3,049	3,163	3,229	3,319	3,389
2,969	3,061	3,167	3,251	3,323	3,391
2,971	3,067	3,169	3,253	3,329	3,407
2,999	3,079	3,181	3,257	3,331	3,413
3,433	3,517	3,581	3,659	3,733	3,823
3,449	3,527	3,583	3,671	3,739	3,833
3,457	3,529	3,593	3,673	3,761	3,847
3,461	3,533	3,607	3,677	3,767	3,851
3,463	3,539	3,613	3,691	3,769	3,853
3,467	3,541	3,617	3,697	3,779	3,863
3,469	3,547	3,623	3,701	3,793	3,877
3,491	3,557	3,631	3,709	3,797	3,881
3,499	3,559	3,637	3,719	3,803	3,889
3,511	3,571	3,643	3,727	3,821	3,907
3,911	4,001	4,073	4,153	4,241	4,327
3,917	4,003	4,079	4,157	4,243	4,337
3,919	4,007	4,091	4,159	4,253	4,339
3,923	4,013	4,093	4,177	4,259	4,349
3,929	4,019	4,099	4,201	4,261	4,357
3,931	4,021	4,111	4,211	4,271	4,363
3,943	4,027	4,127	4,217	4,273	4,373
3,947	4,049	4,129	4,219	4,283	4,391
3,967	4,051	4,133	4,229	4,289	4,397
3,989	4,057	4,139	4,231	4,297	4,409
4,421	4,507	4,591	4,663	4,759	4,861
4,423	4,513	4,597	4,673	4,783	4,871
4,441	4,517	4,603	4,679	4,787	4,877
4,447	4,519	4,621	4,691	4,789	4,889
4,451	4,523	4,637	4,703	4,793	4,903
4,457	4,547	4,639	4,721	4,799	4,909
4,463	4,549	4,643	4,723	4,801	4,919
4,481	4,561	4,649	4,729	4,813	4,931
4,483	4,567	4,651	4,733	4,817	4,933
4,493	4,583	4,657	4,751	4,831	4,937

4,943	5,009	5,099	5,189	5,281	5,393
4,951	5,011	5,101	5,197	5,297	5,399
4,957	5,021	5,107	5,209	5,303	5,407
4,967	5,023	5,113	5,227	5,309	5,413
4,969	5,039	5,119	5,231	5,323	5,417
4,973	5,051	5,147	5,233	5,333	5,419
4,987	5,059	5,153	5,237	5,347	5,431
4,993	5,077	5,167	5,261	5,351	5,437
4,999	5,081	5,171	5,273	5,381	5,441
5,003	5,087	5,179	5,279	5,387	5,443
5,449	5,527	5,641	5,701	5,801	5,861
5,471	5,531	5,647	5,711	5,807	5,867
5,477	5,557	5,651	5,717	5,813	5,869
5,479	5,563	5,653	5,737	5,821	5,879
5,483	5,569	5,657	5,741	5,827	5,881
5,501	5,573	5,659	5,743	5,839	5,897
5,503	5,581	5,669	5,749	5,843	5,903
5,507	5,591	5,683	5,779	5,849	5,923
5,519	5,623	5,689	5,783	5,851	5,927
5,521	5,639	5,693	5,791	5,857	5,939
5,953	6,067	6,143	6,229	6,311	6,373
5,981	6,073	6,151	6,247	6,317	6,379
5,987	6,079	6,163	6,257	6,323	6,389
6,007	6,089	6,173	6,263	6,329	6,397
6,011	6,091	6,197	6,269	6,337	6,421
6,029	6,101	6,199	6,271	6,343	6,427
6,037	6,113	6,203	6,277	6,353	6,449
6,043	6,121	6,211	6,287	6,359	6,451
6,047	6,131	6,217	6,299	6,361	6,469
6,053	6,133	6,221	6,301	6,367	6,473
6,481	6,577	6,679	6,763	6,841	6,947
6,491	6,581	6,689	6,779	6,857	6,949
6,521	6,599	6,691	6,781	6,863	6,959
6,529	6,607	6,701	6,791	6,869	6,961
6,547	6,619	6,703	6,793	6,871	6,967
6,551	6,637	6,709	6,803	6,883	6,971
6,553	6,653	6,719	6,823	6,899	6,977
6,563	6,659	6,733	6,827	6,907	6,983

6,569	6,661	6,737	6,829	6,911	6,991
6,571	6,673	6,761	6,833	6,917	6,997
7,001	7,109	7,211	7,307	7,417	7,507
7,013	7,121	7,213	7,309	7,433	7,517
7,019	7,127	7,219	7,321	7,451	7,523
7,027	7,129	7,229	7,331	7,457	7,529
7,039	7,151	7,237	7,333	7,459	7,537
7,043	7,159	7,243	7,349	7,477	7,541
7,057	7,177	7,247	7,351	7,481	7,547
7,069	7,187	7,253	7,369	7,487	7,549
7,079	7,193	7,283	7,393	7,489	7,559
7,103	7,207	7,297	7,411	7,499	7,561
7,573	7,621	7,687	7,741	7,823	7,879
7,577	7,639	7,691	7,753	7,829	7,883
7,583	7,643	7,699	7,757	7,841	7,901
7,589	7,649	7,703	7,759	7,853	7,907
7,591	7,669	7,717	7,789	7,867	7,919
7,603	7,673	7,723	7,793	7,873	
7,607	7,681	7,727	7,817	7,877	

Table D. Primitive Roots

N	Primitive Roots of N
2	1
3	2
4	3
5	2,3
6	5
7	3,5
9	2,5
10	3,7
11	2,6,7,8
13	2,6,7,11
14	3,5
17	3,5,6,7,10,11,12,14
18	5,11
19	2,3,10,13,14,15
22	7,13,17,19
23	5,7,10,11,14,15,17,19,20,21
25	2,3,8,12,13,17,22,23
26	7,11,15,19
27	2,5,11,14,20,23
29	2,3,8,10,11,14,15,18,19,21,26,27
31	3,11,12,13,17,21,22,24
34	3,5,7,11,23,27,29,31
37	2,5,13,15,17,18,19,20,22,24,32,35
38	3,13,15,21,29,33
41	6,7,11,12,13,15,17,19,22,24,26,28,29,30,34,35
43	3,5,12,18,19,20,26,28,29,30,33,34
46	5,7,11,15,17,19,21,33,37,43
47	5,10,11,13,15,19,20,22,23,26,29,30,31,33,35,38,39,40,41,43,44,45
49	3,5,10,12,17,24,26,33,38,40,45,47
50	3,13,17,23,27,33,37,47
53	2,3,5,8,12,14,18,19,20,21,22,26,27,31,32,33,34,35,39,41,45,48,50,51
54	5,11,23,29,41,47
58	3,11,15,19,21,27,31,37,39,43,47,55
59	2,6,8,10,11,13,14,18,23,24,30,31,32,33,34,37,38,39,40,42,43,44,47,50, 52,54,55,56
61	2,6,7,10,17,18,26,30,31,35,43,44,51,54,55,59

N	Primitive Roots of N
67	2,7,11,12,13,18,20,28,31,32,34,41,44,46,48,50,51,57,61,63
71	7,11,13,21,22,28,31,33,35,42,44,47,52,53,55,56,59,61,62,63,65,67,68,69
73	5,11,13,14,15,20,26,28,29,31,33,34,39,40,42,44,45,47,53,58,59,60,62,68
79	3,6,7,28,29,30,34,35,37,39,43,47,48,53,54,59,60,63,66,68,70,74,75,77
83	2,5,6,8,13,14,15,18,19,20,22,24,32,34,35,39,42,43,45,46,47,50,52,53,54, 55,56,57,58,60,62,66,67,71,72,73,74,76,79,80
89	3,6,7,13,14,15,19,23,24,26,27,28,29,30,31,33,35,38,41,43,46,48,51,54, 56,58,59,60,61,62,63,65,66,70,74,75,76,82,83,86
97	5,7,10,13,14,15,17,21,23,26,29,37,38,39,40,41,56,57,58,59,60,68,71,74, 76,80,82,83,84,87,90,92

g	Primes in [101, 797] With g as the Smallest Primitive Root
2	101,107,131,139,149,163,173,179,181,197,211,227,269,293,317,347, 349,373,379,389,419,421,443,461,467,491,509,523,541,547,557,563, 587,613,619,653,659,661,677,701,709,757,773,787,797
3	113,127,137,199,223,233,257,281,283,331,353,401,449,463,487,521, 569,571,593,607,617,631,641,691,739,751
5	103,157,167,193,263,277,307,383,433,503,577,647,673,683,727,743
6	109,151,229,251,271,367,733,761
7	239,241,359,431,499,599,601
10	313,337
11	643,719,769
13	457,479
15	439
17	311
19	191
21	409

Table E. Indices ($\text{ind}_g N$)

$p = 3, g = 2$

N	0	1	2
0		0	1

$p = 5, g = 2$

N	0	1	2	3	4
0		0	1	3	2

$p = 7, g = 3$

N	0	1	2	3	4	5	6
0		0	2	1	4	5	3

$p = 11, g = 2$

N	0	1	2	3	4	5	6	7	8	9
0		0	1	8	2	4	9	7	3	6
1	5									

$p = 13, g = 2$

N	0	1	2	3	4	5	6	7	8	9
0		0	1	4	2	9	5	11	3	8
1	10	7	6							

$p = 17, g = 3$

N	0	1	2	3	4	5	6	7	8	9
0		0	14	1	12	5	15	11	10	2
1	3	7	13	4	9	6	8			

$p = 19, g = 2$

N	0	1	2	3	4	5	6	7	8	9
0		0	1	13	2	16	14	6	3	8
1	17	12	15	5	7	11	4	10	9	

$p = 23, g = 5$

N	0	1	2	3	4	5	6	7	8	9
0		0	2	16	4	1	18	19	6	10
1	3	9	20	14	21	17	8	7	12	15
2	5	13	11							

$p = 29, g = 2$

N	0	1	2	3	4	5	6	7	8	9
0		0	1	5	2	22	6	12	3	10
1	23	25	7	18	13	27	4	21	11	9
2	24	17	26	20	8	16	19	15	14	

$p = 31, g = 3$

N	0	1	2	3	4	5	6	7	8	9
0		0	24	1	18	20	25	28	12	2
1	14	23	19	11	22	21	6	7	26	4
2	8	29	17	27	13	10	5	3	16	9
3	15									

$p = 37, g = 2$

N	0	1	2	3	4	5	6	7	8	9
0		0	1	26	2	23	27	32	3	16
1	24	30	28	11	33	13	4	7	17	35
2	25	22	31	15	29	10	12	6	34	21
3	14	9	5	20	8	19	18			

$p = 41, g = 6$

N	0	1	2	3	4	5	6	7	8	9
0		0	26	15	12	22	1	39	38	30
1	8	3	27	31	25	37	24	33	16	9
2	34	14	29	36	13	4	17	5	11	7
3	23	28	10	18	19	21	2	32	35	6
4	20									

$p = 43, g = 3$

N	0	1	2	3	4	5	6	7	8	9
0		0	27	1	12	25	28	35	39	2
1	10	30	13	32	20	26	24	38	29	19
2	37	36	15	16	40	8	17	3	5	41
3	11	34	9	31	23	18	14	7	4	33
4	22	6	21							

$p = 47, g = 5$

N	0	1	2	3	4	5	6	7	8	9
0		0	18	20	36	1	38	32	8	40
1	19	7	10	11	4	21	26	16	12	45
2	37	6	25	5	28	2	29	14	22	35
3	39	3	44	27	34	33	30	42	17	31
4	9	15	24	13	43	41	23			

$p = 53, g = 2$

N	0	1	2	3	4	5	6	7	8	9
0		0	1	17	2	47	18	14	3	34
1	48	6	19	24	15	12	4	10	35	37
2	49	31	7	39	20	42	25	51	16	46
3	13	33	5	23	11	9	36	30	38	41
4	50	45	32	22	8	29	40	44	21	28
5	43	27	26							

$p = 59, g = 2$

N	0	1	2	3	4	5	6	7	8	9
0		0	1	50	2	6	51	18	3	42
1	7	25	52	45	19	56	4	40	43	38
2	8	10	26	15	53	12	46	34	20	28
3	57	49	5	17	41	24	44	55	39	37
4	9	14	11	33	27	48	16	23	54	36
5	13	32	47	22	35	31	21	30	29	

$p = 61, g = 2$

N	0	1	2	3	4	5	6	7	8	9
0		0	1	6	2	22	7	49	3	12
1	23	15	8	40	50	28	4	47	13	26
2	24	55	16	57	9	44	41	18	51	35
3	29	59	5	21	48	11	14	39	27	46
4	25	54	56	43	17	34	58	20	10	38
5	45	53	42	33	19	37	52	32	36	31
6	30									

$p = 67, g = 2$

N	0	1	2	3	4	5	6	7	8	9
0		0	1	39	2	15	40	23	3	12
1	16	59	41	19	24	54	4	64	13	10
2	17	62	60	28	42	30	20	51	25	44
3	55	47	5	32	65	38	14	22	11	58
4	18	53	63	9	61	27	29	50	43	46
5	31	37	21	57	52	8	26	49	45	36
6	56	7	48	35	6	34	33			

$p = 71, g = 7$

N	0	1	2	3	4	5	6	7	8	9
0		0	6	26	12	28	32	1	18	52
1	34	31	38	39	7	54	24	49	58	16
2	40	27	37	15	44	56	45	8	13	68
3	60	11	30	57	55	29	64	20	22	65
4	46	25	33	48	43	10	21	9	50	2
5	62	5	51	23	14	59	19	42	4	3
6	66	69	17	53	36	67	63	47	61	41
7	35									

Table F. Quadratic, Cubic, and Quartic Residues

N	$\mathbf{A}_N^{(2)}$	$\mathbf{A}_N^{(3)}$	$\mathbf{A}_N^{(4)}$
2	1	1	1
3	1	1,2	1
4	1	1,3	1
5	1,4	1,2,3,4	1
6	1	1,5	1
7	1,2,4	1,6	1,2,4
8	1	1,3,5,7	1
9	1,4,7	1,8	1,4,7
10	1,9	1,3,7,9	1
11	1,3,4,5,9	1,2,3,4,5,6,7,8,9,10	1,3,4,5,9
12	1	1,5,7,11	1
13	1,3,4,9,10,12	1,5,8,12	1,3,9
14	1,9,11	1,13	1,9,11
15	1,4	1,2,4,7,8,11,13,14	1
16	1,9	1,3,5,7,9,11,13,15	1
17	1,2,4,8,9,13,15,16	1,2,3, . . . ,16	1,4,13,16
18	1,7,13	1,17	1,7,13
19	1,4,5,6,7,9,11,16,17	1,7,8,11,12,18	1,4,5,6,7,9,11,16,17
20	1,9	1,3,7,9,11,13,17,19	1
21	1,4,16	1,8,13,20	1,4,16
22	1,3,5,9,15	1,3,5,7,9,13,15,17, 19,21	1,3,5,9,15
23	1,2,3,4,6,8,9,12,13, 16,18	1,2,3, . . . ,22	1,2,3,4,6,8,9,12,13, 16,18
24	1	1,5,7,11,13,17,19,23	1
25	1,4,6,9,11,14,16,19, 21,24	1,2,3,4,6,7,8,9,11, 12,13,14,16,17,18, 19,21,22,23,24	1,6,11,16,21
26	1,3,9,17,23,25	1,5,21,25	1,3,9
27	1,4,7,10,13,16,19, 22,25	1,8,10,17,19,26	1,4,7,10,13,16,19, 22,25
28	1,9,25	1,13,15,27	1,9,25
29	1,4,5,6,7,9,13,16,20, 22,23,24,25,28	1,2,3, . . . ,28	1,7,16,20,23,24,25
30	1,19	1,7,11,13,17,19,23, 29	1

N	$\mathbf{A}_N^{(2)}$	$\mathbf{A}_N^{(3)}$	$\mathbf{A}_N^{(4)}$
31	1,2,4,5,7,8,9,10,14, 16,18,19,20,25,28	1,2,4,8,15,16,23,27, 29,30	1,2,4,5,7,8,9,10,14, 16,18,19,20,25,28
32	1,9,17,25	1,3,5,7,9,11,13,15, 17,19,21,23,25,27, 29,31	1,17
33	1,4,16,25,31	1,2,4,5,7,8,10,13,14, 16,17,19,20,23,25, 26,28,29,31,32	1,4,16,25,31
34	1,9,13,15,19,21,25, 33	1,3,5,7,9,11,13,15, 19,21,23,25,27,29, 31,33	1,13,21,33
35	1,4,9,11,16,29	1,6,8,13,22,27,29,34	1,11,16
36	1,13,25	1,17,19,35	1,13,25
37	1,3,4,7,9,10,11,12, 16,21,25,26,27,28, 30,33,34,36	1,6,8,10,11,14,23, 26,27,29,31,36	1,7,9,10,12,16,26, 33,34
38	1,5,7,9,11,17,23,25, 35	1,7,11,27,31,37	1,5,7,9,11,17,23,25, 35
39	1,4,10,16,22,25	1,5,8,14,25,31,34,38	1,16,22
40	1,9	1,3,7,9,11,13,17,19, 21,23,27,29,31,33, 37,39	1
41	1,2,4,5,8,9,10,16,18, 20,21,23,25,31,32, 33,36,37,39,40	1,2,3, . . . ,40	1,4,10,16,18,23,25, 31,37,40
42	1,25,37	1,13,29,41	1,25,37
43	1,4,6,9,10,11,13,14, 15,16,17,21,23,24, 25,31,35,36,38,40, 41	1,2,4,8,11,16,21,22, 27,32,35,39,41,42	1,4,6,9,10,11,13,14, 15,16,17,21,23,24, 25,31,35,36,38,40, 41
44	1,5,9,25,37	1,3,5,7,9,13,15,17, 19,21,23,25,27,29, 31,35,37,39,41,43	1,5,9,25,37
45	1,4,16,19,31,34	1,8,17,19,26,28,37, 44	1,16,31
46	1,3,9,13,25,27,29, 31,35,39,41	1,3,5,7,9,11,13,15, 17,19,21,25,27,29, 31,33,35,37,39,41, 43,45	1,3,9,13,25,27,29, 31,35,39,41

N	$\mathbf{A}_N^{(2)}$	$\mathbf{A}_N^{(3)}$	$\mathbf{A}_N^{(4)}$
47	1,2,3,4,6,7,8,9,12, 14,16,17,18,21,24, 25,27,28,32,34,36, 37,42	1,2,3, . . . ,46	1,2,3,4,6,7,8,9,12, 14,16,17,18,21,24, 25,27,28,32,34,36, 37,42
48	1,25	1,5,7,11,13,17,19, 23,25,29,31,35,37, 41,43,47	1
49	1,2,4,8,9,11,15,16, 18,22,23,25,29,30, 32,36,37,39,43,44, 46	1,6,8,13,15,20,22, 27,29,34,36,41,43, 48	1,2,4,8,9,11,15,16, 18,22,23,25,29,30, 32,36,37,39,43,44, 46
50	1,9,11,19,21,29,31, 39,41,49	1,3,7,9,11,13,17,19, 21,23,27,29,31,33, 37,39,41,43,47,49	1,11,21,31,41

Table G. Values of Some Multiplicative Functions

n	$\tau(n)$	$\sigma(n)$	$\phi(n)$	$\lambda(n)$	$\mu(n)$
1	1	1	1	1	1
2	2	3	1	−1	−1
3	2	4	2	−1	−1
4	3	7	2	1	0
5	2	6	4	−1	−1
6	4	12	2	1	1
7	2	8	6	−1	−1
8	4	15	4	−1	0
9	3	13	6	1	0
10	4	18	4	1	1
11	2	12	10	−1	−1
12	6	28	4	−1	0
13	2	14	12	−1	−1
14	4	24	6	1	1
15	4	24	8	1	1
16	5	31	8	1	0
17	2	18	16	−1	−1
18	6	39	6	−1	0
19	2	20	18	−1	−1
20	6	42	8	−1	0
21	4	32	12	1	1
22	4	36	10	1	1
23	2	24	22	−1	−1
24	8	60	8	1	0
25	3	31	20	1	0
26	4	42	12	1	1
27	4	40	18	−1	0
28	6	56	12	−1	0
29	2	30	28	−1	−1
30	8	72	8	−1	−1
31	2	32	30	−1	−1
32	6	63	16	−1	0
33	4	48	20	1	1
34	4	54	16	1	1
35	4	48	24	1	1
36	9	91	12	1	0

n	$\tau(n)$	$\sigma(n)$	$\phi(n)$	$\lambda(n)$	$\mu(n)$
37	2	38	36	−1	−1
38	4	60	18	1	1
39	4	56	24	1	1
40	8	90	16	1	0
41	2	42	40	−1	−1
42	8	96	12	−1	−1
43	2	44	42	−1	−1
44	6	84	20	−1	0
45	6	78	24	−1	0
46	4	72	22	1	1
47	2	48	46	−1	−1
48	10	124	16	−1	0
49	3	57	42	1	0
50	6	93	20	−1	0
51	4	72	32	1	1
52	6	98	24	−1	0
53	2	54	52	−1	−1
54	8	120	18	1	0
55	4	72	40	1	1
56	8	120	24	1	0
57	4	80	36	1	1
58	4	90	28	1	1
59	2	60	58	−1	−1
60	12	168	16	1	0
61	2	62	60	−1	−1
62	4	96	30	1	1
63	6	104	36	−1	0
64	7	127	32	1	0
65	4	84	48	1	1
66	8	144	20	−1	−1
67	2	68	66	−1	−1
68	6	126	32	−1	0
69	4	96	44	1	1
70	8	144	24	−1	−1
71	2	72	70	−1	−1
72	12	195	24	−1	0
73	2	74	72	−1	−1
74	4	114	36	1	1
75	6	124	40	−1	0

Table H. Numbers of Unrestricted Partitions*

N	p(N)	N	p(N)
1	1	38	26,015
2	2	39	31,185
3	3	40	37,338
4	5	41	44,583
5	7	42	53,174
6	11	43	63,261
7	15	44	75,175
8	22	45	89,134
9	30	46	105,558
10	42	47	124,754
11	56	48	147,273
12	77	49	173,525
13	101	50	204,226
14	135	51	239,943
15	176	52	281,589
16	231	53	329,931
17	297	54	386,155
18	385	55	451,276
19	490	56	526,823
20	627	57	614,154
21	792	58	715,220
22	1,002	59	831,820
23	1,255	60	966,467
24	1,575	61	1,121,505
25	1,958	62	1,300,156
26	2,436	63	1,505,499
27	3,010	64	1,741,630
28	3,718	65	2,012,558
29	4,565	66	2,323,520
30	5,604	67	2,679,689
31	6,842	68	3,087,735
32	8,349	69	3,554,345
33	10,143	70	4,087,968
34	12,310	71	4,697,205
35	14,883	72	5,392,783
36	17,977	73	6,185,689
37	21,637	74	7,089,500

N	p(N)	N	p(N)
75	8,118,264	101	214,481,126
76	9,289,091	102	241,265,379
77	10,619,863	103	271,248,950
78	12,132,164	104	304,801,365
79	13,848,650	105	342,325,709
80	15,796,476	106	384,276,336
81	18,004,327	107	431,149,389
82	20,506,255	108	483,502,844
83	23,338,469	109	541,946,240
84	26,543,660	110	607,163,746
85	30,167,357	111	679,903,203
86	34,262,962	112	761,002,156
87	38,887,673	113	851,376,628
88	44,108,109	114	952,050,665
89	49,995,925	115	1,064,144,451
90	56,634,173	116	1,188,908,248
91	64,112,359	117	1,327,710,076
92	72,533,807	118	1,482,074,143
93	82,010,177	119	1,653,668,665
94	92,669,720	120	1,844,349,560
95	104,651,419	121	2,056,148,051
96	118,114,304	122	2,291,320,912
97	133,230,930	123	2,552,338,241
98	150,198,136	124	2,841,940,500
99	169,229,875	125	3,163,127,352
100	190,569,292		

*Computed from the recurrence relation given in Theorem 10.9.

Table I. Special Numbers

1. Lucas and Companion Lucas Sequences

$P = 1, Q = -1$

n	$U_n(P, Q)$	$V_n(P, Q)$	n	$U_n(P, Q)$	$V_n(P, Q)$
0	0	2	13	233	521
1	1	1	14	377	843
2	1	3	15	610	1,364
3	2	4	16	987	2,207
4	3	7	17	1,597	3,571
5	5	11	18	2,584	5,778
6	8	18	19	4,181	9,349
7	13	29	20	6,765	15,127
8	21	47	21	10,946	24,476
9	34	76	22	17,711	39,603
10	55	123	23	28,657	64,079
11	89	199	24	46,368	103,682
12	144	322	25	75,025	167,761

$P = 2, Q = -1$

n	$U_n(P, Q)$	$V_n(P, Q)$	n	$U_n(P, Q)$	$V_n(P, Q)$
0	0	2	13	33,461	94,642
1	1	2	14	80,782	228,486
2	2	6	15	195,025	551,614
3	5	14	16	470,832	1,331,714
4	12	34	17	1,136,689	3,215,042
5	29	82	18	2,744,210	7,761,798
6	70	198	19	6,625,109	18,738,638
7	169	478	20	15,994,428	45,239,074
8	408	1,154	21	38,613,965	109,216,786
9	985	2,786	22	93,222,358	263,672,646
10	2,378	6,726	23	225,058,681	636,562,078
11	5,741	16,238	24	543,339,720	1,536,796,802
12	13,860	39,202	25	1,311,738,121	3,710,155,682

$P = 5, Q = 6$

n	$U_n(P, Q)$	$V_n(P, Q)$	n	$U_n(P, Q)$	$V_n(P, Q)$
0	0	2	13	1,586,131	1,602,515
1	1	5	14	4,766,585	4,799,353
2	5	13	15	14,316,139	14,381,675
3	19	35	16	42,981,185	43,112,257
4	65	97	17	129,009,091	129,271,235
5	211	275	18	387,158,345	387,682,633
6	665	793	19	1,161,737,179	1,162,785,755
7	2,059	2,315	20	3,485,735,825	3,487,832,977
8	6,305	6,817	21	10,458,256,051	10,462,450,355
9	19,171	20,195	22	31,376,865,305	31,385,253,913
10	58,025	60,073	23	94,134,790,219	94,151,567,435
11	175,099	179,195	24	282,412,759,265	282,446,313,697
12	527,345	535,537	25	847,255,055,011	847,322,163,875

2. Stirling's Numbers of the First Kind, $S(n, k)$

$n \backslash k$	0	1	2	3	4	5	6	7	8	9	10
0	1										
1	0	1									
2	0	1	1								
3	0	2	3	1							
4	0	6	11	6	1						
5	0	24	50	35	10	1					
6	0	120	274	225	85	15	1				
7	0	720	1,764	1,624	735	175	21	1			
8	0	5,040	13,068	13,132	6,769	1,960	322	28	1		
9	0	40,320	109,584	118,124	67,284	22,449	4,536	546	36	1	
10	0	362,880	1,026,576	1,172,700	723,680	269,325	63,273	9,450	870	45	1

3. Stirling's Numbers of the Second Kind, s(*n*, *k*)

n \ *k*	0	1	2	3	4	5	6	7	8	9	10
0	1										
1	0	1									
2	0	1	1								
3	0	1	3	1							
4	0	1	7	6	1						
5	0	1	15	25	10	1					
6	0	1	31	90	65	15	1				
7	0	1	63	301	350	140	21	1			
8	0	1	127	966	1,701	1,050	266	28	1		
9	0	1	255	3,025	7,770	6,951	2,646	462	36	1	
10	0	1	511	9,330	34,105	42,525	22,827	5,880	750	45	1

4. Eulerian Numbers, E(*n*, *k*)

n \ *k*	0	1	2	3	4	5	6	7	8	9	10
0	1										
1	1	0									
2	1	1	0								
3	1	4	1	0							
4	1	11	11	1	0						
5	1	26	66	26	1	0					
6	1	57	302	302	57	1	0				
7	1	120	1,191	2,416	1,191	120	1	0			
8	1	247	4,293	15,619	15,619	4,293	247	1	0		
9	1	502	14,608	88,234	156,190	88,234	14,608	502	1	0	
10	1	1,013	47,840	455,192	1,310,354	1,310,354	455,192	47,840	1,013	1	0

5. Bernoulli Numbers, $B_n = N_n/D_n$

n	N_n	D_n	n	N_n	D_n
2	1	6	18	43,867	798
4	−1	30	20	−174,611	330
6	1	42	22	854,513	138
8	−1	30	24	−236,364,091	2,730
10	5	66	26	8,553,103	6
12	−691	2,730	28	−23,749,461,029	870
14	7	6	30	8,615,841,276,005	14,322
16	−3,617	510	32	−7,709,321,041,217	510

6. Carmichael Numbers*

561	8,911	52,633	126,217	294,409	449,065	658,801	997,633
1,105	10,585	62,745	162,401	314,821	488,881	670,033	1,024,651
1,729	15,841	63,973	172,081	334,153	512,461	748,657	1,033,669
2,465	29,341	75,361	188,461	340,561	530,881	825,265	1,050,985
2,821	41,041	101,101	252,601	399,001	552,721	838,201	1,082,809
6,601	46,657	115,921	278,545	410,041	656,601	852,841	1,152,271

*Communicated to the authors by Professor R.G.E. Pinch (University of Cambridge).

7. Mersenne Primes, M_p

p	M_p	p	p	p
2	3	89	4,253	86,243
3	7	107	4,423	110,503
5	31	127	9,689	132,049
7	127	521	9,941	216,091
13	8,191	607	11,213	756,839
17	131,071	1,279	19,937	859,432
19	524,287	2,203	21,703	1,257,787
31	2,147,483,647	2,281	23,209	1,398,269
61	2,305,843,009,213,693,953	3,217	44,497	2,976,221

8. Euler Numbers, E_n

n	E_n	n	E_n
0	1	14	−199,360,981
2	−1	16	19,391,512,145
4	5	18	−2,404,879,675,441
6	−61	20	370,371,188,237,525
8	1,385	22	−69,348,874,393,137,901
10	−50,521	24	15,514,534,163,557,086,905
12	2,702,765	26	−4,087,072,509,293,123,892,361

Table J. Continued Fractions

$\sqrt{n}$	$[d_0; \overline{d_1, d_2, \ldots}]$	$\sqrt{n}$	$[d_0; \overline{d_1, d_2, \ldots}]$
$\sqrt{2}$	$[1; \overline{2}]$	$\sqrt{28}$	$[5; \overline{3, 2, 3, 10}]$
$\sqrt{3}$	$[1; \overline{1, 2}]$	$\sqrt{29}$	$[5; \overline{2, 1, 1, 2, 10}]$
$\sqrt{5}$	$[2; \overline{4}]$	$\sqrt{30}$	$[5; \overline{2, 10}]$
$\sqrt{6}$	$[2; \overline{2, 4}]$	$\sqrt{31}$	$[5; \overline{1, 1, 3, 5, 3, 1, 1, 10}]$
$\sqrt{7}$	$[2; \overline{1, 1, 1, 4}]$	$\sqrt{32}$	$[5; \overline{1, 1, 1, 10}]$
$\sqrt{8}$	$[2; \overline{1, 4}]$	$\sqrt{33}$	$[5; \overline{1, 2, 1, 10}]$
$\sqrt{10}$	$[3; \overline{6}]$	$\sqrt{34}$	$[5; \overline{1, 4, 1, 10}]$
$\sqrt{11}$	$[3; \overline{3, 6}]$	$\sqrt{35}$	$[5; \overline{1, 10}]$
$\sqrt{12}$	$[3; \overline{2, 6}]$	$\sqrt{37}$	$[6; \overline{12}]$
$\sqrt{13}$	$[3; \overline{1, 1, 1, 1, 6}]$	$\sqrt{38}$	$[6; \overline{6, 12}]$
$\sqrt{14}$	$[3; \overline{1, 2, 1, 6}]$	$\sqrt{39}$	$[6; \overline{4, 12}]$
$\sqrt{15}$	$[3; \overline{1, 6}]$	$\sqrt{40}$	$[6; \overline{3, 12}]$
$\sqrt{17}$	$[4; \overline{8}]$	$\sqrt{41}$	$[6; \overline{2, 2, 12}]$
$\sqrt{18}$	$[4; \overline{4, 8}]$	$\sqrt{42}$	$[6; \overline{2, 12}]$
$\sqrt{19}$	$[4; \overline{2, 1, 3, 1, 2, 8}]$	$\sqrt{43}$	$[6; \overline{1, 1, 3, 1, 5, 1, 3, 1, 1, 12}]$
$\sqrt{20}$	$[4; \overline{2, 8}]$	$\sqrt{44}$	$[6; \overline{1, 1, 1, 2, 1, 1, 1, 12}]$
$\sqrt{21}$	$[4; \overline{1, 1, 2, 1, 1, 8}]$	$\sqrt{45}$	$[6; \overline{1, 2, 2, 2, 1, 12}]$
$\sqrt{22}$	$[4; \overline{1, 2, 4, 2, 1, 8}]$	$\sqrt{46}$	$[6; \overline{1, 3, 1, 1, 2, 6, 2, 1, 1, 3, 1, 12}]$
$\sqrt{23}$	$[4; \overline{1, 3, 1, 8}]$	$\sqrt{47}$	$[6; \overline{1, 5, 1, 12}]$
$\sqrt{24}$	$[4; \overline{1, 8}]$	$\sqrt{48}$	$[6; \overline{1, 12}]$
$\sqrt{26}$	$[5; \overline{10}]$	$\sqrt{50}$	$[7; \overline{14}]$
$\sqrt{27}$	$[5; \overline{5, 10}]$	$\sqrt{51}$	$[7; \overline{7, 14}]$

$[0; \overline{d_1}] = x$	x	$[0; \overline{d_1}] = x$	x
$[0; \overline{1}]$	$(\sqrt{5} - 1)/2$	$[0; \overline{11}]$	$(\sqrt{125} - 11)/2$
$[0; \overline{2}]$	$\sqrt{2} - 1$	$[0; \overline{12}]$	$\sqrt{37} - 6$
$[0; \overline{3}]$	$(\sqrt{13} - 1)/2$	$[0; \overline{13}]$	$(\sqrt{173} - 13)/2$
$[0; \overline{4}]$	$\sqrt{5} - 2$	$[0; \overline{14}]$	$\sqrt{50} - 7$
$[0; \overline{5}]$	$(\sqrt{29} - 5)/2$	$[0; \overline{15}]$	$(\sqrt{229} - 15)/2$
$[0; \overline{6}]$	$\sqrt{10} - 3$	$[0; \overline{16}]$	$\sqrt{65} - 8$
$[0; \overline{7}]$	$(\sqrt{53} - 7)/2$	$[0; \overline{17}]$	$(\sqrt{293} - 17)/2$
$[0; \overline{8}]$	$\sqrt{17} - 4$	$[0; \overline{18}]$	$\sqrt{82} - 9$
$[0; \overline{9}]$	$(\sqrt{85} - 9)/2$	$[0; \overline{19}]$	$(\sqrt{365} - 19)/2$
$[0; \overline{10}]$	$\sqrt{26} - 5$	$[0; \overline{20}]$	$\sqrt{101} - 10$

Appendix III
Sample Final Examinations

* * * FINAL EXAMINATION 1 * * *
(Chapters 1–8)
(closed book; 120 min.)

1. (24 pts.) Provide brief but accurate definitions or statements of the following terms:

(a) Fermat number.
(b) Multiplicative function.
(c) Algebraic number.
(d) Euler's Theorem.
(e) Cubic residue.
(f) Prime Number Theorem.

2. (27 pts.) Mark each statement as true (T) or as false (F):

(a) ________ Every integer greater than 8 can be written as the sum of three positive squares.
(b) ________ If $ax \equiv ay \pmod{m}$, then $x \equiv y \pmod{m}$.
(c) ________ If A_1, A_2 are two quartic residues of the modulus m, then so is $A_1 A_2$.
(d) ________ There is an integer in $\mathbf{Z}_{24}$ whose order modulo 24 is 12.
(e) ________ If $x \mid (a + b)$, where $x, a, b \in \mathbf{N}$, then either $x \mid a$ or $x \mid b$.
(f) ________ The number 3185 can be written as the sum of two squares.
(g) ________ If $p > 2$ is a prime, then so is $p^2 + 4$ since the polynomial $x^2 + 4$ cannot be factored over $\mathbf{Z}$.
(h) ________ If g is a primitive root of the prime p, then so is $g^3 \pmod{p}$.
(i) ________ The value of the Jacobi symbol $(14 \mid 75)$ is -1.

3. (18 pts.) Find all incongruent roots of the congruence $4x^2 - x \equiv 7 \pmod{17}$.

4. (20 pts.) Let $\mathbf{S}'$ be a reduced residue system of the modulus n. Prove that $\mathbf{S}'$ is a group.

5. (20 pts.) Solve the system of linear congruences:

$$\begin{cases} 2x \equiv 3 \pmod{11} \\ -3x \equiv 1 \pmod{16}. \end{cases}$$

6. (15 pts.) Suppose that N is a perfect number (odd or even). Show that

$$\sum_{d|N} \frac{1}{d} = 2.$$

7. (20 pts.) Let p, $q = 2p + 1$ be primes. Prove that q divides one, and only one, of M_p, $M_p + 2$, where M_p denotes the Mersenne number.

8. (15 pts.) Show that the congruence $3x^4 + y^2 \equiv 2 \pmod 8$ has no solution (x, y).

9. (18 pts.) Let A be a quadratic residue of the prime p; show that it is not a primitive root of p.

10. (23 pts.) One of the three sides of a right triangle with integral dimensions has length 175. Find two possibilities for the lengths of the other two sides. Support your results with adequate discussion.

* * * FINAL EXAMINATION 2 * * *
(Chapters 1–8)
(closed book; 120 min.)

1. (24 pts.) Provide brief but accurate definitions or statements of the following terms:

(a) Bertrand's Postulate.
(b) Reduced residue system.
(c) Convergent of a continued fraction.
(d) Möbius function.
(e) Waring's Problem.
(f) Well-Ordering Property.

2. (27 pts.) Mark each statement as true (T) or false (F):

(a) ________ Let N_1, N_2 be two positive integers; if their least common multiple is four times their greatest common divisor, then N_1N_2 is a perfect square.

(b) ________ If $x^3 \equiv y^3 \pmod{n}$, then $x^2 \equiv y^2 \pmod{n}$.
(c) ________ Let p be a prime. Then p divides $1 + (p - 1)!$.
(d) ________ Let p be an odd prime. Then

$$\sum_{k=1}^{p-1} k^{p-1} \equiv -1 \pmod{p}.$$

(e) ________ 100 is an abundant number.
(f) ________ The primitive roots of an integer N form a cyclic multiplicative group.
(g) ________ Let $\lambda(n)$, $\mu(n)$ be the Liouville and Möbius functions, respectively. Then $\lambda(n)\mu(n)$ can never be -1.
(h) ________ The prime 61 has exactly 15 quartic residues.
(i) ________ Let $n > 4$ be an arbitrary modulus. Then $x^4 - 1 \equiv 0 \pmod{n}$ can have no more than four incongruent roots.

3. (21 pts.) Compute the following Legendre and Jacobi symbols:

(a) $\left(\dfrac{231}{233}\right)$. (b) $(2 \mid 43)$. (c) $(58 \mid 329)$.

4. (20 pts.) Let $p > 5$ be a prime. Prove that either $10 \mid (p^2 + 1)$ or $10 \mid (p^2 - 1)$.

5. (18 pts.) Show that 561 is a pseudoprime to the base 2.

6. (20 pts.) Let p_1, p_2 be distinct primes and $P = p_1 p_2$. Prove that

$$p_1^{p_2 - 1} + p_2^{p_1 - 1} \equiv 1 \pmod{P}.$$

7. (20 pts.) Let $\sigma(n)$ be the sum-of-divisors function. Describe a general set of circumstances in which for $n, a > 1$ one has $\sigma(n + a) = \sigma(n) + \sigma(a)$.

8. (15 pts.) Prove that the prime p can be represented as the sum of two squares iff the congruence $x^2 + 1 \equiv 0 \pmod{p}$ has a root.

9. (22 pts.) Find the smallest primitive root g of the prime 17. Then construct for this prime the table of indices to the base g. Use the table to find all the incongruent solutions to $x^6 \equiv 13 \pmod{17}$.

10. (13 pts.) A certain integer between 1 and 1150 leaves a remainder of 5 or 10 when divided by 31 or 37, respectively. Find the integer.

* * * FINAL EXAMINATION 3 * * *
(Chapters 1–7, 9)
(closed book; 120 min.)

1. (24 pts.) Provide brief but accurate definitions or statements of the following terms:

(a) Carmichael number.
(b) Algebraic number field.
(c) Fundamental Theorem of Arithmetic.
(d) Group.
(e) Legendre symbol.
(f) Gauss's Law of Quadratic Reciprocity.

2. (27 pts.) Mark each statement as true (T) or as false (F):

(a) ______ For the congruence $2x^2 \equiv 3 \pmod{13}$, the pertinent Legendre symbol to be considered is

$$\left(\frac{8}{13}\right).$$

(b) ______ In $\mathbf{Z}_{18}$, 10 is a divisor of zero.
(c) ______ Let p be an odd prime; then

$$\sum_{k=1}^{p-1} \left(\frac{k}{p}\right) = 1.$$

(d) ______ There are only a finite number of primes that end in 11.
(e) ______ 15 is a pseudoprime to the base 5.
(f) ______ If * indicates Dirichlet convolution, then $(\tau * \sigma)(4) = 16$, where τ is the number-of-divisors function and σ is the sum-of-divisors function.
(g) ______ Let $\mathbf{F} = \mathbf{Q}$ and $\theta = \sqrt[3]{2}$; then $[\mathbf{F}(\theta)\colon \mathbf{F}] = 2$.
(h) ______ Let $n > 4$ be an arbitrary modulus. Then $x^2 - 1 \equiv 0 \pmod{n}$ can have no more than two incongruent roots.
(i) ______ The Jacobi symbol $(2 \mid 45)$ has the value -1.

3. (18 pts.) Consider the sequence $\{F_n\}_{n=1}^{\infty}$ of Fibonacci numbers, defined by $F_1 = F_2 = 1$, $F_n = F_{n-1} + F_{n-2}$ for $n > 2$. Prove that any two consecutive Fibonacci numbers are relatively prime.

4. (15 pts.) Compute the least common multiple of 2431, 2873.

5. (20 pts.) Find the minimal polynomial over $\mathbf{Q}$ for the algebraic number $\theta = 1 + i\sqrt[3]{2}$, and justify the minimality of the polynomial.

6. (22 pts.) Find all the units in $\mathbf{Q}(\sqrt{-3})$. Support your answer with adequate discussion.

7. (20 pts.) Let $T_{163}^{(3)}$ denote the sum of the cubic residues of the prime $p = 163$. Prove that $163 \mid T_{163}^{(3)}$.

8. (16 pts.) One has $e = 2.7182818284\ldots$. Find the 5th-order convergent to the simple continued fraction representation of e.

9. (20 pts.) Let $N > 6$ be an even perfect number. Prove that $N \equiv 1 \pmod 9$.

10. (18 pts.) Solve the system of linear congruences

$$\begin{cases} 5x \equiv 1 \pmod{12} \\ 2x \equiv 3 \pmod{25}. \end{cases}$$

Appendix IV
Hints and Answers to Selected Problems

Chapter 1

1.3. (a) If $\sqrt{x}$ is an irrational number, then x is a rational number.

1.8. 2^n.

1.9. Use facts about congruent triangles.

1.13. Use facts about parallel and perpendicular lines.

1.21. (a) $T_5(x) = 16x^5 - 20x^3 + 5x$.

1.23. (d) Any nonzero real number has a multiplicative inverse.

1.31. Use Axiom III (see Problem 1.29) and the fact that $x < 0 \rightarrow (-x)$ is positive.

1.34. Use the Archimedean Property in Problem 1.33.

1.42. Use Theorem 1.3 twice.

1.43. Write N as the sum of two terms, each of which is divisible by 3.

1.45. Consider $3^n = 10^m$; express m in terms of n.

1.49. $\pi(10{,}000) = 1229$.

1.53. Use Theorem 1.4 (b).

1.60. No.

1.71. Diverges; converges.

1.76. Consider Theorem 1.10.

1.78. Factoring.

1.82. (b) B must be odd and A must be even. Use proof by contrapositive: Begin by supposing $5 \nmid A$.

1.85. $\pi(13) = 6$.

1.86. $\mathrm{li}(10{,}000) \approx 1247$.

1.97. One approach is to use the conjecture in Problem 1.93.

Chapter 2

2.2. (c) $2^6 \cdot 3 \cdot 73$.

2.3. False.

2.9. Begin with the idea that a single power of p will certainly divide the product of any p consecutive positive integers.

2.12. (a) Let r denote the number of digits in N; then $N \geq 10^{r-1}$.
(d) Three.

2.19. Proof by contradiction.

2.25. Consider p, the largest prime not exceeding n, and use Theorem 2.3.

2.30. 27.

2.43. Just count the number of elements whose order is greater than 2, and use Problem 2.41.

2.47. Yes.

2.49. (d) 7623.

2.57. Use Theorem 1.4 (b).

2.60. No, no, no, no, ...

2.72. (a) $x = -5 - 23k,\ y = -17 - 77k$.

2.76. Yes.

2.79. (d) 4,284,466.

2.83. There are six values of y.

2.85. Think about intersections of sets.

Chapter 3

3.3. No.

3.5. (a) $x \equiv 4 \pmod{24}$.

3.8. No.

3.12. Let p be of the form $3k + 1$ or $3k + 2$.

3.14. Factoring.

3.16. Begin by letting $N = \sum_{i=0}^{r} a_i 9^i$.

3.19. Factoring.

3.20. Use mathematical induction to show that $7 \mid (a^7 - a)$.

3.21. The analogous statement does hold for certain other primes.

3.28. Use mathematical induction twice.

3.29. (d) $k = 8$.

3.40. Careful: **R** may not contain 1 and may not be commutative.

3.48. Assume, to the contrary, that there is a nonzero element $y \in \mathbf{R}$ such that $y^n = 0$ for some minimal $n > 2$.

3.54. No.

3.61. Try to construct the five-element addition table.

3.62. α satisfies a quartic equation.

3.68. (a) $x \equiv 59 \pmod{113}$.

3.75. Yes.

3.83. Show that each mu is coprime to n, and that the mu's are incongruent modulo n.

3.84. The proposition is not true if p is composite.

3.85. It is sufficient to show that no two integers $r + mu + nv$ are congruent modulo mn.

3.91. (a) Use Theorem 2.12.

3.92. (f) 171.

3.93. (b) Factoring.

3.97. Use Fermat's Theorem.

3.101. Just multiply congruences.

3.106. $1000 < \text{spsp}(2) < 3000$.

3.110. (d) $k = 21$.

3.112. (f) $x \equiv 12 \pmod{51}$.

3.114. Show that the powers $a^0, a^1, \ldots, a^{\phi(m)-1}$ are incongruent modulo m, then use Problem 3.91.

3.123. (b) Just use the result from part (a).

3.125. NUMBER THEORY.

3.127. (b) $j \equiv 1243 \pmod{1740}$.

3.129. (c) Use parts (a), (b).

Chapter 4

4.4. (c) $q(x) = 2x^2$, $r(x) = -2x^2 + x + 1$.

4.5. Proof by contradiction.

4.11. Any solution x must be odd, so consider the set of values of $f(x) = 2x^2 + 3x + 3$ obtained by letting $x = 1, 3, 5, \ldots, 2^n - 1$.

4.16. (a) $x \equiv 13, 14 \pmod{17}$.

4.17. Let $p - 1 = dk$, and use Lagrange's Theorem.

4.25. (b) Use part (a).

4.28. (a) Use Problem 4.25 (b).

4.30. Use Problems 4.23, 4.29.

4.35. (a) $(2^6 - 1, 2^8 - 1) = 2^{(6,8)} - 1 = 2^2 - 1$.

4.40. (b) 1.

4.45. If $M(n)$ is the nth Mersenne prime and p_n is the nth prime, then is $M(n) \geq 2^{p_n} - 1 > 2^n$?

4.49. 301.

4.50. 0.0285.

4.51. (a) $x \equiv 13084 \pmod{27559}$.

4.59. (c) There are two solutions modulo 2401.

4.61. (a) $x \equiv 1049713 \pmod{2072070}$.

4.63. ($\rightarrow$) Show that if $f(x)$, $g(x)$ are polynomials over $\mathbf{Z}$ and $f(x) \equiv g(x) \pmod{p}$ identically, then $f'(x) \equiv g'(x) \pmod{p}$ identically.

Chapter 5

5.1. (e) Order $= 18$.

5.9. Assume, to the contrary, that n is composite, and make use of Theorem 5.1.

5.10. $\mathbf{Z}_{13}^{\times}$ is cyclic.

5.15. (c) $\mathbf{G}$ has only one element of order 2.

5.16. (c) All the generators for $\mathbf{Z}_{17}^{\times}$ lie outside of $\mathbf{Z}_{17}^{\times 2}$.

5.20. Sometimes.

5.22. Use Wilson's Theorem.

5.28. (b) This is a pairing argument.

5.33. (a) $x \equiv 19 \pmod{43}$.

5.35. (c) $t \equiv 7 \pmod{20}$.

5.42. Use Theorems 2.7(b), 2.16.

5.46. Examine Table G in Appendix II.

5.50. Show that g, a primitive root of p^2, is also a primitive root of p.

Chapter 6

6.1. (d) $x \equiv 0, 26, 71 \pmod{97}$.

6.2. (b) $T_{37}^{(2)} = 333 = 9 \cdot 37$.

6.7. (c) Use part (b).

6.10. (b) 96.55%.

6.12. (a) Begin by letting $\mathbf{H} = \{g^a, g^b, \ldots, g^n\}$ be a subgroup of the cyclic group $\mathbf{G} = \{g, g^2, \ldots, g^{|\mathbf{G}|}\}$; then look at the smallest exponent from the set $\{a, b, \ldots, n\}$.

6.16. This is a pairing argument.

6.22. (a) $\left(\frac{2}{19}\right)\left(\frac{7}{19}\right) = -1$.

6.23. (b) Make use of Wilson's Theorem.

6.25. 0.

6.27. (b) Use Problem 6.25.

6.41. (a) i) $\{r_i\} = \{3, 6, 9, 1, 4\}$.

6.44. (a) -1.

6.46. Look at the allowed values for the order of 2, and rule out all of them except one.

6.47. $p \equiv 1, 9, 11, \text{or } 19 \pmod{20}$.

6.50. (c) There are four solutions.

6.63. Pairing argument.

6.70. (c) Consider, successively, DS(N) $= 1, 2, 3, 4$.

6.73. ($\leftarrow$) Let the integer $A = 8k + 1$, $k \geq 0$. Assume $n \geq 6$, as the cases $n = 1$ through 5 can be dealt with individually. Show how to construct a solution to $x^2 \equiv 8(k + 1) + 1 \pmod{2^n}$ if you have a solution to $x^2 \equiv 8k + 1 \pmod{2^n}$.

6.76. There are three such primes.

6.77. (a) Use Lemma 6.4.1 and Corollary 4.1.1.

6.78. (c) 1, 13, 47.

6.85. (b) $x \equiv 19, 61 \pmod{448}$.

6.86. The set of residues is a cyclic multiplicative group.

6.90. Begin by rationalizing $(a + b\omega)^{-1}$.

6.91. Only two are composites.

6.93. (a) Use proof by contradiction.

6.94. (f) $(6 + \omega)(5 - \omega)$, or any equivalent of this.

6.95. (a) ω^2.

6.96. (c) The group has order 6.

6.98. (c) i.

6.99. $\left(\dfrac{-1 - 2i}{3 - 2i}\right)_4 = 1$.

Chapter 7

7.11. $\sigma(10^6) = 2{,}480{,}437$.

7.15. Use mathematical induction on the number of distinct prime powers in the factorization of n.

7.20. 1.

7.21. There are only two such functions.

7.22. Show first that σ_k is multiplicative.

7.25. Use proof by contradiction.

7.26. (a) Compare $p_j^{\alpha_j/2}$ with $p_j^{\alpha_j-1}(p_j - 1)$. Yes, for even $n > 6$.

7.30. (a) $r(7) = -16{,}744$.

7.40. Use proof by contradiction, as in Theorem 1.6.

7.43. The first two such integers are $n = 10, 14$.

7.49. (c) Both partners are less than 20,000.

7.55. Suppose, to the contrary, that n is a superabundant number and that for some pair of primes $p_j < p_k$ one has $\alpha_j < \alpha_k$. Now consider $n' = (p_j/p_k)n$.

7.56. (a) Yes.

7.59. (a) There are four solutions.

7.62. There are three solutions of the form $n = p^\alpha$.

7.67. (a) Show that $2p_1p_2 \equiv 2 \pmod 3$.

7.71. Corollary 7.12.2 dominates in the interval [1, 1200] for integers x of the form $2(6k + 1)$.

7.73. (c) 480.

7.74. (b) 112.

7.79. (b) In Theorem 7.13 let $f(d) = \Lambda(d)$ and $g(n) = \ln n$.

7.83. (a) 48.

7.84. (a) 2.

7.87. (b) Look at $(f^{-1} * f)(p^{\alpha})$.

7.89. (e) $(\tau_u \otimes \phi)(24) = 24$.

7.91. (a) $(\phi ** \phi)(12) = 96$.

Chapter 8

8.7. In $w^2 + x^2 + y^2 = z^2$, where w, x, y, z are primitive, consider among $\{w, x, y\}$ the cases: No even integers are present, one even integer is present, three even integers are present.

8.12. (a) $M = 4, x = 3, y = 1, 17y \equiv -x \pmod{10}$.

8.13. Use Theorem 8.6.

8.14. $B(10{,}000) = 2749$.

8.21. Let $\mathbf{M} = \begin{pmatrix} a + bi & c + di \\ A + Bi & C + Di \end{pmatrix}$ and work out det $\mathbf{M}^*$, $(\det \mathbf{M})^*$.

8.23. 96 representations.

8.28. The argument is analogous to that in Problem 8.27.

8.31. (b) Use a parity argument.

8.35. Each summand $< (159)^4$.

8.37. (b) $N > 20$.

8.38. (b) $N > 50$.

8.40. Use an argument like that in Problem 8.39: Seek a class of linear forms $N = ak + b$ such that $N = x^3 + y^3 + z^3$ is impossible.

8.44. (b) Combine the identity in part (a) with Theorem 8.8.
(c) Write the indicated residues as sums of biquadrates, and then combine this with part (a).

Chapter 9

9.3. (c) $\sqrt{2} + \sqrt[3]{2}$ satisfies $x^6 - 6x^4 - 4x^3 + 12x^2 - 24x - 4 = 0$.

9.6. (c) Use proof by contradiction.

9.7. Model your proof after analogous material in Chapter 2, together with Theorem 4.1.

9.17. $(\alpha^2 - \alpha - 2)^{-1} = -[5\alpha^4 - 5\alpha^3 + 6\alpha^2 - 4\alpha + 13]/16.$

9.24. Use the fact that π is transcendental.

9.30. Use Problem 9.28.

9.32. Look at a typical term of the minimal polynomial for $a + b$.

9.36. Use Theorem 9.11.

9.43. (a) $N\alpha = 19$.

9.44. (b) $\mathbf{C} = \begin{pmatrix} 1/2 & 0 & 0 \\ 0 & 1/3 & -1/3 \\ 0 & 2/3 & 1/3 \end{pmatrix}$. (c) In each base, $N\alpha = 223/16$.

9.47. Use Theorem 9.15 and Corollary 9.16.1.

9.52. (a) α satisfies a minimal polynomial of degree 3; use Theorem 9.16. (b) Use part (a).

9.57. (b) $N\alpha = 2$. (d) The minimum polynomial for β is a quartic.

9.58. (a) Use mathematical induction. (c) Discover an analog to the results in parts (a), (b).

9.59. (b) Start by finding one unit.

Chapter 10

10.10. $k = 5, N_0 = 54, K = 1.98$.

10.15. $g_1(x)g_2(x) = 2x + 6x^2 + \left(\frac{38}{3}\right)x^3 + 22x^4 + \left(\frac{172}{5}\right)x^5 + \left(\frac{748}{15}\right)x^6 + \ldots + \ldots + .$

10.16. 286; 84.

10.20. $N_{13} = 6$.

10.21. Count the number of solutions in positive integers and the number of solutions that contain one or more zeros. Make use of Problem 10.17.

10.23. $B_6 = \frac{1}{42}$.

10.27. $G_n - G_{n-1} - G_{n-2} = F_{n-1}$.

10.29. One approach is to try plausible experimental forms for the generating function. $\lim_{n\to\infty} f_n\left(\frac{1}{4}\right) \approx 2.00092$.

10.32. $\pi_4(10) = 23$.

10.37. $p(10) = 42$.

10.39. (c) Make use of $\Sigma_{n=1}^{\infty} 1/n^2 = \pi^2/6$.

10.55. Example: The number of partitions of 18 into summands that are congruent to either 1 or 5 (mod 6) is 14.

10.61. Use mathematical induction.

10.62. (c) Assume a plausible polynomial form in k with undetermined coefficients.

10.65. $k = 8$.

10.67. $p_1(20, 4) = 23$.

10.68. $p(20) = 627$.

10.69. Two more suggestions: (1) Work with $\ln \Theta(x)$; (2) The general strategy is, after differentiating $F(x) = 1/\Theta(x)$, to ultimately express each side as a power series in x, and then to equate coefficients of like powers of x (i.e., of x^N, where $1 \leq N < \infty$).

10.70. $p(50) = 204{,}226$.

Chapter 11

11.5. $s_n = 2s_{n-1} + 1, s_0 = 0, n \geq 1$.

11.6. (c) $s_n = \begin{cases} 2 & \text{if } n = \text{odd} \\ 2 - (-1)^k & \text{if } n = 2k. \end{cases}$

11.7. $k = 4$.

11.9. (c) $H_5(x) = 32x^5 - 160x^3 + 120x$.

11.11. (a) $g(x) = (1 - 3x)/(1 - 5x + 3x^2)$.

11.13. (e) $s_n = \frac{8}{5}\left(\frac{1}{2}\right)^n + \frac{1}{5}(1 - 7i)(1 + i)^n + \frac{1}{5}(1 + 7i)(1 - i)^n$.

11.19. $(V_n, V_{n+1}) = 1$ is also true.

11.20. Use mathematical induction for parts (b), (c).

11.22. Use mathematical induction.

11.27. Use the Binet formulas, and let $x = \beta/\alpha$.

11.28. Use the Binet formulas throughout. In part (c) couple this with mathematical induction.

11.30. (a) $U_n(x) = \{[(x + \sqrt{x^2 + 4})/2]^n - [(x - \sqrt{x^2 + 4})/2]^n\}/\sqrt{x^2 + 4}$.
(e) $x = \pm 2i \cos(k\pi/n)$, $k = 0, 1, \ldots, n - 1$.

11.35. (b) Use mathematical induction, and call on one of the results from part (a).

11.41. $3!\, s(6, 3) - 3!\, s(5, 3) - 2!\, s(5, 2) = 360$ ordered subsets.

11.43. Make repeated use of the identity $j\binom{k}{j} = k\binom{k-1}{j-1}$. Some computed values: $s(10, 3) = 9330$, $s(10, 7) = 5880$.

11.45. Use Theorems 11.5, 11.7.

11.46. $S = n(n + 1)(n - 1)(n - 2)/24$.

11.48. $B_7(x) = x^7 - \frac{7}{2}x^6 + \frac{7}{2}x^5 - \frac{7}{6}x^3 + \frac{1}{6}x$.

11.51. ~~No.~~ YES.

11.55. Use Theorem 11.9 with an obvious choice for x.

11.60. If $\Sigma_{n=0}^{\infty} B_n x^n/n!$ does converge, then it can do so only for $|x| < 2\pi$.

11.62. Begin by letting $g(x, t) = \Sigma_{n=0}^{\infty} B_n(x)t^n/n!$.

11.65. Use mathematical induction together with the relation $B'_{k+1}(x) = (k + 1)B_k(x)$.

11.66. Use Problem 11.65.

11.67. Use Rolle's Theorem from the calculus.

11.72. Yes.

11.74. Make a comparison with approximating rectangles.

11.75. (b) Make use of part (a) and of Theorem 11.9.

11.77. If $n = 15$, there is agreement out to seven decimal places.

11.78. (c) $7\pi^3/216$.

Name Index

K

L

M

N

O

P

R

Subject Index

H

I

Q

R